pringer Series on Atomic, Optical, and Plasma Physics covers in a
hensive manner theory and experiment in the entire field of atoms and
les and their interaction with electromagnetic radiation. Books in the series
e a rich source of new ideas and techniques with wide applications in fields
as chemistry, materials science, astrophysics, surface science, plasma
logy, advanced optics, aeronomy, and engineering. Laser physics is a
lar connecting theme that has provided much of the continuing impetus for
evelopments in the field, such as quantum computation and Bose-Einstein
msation. The purpose of the series is to cover the gap between standard
graduate textbooks and the research literature with emphasis on the
mental ideas, methods, techniques, and results in the field.

Springer Series on Atomic, Optica and Plasma Physics

Volume 125

Vadim Dudnikov

Development
and Applications
of Negative Ion Sources

Second Edition

 Springer

Vadim Dudnikov
Muons, Inc (United States)
Batavia, IL, USA

ISSN 1615-5653   ISSN 2197-6791   (electronic)
Springer Series on Atomic, Optical, and Plasma Physics
ISBN 978-3-031-28407-6   ISBN 978-3-031-28408-3   (eBook)
https://doi.org/10.1007/978-3-031-28408-3

This Springer imprint is published by the registered company Springer Nature Switzerland AG
The registered company address is: Gewerbestrasse 11, 6330 Cham, Switzerland

Springer Series on Atomic, Optical, and Plasma Physics

Volume 125

The Springer Series on Atomic, Optical, and Plasma Physics covers in a comprehensive manner theory and experiment in the entire field of atoms and molecules and their interaction with electromagnetic radiation. Books in the series provide a rich source of new ideas and techniques with wide applications in fields such as chemistry, materials science, astrophysics, surface science, plasma technology, advanced optics, aeronomy, and engineering. Laser physics is a particular connecting theme that has provided much of the continuing impetus for new developments in the field, such as quantum computation and Bose-Einstein condensation. The purpose of the series is to cover the gap between standard undergraduate textbooks and the research literature with emphasis on the fundamental ideas, methods, techniques, and results in the field.

Vadim Dudnikov

Development and Applications of Negative Ion Sources

Second Edition

 Springer

Vadim Dudnikov
Muons, Inc (United States)
Batavia, IL, USA

ISSN 1615-5653   ISSN 2197-6791   (electronic)
Springer Series on Atomic, Optical, and Plasma Physics
ISBN 978-3-031-28407-6   ISBN 978-3-031-28408-3   (eBook)
https://doi.org/10.1007/978-3-031-28408-3

This Springer imprint is published by the registered company Springer Nature Switzerland AG
The registered company address is: Gewerbestrasse 11, 6330 Cham, Switzerland

Foreword

My first introduction to Vadim Dudnikov was at the International Conference on Ion Sources held in Berkeley, California, in 1989. The following year, I had the pleasure of meeting him again at his own laboratory at the Budker Institute of Nuclear Physics in Novosibirsk, Russia. At that time, Vadim was already well recognized internationally as a leader in the areas of charge exchange and negative ion production. In 1971, he had discovered the enhancement of negative ion generation in gas discharge plasmas that is brought about by a small admixture of cesium (or other low ionization potential material) to the plasma – the cesiation effect. This method, which dramatically enhances negative ion production, is now a widely used technique for the production of intense, high brightness, negative ion beams. The early work on charge exchange and cesium catalysis carried out at the Budker Institute is well recorded in the literature, and includes the ground-breaking work of Vadim Dudnikov and other key researchers at Novosibirsk – Belchenko, Dimov, Derevyankin, Klenov, Davydenko, and others.

There have evolved over the years a great number of embodiments of surface plasma negative ion sources utilizing cesium catalysis. In surface plasma sources, a flux of negative ions is produced when electrons are captured from the cesiated electrode surface by sputtered and reflected particles. Discovery of the physical basis of cesium catalysis has led to the development of surface plasma sources (also called surface production sources, referring to the cesiated surface at which the negative ions are formed) in many laboratories in the USA, France, Germany, Italy, Japan, Russia, UK, and other countries. Cesium catalysis is used in neutral beam injectors for experimental thermonuclear fusion devices, and the negative ion sources that produce the H^- ion beams from which the immensely intense neutral beams are subsequently formed by charge exchange have reached hugely impressive parameters. The International Thermonuclear Experimental Reactor (ITER) presently being constructed at Cadarache, France, calls for two cesium-seeded negative ion sources each producing beams of H^- or D^- ions at energy 1 MeV and current 73 A (H^-) or 40 A (D^-) for an on-time of one hour. It is remarkable—the beam current that can be generated has increased by a factor of about 10,000 in the past several decades. To paraphrase Darwin, there is grandeur in this view of science.

Vadim's book describes the origin, history, development, and applications of negative ion sources, with emphasis on cesium seeding and those kinds of sources that utilize this phenomenon. It forms an impressive account, including sources that have been developed and used for accelerator injection, ion implantation, accelerator mass-spectrometry, neutral beam injectors, and more. The book will make a valuable contribution to the ion source literature, and will be particularly valuable to the negative ion source community. I hope you enjoy and appreciate it as much as I have.

Berkeley, CA, USA   Ian Brown
June 2019

Preface to Second Edition

This Second Edition comes 4 years after publication of the First Edition in 2019.

Many things have evolved in the field of negative ion beams and sources development and applications. Understanding of negative ion source behavior has grown, new negative ion sources have been developed, new applications have evolved, and generally the achievable negative ion beam parameters have improved greatly. Many new reports were presented in two occurrences of the biennial International Symposium on Negative Ions, Beams and Sources (NIBS'20 and NIBS'22), two International Ion Source Conferences (ICIS'19 and ICIS'21), three International Particle Accelerator Conference (IPAC'19, IPAC'20 and IPAC'22), two LINAC Conferences (LINAC'20 and LINAC'22), and the North American Particle Accelerator Conference (NAPAC'22) and published in different journals. Chapter 7 "Development of Conversion Targets for High-Energy Neutral Beam Injectors" has been added, along with more than 100 figures.

This book is designed to serve as a review and reference. The target readers are researchers actively involved in ion sources development and applications. The objective of the book is to provide a comprehensive, easily understood introduction to and review of the field.

Negative ion source research is a very empirical field, and for the most part the emphasis presented in this book is experimental. It has overwhelmingly been the case that theoretical understanding of a negative ion source has followed after its construction and after its experimental performance has been characterized. This situation is reflected in the presentation here.

The academic background assumed of the readers is roughly physics graduate level. A working knowledge of plasma physics, atomic physics, and electromagnetic theory will definitely aid the digestion, and some familiarity with electronic and electrical systems will help too. I am most grateful to my friends and colleagues who help me in development and preparation of this book.

Batavia, IL, USA   Vadim Dudnikov

Preface

Increased interest in negative ion sources in recent years is strongly associated with the growing importance of negative ion beams for a range of scientific applications, including for tandem accelerators, high-energy ion implantation, accelerator mass-spectrometry, super-collimated beams, charge-exchange injection into cyclic accelerators and storage rings, charge-exchange extraction of beams from cyclotrons, injection of high energy neutrals into plasma traps, charge-exchange beam distribution, and more. This book describes the development of negative ion sources and their applications in research and industry. Applications of charge-exchange technologies for controlling beams of accelerated particles are considered. A description is given of the physical foundations and design of negative ion sources utilizing charge-exchange, plasma volume, thermionic, secondary emission (sputtering) and surface-plasma sources, as well as the history of their development, with much of the work described having been realized with the direct participation of the author. Negative ion beam transport and the development of beam-plasma instabilities are considered. An extensive bibliography (over 600 titles) provides good reference material. The anticipated readership of this book includes professional researchers involved in the development and use of ion sources, as well as graduate and undergraduate students entering the field or an application area.

This book is based on the Russian text, "Источники отрицательных ионов" ("Negative Ion Sources", V. Dudnikov, Novosibirsk State University, 2018), with broad extension. The English manuscript was extensively reviewed and re-edited by Ian Brown, who made a great contribution to the book quality. Many thanks are due to many publishers, organizations, and authors for permission to reproduce material and figures from various books and journals. Many thanks to my collaborators from different institutes around the word.

Batavia, IL, USA   Vadim Dudnikov

Contents

Acronyms

AGS	Alternating Gradient Synchrotron
AMFC	Axial Magnetic Field Correction
ANL	Argonne National Laboratory
BATMAN	Bavarian Test Machine for Negative ions (Garching, Germany)
BINP	Budker Institute of Nuclear Physics (Novosibirsk, Russia)
BNL	Brookhaven National Laboratory
CELSIUS	Cooling with ELectrons and Storing of Ions from the Uppsala Synchrocyclotron
COSY	Cooler Synchrotron (Jülich Research Center, Germany)
CSNS	Chinese Spallation Neutron Source
DESY	German Electron Synchrotron (Deutsches Elektronen-SYnchrotron) (Hamburg)
ECR	Electron Cyclotron Resonance
ELISE	Extraction from a Large Ion Source Experiment (Garching, Germany)
FNAL	Fermi National Accelerator Laboratory
HERA	Hadron Elektron Ring Anlage (Electron Proton Collider) (DESY, Germany)
HOPG	Highly Oriented Pyrolytic Graphite (a highly pure and ordered form of graphite)
IHEP	Institute of High Energy Physics (Protvino, Russia)
INP	Institute of Nuclear Physics (Novosibirsk, and also other locations)
INR	Institute of Nuclear Research (Troitsk, Russia)
IOTA	Integrable Optics Test Accelerator (Fermilab)
ISIS[1]	A spallation neutron source at the Rutherford Appleton Laboratory (UK)
ISR	Intersecting Storage Rings (CERN)

---

[1] The name ISIS is not an acronym. It refers to the Ancient Egyptian goddess and the local name for the River Thames. The name was selected for the official opening of the facility in 1985, prior to this it was known as the SNS, or Spallation Neutron Source. The name was considered appropriate as Isis was a goddess who could restore life to the dead, and ISIS made use of equipment previously constructed for the Nimrod and NINA accelerators.

ITEP	Institute of Theoretical and Experimental Physics (Moscow)
ITER	International Thermonuclear Experimental Reactor (Cadarache, France)
JAEA	Japan Atomic Energy Agency
J-PARC	Japan Proton Accelerator Research Complex
KEK	High Energy Accelerator Research Organization (Japan)
LANL	Los Alamos National Laboratory
LANSCE	Los Alamos Neutron Science Center
LBNL	Lawrence Berkeley National Laboratory
LEBT	Low Energy Beam Transport
LHC	Large Hadron Collider (CERN)
LHD	Large Helical Device (Japan)
MACOR	Machinable ceramic
MITICA	Megavolt ITER Injector and Concept Advancement (Padua, Italy)
NBI	Neutral Beam Injector
NIFS	National Institute for Fusion Science
NNBI	Negative ion based Neutral Beam Injector
OPPIS	Optically-Pumped Polarized H⁻ Ion Source (Brookhaven)
ORNL	Oak Ridge National Laboratory
PEEK	PolyEther Ether Ketone (a colorless organic thermoplastic polymer)
RAL	Rutherford Appleton Laboratory (UK)
RF	Radio Frequency
RFQ	Radio Frequency Quadrupole (a kind of linear accelerator used at low energies)
RFX	Reversed Field Experiment
RHIC	Relativistic Heavy Ion Collider (Brookhaven)
SB RAS	Siberian Branch of the Russian Academy of Science
SBAS	Siberian Branch of the Academy of Science
SCC	Space Charge Compensation
SIMS	Secondary Ion Mass Spectroscopy
SNS	Spallation Neutron Source (Oak Ridge)
SPG	Surface Plasma Generation
SPIDER	Source for Production of Ions of Deuterium Extracted from RF plasma (Padua, Italy)
SPM	Surface Plasma Method
SPP	Surface Plasma Production
SPS	Surface Plasma Source
TRIUMF	TRI-University Meson Facility, Canada's national particle accelerator center
VAPP	Proton-Antiproton Collider
VITA	Vacuum Insulation Tandem Accelerator
XiPaf	Xi'an Proton Application Facility (China)
ZGS	Zero Gradient Synchrotron

Chapter 1
Introduction

Abstract The motivation for the development of negative ion sources and the main methods of negative ion production and applications of negative ion source are described and briefly discussed.

Many important achievements in atomic physics, nuclear physics, and elementary particle physics have been inextricably linked to the development and improvement of methods for the production and acceleration to high energies of particle fluxes [1, 2]. For many years the growth of the science and technology of energetic particle beams was stimulated by the needs of these physics subfields. More recently however, energetic particle beams have become an indispensable tool for a substantially broader range of scientific applications, including plasma physics, solid state physics, chemistry, biology, medicine, and a number of important areas of technology and industry. As yet, we can accelerate to high-energies-only particles that are electrically charged, and thus the starting point for the production of energetic particle beams is the generation of charged particles of the desired kind in an appropriate ion or electron source and the subsequent formation of a flow of these particles as an ordered beam.

Since the characteristics of the charged particle source have a strong influence on the details of the accelerated particle flux, considerable attention has been paid to the development of ion sources. In many cases, the choice of particle species to be employed in a particular application is determined by the capability of generating the original beam with appropriate and desirable features.

The relative simplicity of methods for electron beam production has benefited greatly the relatively rapid transformation of electron accelerators from unique instruments for scientific research into reliable high-performance equipment for industrial processing lines.

The basis for the widespread use of positive ion beams in scientific research and their introduction into industrial technology lies largely with the kinds of ion sources developed in the past that have now become traditional methods for the production of beams of positive ions. The development of positive ion sources and negative ion

V. Dudnikov, *Development and Applications of Negative Ion Sources*, Springer Series on Atomic, Optical, and Plasma Physics 125, https://doi.org/10.1007/978-3-031-28408-3_1

sources prior to 1970 has been well described by Gabovich in his book, *Physics and Technology of Plasma Ion Sources* [3], and in more recent years by a number of authors. A powerful driver for the development of ion beam sources was provided by the needs of accelerator technology, large-scale isotope separation, space thrusters, ion implantation, neutron generators, and neutral beam injectors. Ion source know-how in the early years facilitated the production of beams of positive ions with beam current up to 1 A and negative hydrogen ions with beam current up to 5 mA.

For space technology, ion implantation, thermonuclear fusion research, and a number of other important applications, ion beams with increasingly greater beam currents were needed. Progress in research programs associated with these areas engendered progressively improved understanding of the physical principles of ion beam production, leading to new techniques, improved technology, and the invention of fundamentally new methods for intense ion beam production, and with these drivers, work has led to beams that until recently seemed far-fetched indeed: sources have now been created that can generate quasi-steady beams of positive ions and hydrogen atoms with beam currents up to hundreds of Amperes and short-pulsed ion beams up to hundreds of kiloamperes. Recent achievements are reflected in a number of texts [4–8]. These accomplishments have led to some major effects in their field of application. An excellent and impressive example is the situation with neutral beam heating of thermonuclear plasmas. In his book *Controlled Thermonuclear Reactions* [9], Lev Artsimovich wrote "…it is possible to use the nuclear energy of fusion due not only to thermal collisions of plasma ions but also when passing a beam of fast particles through the plasma with sufficiently high electron temperature." However later it was necessary to state "It can hardly be expected that such a process will be of practical interest in the case when a beam of deuterons produced by an accelerator is injected into a plasma…This possibility is almost certainly of no practical significance but nevertheless should be keep in mind…." That is, at the state-of-the-art in those early years, it seemed rather implausible that ion beams of sufficiently high current could be produced to be of any practical interest for fusion application. But developments in intense ion beam production have rendered feasible the heating and fueling of thermonuclear fusion plasmas using intense, energetic beams of neutral hydrogen atoms [10]. Equally crucially, the achievements of recent years have affected many other areas of application of energetic particle fluxes. The generation of positive ions in a gas-discharge plasma, a straightforward approach used in early ion sources, has turned out to be an effective base method for producing beams of positive ions. Modern adaptations of this approach provide highly efficient positive ion sources, close to the theoretical limit. In most cases, the intensity of the beam is limited not by ion generation capability but by constraints of the ion-optical beam formation systems. In consequence of this, increasing effort is being concentrated on problems of the formation and acceleration of beams of positive ions; positive ion sources with acceleration in a plasma with closed electron drift and space charge compensation have been developed [11].

Methods for the production of negative ion beams constitute a significant and substantial body of work. Development of negative ion sources lagged historically behind the development of positive ion sources and grew from the needs of important applications. Perhaps the most important application of negative ions is their use for the efficient conversion of energetic negative ion beams into intense, energetic beams of neutral atoms (and also of positive ions). These applications are inherent to the use of charge-exchange methods in the general field of energetic particle flows, enabling them to be combined in charge-exchange technology for the production and use of beams of energetic particles [12].

Prior to 1968, the intensity of negative ion beams that could be achieved by sources of various kinds was limited to a level of several milliamps, a limitation brought about largely by the constraints of negative ion formation. But by 1971, the primary limitations to the achievable intensity of negative ions beams were eliminated, and the production of H$^-$ ion beams with current of 1 A was accomplished. Thereafter, the generation of negative ion beams with much greater intensity became, to a large extent, a technical problem [13].

Progress in negative ion beam production has been associated with developments in two directions: a significantly enhanced charge-exchange method using a positive ion source with multi-aperture extraction for beam formation together with effective targets of alkali and alkaline earth metals and the development of a fundamentally new surface-plasma method (SPM) of negative ion formation. Following wide exploration for new methods of negative ion beam production at the Institute of Nuclear Physics of the Siberian Branch of the Russian Academy of Science at Novosibirsk, it was found in 1971 by V. Dudnikov that addition of cesium to the plasma device substantially intensifies the formation of negative ions in the plasma [13]. Subsequent studies have shown that the increased emission of negative ions from the discharge is due to increased negative ion generation at the surface of electrodes bombarded by plasma particles [14–16]. The adsorption of cesium on the electrode surface lowers the surface work function from 4–5 eV down to 1.2–2 eV. In this case, negative ions are formed due to electron capture from the electrodes, and an increased flux of back-scattered and sputtered particles are negative ions. In the same way, the emission of negative ions of positronium is intensified by the deposition of alkali metals on the surface of single-crystal tungsten [17, 18], and similarly for negative muonium ions. A method for monitoring the work function of the plasma electrode of a surface production source (SPS), based on the formation of negative positronium ions, has been proposed [16]. A related method has been proposed [19, 20] for obtaining cold negative muonium ions by electron capture by muonium atoms on the cesiated surface of a monocrystalline foil of tungsten or palladium or silica gel.

The processes of negative ion formation when a solid surface is bombarded by fast particles were explored experimentally by Woodcock in 1931 [21]. Ayukhanov and coworkers in 1961 [22] and Kron in 1962 [23] found that the emission of secondary negative ions from the surface is intensified upon adsorption onto the surface of alkali metal films. However, the resulting efficiency for the formation of negative ions with small electron affinity was low, and the intensity of negative ion beams produced were so small that these processes were not taken into account later in analyzing the processes of formation of negative ions in ion sources and were not used for generating intense beams of negative ions [3].

Fortunately, under certain conditions the formation efficiency of negative ions by electron capture from a solid surface can be quite high and can be maintained at a high level up to significant flux density of the bombarding particles, sufficient to produce H⁻ ion fluxes with current density of up to several Amperes/cm². The physical mechanism responsible for the generation of intense negative ion fluxes by the interaction of plasma with the surface of a solid body was called surface plasma generation (SPG). Optimized surface plasma generation of negative ions forms the foundation of surface plasma methods for the production of negative ion beams (SPM). As already noted, the formation of negative ions in the interaction of particles with a solid surface has been studied for nearly 90 years, over which timespan a huge amount of factual material about the processes involved has been accumulated and concentrated in a truly immense number of publications.

Although at present some important processes involved with surface plasma production remain not well understood, the main features of this method of negative ion production have been explored and described quite clearly. In recent years, the development of methods of negative ion beam production has intensified considerably, as manifested by the increasing attention to negative ions at conferences on accelerators, ion sources, controlled thermonuclear fusion, and other topics, and particularly by the ongoing biennial International Symposium on Negative Ions, Beams and Sources (NIBS) and International Conference of Ion Sources (ICIS).

References

1. V. Dudnikov, *Negative Ion Sources* (NSU, Novosibirsk, 2018). В. Дудников, *Источники отрицательных ионов*, НГУ, Новосибирск, 2018
2. V. Dudnikov, *Development and Applications of Negative Ion Sources* (Springer, Switzerland AG, 2019)
3. M.D. Gabovich, *Physics and Technology of Plasma ion Sources* (Atomizdat, Moscow, 1972). М.Д. Габович, *Физика и техника Плазменных источников ионов*, Москва, Атомиздат, 1972
4. H. Zhang, *Ion Sources* (Springer, Switzerland AG, 1999)
5. I. Brown (ed.), *The Physics and Technology of ion Sources* (Wiley-VCH Verlag GmbH & Co. KGaA, New York, 2004)
6. A.T. Forrester, *Large Ion Beams* (Wiley, New York, 1988)
7. B. Wolf, *Handbook of Ion Sources* (CRC Press, Boca Raton, 1995)
8. D.F. Hunt, F.W. Crow, Electron capture negative ion chemical ionization mass spectrometry. Anal. Chem. **50**(13), 1781 (1978)
9. L.A. Arcymovich, *Controlled Nuclear Reactions* (Fismatgis, Moscow, 1961). Л.А. Арцимович, *Управляемые термоядерные реакции*, Москва, Физматгиз, 1961
10. F. Santini, Non-thermal fusion in a beam plasma system. Nucl. Fusion **46**, 225 (2006)
11. V. Dudnikov, A. Westnern, Ion source with closed drift anode layer plasma acceleration. Rev. Sci. Instrum. **73**, 729 (2002)

12. G.I. Dimov, V.G. Dudnikov, Charge-exchange method for controlling particle beams. Sov. J. Plasma Phys. Engl. Transl. **4**(3), 388–396 (1978)

13. V. Dudnikov, Method of negative ion obtaining, Patent cccp, 411542, 10/III. 1972.; http://www. findpatent.ru/patent/41/411542.html В.Г. Дудников, "Способ получения отрицательных ионов", Авторское свидетельство, М. Кл.Н 01 J 3/0,4, 411542, заявлено 10/III, 1972

14. Y. Belchenko, G. Dimov, V. Dudnikov, Physical principles of surface plasma source operation, in *Symposium on the Production and Neutralization of Negative Hydrogen Ions and Beams, Brookhaven, 1977*, (Brookhaven National Laboratory (BNL), Upton, NY, 1977), pp. 79–96. Yu Belchenko, G. Dimov, V. Dudnikov, Physical principles of surface plasma source method of negative ion production, Preprint IYaF 77-56, Novosibirsk 1977. Ю. Бельченко, Г. Димов, В. Дудников, «Физические основы поверхностно плазменного метода получения пучков отрицательных ионов», препринт ИЯФ 77-56, Новосибирк 1977. http://irbiscorp.spsl.nsc. ru/fulltext/prepr/1977/p1977_56.pdf

15. V.G. Dudnikov, Surface-plasma method for the production of negative ion beams. Physics-Uspekhi **62**(12), 1233 (2019)

16. Y.I. Belchenko, V.I. Davydenko, P.P. Deichuli, et al., Studies of ion and neutral beam physics and technology at the Budker Institute of Nuclear Physics, SB RAS. Physics - Uspekhi **61**(6), 531–581 (2018)

17. Y. Nagashima, T. Hakodate, A. Miyamoto, K. Michishio, New J. Phys. **10**, 123029 (2008).; Y. Nagashima, T. Hakodate, A. Miyamoto, K. Michishio, H. Terabe, J. Phys. Conf. Ser. **194**, 012039 (2009); Y. Nagashima, K. Michishio, T. Tachibana et al., Positronium negative ion experiments - formation, photodetachment and production of an energy tunable positronium beam, in *XXVII International Conference on Photonic, Electronic and Atomic Collisions (ICPEAC 2011)*, IOP Publishing Journal of Physics: Conference Series **388**, 012021 (2012)

18. V. Dudnikov, A. Dudnikov, "Positronium negative ions for monitoring surface plasma source work functions", NIBS 2016, Oxford, UK, 2016. AIP Conf. Proc. **1869**, 020007 (2017)

19. V. Dudnikov, M.A.C. Cummings, R.P. Johnson, A. Dudnikov, Cold muonium negative ion production, IPAC 2017, Copenhagen, Denmark, 2017. V

20. V. Dudnikov, A. Dudnikov, Ultracold muonium negative ion production. AIP Conf. Proc. **2052**, 060001 (2018)

21. K. Woodcock, The emission of negative ions under the bombardment of positive ions. Phys. Rev. **38**, 1696 (1931)

22. U. A. Arifov, A. H. Ayukhanov, Isvestiya AN, Us. CCR, seriya Fis-mat. nauk, 6, 34 (1961). У.А. Арифов, А.Х. Аюханов, Изв. АН Уз. ССР, серия физ-мат. наук, 6, 34, (1961)

23. V.E. Kron, J. Appl. Phys. **34**, 3523 (1962)

Chapter 2
Charge-Exchange Technologies

Abstract Topics covered in this chapter include: Features of charge-exchange technology for the production and use of accelerated particles; regularities in the redistribution in mass and charge of accelerated particles; charge-exchange tandem accelerators; super-collimated beam production; charge-exchange extraction of particles from accelerators; charge-exchange distribution of accelerated particle beams; charge-exchange injection into accelerators and accumulator rings; charge-exchange injection into magnetic plasma traps.

2.1 Features of Charge-Exchange Method for the Production and Use of Accelerated Particles

Progress in the production and applications of negative ions provides opportunity for a significant increase in the efficiency of many energetic ion beam applications [1–3]. Awareness of these opportunities is due to an increased interest in developing methods for obtaining high-intensity, high-quality beams of negative ions.

Negative ions were discovered practically simultaneously with positive ions—H^-, O^-, and C^- ions were observed on the first Thomson mass spectrograms of "channel rays" together with H^+, H_2^+, O^+, C^+, etc. ions [4, 5]. Nevertheless, until recently almost all applications of accelerated particle fluxes exclusively employed positive ions. Lack of efficient methods of negative ion generation has driven the use of positive ions in some disadvantageous situations when the use of negative ions would be preferable. Systematic exploration for effective methods for generating negative ions have been developed in connection with the special role of negative ions in charge-exchange technology for obtaining and using high-energy particle fluxes. In most of the many applications of ion beams with constant charge and mass, electric and magnetic fields are used to control (steer, focus, etc.) the beam. Random changes in the charge or mass of the beam particles, caused by the loss or capture of electrons and dissociation in interaction with fields, with residual gas, with collimator walls, etc., are harmful. On the other hand, controlled change in the charge or mass when interacting with special targets is an

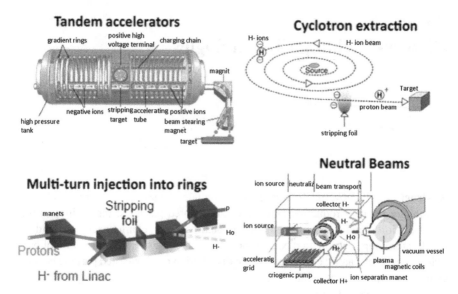

Fig. 2.1 Examples of the use of charge-exchange technologies. (Reproduced from Dudnikov [6])

effective method for influencing the parameters that determine the motion of the particles in macroscopic electric and magnetic fields. Thus, due to the action of the target on the electron shells of the beam particles, which changes their charge-to-mass ratio but does not significantly change their velocity, it is possible to change the acceleration that determines the further motion of the particles. This makes it possible to relatively easily realize phenomena that cannot be realized using only traditional methods of influencing accelerated particle flows. Some examples of the application of charge-exchange technologies are shown in Fig. 2.1. In tandem accelerators, negative ions are accelerated to a positive electrode, where they are stripped by a suitable target to become positive ions and are again accelerated to a grounded electrode. In a cyclotron, accelerated negative ions are passed through a thin foil, converted to positive ions, and extracted from the accelerator. During charge-exchange injection, accelerated negative ions are introduced into orbit and stripped to positive ions (multiply charged) which are then captured in an equilibrium orbit. In thermonuclear installations, accelerated negative ions are converted by charge exchange to neutral atoms and injected into the plasma through the confining magnetic field, where they are ionized and captured in the plasma, increasing its temperature and maintaining the current in the plasma.

2.2   Regularities in the Redistribution of Accelerated Particles in Mass and Charges

An important application of energetic particle beams has to do with streams of accelerated hydrogen isotopes, and in this connection, charge-exchange technology for handling hydrogen isotopes has been developed. At the same time, this simplest case reflects the main features of charge-exchange technology for obtaining and using accelerated particle fluxes. In what follows we shall speak of charged and neutral hydrogen particles. Since we are interested in the behavior of electron shells, all that is said can be transferred to other isotopes of hydrogen, with an appropriate velocity recalculation. A flow of accelerated hydrogen particles can be realized as positive ions H^+, H_2^+, and H_3^+; neutral particles H^0 and H_2; negative ions H^-; or a combination of these mass and charge states. Metastable negative molecular ions H_2^- (negative ion of molecular H_2) have been observed [7] and confirmed [8]. There are also doubly charged negative ions. The number of possible states increases because of the presence of internal degrees of freedom in complex particles: vibrational, rotational, and electronically excited states. The H^- ion is realized in the form of a single-bound state of a proton and two electrons. The binding energy of the additional electron in the H^- ion is the affinity of the hydrogen atom to the electron, $S = 0.754$ eV. The probability of removing an electron from a H^- ion remains significant at distances up to 5 atomic radii from the proton. This corresponds to a large cross-section for the detachment of an electron from the H^- ion when interacting with other systems. Information on the properties of H^- ions has been systematized in a number of monographs [9–16]. Positive ions with different masses, negative ions, and neutral particles experience different acceleration in the same fields, so redistributing particles over these states can control their behavior. In principle, transitions are possible between almost all the states mentioned, so that the complete set of possible transitions is difficult to see even for the simplest set—the hydrogen particles. Excitation of internal degrees of freedom does not affect the motion of particles in macroscopic fields, but it does affect the patterns of their redistribution over mass and charge states.

The basis of charge-exchange technology is the physics of atomic collisions, and an extensive literature has accumulated on the elementary processes responsible for the redistribution of accelerated particles by mass and charge state when interacting with targets. Data on these processes has been reported both in monographs [2, 3, 9–16] and reviews [17–20]. The available data do not provide all the requirements for charge-exchange technology, but the methods developed provide the necessary information in most cases. It is important for charge-exchange technology that the transfer of particles into another specific mass or charge state can be effectively accomplished by appropriate choice of the initial state of the accelerated particles and the properties of the target. Redistribution of mass and charge states occurs when accelerated particles interact with targets consisting of neutral gases and vapors, plasma, solid-state films; with electron and photon fluxes; etc. The interaction processes of energetic particles with targets are characterized by a number of laws as listed below.

1. At the comparatively high-energy characteristic of charge-exchange technology, one can neglect the capture by fast particles of heavy target particles or their fragments.
2. In all targets, fast molecular particles dissociate into their charged and neutral components, which subsequently behave as corresponding atomic particles. As a rule, fast molecular particles will dissociate until an equilibrium distribution over charge states is established.

3. Electron capture by fast particles is determined by a charge-exchange process, and the loss of electrons by inverse charge exchange and the ionization of fast particles in collisions with electrons, photons, neutral, and charged particles. When the velocity of the energetic particle is comparable to the velocity of electrons in the outer shell of neutral target particles, the capture of electrons by positive ions predominates over the loss of electrons by fast neutral particles. In this energy range (of order 105 eV for protons), it is possible to convert positive ions into fast neutral atoms using targets of suitable gases or vapors.

4. In targets of highly ionized plasma and photons, the detachment of electrons from negative ions and neutral particle stripping prevails. Such targets efficiently convert fast negative ions into neutral atoms and positive ions and fast atoms into positive ions.

5. The detachment of electrons from negative ions always prevails over the capture of electrons from the target by fast neutral particles, and thus negative ions are converted into neutral particles with high efficiency at all energies, but at high energy the efficiency of this conversion is substantially greater for plasma and photon targets.

6. With increasing energy, the cross-section for the capture of electrons from the target by fast positive ions decreases much faster than the cross-section for electron loss by fast neutral particles. Thus, the interaction of high-energy particles with targets causes, with a predominant probability, transitions to states with a smaller mass and a smaller number of electrons in the envelope. For atomic hydrogen particles accelerated to an energy of more than 10^5 eV, only the following transitions between charge states are realized with high probability: $H^- \rightarrow H^o \rightarrow H^+$, $H^o \rightarrow H^+$. Protons are the final product of these almost irreversible transformations, and they are not prone to [21] transformation into other charge states. Because of this, high-energy protons cannot be used as the initial state of particles used in charge-exchange technologies.

7. In a number of applications of charge-exchange technology, positive molecular ions can be used as the initial state of the particles. High-energy positive molecular ions are converted to neutral atoms with high efficiency; however, the atoms produced have only a half or a third of the kinetic energy of the molecular ion. Since progress in the higher-energy region involves great difficulties, such a significant shift down the energy scale makes the use of molecular ions not always beneficial.

8. An effective and often-used technique in charge-exchange technology is the transformation of charged particles into neutrals, which switches off any effect of macroscopic fields on the particle motion. Generalized data on the achievable conversion efficiency of various ions into neutral particles when interacting with targets of plasma and neutral particles, shown in Fig. 2.2, clearly demonstrate the advantages of using negative ions H^- to produce beams of neutral particles.

In targets of gases or vapors, the fraction of fast atoms formed from energetic protons increases with increasing target thickness to an equilibrium value, η^{+o} (without scattering). If at the exit from the target the proton beam can be separated from

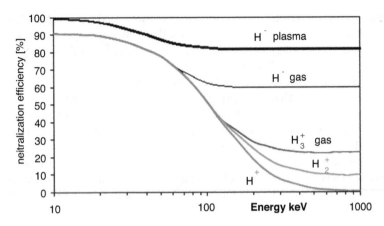

Fig. 2.2 Generalized data on the achievable efficiency of conversion of deuterium ion beams into fast neutral atom beams as a function of energy of the neutral atoms produced. (Reproduced from Dudnikov [6])

the fast atoms and again passed through the target, then it is possible to convert the residual proton fraction $(1 - \eta^{+o})$ into neutrals, again with efficiency η^{+o}. A sequence of such manipulations can be used to complete the transformation of protons into fast neutral atoms and even into negative ions, but the phase volume of the beams of particles produced with the required charge inevitably increases many times over the phase volume of the initial beam. A number of ingenious schemes designed to increase the efficiency of particle redistribution over charge states by multiple transmission of beams through the target have been described, but as far as we know, these methods have not been used yet [22]. The ratio of the cross-section for dissociation of molecular ions into neutral particles to the cross-section for the loss of electrons by fast neutral particles depends weakly on energy. Estimates allow us to hope that it is possible to convert up to 10–30% of the power transferred by molecular ions to the power of the neutral particle flux at a molecular ion energy up to several MeV, but in this case the decrease in kinetic energy of the particles produced makes such a process unprofitable. The efficiency of the formation of neutral atoms from negative ions is much greater because the electron detachment cross-section from the H^- ion is much greater than the electron detachment cross-section from H^o. A stripping cross-section for high-energy H^- and H^o and efficiency of H^o formation (up to 55%) was measured by Dimov and Dudnikov [23]. The ratio of these cross-sections is especially great when interacting with a target of highly ionized plasma. Estimates indicate that fast H^o neutrals can be formed from H^- ions at an efficiency of 85% using a plasma target [24]. A conversion efficiency of 82% has been observed [25] at energy up to 10^6 eV.

In addition, the $H^- \rightarrow H^o$ conversion ensures the production of neutral particles with minimal impact on the beam velocity. Scattering upon detachment of electrons from accelerated negative ions in the process of interaction with target particles has been considered [26, 27]. In calculating scattering angle, it was assumed that in the scattering event, the kinetic energy of transverse motion ΔW is comparable to the electron affinity S. From these considerations, the scattering angle α of fast atoms formed with kinetic energy W was obtained as $\alpha = (S/W)^{1/2}$, which is a large value. In fact, in fast particle collisions, an electron is detached as a result of the short electric field pulse produced in the fast particle system by the field of the target particle. As shown by Dudnikov [6], when an electron is separated from its atom, the electron is torn away from the negative ion, and the remaining neutral atom, comparable transverse momenta, should be transmitted, and the transferred energies should differ by a factor of M/m, where M is the mass of the heavy particle an m is the electron mass. The neutrality of the remainder and the short-range action of the polarization forces in the case of electron detachment from a negative ion should provide an additional decrease in the momentum transfer, so that for the scattering angle we obtain the relation:

$$\alpha = \left[mS / \left(MW \right) \right]^{1/2}.$$

According to measurements [26], when neutralizing H^- ions with an energy of 10^5 eV, the angular spread of the resulting fast atoms increases by 10^{-4} rad. The scattering angle decreases proportionally to $W^{-1/2}$ and depends weakly on the target material and its thickness to almost the maximum neutralization efficiency. When bound electrons are captured by protons and when molecular ions are dissociated, the angular spread of the fast atoms formed is found to be significantly greater [28].

Photonic-neutralizing targets have been discussed recently [29, 30]. Stripping of H^- ions with energy 10^3–10^4 eV has been tested experimentally [29, 31], and one can hope to achieve a conversion coefficient close to unity. At sufficiently high-ion current density, high-energy efficiency of these targets is assured. The technical difficulties involved in creating effective photon targets at the present time seem surmountable. The use of these regularities of the redistribution of accelerated particles over charge and mass states will be considered using examples of specific applications of charge-exchange technology. For the time being, we will not deal with methods for obtaining negative ions, but will dwell only on the requirements imposed on negative ion beams in specific applications of charge-exchange technology.

The cross-section for H^- "intrabeam stripping": $H^- + H^- \rightarrow H^- + H^0 + e^-$ was measured by analyzing the decay of a stored H^- beam. Results ($\sigma_{max} = 3.6 \times 10^{-15}$ cm^2 ± 30%) [32, 33] agree with recently published classical trajectory Monte Carlo calculations but suggest a smaller cross-section than obtained from earlier theoretical models. This sets more favorable conditions for the storage of H^- beams than previously assumed. Comparison of predictions intrabeam stripping cross-sections (green solid line) to the numerical simulations is shown in Fig. 2.3.

For a megawatt scale H^- linacs, the intrabeam stripping, if not addressed, can result in a power loss in excess of 1 W/m, creating considerable residual radiation in the high-energy part of the linac. Therefore, mitigation measures should be

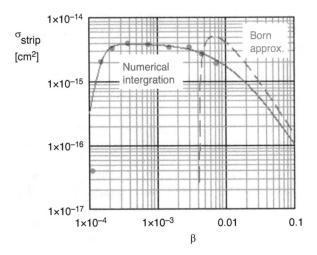

Fig. 2.3 Comparison of predictions of intrabeam stripping cross-sections (green solid line) to the numerical simulations of Ref. [34] (red dots) and to the results of Born approximation (dashed blue line)

considered. As one can see for a fixed bunch population an increase of beam sizes is the most effective way. At fixed emittance, it also reduces the velocity spread, resulting in an additional decrease of particle loss. However, there is quite limited potential for bunch size increase for both transverse and longitudinal degrees of freedom.

2.3 Charge-Exchange Tandem Accelerators

Charge-exchange accelerators (see Fig. 2.1), which provide repeated acceleration of particles in electrostatic fields, are one of the oldest and most established yet still vigorously developed applications of charge-exchange technology. The use of controlled charge exchange of particles for their multiple acceleration in electrostatic fields was suggested by Gerthsen [35] and by Dempster [36] in 1932 [37]. In their accelerator scheme, a spatially periodic distribution of electrostatic potential along the acceleration path was established. Positive ions (protons) were accelerated in the electric field and converted into neutral particles by charge exchange in a suitable target (hydrogen cell) before entering the decelerating region. Proton energy up to 145 keV was achieved [38], but particles with higher energy were also observed. Their appearance was explained by the transformation of a part of the proton beam into H⁻ ions which were then accelerated in the field, while positive protons were retarded. The present-day multistage acceleration system using special methods for producing negative ions was proposed by Peter [39] and Bennet [40] in 1936. Negative ions are accelerated from ground potential toward an electrode at high-positive potential; at the maximum potential, the particles are passed through a target so as to convert the negative ions into positive ions, which are then accelerated further by falling from high-positive potential back to ground potential. Such a two-step acceleration makes it possible to accelerate hydrogen particles to twice the potential of the high-voltage electrode. The negative ions can be accelerated in yet a further accelerator; in this way, a three-step acceleration is realized that generates hydrogen particles with energy tripled in comparison with the potential of the high-voltage terminal. When ions of heavier elements are used, the energy boost can be significantly greater, since the heavy negative ions accelerated to high energy are converted into multiply-charged positive ions upon interacting with the charge-exchange target.

Increasing the potential of the high-voltage electrode is fraught with considerable difficulty, and a practical limit to what is reasonably feasible is around tens megavolts. Given this situation, the opportunities provided by charge-exchange technology have proved to be very attractive. Their implementation has made it possible to move far beyond the energy limit of traditional electrostatic accelerators. Alvarez turned to the idea of tandem acceleration in 1951 [41], when the practicality of producing the necessary beams of negative ions was starting to look realistic; for real-world charge-exchange accelerators, beams of negative ions with an intensity of several microamperes were required, but even such modest needs could be met only in the late 1950s [42, 43]. Tandem accelerators can accelerate ions with micro-ampere beam current. Beams of negative ions with the necessary intensity and of acceptable quality were obtained at a relatively early stage of development of negative ion sources. In charge-exchange accelerators, the beam is transported through long accelerating tubes with long-focus optics and through long charge-exchange tubes. It is desirable to make the tube aperture as small as possible, since this simplifies and facilitates the construction of the tube and increases its electrical strength; however, the reduction of the accelerator tube aperture is limited by the minimum attainable beam emittance. In 1965, a two-stage acceleration of protons with current up to 350 μA was achieved [44]. At present, three-stage charge-exchange accelerators produce protons and deuterons with energy up to 30 MeV at beam current of several μA [45, 46]. Charge-exchange accelerators with high-voltage electrode potential up to 30 MV have been implemented [45, 47], and the possibility of increasing the voltage yet further, up to 50 MV, has been considered [45]. At the same time, the quality of the negative ion beams used sets a constraint on further extension of the capabilities of charge-exchange accelerators.

The transition to a two-stage acceleration system greatly facilitated the development of ion sources for electrostatic accelerators. This scheme allows the ion source to be located outside the tank with compressed gas, providing increased electrical strength of the high-voltage gaps; also, restrictions on the dimensions of the source and the necessary power, cooling, and vacuum pumping are then largely eliminated. Nevertheless, improving the technical characteristics of the source remains an urgent task. Particularly complex problems arise when the source must be situated in high-voltage terminals, because of the inevitably limited resources available there [48]. Recently, new high-voltage generators have been developed for direct-action accelerators, which make it possible to accelerate beams with current up to several

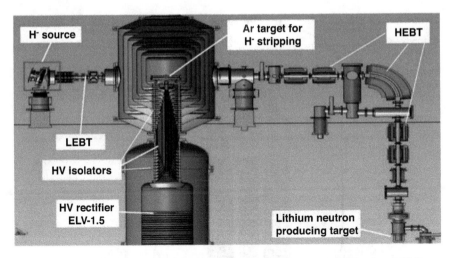

Fig. 2.4 Scheme of accelerator-based neutron source with vacuum-insulated tandem (VITA)

milliamperes and even tens and hundreds of milliamperes [49–51]. It has been possible to increase the proton current from a tandem with vacuum insulation (VITA shown in Fig. 2.4) to 10 mA [52, 53]. The limitation of the voltage stress in this case also allows the use of charge-exchange technology to increase the accelerated particle energy. The development of appropriate sources of negative deuterium ions with average beam current up to hundreds of milliamperes would allow the construction of impressively powerful neutron sources [28].

The specific application of tandems is accelerating mass spectrometry [54]. This analysis technique uses hypersensitive detection of the radioactive carbon isotope ^{14}C, based on the absence of stable negative nitrogen ions. Negative ions with mass $M = 14$ (such as $^{14}C^-$, $^{12}CH_2^-$, $^{13}CH^-$, $^{12}CD^-$) are separated by a mass separator and accelerated to a positive electrode. After acceleration they are passed through a gas or film target in which the ions lose electrons and molecular ions dissociate, and again the masses are analyzed and accelerated to the ground electrode. After that, the allocated ions are registered in and counted. Thus, it is possible to register ^{14}C ions with sensitivity up to 10^{-15} (one ^{14}C ion in 10^{15}, ^{12}C ions). A schematic of accelerator mass-spectrometer of BINP is shown in Fig. 2.5.

High-voltage ion implantation facilities using the tandem acceleration principle have been successfully employed in microchip fabrication [55]. For time-of-flight neutron studies, a pulsed mode of operation of charge-exchange accelerators has been used, where pulse duration of about 10^{-9} sec and repetition rate up to several MHz are typical. It is desirable to have the greatest possible negative ion pulse beam current [56]. Presently available negative ion sources do not allow the full use of the capabilities of the accelerators, and the development of suitable sources of negative ions remains an urgent task.

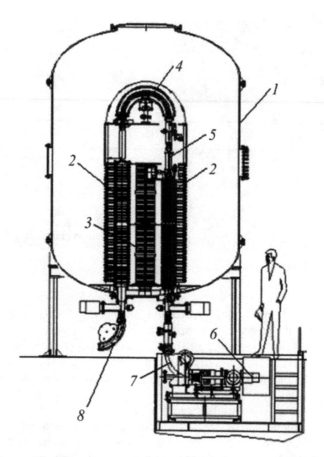

Fig. 2.5 Schematic of accelerator mass-spectrometer of BINP. (1) Tank with SF$_6$, (2) accelerating tubes, (3) electrostatic power supply, (4) magnetic mass-spectrometer, (5) magnesium stripping target, (6) sputtering cesiated negative ion source, (7) magnetic mass spectrometer, (8) exit magnetic mass spectrometer

2.4 Super-Collimated Beam Production

For precision experiments on the interaction of energetic particles with material in nuclear and atomic physics, solid-state physics, biology, and medicine, it is often necessary to form beams that are of very small cross-section and of minimum angular divergence. Producing such beams by collimating positive ion beams is difficult because of the scattering of particles at the edges of the collimating diaphragms, a situation that is amplified with decreasing ratio of hole diameter to collimator thickness, and reducing the thickness is constrained by the penetration of particles through the diaphragm material. The use of negative ions makes it possible to overcome these difficulties, since the interaction of negative ions with matter is accompanied by the detachment of the additional electron, and unscattered negative ions can be separated from the interacting particles by a magnetic field. Thus, it was possible to obtain "super-collimated" negative hydrogen ion beams of energy of 4 MeV with a diameter of 30 microns and angular divergence of 2.6×10^{-5} rad [25]. The angular divergence of the collimated proton beam was 100 times greater. To obtain intense super-collimated beams, beams of negative ions are needed with little intensity, but with the greatest possible brightness.

2.5   Charge-Exchange Extraction of Particles from Accelerators

Charge-exchange technology is used to extract accelerated particles from cyclotrons (see Fig. 2.1). In high-energy cyclotrons, and especially in isochronous cyclotrons, the increments in the spiral particle trajectory become very small toward the end of the particle acceleration. This makes it difficult to use traditional methods for efficient particle extraction. But if the particles are caused to interact with a target, the change in their mass or charge leads to a significant change in the trajectory curvature and ensures reliable particle separation. Negative hydrogen and hydrogen isotope ions have been used in some systems. Typically, negative ions are injected into the cyclotron into an internal orbit either from an internal source or from a source located outside the cyclotron; in the latter case, negative ions are passed through a channel in the yoke and bent through 90° by an electrostatic deflector [57]. To extract the accelerated particles, a target is introduced in the form of a thin foil, at which target the negative ions are converted to positive ions that are then readily extracted outside the magnet. This method of particle extraction from the cyclotron by charge-exchange provides a number of advantages, including high efficiency, higher extracted beam quality, the possibility of smooth variation of extracted beam energy by target movement, and the option of simultaneous extraction of beams at different energies via the incomplete overlap of beams at different orbits by targets. Negative ions were first used in the University of Colorado cyclotron in 1962 [58] and were subsequently used in many (>3000) cyclotrons for the production of medical isotopes. In 1974, an isochronous cyclotron was launched at TRIUMF, Canada, with projected average extracted beam power of up to 50 kW at proton energy up to 520 MeV and with energy spread of less than 0.1%. Accelerated particles are extracted from the cyclotron by the conversion to protons of accelerated H^- ions. Simultaneously extracted proton beams were obtained in the energy range 180–520 MeV [59]. By using a sufficiently thin target, accelerated negative ions can be extracted with high efficiency in the form of beams of neutral particles. Such beams can easily and without additional loss be split into a plurality of beams using stripping targets and magnets ("charge-exchange beam distribution") [23, 60]. For cyclotron application, negative ion beams with average beam current of several milliamperes to tens of milliamperes are needed. The quality of the beam used influences the design and efficiency of the cyclotrons.

2.6   Charge-Exchange Distribution of Accelerated Particle Beams

Acceleration of negative ions instead of protons, or even together with protons, in high-power linear accelerators allows the use of charge-exchange technology of operating with accelerating beams. Negative ions can be accelerated along with positive ions in the antiphase high-frequency accelerating field, allowing doubling the power and increasing the efficiency of the accelerator for the case when the accelerator is limited to the peak value of beam current. Negative ion beams can be divided into an arbitrarily large number of proton beams by stripping targets and bending magnets without loss of intensity and deterioration of quality [59], whereas using traditional methods of impact separation of beams is a challenge. Other manipulations with beams that are necessary for experiments are also facilitated. H^- ions are used in the powerful LANSCE linear accelerator, with particle energy 800 MeV at projected average beam power up to 800 kW [58]. For these applications, it is desirable to have H^- ion beams with parameters close to the parameters of the proton beams used: pulse beam current ~100 mA, pulse duration 10^{-3} sec, repetition frequency of 120 Hz, and normalized emittance 0.2–0.5 mm.mrad.

2.7   Charge-Exchange Injection into Accelerators and Accumulator Rings

The charge-exchange injection of particles into accelerators and storage rings constitutes an important applications of charge-exchange technology (see Fig. 2.1) [61].

Particles entering a region of steady magnetic field from a region without magnetic field move along infinite trajectories that intersect the magnetic field region. Particles irreversibly trapped in the magnetic field move along finite quasi-periodic trajectories. Particles starting their motion along finite trajectories from a particle injector located on this trajectory again come to the injector after one or a few periods, and, if the injector is not sufficiently "transparent," they are lost. For transformation of "lossy" trajectories into captured trajectories and for moving trajectories away from the "opaque injector," pulse techniques are required to control the particle motion, such as by changing the magnetic field or by influencing the particles with additional electric and/or magnetic fields. Usually these effects are not selective—in providing irreversible capture of injected particles, they also cause loss of previously captured particles. Many substantial methods have been developed for the capture of accelerated particles into stationary orbits in the leading magnetic field of accelerators and storage devices, including single-turn injection with an inflector, injection into an increasing field, spiral accumulation, a phase displacement method, and others. These traditional methods of injection allow the accumulation of particles up to hundreds of circulation periods. In this case, newly arrived particles are placed in a region of the system phase space not yet occupied by earlier

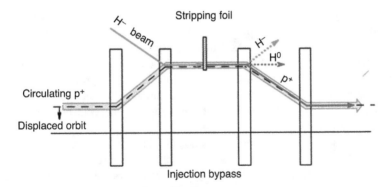

Fig. 2.6 Schematic representation of charge-exchange injection

particles, so the brightness of the accumulated beam (particle density in the space of transverse coordinates and momentum) cannot be higher than the brightness of the injected beam. Injector technology that has been developed allows filling accelerator paths filled at low-magnetic-field strength, in accordance with their phase capacitance. However, the brightness of available proton beams is insufficient to completely fill the storage ring tracks, which have a greater space charge limit. A schematic representation of charge-exchange injection is shown in Fig. 2.6.

Charge-exchange technology allows effecting an injector that is located on an equilibrium finite orbit and is transparent to irreversibly trapped particles. Such an injector is the transforming target, which "generates" protons. The "input transformation particles" H^-, H^o, H_2^+, or H_3^+ are accelerated to the required energy and brought to the target in such a way that protons formed there move along the necessary trajectories. At the end of the injection phase, the target can be "removed," thereby completely eliminating its effect on particle motion. It is important in this case that new particles fall into the region of phase space already filled with trapped particles; in this way the brightness of the accumulated beam can exceed the brightness of the injected beam by orders of magnitude. In doing so, it is possible to circumvent the restrictions imposed by Liouville's theorem. The charge-exchange method of trapping particles into stable orbits was discussed by Alvarez in 1951 [40]. However, at that time the level of technology for obtaining negative ions was so low that one could only talk about studying orbits in a steady magnetic field with no expectation for accumulating beams of appreciable intensity.

Later, the attractiveness of charge-exchange injection of protons into accelerators was independently noted by a number of authors [23]. A focused study of the problems associated with charge-exchange injection of protons into accelerators was begun in 1960 at the Novosibirsk Institute of Nuclear Physics (INP) of the SB AS USSR, on the suggestion of G.I. Budker [63] for the development of the proton-antiproton collider VAPP (later on, this program was realized at FNAL, USA, with INP developments in charge-exchange injection, antiproton production, and electron cooling, with important participation of former BINP members [64]). Since we were talking about the accumulation of beams with maximum intensity, and the record intensity of H^- ion beams that time was only 70 µA, an important part of the program was the development of methods for obtaining intense negative hydrogen ion beams (several mA). Charge-exchange injection of protons into the storage ring was carried out experimentally at the Novosibirsk Institute of Nuclear Physics in 1964 [65]. It was then possible to increase the intensity of the proton beam accumulated in the storage ring by the charge-exchange method up to the space charge limit [62, 66–68]. A simplified schematic of the setup for studying charge-exchange injection is shown in Fig. 2.7a.

This small circular proton storage ring consisted of a continuous, cyclotron-type magnetic field with weak focusing, radius $R = 42$ cm, aperture 4×8 cm^2, field index $n = 0.6$, vertical betatron frequency $Q_z = (0.6)^{1/2}$, radial betatron frequency $Q_r = (1–0.6)^{1/2}$, and started experiments with a bunched beam.

In these experiments, a beam of 1 MeV H^o atoms was produced at the charge-exchange (stripping) target 1, by conversion in a gas target from H^- ions accelerated to an energy of 1 MeV. The conversion efficiency is high (>50%), as shown in Fig. 2.2. At a second stripping target for conversion of H^o atoms to protons, a supersonic jet of hydrogen 3 was used, which was switched on for the time of injection. A residual gas ionization beam profile monitor 8 (Fig. 2.7a, b) and a fluorescent beam profile monitor 6 were used for the first time in this storage ring [69], devices now routinely used in all proton and ion accelerators and storage rings. A drift tube RF accelerator Sect. 2.5 was used for the compensation of ionization energy loss. The experiments completely confirmed the initial premises. With the compensation of ionization energy losses by RF voltage, the capture efficiency over a span of 2000 revolutions was 75%, in accordance with the separatrix area, and when injection was for 4000 revolutions, the efficiency was reduced only by 20%. In these experiments, the electron-proton instability (electron cloud effect), limiting the intensity of beams in meson factories and in other large accelerators and accumulators, was observed, explained, and feedback-suppressed for the first time [70]. The accumulated beam lives for 1.5–5 msec and then betatron oscillations develops, and the beam is dropped from the orbit in several dozen revolutions. With a high RF, voltage is lost only at the central (coherent) part of the beam.

Subsequently, in 1967, experiments were carried out to achieve a circulating proton beam with space charge compensation of the beam by electrons and with compensation of ionization energy losses by an induction electric field. A schematic of the storage ring with betatron compensation of ionization losses is shown in Fig. 2.8 [71–74].

Hollow "donut" (donut made from copper with a channel inside, with radius 42 cm and with aperture 3×4 cm^2) 3 was located between the poles of the electromagnet 2, into which a beam of high-energy (1 MeV) neutrals was injected. The induction field was produced by the discharge of a capacitor bank to donut cut 5 with a labyrinth (a coaxial copper structure) 5 which prevents the penetration of pulsed magnetic field into the donut. A selection of oscillograms characterizing the accumulation of the circulating proton beam in the donut is shown in Fig. 2.9. The circulating beam accumulates to the equilibrium level (first oscillogram, trace 1 in Fig. 2.9). The potential of the beam, measured by an annular pickup electrode,

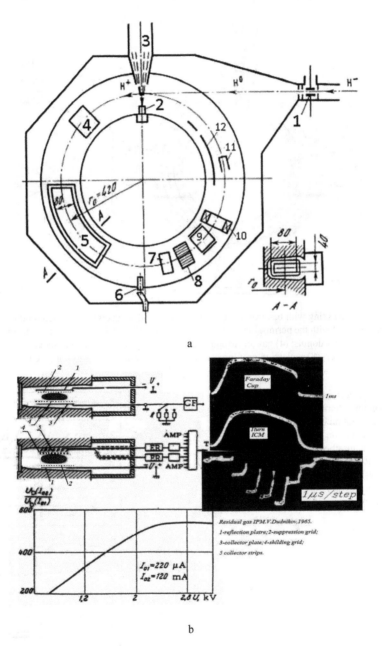

a

b

Fig. 2.7 (**a**) Schematic of installation for the study of charge-exchange injection. (Reproduced from Dudnikov [62]). (1) First stripping target, (2) supersonic jet nozzle, (3) jet receiver, (4) annular pickup electrode, (5) drift tube RF accelerator, (6) collimator of fluorescent beam profilometer, (7) ionization beam intensity meter, (8) ionization profilometer, (9) beam position pickup, (10) Rogowski coil, (11) Faraday cup, (12) deflector for suppressing e-p instability. (**b**) Residual gas ionization beam current and beam profile monitors (ICM, IPM). (Reproduced from Dudnikov [62]). (1) Reflection plate, (2) suppression grid, (3) collection plate, (4) screening grid, (5) collector strips for separate collection of electrons along magnetic field lines, connected to oscilloscope by fast commutator for produce histograms of beam profiles

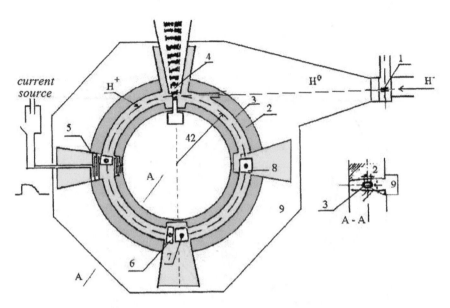

Fig. 2.8 Storage ring with betatron compensation of ionization energy losses. (Reproduced from Shamovsky [22] with the permission of V. Shamovsky). (1) First stripping target; (2) magnet pole; (3) hollow copper donuts; (4) gas jet stripping target; (5) labyrinth; (6) annular pickup; (7) ionization current meter; (8) ionization profilometer; ion-electron collectors; (9) vacuum chamber

Fig. 2.9 Selection of oscillograms characterizing the accumulation of a circulating proton beam in a donut in quasibetatron mode. (Reproduced from Shamovsky [71] with the permission of V. Shamovsky)

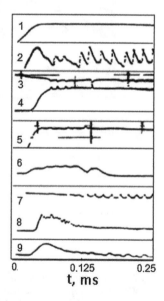

increases and then slowly decreases due to the accumulation of compensating electrons for ~10 microseconds. Then the potential increases sharply due to the development of transverse instability of proton beam and removal of electrons, and the

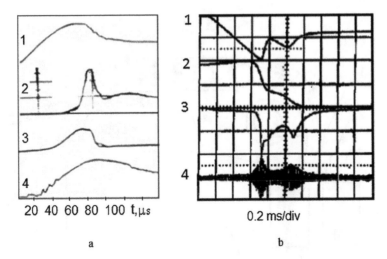

Fig. 2.10 (**a**) Oscillograms showing beam accumulation with ionization loss compensation by an induction field. (Reproduced from Shamovsky [71] with the permission of V. Shamovsky). (1) Signal from the current meter, (2) signal from the vertical loss probe, (3) signal from the horizontal loss probe, (4) signal showing vertical coherent oscillations of the beam. (**b**) Oscillograms showing beam accumulation with ionization loss compensation by an induction field. (1) Signal from the current meter, (2) signal from the horizontal loss probe, (3) signal from the vertical loss probe, (4) signal showing vertical coherent oscillations of the beam

electron accumulation process is repeated (trace 2). The collector, screened by a grid, registers the emission of electrons and ions simultaneously over the entire orbit (traces 3, 4, 5, 7). When bunching the beam due to the negative mass effect (trace 8), electrons do not accumulate and the instability is suppressed (traces 6, 9).

To suppress the negative mass effect, poles with strong focusing were installed in the electromagnet. With these poles, the beam accumulation with compensation of ionization losses by an induction field was investigated. Oscillograms characterizing the beam accumulation with ionization energy loss compensation by the induction field are shown in Fig. 2.10a, b. The beam current accumulates and then saturates (trace 1), with an increase in the signal from the horizontal loss probe (trace 3). At the same time, the signal from the vertical loss probe (trace 2) increases. The vertical position monitor of the beam fixes the growth of vertical betatron oscillations (trace 4) before the beam is dropped vertically [68]. This instability, associated with the oscillation of compensating particles in the beam potential well, is well described by the instability theory developed by Chirikov [75] for an electron beam compensated by ions. A more detailed theory of this instability was developed by Koshkarev and Zenkevich [76] and later by Bosch [77].

The examination of collective effects in circulating beams with extreme space charge intensity in combination with charge-exchange injection helped to create this "super-unstable" circulating proton beam, compensated by an electron cloud, with an intensity almost an order of magnitude greater than the space charge limit

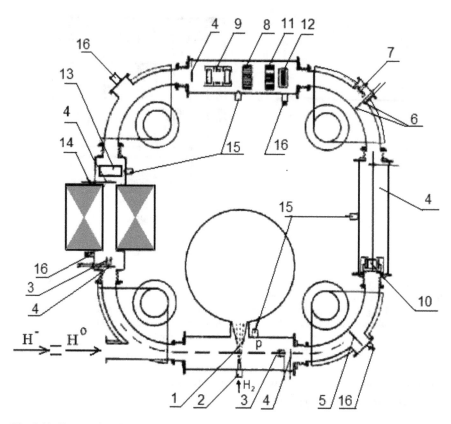

Fig. 2.11 Storage ring for producing a circulating proton beam with intensity above the space charge limit. (Reproduced from Shamovsky [71] with the permission of V. Shamovsky). (1) Supersonic jet stripping target, (2) pulsed jet valve, (3) beam collector, (4) quartz screen, (5, 6) mobile target, (7) ion collector, (8) Rogowski coil, (9) beam position monitor, (10) electrostatic pickup of quadrupole beam oscillations, (11) electromagnetic transverse beam oscillation sensor, (12) vertical beam sensor, (13) secondary charged particle meter, (14) induction core, (15) pulsed gas inlet, (16) gas inlet valve

[78–80]. The installation developed for producing this circulating proton beam with intensity above the space charge limit is shown in Fig. 2.11.

This facility consists of a race-track-type magnetic system with a large circumference of about 6 m, radius of bending magnets $R = 42$ cm, magnetic field 3.5 kG, index $n = 0.2$–0.7, straight sections 106 cm, aperture 4×6 cm^2, and revolution frequency 1.86 MHz. An inductive core was used for the compensation of the ionization energy loss of ~200 eV per turn, which produced some ionization cooling.

A super-intense proton beam with intensity ~1 A, corresponding to a calculated vertical betatron tune shift $\Delta Q = 0.85 \times 6 = 5.1$ with $Q = 0.85$, was accumulated with e-p instability self-stabilization by fast accumulation of high-circulated beam current and accumulation of plasma from residual gas ionization. It also suffered an electron cloud instability. In this case, the threshold corresponded to 1.2×10^{10}

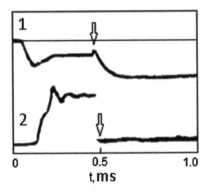

Fig. 2.12 Oscillograms showing the first observation of beam accumulation with intensity exceeding the space charge limit. (Reproduced from Shamovsky [71] with the permission of V. Shamovsky). (1) Accumulated beam intensity, (2) beam vertical position. The arrow indicates the time at which electron removal is switched off

protons. The coasting beam instability was suppressed (self-stabilized) by increase in the injection beam current and the gas density. The existence of an "island of stability" above threshold was consistent with an analysis based on the prediction by Chirikov [75]. This self-stabilization of the transverse e-p instability in the PSR was explained by increasing the beam density and increasing the rate of secondary particle generation above a threshold level with fast decrease of the unstable wavelength λ below the transverse beam size a. (i.e., the sum of beam density n_b and ion density n_i are above a threshold level):

$$\left(n_b + n_i\right) > \beta \, / \, 2\pi r_e a^2. \quad \left(r_e = e^2 \, / \, mc^2\right).$$

Oscillograms confirming the first observation of the accumulation of a beam with intensity exceeding the space charge limit are shown in Fig. 2.12. The beam was accumulated up to the space charge limit with the removal of electrons from the beam by a transverse electric field (trace 1). In this case the Hervard instability, connected with the beam interaction with the ion trace, develops, which does not lead to beam loss (trace 2). After electron removal is switched off, the oscillations rapidly decay (trace 2), and the intensity of the beam rises above the space charge limit (trace 1) [81] if the intensity of the injected beam exceeded the critical value of ~5 mA.

If the injected beam intensity was below the critical value, intense oscillations developed, and the accumulated beam intensity decreased significantly after switching off the electron-removing field. After increasing the intensity of the injected H⁻ ion beam to 15 mA (8 mA injected atom current), it was possible to accumulate a beam with intensity above the space charge limit without electron removal. Oscillograms showing the accumulation of a beam with intensity above the space charge limit is shown in Fig. 2.13 [78–81]. If the intensity of the injected beam is above the critical level (~5 mA), many positive ions accumulate in the beam, and the

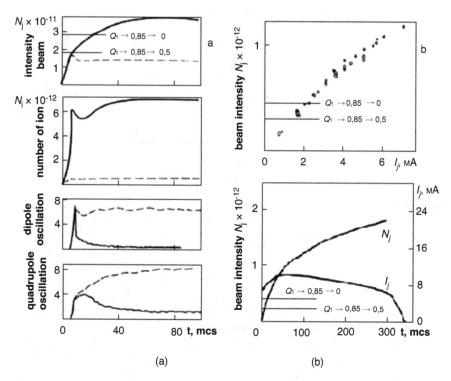

Fig. 2.13 (**a**) Oscillograms showing the accumulation of a beam with intensity above the space charge limit (solid) and below (dashed line) and (**b**) dependence of circulating beam intensity on injected beam intensity. (Reproduced from Belchenko et al. [28], with the permission of AIP Publishing)

circulating beam accumulated is above the space charge limit (oscillogram 2 in Fig. 2.12a), and dipole and quadrupole oscillations are rapidly damped (oscillograms 3 and 4 in Fig. 2.12a). If the intensity of the injected beam is below the critical level, the intensity of the circulating beam is limited to a low level (dashed line in Fig. 2.12a), positive ions do not accumulate in the beam, and dipole and quadrupole oscillation remain high. Stabilization of oscillations in this case is due to the decrease of the unstable oscillation wavelength to the transverse beam size, when they become surface waves and cannot influence the beam volume. The intensity of the accumulated beam can reach up to 1 A, a factor of nine greater than the absolute space charge limit, when the space charge reduces the focusing effect of the magnetic system to zero, and a factor of 150 greater than the threshold for electron-proton instability (electron cloud effect). Plasma generators were used for the stabilization of e-p instability at lower injection current and lower residual gas density [82, 83]. An attempt to obtain a circulating beam with compensated space charge has been made with the Fermi National Accelerator Laboratory installation IOTA [84].

Results of investigations of the charge-exchange method of injection of protons into accelerators have been described in the literature [22, 72, 73, 76, 82, 85]. The inverse charge-exchange of protons into atoms limits the injection time at energies lower than 10^6 eV. At higher energies, only multiple scattering of circulating protons is essential, limiting the injection time to a level of 10^4 revolutions. By placing a target at the local minimum of the β-function of the focusing system, it is possible to reduce the effect of multiple scattering and even ensure the attenuation of incoherent betatron oscillations due to ionization energy losses in the target [35, 83]. Such a large permissible duration of highly effective capture of protons allows us to sharply reduce the injected beam intensity requirements, and, especially important, when the charge is injected the injected beam brightness requirements are reduced. While traditional injection methods necessitate proton beam injection with the maximum allowable intensity and brightness, the same results can be obtained using H⁻ beams with intensity and brightness less than a hundredfold.

The sources available and being developed do not allow injection into synchrotrons of sufficiently intense beams of nuclear spin polarized ions. Sources with an intensity of 10^{-3} A have been developed. With charge-exchange injection, however, this intensity is sufficient to fill the booster synchrotron to the space charge limit [86].

Charge-exchange injection can also be used for injection into cyclotrons. The injected particles can be passed to the center of the cyclotron as neutral particles and recharged (stripped) at a target located at the first turn. This injection method has been used to inject polarized protons into cyclotrons [87, 88].

In 1968, Ron Martin, director of the Zero Gradient Synchrotron (ZGS) at Argonne National Laboratory, visited the INP, Novosibirsk, and became acquainted with charge-exchange injection developments there. He decided that charge-exchange injection would allow the Zero Gradient Synchrotron to compete in intensity with the AGS at Brookhaven. In 1969, charge-exchange injection was successfully tested on the 12 GeV ZGS proton synchrotron at Argonne at an injection energy of 50 MeV [89]. Protons were captured into orbit upon charge exchange of H⁻ ions in a thin organic film target. At Martin's suggestion, charge-exchange injection of protons into the ZGS synchrotron booster was developed [90]. In 1971, charge-exchange injection of protons into a 200 MeV synchrotron was performed— a prototype of the ZGS booster [85]. The booster served for many years since 1977 as an intense-pulsed neutron generator [91]. In 1978, charge-exchange injection was developed for the Fermi National Accelerator Laboratory (FNAL) booster with 200 MeV injection energy [92]. In 1982, the AGS synchrotron at BNL was switched over to charge-exchange injection [93]. In 1984, charge-exchange injection was implemented on the ISIS synchrotron at the Rutherford Appleton Laboratory, England [94], and on the proton accumulator ring at Los Alamos National Laboratory [95, 96]. In 1980–1984, charge-exchange injection was realized at KEK, Japan [97], at DESY (Deutsch's Electron Synchrotron, Hamburg, Germany) [98] and in Chinese Spallation Neutron Source (CSNS) [99]. Charge-exchange injection is used on the CELSIUS ring (Uppsala, Sweden) [100], in the COSY ring (Jülich Research Center, Germany) [101, 102], and in the booster at the Institute of Theoretical and Experimental Physics (ITEP, Moscow) for the accumulation of carbon ions [103].

The transition to charge-exchange injection in the CERN booster (Geneva, Switzerland) [104] is finished [105] and in the IHEP booster (Protvino, Russia) [106] is now being prepared. It is planned to use charge-exchange injection for proton injection into the storage ring of European Spallation Source (ESS) [107].

In 1975, a patent was issued for a pulsed neutron generator [108] in which a ring proton accumulator is used to convert the time structure of the beam from a powerful linear accelerator. The negative-ion beam macro-pulses of duration of 1 msec and frequency up to 120 Hertz are assumed to be captured in the storage ring by a charge-exchange method. By compressing the accumulated beam into a short bunch and using one-turn extraction, it is possible to convert the macro-pulses into short pulses of very-high proton current, as is necessary for time-of-flight studies. Using a slow output, it is possible to convert the macro-pulses into a quasi-continuous proton beam, as is necessary for uniform loading of recording equipment. Stripping was supposed to be carried out using carbon or beryllium foil targets. To avoid unnecessary interaction with the target, it was intended to use a local orbit offset (bypassing the target). The complex of problems associated with implementation of this project is being investigated deeply. The project was implemented at LANSCE in 1984 [109]. The pulse intensity of the beam in the storage ring is limited by the development of the electron-proton instability [61, 62, 66, 67, 73, 78, 80, 110–117].

In 2006, work started on the Spallation Neutron Source (the SNS), an intense source of neutrons with charge-exchange injection at Oak Ridge National Laboratory [111]. Ions of H^- with current ~40 mA are accelerated in a superconducting linear accelerator to an energy of 1 GeV and accumulate for ~10^3 revolutions in a compact storage ring with revolution period ~1 μs (accumulated current up to ~50 A, power up to ~50 GW).

Let us consider the implementation of charge-exchange injection using the example of charge-exchange injection into the SNS storage ring. A stripping foil is placed in the second magnet fringe field of strength of 0.25 T, and a thick stripping foil is positioned after a third magnet for stripping residual H^- ions and atoms (including excited) to protons, which are sent to a shielded beam collector (see Fig. 2.1). A stripping foil is attached to the side of a holder, as shown in Fig. 2.14. The thickness of the small crystalline carbon foil is 350 μg/cm^2 and of size 12×30 mm^2.

The foil holder is attached to a bicycle chain and is inserted into the beam by moving the chain, as can be seen in Fig. 2.14. The stripped electrons have a kinetic energy of 0.5 MeV and a power of 3 kW at an average H^- ion beam power of 1 MW. They move in spirals around the magnetic field lines to a graphite electron trap at the lower pole of the second magnet after the foil. Reflected electrons can damage the foil holder. With an average beam power of up to 1.4 MW, the foil can withstand stripping up to 5×10^3 Coulombs (10^5 pulses). This intensity is almost the limit for a carbon foil, since the foil overheats and the carbon begins to sublime. The complex stability problems of charge-exchange foils in high-power beam injection have been discussed in the literature [112].

Fig. 2.14 Photo of the stripping foil on the SNS holder. (Reproduced from Plum [112] with permission of M. Plum)

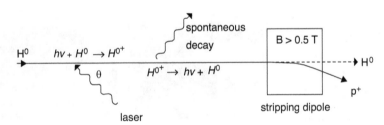

Fig. 2.15 Schematic of the system for ionization of hydrogen atoms with energy 1 GeV by laser irradiation for laser charge-exchange injection. (Reproduced from Cousineau et al. [113], with the permission of APS Publishing)

The laser ionization of H° atoms into protons for laser charge-exchange injection has been described [113, 114]. A schematic of the experimental system is shown in Fig. 2.15.

The first electron is detached by Lorentz ionization in a strong magnetic field. Laser radiation is encountered with the H° beam at an angle θ in the laboratory reference frame. The laser radiation frequency in the rest frame of the atoms, ω_0, is related to the frequency of the laser radiation in the laboratory frame, ω, by $\omega_0 = \omega\gamma(1 + \beta\cos\theta)$, where θ is the angle between the direction of laser radiation and the direction of atom motion, $\gamma = 1/(1 - \beta^2)^{1/2}$, and $\beta = v/c$. For the upper excited state with $n = 3$, which can be stripped in the magnetic field by Lorentz ionization, $\omega_0 = 1.84 \times 10^{16}$ Hz and $\lambda_0 = 102.6$ nm. This wavelength corresponds to the third harmonic of 355 nm radiation, with wavelength 1064 nm at an angle $\theta = 1.064$ radians, for an atom beam energy of 1 GeV. The time structure of the laser radiation is consistent with the time structure of the linac beam: pulses of 50 psec at a frequency of 402.5 MHz. Atoms with greater energy have a slightly smaller angle to the radiation beam, which allows them to be in resonance with the radiation. The pulse radiation power is ~1 MW, and the duration of the macro-pulse is 10 microseconds. This power is enough for Lorentz stripping of 90% of the excited atoms by the second magnet (Fig. 2.15). The next step should be the use of an optical resonator for the accumulation of laser radiation so as to increase the duration of atom stripping up to 1 ms.

This scheme of charge-exchange injection permits the production of polarized nuclear by optical pumping by a laser with circular polarization.

The history of the observation of the electron-proton instability (electron cloud effect) is shown in Fig. 2.16 [115]. In 1965, as is now clear, the e-p instability was observed in the ZGS at ANL and in the AGS at BNL, but only at the INP, Novosibirsk, was it correctly interpreted, explained, and suppressed [65–67]. The instability was damped by negative feedback in the Bevatron, Berkeley, in 1971. At the Intersecting Storage Rings facility (ISR, CERN) in 1972, this instability formed a background for the detector. At the Los Alamos National Laboratory, this instability has limited the pulse intensity of the beam since 1988, and since 2000, it has limited beam intensity at B-factories, intense beams in positron storage rings, and proton beams

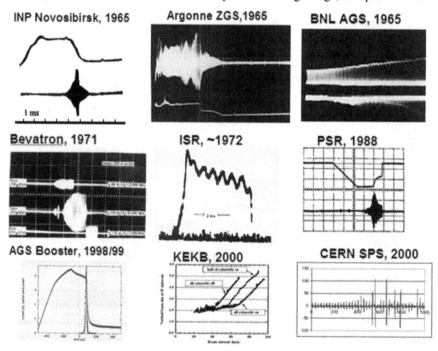

Fig. 2.16 The history of observation of the electron-proton instability. (Reproduced from Zimmermann [115] with permission from F. Zimmermann)

in the Large Hadron Collider (LHC, CERN). The review of electron cloud effect in present and future high-intensity hadron machines was presented in [116].

Charge-exchange injection calls for reliable H^- ion sources with pulse intensity up to 100 mA and high-duty cycle. Such sources became possible after the discovery and development at the Institute of Nuclear Physics, Novosibirsk, of the surface plasma method for producing negative ions using the cesiation effect [118–120], and the development of surface plasma sources of intense, bright beams of negative ion [2, 3, 121].

We may summarize the principal features of charge-exchange injection into accelerators and accumulator rings as follows.

- The successful development of charge-exchange injection at the BINP, Novosibirsk, opened opportunities for the widespread use of charge-exchange injection for all ion accelerators. Without charge-exchange injection, the loss of injected beam is ~10%, while charge-exchange injection allows decreasing the loss to 0.002% [117]
- The discovery and explanation of the electron-proton instability (electron cloud effect) was an important achievement in the physics of intense ion beams. This instability still limits the intensity of the largest accelerators and storage rings.
- Obtaining circulating beams with an intensity much greater than the space charge limit, and the development of a high-frequency acceleration system with space charge compensation, [122] presents optimism for a significant increase in the intensity of cyclic accelerators.
- The requirements for negative ion sources for charge-exchange injection have been summarized [123].

2.8   Charge-Exchange Injection into Magnetic Plasma Traps

Charge-exchange injection of powerful beams of accelerated neutral particles is one of the most promising methods for producing plasma with thermonuclear parameters in magnetic traps (Fig. 2.1). Plasma confined in the magnetic field is used as the target, which converts the injected energetic neutral particles into magnetically trapped energetic ions. For primary accumulation of plasma, one has to resort to tricks. One of the first applications of charge-exchange technology in thermonuclear research, as proposed by Budker in 1958 [124] and implemented under the leadership of Golovin, was the accumulation of plasma in the OGRA magnetic mirror trap in 1961 [125]. The primary accumulation of the plasma was by the formation of fast protons in collisions of accelerated H_2^+ ions with residual gas. At the DCX installation, a magnetic mirror trap at Oak Ridge, an injected molecular H_2^+ ion beam, was dissociated into H^+ ions and H^o atoms by collisions with a high-current arc that intersected the beam [126]. Later, Lorentz ionization in the magnetic field of highly excited atoms formed during the charge exchange of protons in special targets was used [127]. Charge-exchange technology for the production of hot plasma in magnetic traps was proposed a long time ago [128, 129] and has a long history of development, but this direction began to develop particularly intensively in recent years [130]. Using the high conversion efficiency of protons into atoms at low energies, it was possible to create injection systems to produce H⁰ atoms with energy tens of keV and intensity up to hundreds and thousands of amperes [128, 131]. It was also possible to effectively trap particles in the magnetic field of the traps, using the ionization of fast atoms in the pre-created highly ionized Tokamak plasma [132] or in the pre-injected plasma of mirror traps [133]. Charge-exchange injection allows one to create arbitrary particle distributions in terms of energy and space, to effect volumetric influences on the state of the plasma, and to maintain a current in the plasma. To further advance toward reactor parameters and under reactor conditions, atoms with energy hundreds of keV and up to about 1 MeV are needed. At such energies, the conversion efficiency of positive hydrogen isotope ions into neutral atoms becomes unacceptably low (see Fig. 2.2). The only alternative is to obtain high-energy atoms from accelerated negative ions. To meet these needs, injectors are needed that generate and accelerate beams of negative ions with an intensity of many tens of amperes.

A closed configuration of field lines with an inverted magnetic field in a circulating electron beam, as proposed by Christoffilos for thermonuclear fusion, was achieved by injecting an electron beam of current tens of kiloamperes into a magnetic field [134]. The instabilities that destroy the circulating beam were overcome by injecting beams for which focusing by the self-magnetic field is already needed for injection. These advances have stimulated the development of systems for the implementation of thermonuclear fusion using configurations with closed field lines in circulating closed ion beams [135]. To obtain such ion beams, it is proposed to use a surface-plasma method for producing beams of negative ions. Thus, well-formed fluxes of H⁻ ions are required with an intensity of order hundreds and thousands of amperes at an energy of hundreds of keV and MeV.

Negative ion NBI systems are currently in operation at two fusion experiments in Japan: at LHD (Large Helical Device, in Toki) with a nominal beam energy of 180 keV and an achieved maximum deposited power of 5.7 MW per beamline and at JT-60 U (Japan Torus-60 Upgrade, in Naka) with an achieved beam energy of 416 keV and an injected power of 5.8 MW. Furthermore, the MeV accelerator located in Naka has shown the feasibility to reach a negative ion acceleration energy of 1 MeV for pulse lengths of up to 60 s and with extracted current densities below the value required for ITER.

The experience gained with the neutral beam injection system at ITER will be extremely important in view of the NNBI system for the demonstration reactor DEMO, whose requirements might even be more demanding with respect to ITER. The final design of the NNBI for DEMO is not yet defined: research and development are aiming to fulfill the ITER NNBI requirements since at present no NNBI system has yet achieved all the ITER requirements simultaneously. At the same time, research and development for the NNBI for DEMO are already started, considering in particular the RAMI requirements.

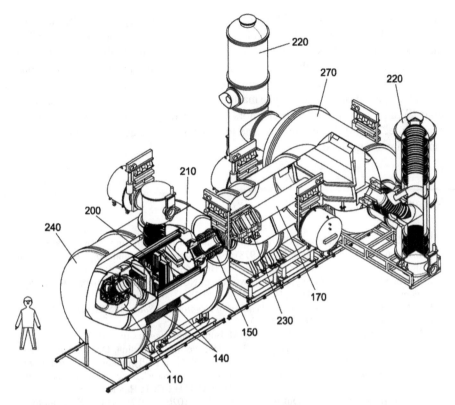

Fig. 2.17 Schematic of NBI with negative ion source of BINP, Novosibirsk [136]. An isometric cutaway view of the discharge external magnets, the absence of embed JET neutral beam injector, (110) RF negative ion source with cesiation, (140) an insulator: support, (150) accelerating tube, (170) neutralizer, (200) pumping panel, (210) vacuum tank, (220) recuperators, (230) quadrupoles, (240) a larger-diameter tank filled by SF_6 gas, and (270) a plasma chamber

Schematic of NBI with RF cesiated Surfacer Plasma negative ion source (SPS) 10 A, 1 MeV in BINP, Novosibirsk is shown in Fig. 2.17 [135].

2.9   Summary

Thus the development of effective methods for obtaining negative ion beams is the starting point for realizing the attractive possibilities provided by charge-exchange technology for developing energetic particle beams. Negative ions can be used in most positive ion applications; their indispensability for charge-exchange technology has driven efforts to develop methods for developing negative ion beams. On the other hand, achievements in negative ion technology have driven and will continue to drive the development of new effective applications of negative ions. With the development of the surface-plasma method for obtaining negative ions, previous limitations on the intensity and brightness of the resulting negative ion beams have been virtually eliminated. Future development of VITA tandem for BNCT in Budker Institute of Nuclear Physics is presented in [137]. There is no doubt that the creation of surface plasma sources that generate high-quality intense beams of negative ions will affect the emergence of many new applications of accelerated particle fluxes. The development of the surface plasma source allows a real discussion of the question: "Will proton sources in accelerators be replaced by negative ion sources in the near future?" [138]. Recent results tend to make a positive answer to this question more and more justified.

References

1. G.I. Dimov, V.G. Dudnikov, Charge-exchange method for controlling particle beams. Sov. J. Plasma Phys. Engl. Transl. **4**(3) (1978). Г.И. Димов, В.Г. Дудникшв, «Перезарядный метод управления пучками ускрренных частиц», Физика плазмы, 4(3) (1978)
2. V. Dudnikov, *Negative Ion Sources* (Novosibirsk, NSU, 2018). В. Дудников, *Источники отрицательных ионов*. НГУ, Новосибирск, (2018)
3. V. Dudnikov, *Development and Applications of Negative Ion Sources* (Springer, 2019)
4. J. Thomson, Phyl. Mag. **21**, 225 (1911)
5. M. Born, *Atomic Physics* (M. Mir, 1965). М. Борн, *Атомная физика*, М. Мир (1965)
6. V.G. Dudnikov, Surface Plasma Method of Negative Ion Production, Dissertation for doctor of Fis-Mat. nauk, INP SBAS, Novosibirsk, (1976). В.Г. Дудников, "Поверхностно-плазменный метод получения пучков отрицательных ионов", Диссертация на соискание учёной степени доктора физ.-мат. Наук, ИЯФ СОАН СССР, Новосибирск, (1976)
7. V.I. Khvostenko, V.M. Dukelsky, JETF **34**, 1026 (1957). В.И. Хвостенко и В.М. Дукельский, ЖЭТФ, 34, 1026 (1957)
8. B. Jordon-Thaden, H. Kreckel, R. Golser, et al., Structure and stability of the negative hydrogen molecular ion. Phys. Rev. Lett. **107**, 193003 (2011)
9. T. Bete, E. Solpiter, *Quantum Mechanics of Atoms with One and Two Electrons* (Fismatgis, Mosow, 1960). Т. Бете, Э. Солпитер, *Квантовая механика атомов с одним и двумя электронами*, Москва, Физматгиз, (1960)
10. B.M. Smirnov, *Atomic Collisions and Elemental Process in Plasma* (Atomizdat, Moscow, 1968). Б.М. Смирнов, *Атомные столкновения и элементарные процессы в плазме*, Москва, Атомиздат, (1968)
11. B.M. Smirnov, *Ions and Excited Atoms in Plasma* (Atomizdat, Moscow, 1974). Б.М. Смирнов, *Ионы и возбуждённые атомы в плазме*, Москва, Атомиздат, (1974)
12. H.S.W. Messy, *Negative Ions* (Cambridge U.P, 1976). Г. Месси, *Отрицательные ионы*, Москва, Мир, (1979)
13. D.J. Hasted, *Physics of Atomic Collisions* (M. Mir, 1965). Дж. Хастед, *Физика атомных столкновений*, Москва, Мир, (1965)
14. I. Mac Daniel, *Collision Processes in Ionized Gases* (Mir, Moscow, 1967). https://doi.org/10.1063/1.555524. И. Мак-Даниель, *Процессы столкновений в ионизованных газах*, Москва, Мир, (1967)
15. H. Hotop, Binding energies in atomic negative ions. J. Phys. Chem. Ref. Data **4**, 539 (1975)
16. B.M. Smirnov, *Negative Ions* (Mc Graw-Hill, New York, 1976)
17. S.K. Allison, Rev. Mod. Phys. **30**, 1137 (1958)
18. N.V. Fedorenko, Н.В. Федоренко, ЖТФ, 40, 2481 (1970). ZTF **40**, 2481 (1970)
19. A.R. Tawara, Charge changing processes in hydrogen beams. Rev. Mod. Phys. **45**, 178 (1973)

20. T. Andersen, Atomic negative ions: Structure, dynamics and collisions. Phys. Rep. **394**, 157–313 (2004). www.elsevier.com/locate/physrep

21. N.N. Semashko, Investigation and Development of Injectors of fast ions and atoms of Hydrogen for stationary Magnetic trap, Dissertation for Doctor of Fis-Mat nauk, IAE im. Kurchatova, Moscow (1974); Н.Н. Семашко, "Исследование и создание инжекторов быстрых ионов и атомов водорода для стационарных магнитных ловушек", Диссертация на соискание учёной степени доктора физ.-мат. наук, ИАЭ им. И.В. Курчатова, Москва, (1974)

22. M.D. Gabovich, N.V. Pleshivtsev, N.N. Semashko, *Beams of Ions and Atoms for Controlled Nuclear Fusion and Technologies* (Energoatom izdat, Moscow, 1986). М. Д. Габович, Н.В. Плешивцев, Н.Н. Семашко, "Пучки ионов и атомов для управляемого термоядерного синтеза и технологических целей", еМосква, Энергоатом издат, (1986)

23. G.I. Dimov, V.G. Dudnikov, Cross sections for stripping of-1-MeV negative hydrogen ions in certain gases.", Zhur. Tekh. Fiz, 36, (1966) 1239;. Sov. Phys. Tech. Phys **11**, 919 (1967)

24. G.I. Dimov, Charge exchange method of protons in accelerators and storage rings, Preprint 304, INP SBAS, Novosibirsk, (1969). Г.И. Димов, "Перезарядный метод инжекции протонов в ускорители и накопители", Препринт 304 ИЯФ СО АН СССР, Новосибирск, (1969)

25. G.I. Dimov, G.V. Roslyakov, Nucl. Fusion **15**, 551 (1975)

26. D.D. Armstrong, H.E. Wegner, Rev. Sci. Instrum. **44**, 44 (1971)

27. B.A. Dyachkov, V.I. Zinenko, V. Kazantsev, ZTF **47**, 416 (1977). Б.А. Дьячков, В.И. Зиненко, Г.В. Казанцев, ЖТФ, 47, 416 (1977)

28. M.A. Abroyan, V.P. Golubev, V.L. Komarov, Negative Ion Sources, NIIEFA, Review OD-4, Leningrad (1976). М.А. Аброян, В.П. Голубев, В.Л. Комаров, Г.В. Чемякин, Источники отрицательных ионов, НИИЭФА, Обзор ОД-4, Ленинград, (1976)

29. I.E. Fink, W.L. Barr, G.W. Hamilton, LLL Rep. UCRL-52173 (1967)

30. S.S. Popov, M.G. Atluhanov, A.V. Burdakov, A.A. Ivanov, A.V. Kolmogorov, M.Y. Ushkova, High effective neutralizer for negative hydrogen and deuterium ion beams on base of non-resonance adiabatic trap of photons, Report MonP1, AIP Conference Proceedings 1869, 070002 (2017)

31. B. Van Zyl, N.G. Utterback, R.C. Amme, Rev. Sci. Instrum. **47**, 814 (1976)

32. M. Chane, R. Giannini, P. Lefévre, R. Ley, D. Manglunki, D. Möhl, Measurements of H⁻ intra-beam stripping cross section by observing a stored beam in lear. Phys. Lett. B **192**(3–4), 475–477 (1987)

33. V. Lebedev, N. Solyak, J.-F. Ostiguy, Fermilab, Batavia, IL 60510, U.S.A. A. Alexandrov, A. Shishlo, INTRABEAM STRIPPING IN H- LINACS, Proceedings of Linear Accelerator Conference LINAC2010, Tsukuba, Japan THP080 (2010)

34. J.S. Cohen, Phys. Rev. A **33**(3) (1986)

35. C. Gerthsen, Naturwissenshaften **20**, 743 (1932)

36. A.J. Dempster, Phys. Rev. **42**, 901 (A.4) (1932)

37. A.P. Grinberg, UFN **117**, 333 (1975). А.П. Гринберг, УФН, 117, 333 (1975)

38. C. Gerthsen, W. Rensse, Phys. Zs. **34**, 478 (1933)

39. O. Peter, Ann. D. Phys. (Lpz) **27**, 294 (1936)

40. W.H. Bennet, B.F. Parby, Phys. Rev. **49**, 422 (1936)

41. L.W. Alvarez, Rev. Sci. Instrum. **22**, 705 (1951)

42. R.I. Van de Graaf, Nucl. Instrum. Methods **8**, 195 (1960)

43. D.A. Brodley, Nucl. Instrum. Methods **122**, 1 (1974)

44. N.B. Brooks, R.P. Bastide, K.H. Purser, et al., IEEE Trans. Nucl. Sci. **NS-12**(3), 313 (1965)

45. H.E. Wegner, IEEE Trans. Nucl. Sci. **NS-18**(3), 68 (1971)

46. H.E. Wegner, IEEE Trans. Nucl. Sci. **NS-22**(3), 946 (1975)

47. L.A. Zubkov, Atomic Technic Abroad **11**, 26 (1975). Л.А. Зубков, Атомная техника за рубежом, №11, 26 (1975

48. G.I. Dimov, I. Ya, Timoshin, V.V. Demidov, V.G. Dudnikov, PTE **4**, 30 (1967). Г. И. Димов, И, Я. Тимошин, В. В. Демидов, В. Г. Дудников, ПТЭ, 4,30 (1967)

49. G. Ryding, T. Smick, B. Park, et al., A 1.2MeV, 100mA Proton Implanter, http://how-canihelpsandiego.com/wp-content/uploads/2012/02/Twin-Creek-Technologies-Hyperion-Proton-Implantar.pdf

50. N. Smick, Hyperion™ Accelerator Technology for BNCT, https://agenda.infn.it/getFile.py/access?contribId=32&resId=0&materialId=slides&confId=7214

51. Hyperion-MeV implanter for solar, http://www.uspvmc.org/proceedings/c-Si_feedstock-wafering_workshop/Talk%206%20Bill%20Park.pdf

52. A. Ivanov, D. Kasatov, A. Koshkarev, A. Makarov, Y. Ostreinov, I. Shchudlo, I. Sorokin, S. Taskaev, Suppression of an unwanted flow of charged particles in a tandem accelerator with vacuum insulation, Published by IOP Publishing for Sissa Medialab, (2016), JINST 11 P04018, http://iopscience.iop.org/1748-0221/11/04/P04018

53. A.A. Ivanov, A. Sanin, Y. Belchenko, Recent achievements in studies of negative beam formation and acceleration in the tandem accelerator at Budker Institute. AIP Conf. Proc. **2373**, 070002 (2021). https://doi.org/10.1063/5.0057441

54. A.J.T. Jull, AMS method, in *Encyclopedia of Quaternary Science*, vol. 4, (Elsevier B.V., Amsterdam), p. 29112007

55. J.P. O'Connor, M.S. Chase, S.L.F. Richards, N. Tokoro, Performance Characteristics of the Genus Inc. Tandetron 1520 Ion Implantation System, Proc. Of the Eleventh International Conf. on Ion Implantation Technology, pp. 454–457, Austin, TX (1996)

56. A. Glotov, G. Lavochkin, R. Romanov, et al., Proc. Allunion workshop for accelerators of charged particles, Dubna, 1967, M. Nauka, (1977). А. Глотов, Г. Лавочкин, Р. Романов и др., Труды Всесоюзного совещания по ускорителям заряженных частиц, Дубна, 1976, М. Наука, 1977

57. K.W. Ehlers, Nucl. Instrum. Meth. **32**, 309 (1965)

58. M.E. Rickey, R.J. Smyth, Nucl. Instrum. Meth. **19**, 66 (1962)

59. J.R. Richardson, E.W. Blachmore, G. Dutto, et al., IEEE Trans. Nucl. Sci. **NS-22**(3), 313 (1975)

60. E.G. Michaelis, IEEE Trans. Nucl. Sci. **NS-22**(3), 1385 (1975)

61. V.G. Dudnikov, Charge exchange injection into accelerators and storage rings. Physics-Uspekhi **62**(4), 405 (2019)

62. V. Dudnikov, Production of intense proton beam in storage ring by charge exchange injection method, Dissertation for candidate of Fis-Mat. Nauk, INP SBAS, Novosibirsk, (1966). В. Дудников, "Получение интенсивного протонного пучка в накопителе методом перезарядной инжекции", Диссертация, представленная на соискание учёной степени кандидата физ.-мат. Наук, ИЯФ СОАН СССР, Новосибирск, (1966)

63. G.I. Budker, G.I. Dimov, Proc. International Conference on Accelerators, Dubna, 1963, 933. М. (1964). Г.И. Будкер, Г.И. Димов, Труды Международной конференции по ускорителям, Дубна, 1963. 933. М. (1964)

64. V. Shiltsev, V. Lebedev, *Accelerator Physics and Tevatron Collider* (Springer, 2014)

65. G.I. Budker, G.I. Dimov, A.G. Popov, et al., Sov. Atomic Energy **19**, 507 (1965). Г.И. Будкер, Г.И Димов, А.Г. Попов, и др., Атомная энергия, 19, 507 (1965)

66. G. Budker, G. Dimov, V. Dudnikov, Experimental investigation of the intense proton beam accumulation in storage ring by charge-exchange injection method, Proc. Int. Symp. on Electron and Positron Storage Ring, France, Sakley, 1966, rep. VIII, 6.1 (1966)

67. G. Budker, G. Dimov, V. Dudnikov, Experimental investigation of the intense proton beam accumulation in storage ring by charge-exchange injection method. Sov. Atomic. Energy **22**, 348 (1967).; Г. Будкер, Г. Димов, В. Дудников, "Экспериментальные исследования накопления интенсивного протонного пучка в накопителе методом перезарядной инжекции", Атомная энергия, 22, 348 (1967)

68. G.I. Dimov, Charge exchange Injection in accelerators and storage rings, Dissertation for Doctor of Fis-Mat. Nauk, INP SBAS, Novosibirsk (1968). Г. И. Димов, "Перезарядная инжекция в ускорители и накопители", докторская диссертация, ИЯФ, (1968)

69. G.I. Dimov, V.G. Dudnikov, Determination of current and its distribution in storage ring. Instrum. Exp. Tech. **3**, 553 (1969). Г.И. Димов, В.Г. Дудников, "Определения тока и его распределения в накопителе", ПТЭ 3, 553 (1969)

70. G. Rumolo, A.Z. Ghalam, T. Katsouleas, et al., Electron cloud effects on beam evolution in a circular accelerator. Phys Rev Special Topics Accel Beams **6**, 081002 (2003)

71. V.G. Shamovsky, Investigation of the interaction of a beam with an ionized gas, Thesis for candidate Phys-Math. nauk, INP, Novosibirsk (1972). В. Шамовкий, «Исследование взаимодествия пучка с ионизованным газом», кандидатская диссертация, ИЯФ СОАН, Новосибирск, (1972)

72. V. Dudnikov, Some features of transverse instability of partly compensated proton beams, PACS2001. Proceedings of the 2001 Particle Accelerator Conference (2001)

73. G. Budker, G. Dimov, V. Dudnikov, V. Shamovsky, Experiments on electron compensation of proton beam in ring accelerator, Proceedings of the sixth Internat. Conf. On High energy accelerators, Harvard University (1967)

74. G.I. Dimov, V.G. Dudnikov, V.G. Shamovskii, Investigation of the effect of secondary-charged particles on a proton beam in a Betatron mode. Sov At Energy **29**(5), 356–361 (1970)

75. B.V. Chirikov, Stability of partly compensated electron beam. Sov At Energy **36**, 1239 (1965). Б. В. Чириков, "Устойчивость частично компенсированного пучка электронов", Атомная Энергия, 36, 1239 (1965)

76. D. Koshkarev, P. Zenkevich, Part Accel **3**, 1 (1972)

77. R.A. Bosch, Suppression of two-stream hose instabilities at wavelengths shorter than the beam's transverse size. Phys. Rev. ST **6** (2003)

78. Y. Belchenko, G. Dimov, V. Dudnikov, Development of high current proton beams in Novosibirsk, Proc. International Accelerator Conference, Protvino, (1977), http://inspire-hep.net/record/127164/files/HEACC77_I_291-298.pdf: Ю.И. Бельченко, Г.И. Будкер. Г.И. Димов, В.Г. Дудников и др. «Работы по сильноточным протонным пучкам в Новосибирске», Труды Международной конференции по ускорителям, Протвино, 1977; Препринт ИЯФ 77–59, Новосибирск, (1977)

79. M. Reiser, *Theory and Design of Charged Particle Beams*, 2nd edn. (Wiley-VCH Verlag, 2009)

80. V. Dudnikov, Development of charge-exchange injection at the Novosibirsk Institute of Nuclear Physics and around the World. ArXiv, 1808.06002 (2018)

81. G. Dimov, V. Chupriyanov, V. Shamovsky, Sov. Phys., Tech. Phys. **16**(10), 1662 (1971)

82. G.I. Dimov, V.E. Chupriyanov, Compensated proton-beam production in an accelerating ring at a current above the space-charge limit. Part Accel **14**, 155–184 (1984)

83. V. Dudnikov, R. Johnson, L. Vorobiev, C. Ankenbrandt, *Beam Brightness Booster with Ionization Cooling* (NA PAC, 2013)

84. C.S. Park, V. Shiltsev, G. Stancari, et al., Space charge compensation using IOTA, Proceedings of NAPAC2016, Chicago, IL, (2016)

85. G.I. Dimov, Use of hydrogen negative ions in particle accelerators. Rev. Sci. Instrum. **67**, 3393–3404 (1996)

86. A.S. Belov, V.G. Dudnikov, V.E. Kuzik, et al., A source of polarized negative hydrogen ions with deuterium plasma ionizer. Nucl. Instrum. Methods Phys. Res **A333**, 256 (1993)

87. Y. Plis, A. Soroko, UFN **107**, 281 (1972).; Н. Ю. Плис, А. Сороко, УФН, 107, 281 (1972)

88. N.I. Zaika, A.V. Mochnach, P.L. Shmarin, *Physics and Technik of Polarized Beams of Ions* (Noukova Dumka, Kiev, 1987). И. Заика, А. В. Мохнач и П. Л. Шмарин, *Физика и техника поляризованных пучков ионов*, Киев, Наукова думка, (1987)

89. R. Martin, Proc. VIII Internat. Conf. on High Energy accel. CERN, p. 540 (1971)

90. I.D. Simpson, IEEE Trans. Nucl. Sci. **NS-20**(3), 198 (1973)

91. J. Simpson, R. Martin, R. Kustom, History of the ZGS 500 MeV booster http://inspirehep.net/record/1322000

92. C. Hojvat, C. Ankenbrandt, B. Brown, et al., The multiturn charge exchange injection system for the Fermilab booster accelerator. IEEE Trans. Nucl. Sci. **NS-26**(3), 3149 (1979)

93. D.S. Barton, Charge Exchange Injection at the AGS, BNL Int. Rep. 32784 PAC, Santa Fe NM, March 3 (1983)
94. V.C. Kempson, C.W. Planner, V.T. Pugh, Injection dynamics and multiturn charge exchange injection into the fast cycling synchrotron for the SNS. IEEE Trans. Nucl. Sci. **NS-28**(3), 3085 (1981)
95. D.H. Fitzgerald, H. Ahn, B. Blind, et al., *Overview and Status of the Los Alamos PSR Injection Upgrade Project* (PAC 1997, 1997)
96. R.K. Cooper, G.P. Lawrence, *Beam Emittance Growth in a Proton Storage Ring Employing Charge Exchange Injection*, vol 22 (PAC, 1975), p. 1916
97. T. Kawakubo, I. Sakai, H. Sasaki, M. Suetake, The H⁻ Charge-Exchange Injection System in the Booster Of the KEK 12 Gev Proton Synchrotron, http://inspirehep.net/record/234778/files/HEACC86_II_287-290.pdf
98. L. Criegee, H. Dederichs, H. Ebel, G. Franke, D.M. Kong, et al., The 50 MeV H⁻ linear accelerator for HERA LINAC 3 collaboration. Rev. Sci. Instrum. **62**, 867 (1991)
99. First neutrons for Chinese spallation source, Physics Word, https://physicsworld.com/a/first-neutrons-for-chinese-spallation-source/
100. L. Hermanson, et al., in *Proc. of workshop on Beam cooling and Related Topics* (Montreux, 4–8 October, 1993, CERN 94–03,26 April 1994), p. 235
101. A.M. Baldin, A. Kovalenko. *JINR Rapid Communications*, 377.-96, Dubna, (1996), p.5
102. A.O. Sidorin, Formation of Intense ion beams in storage rings with multiturn charge exchange injection and electron cooling, Dissertation for candidate Fis-Mat. nauk, Dubna, (2003). Сидорин А О "Формирование интенсивных ионных пучков в накопителях с многооборотной перезарядной инжекцией и электронным охлаждением", Дисс. ... кандидата физ.-мат. наук (Дубна: 2003)
103. N.N. Alekseev, D.G. Koshkarev, B.Y. Sharkov, JETP Lett. **77**, 123 (2003)
104. J. Lettry, D. Aguglia, J. Alessi, et al., CERN's Linac4 H⁻ sources: Status and operational results. AIP Conf Proc **1655**, 030005 (2015)
105. https://home.cern/news/news/accelerators/first-accelerators-are-back-action
106. B.A. Frolov, V.S. Klenov, V. N. Mihailov, Simulation and optimization of ion optical extraction, acceleration and H⁻ ion beam matching systems, Proceedings of RuPAC2014, Obninsk, Kaluga Region, Russia, (2014)
107. A. Alekoue, R.E. Baussano, N. Blaskovic Kraljevici, et al., The European Spallation Source neutrino Super Beam, White Paper to be submitted to the Snowmass (2021) USA Particle Physics Community Planning Exercise, https://arxiv.org/pdf/2203.08803.pdf
108. A.A. Vasil'ev, R.A. Mescherov, B.P. Murin, J.Y. Stavissky, US Patent, 3,860,828, filed August 30, 1972, issued Jan. 14, (1975)
109. P. George, Lawrence, PAC 1987, p. 825 (1987)
110. R.J. Macek, et al., PAC 1993, p. 3739 (1993)
111. J. Wei, et al., PRST-AB, 3, 080101(1999)
112. M. Plum, AAC – HEBT/Ring/RTBT Overview, Plum-AAC-Feb10_r1.pptx (2010)
113. S. Cousineau, A. Rakhman, M. Kay, A. Aleksandrov, et al., First demonstration of laser-assisted charge exchange for microsecond duration H⁻ beams. PRL **118**, 074801 (2017). S. Cousineau, A. Aleksandrov, T. Gorlov, et al., "The SNS laser stripping experiment and its implications on beam accumulation", thwcr02, Proceedings of cool 2015, newport news, va, usa (2015)
114. V. Danilov, K. Beard, R. Johnson, V. Dudnikov, SNS Laser Stripping for H⁻ injection, Proceedings of PAC09, Vancouver, BC, Canada TU6RFP039 (2009)
115. F. Zimmermann, Review of single bunch instabilities driven by an electron cloud. Phys. Rev. ST Accel. Beams **7**, 124801 (2004). https://doi.org/10.1103/PhysRevSTAB.7.124801. http://link.aps.org/doi/10.1103/PhysRevSTAB.7.124801. F. Zimmermann, Electron-Cloud Effects in past and future machines – walk through 50 years of Electron-Cloud studies, arXiv: 1308.17. https://arxiv.org/abs/1308.1274
116. M. Blaskiewic, Electron Cloud Effects in Present and Future High Intensity Hadron Machines, Proceedings of the 2003 Particle Accelerator Conference (2003)

117. M. Plum, H⁻ Charge Exchange Injection Issues at High Power, HB2016, July 3–8, 2016, Malmo

118. V. Dudnikov, Method of negative ion obtaining, Patent cccp, 411542, 10/III, (1972.); http://www.findpatent.ru/patent/41/411542.html В.Г. Дудников, "Способ получения отрицательных ионов", Авторское свидетельство, М. Кл.Н 01 J 3/0,4, 411542, заявлено 10/III, (1972)

119. Y. Belchenko, G. Dimov, V. Dudnikov, Physical principles of surface plasma source operation, in *Symposium on the Production and Neutralization of Negative Hydrogen Ions and Beams, Brookhaven, 1977*, (Brookhaven National Laboratory (BNL), Upton, NY, 1977), pp. 79–96. Yu Belchenko, G. Dimov, V. Dudnikov, "Physical principles of surface plasma source method of negative ion production", Preprint IYaF 77-56, Novosibirsk (1977). Ю. Бельченко, Г. Димов, В. Дудников, «Физические основы поверхностно плазменного метода получения пучков отрицательных ионов», препринт ИЯФ 77-56, Новосибирк (1977). http://irbiscorp.spsl.nsc.ru/fulltext/prepr/1977/p1977_56.pdf

120. V. Dudnikov, *Surface Plasma Method of Negative Ion Production* (Pysics, Uspekhi, 2019)

121. G.I. Dimov, G.E. Derevyankin, V.G. Dudnikov, IEEE Trans. Nucl. Sci. **NS24**(3) (1977)

122. V. Dudnikov, Dudnikov A., Acceleration system of beam brightness booster, WEPOA44 report, Proceedings of NAPAC2016, Chicago, IL (2016)

123. J.R. Alonso, High-current negative-ion sources for spallation neutron sources: LBNL Workshop, October 1994. Rev. Sci. Instrum. **67**, 1308 (1996)

124. G.I. Budker, Plasma Physics and Problems of Controlled Thermonuclear Reactions, Edit. M. A, Leontovich, 3,3 M., (1958). Г.И. Будкер, в сборнике *"Физика плазмы и проблема управляемых термоядерных реакций"* (под редакцией М.А. Леонтовича) т.3, стр.3, М. (1958)

125. I.N. Golovin et al., UFN **73**, 685 (1961). И.Н. Головин и др. УФН,73, 685 (1961)

126. Barnett, Bely, Lyuis, et al., Plasma Physics and Thermonuclear Reactions, Selected reports of II Geneva Conference for picfull applications of atomic energy, p. 302 M. (1959). Барнетт, Бели,Льюис и др. *"Физика горячей плазмы и термоядерные реакции"*, Избранные доклады иностранных учёных на II Женевской конференции по мирному использованию атомной энергии, стр.302, М. (1959)

127. D. Sweetman, Nucl. Fusion, Supplements Conferenc Proceedings, Saltsburg, (1962)

128. P.F. Post, in [117}, 1, 548, (1959). Р.Ф. Пост, в сборнике 117, том 1, 548, (1959)

129. S.Y. Lukyanov, *Hot Plasma and Controlled Nuclear Fusion* (M. Nauka, 1975). С.Ю. Лукьянов, *Горячая плазма и управляемый ядерный синтез*, М. Наука, (1975)

130. V. Davydenko et al., Multi-slit triode ion optical system with ballistic beam focusing. Rev. Sci. Instrum. **87**(02B303), 02B303 (2016)

131. G.U. Hamilton, Atomic Technic in Abroad, 12, 20 (1975). Г.У. Гамильтон, Атомная техника за рубежом, №12, 20, (1975)

132. L.A. Barry, Symposium on Plasma Heating in for Devices, Varenna-Italy, p. 151, Bologna, (1974)

133. F.N. Coensgen, W.F. Cummins, D.G. Logan, A.W. Molvik, et al., *Recent Experiments on the 2XIIB Mirro Machine, VII Europ. Conf. On Contr. Fusion and Plasma Physics* (Losanne, 1975)

134. M.N. Andrews, H. Davitian, H. Fleishman, et al., Phys. Rev. Lett. **27**, 1428 (1971)

135. H. Gota, M. Tuszewski, A. Smirnov, et al., A high performance field-reversed configuration regime in the C-2 device. J Fusion Sci Technol **63**, 139 (2013)

136. Y. Belchenko, et al., Negative Ion bused neutral beam Injector, US patent 10887976 B2

137. A.A. Ivanov, A.N. Smirnov, S.Y. Taskaev, et al., Accelerator-based neutron source for boron neutron capture therapy. Physics-Uspekhi **65**, 834–851 (2022)

138. T. Slayers, K. Prelec, IX Internat. Conf. on H. E. Accel., Polo Alto, (1974)

Chapter 3
Methods of Negative Ion Production

Abstract Topics covered in this chapter include: Formation and destruction of negative ions; charge-exchange methods for negative ion production; charge-exchange negative ion sources; charge-exchange polarized negative ion sources; cold muonium negative ion production; negative ion beam formation from gaseous plasmas; formation and destruction of negative ions in a gaseous plasma; beam formation from negative ions generated in the plasma volume; plasma volume sources of negative ions; thermionic production of negative ion beams; secondary emission (sputtering) production of negative ion beams.

3.1 Introduction

Production and application of negative ions are discussed in books [1–8].

The properties of negative ions, their interaction with other systems, and processes of their formation and destruction have been studied for over 100 years. In Table 3.1 is presented an electron affinity (binding energy of addition electron in negative ions) for different elements in Mendeleev table.

To date, more than 50 atomic species of negative ions have been experimentally formed and detected, and it has been found that they can possess a long lifetime, since they have positive electron affinity, almost all binary molecular negative ions, and an immense set of complex molecular negative ions [9–26]. Recently, electronically excited states of negative ions have been observed, and their binding energies have been measured [27]. Doubly-charged negative ions O^{2-}, F^{2-}, Cl^{2-}, Br^{2-}, I^{2-}, Te^{2-}, and Bi^{2-} have been observed experimentally also [5, 28]. Some elements, such as N, Be, and Mg, have very low or negative electron affinity and cannot create negative ions. To obtain high-energy nuclei through tandem accelerator, their negative compound ions can be used, such as BeH-, BH-, NH-, MgH-, AlH-, WO_3^-, and ArF-. The charge transfer process (discussed later in this chapter) is particularly well suited for metastable-ion formation. The subsequently formed negative ion may live for extended periods of time if the excited compound state is forbidden to decay. Classic examples of

© The Author(s), under exclusive license to Springer Nature Switzerland AG 2023   41
V. Dudnikov, *Development and Applications of Negative Ion Sources*, Springer
Series on Atomic, Optical, and Plasma Physics 125,
https://doi.org/10.1007/978-3-031-28408-3_3

Table 3.1 Electron affinity (eV) for different elements in Mendeleev table

1	2	3	4	5	6	7	8	9	10	11	12	13	14	15	16	17	18
0.754 H																	0.078 He
0.62 Li	0.19 Be											0.28 B	1.26 C	- N	1.46 O	3.4 F	- Ne
0.55 Na	- Mg											0.41 Al	1.38 Si	0.76 P	2.08 S	3.61 Cl	- Ar
0.50 K	0.02 Ca	0.19 Sc	0.08 Ti	0.53 V	0.67 Cr	- Mn	0.15 Fe	0.65 Co	1.15 Ni	1.24 Cu	- Zn	0.3 Ga	1.23 Ge	0.81 As	2.02 Se	3.36 Br	- Kr
0.49 Rb	0.05 Sr	0.31 Y	0.43 Zr	0.89 Nb	0.75 Mo	0.55 Tc	1.05 Ru	1.137 Rh	0.45 Pd	1.30 Ag	- Cd	0.3 In	1.11 Sn	1.07 Sb	1.97 Te	3.06 I	- Xe
0.47 Cs	0.14 Ba	0.5 La	~0 Hf	0.32 Ta	0.82 W	0.15 Re	1.1 Os	1.56 Ir	2.13 Pt	2.31 Au	- Hg	0.2 Tl	0.36 Pb	0.95 Bi	1.9 Po	2.8 At	- Rn
Fr	Ra	Ac	Rf	Db	Sg	Bh	Hs	Mt	Ds	Rg	Cn	Nh	Fl	Mc	Lv	Ts	Og

metastable negative ions that can be formed only through attachment to excited states of the parent atoms are He− [29] and Be−, both forming with high probability in the He(1s2s)^{1}S$_1$ and He(1s2s)^{3}S$_1$ states through sequential charge exchange with a low ionization-potential charge-exchange vapor. Although much remains to be done in this field of atomic physics, the properties and behavior of many negative ion processes are now fairly well understood. Negative ions play an important role in astrophysical phenomena and in the physics of the upper layers of the atmosphere. In particular, the radiation processes of formation and destruction of H$^-$ ions exert a decisive influence on the solar radiation spectrum and also on the spectra of some other stars. The role of negative ions in gas-discharge physics, in chemical processes, in mass spectrometry, in ion sources, in the electrical strength of gas and vacuum insulation, in the generation of coherent radiation, and in many other phenomena is well recognized. The accumulated data serve as the basis for the applications of negative ions and the development of methods for their production.

3.2 Formation and Destruction of Negative Ions

As already noted, in charge-exchange technology negative ions of hydrogen isotopes are the most widely used, and as a consequence, methods for producing beams of these ions are the most intensively developed. Although the binding energy of an electron to the atom in the H$^-$ ion is much lower than in many other negative ions, H$^-$ beams have been produced that are substantially greater in intensity than beams of other negative ions. Over the past ~60 years of development of negative ion generation technology, from 1956, achievable H$^-$ ion beam currents have increased from about a microampere up to tens of amperes.

The early stages of development of negative ion sources have been reported in a number of review papers [1, 30–37], and a number of new reviews of the development of methods for obtaining negative ions have appeared recently [38–46]. Up-to-date research results are reflected in the proceedings of ion source conferences

[47–49] and in symposia specifically on negative ion sources [50–59] and books [1, 2, 60]. Important data on the processes of formation and destruction of negative ions and the characteristic features of the sources developed are systematized in these papers and proceedings.

Known formation processes of H^- ions include capture of free electrons at the electron affinity level of the H° atom, radiative capture; three-body collisions; capture of electrons in charge exchange with gas molecules to form an H^- ion and an H° atom; formation of H^- ions in the destruction of molecules and molecular ions containing hydrogen; in collisions with electrons and molecules, dissociative attachment; and electron capture from condensed bodies on the electron affinity level of particles ejected from condensed phase or reflected from condensed bodies.

The simplest negative hydrogen production reaction, the radiative attachment of an extra electron to the hydrogen atom, played an important role in the origin of structures in the early universe,

$$H + e = H^- + h\nu, \tag{3.1}$$

but it gives a negligible increase of negative ion production in the ion sources due to the low value of cross sections. The formed H^- ions could be then lost in an atomic hydrogen-rich environment by associative detachment to form hydrogen molecules,

$$H^- + H = H_2 + e. \tag{3.2}$$

The isotopic counterpart of reaction (3.2),

$$D^- + D = D_2 + e. \tag{3.3}$$

Reactions (3.2) and (3.3) are the time-reversed process of the dissociative electron attachment (DEA),

$$e + H_2 = H^- + H \tag{3.4}$$

and

$$e + D_2 = D^- + D, \tag{3.5}$$

which are the fundamental processes producing H^- and D^- in ion source plasmas. Negative hydrogen and deuterium ions are also produced in the plasma volume through the processes of dissociative recombination of H_2^+ and D_2^+ and H_3^+ and D_3^+ ions, which lead to the production of an ion pair,

$$H_2^+ + e = H^- + H^+, \tag{3.6}$$

$$D_2^+ + e = D^- + D^+, \tag{3.7}$$

$$H_3^+ + e = H^- + H_2^+ \tag{3.8}$$

$$D_3^+ + e = D^- + D_2^+ \tag{3.9}$$

Other formation reactions requiring higher electron energy are polar dissociation of excited molecules,

$$H_2(v,J) + e = H^- + H^+ + e \tag{3.10}$$

$$D_2(v,J) + e = D^- + D^+ + e \tag{3.11}$$

and double-charge-exchange of positive hydrogen/deuterium ions,

$$H^+ + H_2 = H + H_2^+, \quad H + H_2 = H^- + H_2^+ \tag{3.12}$$

$$D^+ + D_2 = D + D_2^+, \quad D + D_2 = D^- + D_2^+. \tag{3.13}$$

H^- and D^- ions can also be produced in ion-molecule collisions such as

$$H_2^+ + H^2 = H^- + H + 2H + \tag{3.14}$$

$$D_2^+ + D_2 = D^- + D + 2D^+ \tag{3.15}$$

$$H_2^+ + H_2 = H^- + H + 2H^+ \tag{3.16}$$

$$D_2^+ + D_2 = D^- + D + 2D^+ \tag{3.17}$$

when molecular ions with high enough energies are present. In the hydrogen and deuterium plasma volume, the negative hydrogen ions (H^- and D^-) are formed mainly by DEA of a low-energy (cold) electron to an excited molecule at vibrational level v' and rotational level J,

$$H_2(v',J) + e = H + H^- \tag{3.18}$$

$$D_2(v',J) + e = D + D^-. \tag{3.19}$$

We denote here "low-energy electrons" the electrons with the energy below 4 eV. The reason why low-energy electrons are required for the formation of negative hydrogen and deuterium ions by DEA, reactions (3.18) and (3.19) can be understood from Fig. 3.17, where the electron energy dependence of the DEA cross sections for H_2, calculated by Bardsley and Wadhera, is shown. These calculations have been done for vibrational excitation only. This DEA cross section is maximum at the threshold energy. The threshold energy is highest for $v' = 0$ and goes down with increasing v'. The comparison of these theoretical results with the experimental data of Allan and Wong is shown in Fig. 3.19. On this figure, the peak DEA cross sections are plotted as a function of the internal energy of each vibrationally and rotationally excited state. These cross sections were derived for individual vibrational and rotational states from experimental spectra (i.e., H^- ion current vs electron energy) of electron-impact mass spectrometer. Vibrationally and rotationally excited molecules were generated in a hot iridium collision chamber. The experiment consisted of measuring the energy dependence in the 1–5 eV region for the dissociative attachment in H_2 and D_2 at temperatures ranging from 300 to 1600 K. It was assumed that the populations of vibrational and rotational states are given by the Maxwell-Boltzmann distribution. The cross sections for H_2, D_2, and HD were measured by Schulz and Asundi which included contributions from both vibrational and rotational excitations to their cross-sectional calculations. Figure 3.17 clearly indicates that the difference in cross section between ground-state molecule and molecules excited to higher vibrational states exceeds several orders of magnitudes. Thus, the reaction channels (3.18) and (3.19) should be the main negative ion formation mechanism, if a sufficient density of vibrationally/rotationally excited molecules is created in the volume. Ionizing electrons present in the plasma have enough energy to excite hydrogen molecules to singlet electronically excited states B and C states of the hydrogen molecule. Hiskes calculated the cross sections for ground-state hydrogen molecules for being excited to vibrational levels by electron impact through B and C states. Electrons should have more than 20 eV kinetic energy to produce vibrationally excited molecules through the mentioned mechanism, according to Hiskes' results shown in Fig. 3.18. These cross sections correspond to vibrational excitation of molecules non-rotationally excited. The H^- and D^- ions can be destroyed in the plasma volume by two-body recombination or mutual neutralization,

$$H^- + H^+ \left(or\ H_2^+, H_3^+ \right) = neutrals \tag{3.20}$$

$$D^- + D^+ \left(or\ D_2^+, D_3^+ \right) = neutrals \tag{3.21}$$

by photodetachment,

$$H^- + h\nu = H + e \tag{3.22}$$

$$D^- + h\nu = D + e, \tag{3.23}$$

or by collision with an electron,

$$H^- + e = H + 2e \tag{3.24}$$

$$D^- + e = D + 2e. \tag{3.25}$$

Atoms are present in the plasma volume, and a H^-/D^- ion is destroyed through collision with an atom,

$$H^- + H = 2H + e \tag{3.26}$$

$$D^- + D = 2D + e, \tag{3.27}$$

and together with associative detachment (3.5) with specifying the vibrational level v'',

$$H^- + H = H_2\left(v''\right) + e \tag{3.28}$$

$$D^- + D = D_2\left(v''\right) + e. \tag{3.29}$$

Reactions (3.28) and (3.29) are also named associative detachment. The inverse process (3.18), and the dissociative electron attachment (DEA) (3.19), should be also dependent upon the initial vibrational level as well as the rotational level, J, of the initial molecular state. A collision with a molecule can also detach electron from H^-/D^-,

$$H^- + H_2 = H + H_2 + e, \tag{3.30}$$

$$D^- + D_2 = D + D_2 + e. \tag{3.31}$$

The efficiency of formation of negative ions strongly depends on the binding energy of electrons in interacting systems and, in particular, on the state of excitation of interacting complex particles. The maximum probability of formation of negative ions, as a rule, constitutes only a very small fraction of the maximum probability of formation of positive ions (Fig. 3.1). The interaction of the negative ions with other systems—electrons, photons, ions, neutral particles, and a solid surface—causes the detachment of an additional electron from the negative ion. Cross sections for the destruction of H^- ions are shown in Figs. 3.2 and 3.3. The reaction rate for destruction of negative ions in general considerably exceeds the reactions rate for their formation. This circumstance greatly complicates the production of intense beams of negative ions. The production of beams involves the following processes: generation of the necessary flux of the desired kind of ions, their transportation to the formation system, ion beam formation, and primary acceleration. Special attention must be paid to minimizing destruction of negative ions at the application region.

The cross section σ_5 is the electron destruction cross section, σ_6 is the destruction cross section with H^+ ions, σ_7 is the H^-/H^0 resonant charge-exchange cross section, σ_8 is the resonant charge-exchange cross section for protons, σ_9 is the ion destruction cross section for collisions with atoms, σ_{10} is the destruction cross section of H^- ions in collisions with H_2 molecules, and σ_{11} is the cross section for charge exchange of protons on hydrogen molecules H_2.

Fig. 3.1 Cross sections for the formation of H⁻ ions by electron collision in hydrogen gas and in hydrogen plasma as a function of electron energy. (Dudnikov [61])

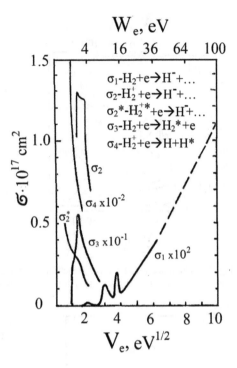

$$W_e, eV$$

$\sigma_1\text{-}H_2+e\rightarrow H^-+...$
$\sigma_2\text{-}H_2^i+e\rightarrow H^-+...$
$\sigma_2^*\text{-}H_2^{+*}+e\rightarrow H^-+...$
$\sigma_3\text{-}H_2+e\rightarrow H_2^*+e$
$\sigma_4\text{-}H_2^++e\rightarrow H+H^*$

σ_2

$\sigma_4 \times 10^{-2}$

σ_2^*

$\sigma_3 \times 10^{-1}$

$\sigma_1 \times 10^2$

$\sigma \cdot 10^{17}\ cm^2$

$$V_e, eV^{1/2}$$

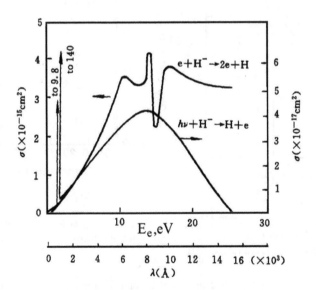

$e+H^-\rightarrow 2e+H$

$h\nu+H^-\rightarrow H+e$

$\sigma(\times 10^{-15} cm^2)$

$\sigma(\times 10^{-17} cm^2)$

to 9.8

to 140

E_e, eV

$\lambda(Å)$

Fig. 3.2 Cross sections for destruction of H⁻ ions by electrons and by photons as a function of electron energy or photon wavelength. (Reproduced from Dudnikov [61])

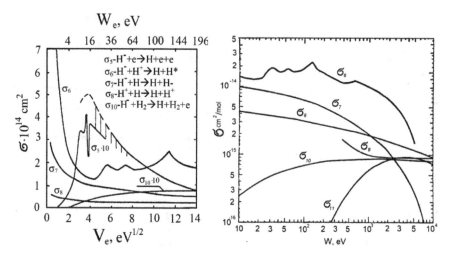

Fig. 3.3 Cross sections for the destruction of H⁻ ions in a hydrogen plasma as a function of electron energy or H⁻ ion energy, in two different scales. (Reproduced from Dudnikov [61])

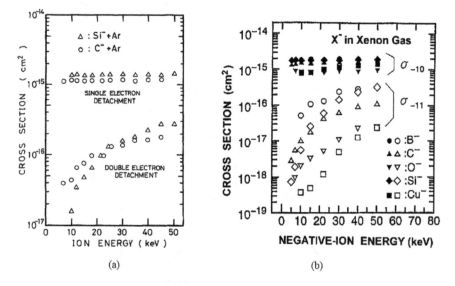

Fig. 3.4 (a) Cross section for the destruction of C⁻ and Si⁻ ions in Ar gas as a function of energy. (b) Cross section for the destruction of a number of different negative ions in Xe gas as a function of energy (Reproduced from Graham [65])

The cross section for the destruction of C^- and Si^- ions in argon as a function of energy is shown in Fig. 3.4a. For other heavy ions, the cross sections for the loss of one and of two electrons have similar dependencies. The cross sections for the destruction of various negative ions in xenon gas as a function of energy are shown in Fig. 3.4b.

Methods for forming beams depend on the ion generation method used. The most important characteristics of the beam are its intensity (ion current) and the volume occupied by the beam in phase space (coordinates and momenta). These characteristics are primarily determined by the density of the generated ion beam, the spread of ion velocities, and the cross sectional area of the ion beam. Methods for generating negative ion beams include by charge-exchange, plasma, secondary emission (sputtering), thermionic, and surface-plasma production, identified according to the negative ion formation mechanism.

3.3   Charge-Exchange Methods for Negative Ion Production

The charge-exchange method for obtaining beams of negative ions is based on the conversion into negative ions of a part of an accelerated beam of positive ions when it interacts with a gas (vapor) target. To obtain H^- ions, beams of H^+, H_2^+, or H_3^+ ions or their superposition, generated by special ion sources, as a rule plasma-based ion sources, are transmitted through the target. H^- ions are formed in collisions of fast ions with target particles as a result of the two-step capture of bound electrons to form first H^o atoms and then H^- ions. This results in the rapid destruction of molecular H_2^+ and H_3^+ ions into H^- ions and H^o atoms by electron capture, as well as directly in the destruction of fast H_2^+ and H_3^+ ions. In collisions of H^- ions with target particles, they can lose an electron and become fast H^o atoms or H^+ ions. In some cases, the positive ion beam is converted in the target into a beam of metastable atoms, and negative ions are obtained by interaction of the fast metastable atom beam with an additional target. The conversion efficiency of atoms or positive ions into negative ions is characterized by a conversion factor $K^- = I^-/I_1$, where I^- is the negative ion beam current at the target output and I_1 is the current of the beam of positive ions or of atoms at the entrance to the target. The value of the conversion coefficient for particles at a given energy is determined by the ratio of cross sections for formation and destruction of negative ions in interaction with the target particles. Naturally, the additional loss of particles in the target by scattering and other causes reduces the conversion coefficient, the intensity, and the brightness of the resulting negative ion beam. Data on the elementary processes of transformation of particles in targets have been systematized in a number of books [12–18, 66]. The dependence of the conversion coefficient on energy has a maximum at a certain value of energy $E = E_{opt}$. For materials with lower ionization potential, E_{opt} is lower, and the corresponding K^- is, as a rule, larger. The achievable conversion efficiency of protons into H^- ions is shown as a function of proton energy in Fig. 3.5a, where the $K^-(E)$ values are given for targets of different materials. The dependence on energy of the maximum efficiency of transformation of different positive ions to negative ions in a Mg target is shown in Fig. 3.5b.

Unfortunately, positive ion sources that have been developed do not allow using the high value of conversion coefficient for cesium. The positive ion beam current generated by electrostatic formation systems increases with increasing extraction

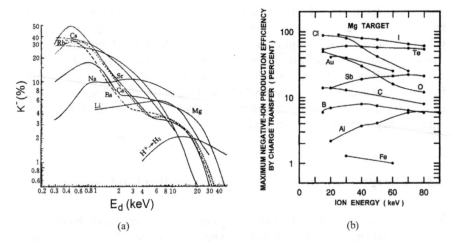

(a)   (b)

Fig. 3.5 (**a**) Energy dependence of the yield of H⁻ ions from protons for targets of various substances. (Reproduced from Schlachter [67]). (**b**) Energy dependence of the maximum efficiency of transformation of different positive ions into negative ions in a Mg target. (Reproduced from Schlachter [67])

voltage U as $U^{3/2}$. Even in ion sources with multi-aperture beam formation systems, after ion extraction at energy $E = eU_{opt}$, the H⁻ ion current density at the exit of the cesium target turns out to be less than for targets of substances with lower values of conversion coefficient but with greater optimal energies. For H⁻ ion beam generation by a charge-exchange method and with maximum current and current density, it is necessary to use an extraction voltage U_{opt}, providing a maximum or, if possible, a larger value of the function $\Phi = K^-(eU)U^{3/2}$. The corresponding values $eU_{opt} > E_{opt}$ and $K^-(eU_{opt}) < K^-(E_{opt})$.

The data show that when using ion sources with modern beam formation systems, only sodium provides some advantages over other non-alkaline targets. There remain hopes that the advantages of cesium can be realized by transforming a high current density plasma stream generated by a plasma accelerator into H⁻ ions [68]. Using the molecular ions H_2^+ and H_3^+ instead of protons, it is possible to obtain H⁻ flows with the same velocity at an intensity of 5.4 and 10.7 times greater, since the number of atomic particles with a given energy in the stream formed by the electrostatic system increases by a corresponding factor. Thus, only a very small fraction of the accelerated particle flux is converted into negative ions. The intensity and brightness of negative ion beams obtained by the charge-exchange method turns out to be tens and hundreds of times less than the intensity and brightness of the converted beams, only because the conversion coefficient K^- is small. This creates an ineradicable gap between the brightness and intensity of negative ion beams obtained by the charge-exchange method and the corresponding proton beams. The scattering of particles during charge exchange and dissociation increases this gap even further. To reduce the effect of scattering, it is desirable to locate the target in the crossover of a beam formed by short-focus optics.

For high-energy ion implantation, beams of heavy negative ions are needed, which can be obtained by charge-exchange methods. The energy dependence of the yield of negative ions with a magnesium target is shown in Fig. 3.5b.

3.4   Charge-Exchange Negative Ion Sources

Alvarez's conclusion [69] on the technical feasibility of charge-exchange accelerators was supported by data on the high efficiency of conversion of positive ions into H^- ions using solid-state film targets [70]. The values obtained for conversion coefficient K^-, up to 10–20% in the presence of comparatively perfect proton ion sources with intensity up to tens of milliamperes, allowed hope for the rapid production of H^- beams with the necessary intensity. However, it was not possible to create solid film targets suitable for use in charge-exchange sources: the films that could be obtained were destroyed at a current density of more than 10 µA/cm². At the same time, the results obtained by Fite [71] did not leave any hope for the competitiveness of other methods of charge-exchange production of H^-. In this connection, the main efforts were concentrated on the development of charge-exchange sources of negative ions using gaseous targets. Such targets provide significantly lower conversion factors, but allow the use of more intense primary ion beams, and provide less scattering and less energy spread.

The first H^- ion beam sources, with microampere intensity, were developed in 1956. These sources used plasma-based positive ion sources with single-aperture beam formation systems in conjunction with different types of charge-exchange targets.

- Fogel and coworkers [72] formed a beam of H^- ions with current up to 2 µA by charge exchange of H^+ and H_2^+ ions, with an energy up to 25–33.5 keV generated by a high-frequency source or a Keller source with a Penning discharge, in a mercury vapor jet.
- Phillips and Tuck [73] produced an H^- ion beam with current up to 2.2 µA by passing positive hydrogen ions extracted from a high-frequency source through an elongated channel in the extractor electrode. Positive ions were charge-exchanged into negative ions in hydrogen, leaving through the channel from the gas-discharge bulb of the source.
- Weinman and coworkers [74] demonstrated up to 30 µA of H^- ions by converting, in a target in the form of a tube with additional hydrogen inlet, a beam of H^+, H_2^+, and H_3^+ ions with an energy of 12 keV generated by a source of the Kistemaker type with heated cathode and reflexing electrons.

Over the first 20 years of development of charge-exchange sources, the H^- ion beam intensity was increased from 10^{-6} to 10^{-3} A. The designs used targets of the three types listed above. In charge exchange in the channel of the working gas source, it is difficult to independently optimize the conditions for the formation of positive ions in the source and the conditions for charge exchange. However, in some cases, such sources can be quite efficient, so that deviation from the optimum is completely compensated by their simplicity of design, compactness, and the absence of additional gas admission into the charge-exchange target. These properties are especially important for sources installed in the high-voltage terminals of electrostatic accelerators. The current of H⁻ ions from such sources could be brought to several tens of µA [75, 76]. By 1969, Dyachkov and coworkers obtained 150 µA of H⁻ ions from this source [77]. In nearly all cases, the primary ions were generated by RF sources with low gas efficiency. With additional gas injection into the charge-exchange target, independent optimization of the conditions for the formation of positive ions in the plasma source and optimization of the conditions for charge exchange in the target are possible. At the same time, flow targets (with gas flow in a charge-exchange gas-cell target) using nonaggressive gases are still quite simple. Such targets are used with sources having high gas efficiency with high extraction voltage, which do not affect the advantages of effective targets from alkali metals. From a source with a single-aperture beam formation system and a flow target, it was possible to obtain a H⁻ ion beam with current of about 1–2 mA [78].

The intensity of H⁻ ion beams produced by the charge-exchange method was substantially increased by the use of multi-aperture primary beam formation systems could form intense beams of positive ions at relatively low extraction voltage. In 1968, Dimov and coworkers produced a pulsed H⁻ ion beam with current 15 mA by charge exchange with a flowing hydrogen target of positive hydrogen ions produced by a source with a high current arc in a diaphragmed channel with multi-aperture extractor system [79, 80]. A schematic of this source is shown in Fig. 3.6. Protons are generated by a cold cathode discharge in a diaphragm channel. The

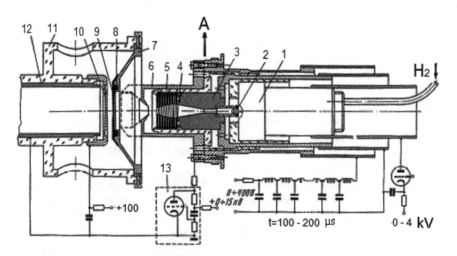

Fig. 3.6 Arc source with diaphragmed gas-discharge channel and multi-slit beam formation system. (Reproduced from Dimov [81], with the permission of G. Dimov). (1) pulsed gas valve, (2) ignition, (3) cold cathode, (4) barrier diaphragm, (5) diaphragmed channel, (6) anode, (7) plasma grid holder, (8) plasma grid shell, (9) plasma grid, (10, 11) extractor grid holder, (12) charge-exchange tube, (13) protective circuit

plasma flows into an expander with the plasma grid. The beam formation system comprises two grids spaced by 2.4 mm and consists of wires 0.05 mm in diameter with mesh size 0.3 mm. This gap could be conditioned up to 15 kV, but during breakdowns, the cathode grid was damaged. Later, the wires of the cathode grid were replaced with metal strips of 0.05×0.5 mm, which made them viable. A protective system 13 with thyratron for protection extraction electrode from breakdown was used. Further improvements of this source made it possible to increase the current of H^- ions to 54 mA [82]. Note that the intensity of the primary beam exceeded 2.5 A.

On this basis, an H^- source for accelerators was developed as shown in Fig. 3.7a [84, 85]. At a proton current density of 1 A/cm^2, the H^- ion current density is 10–20 mA/cm^2. The extraction voltage is applied to the charge-exchange tube (item 5 in Fig. 3.7a). The positive ion beam is accelerated upon entry to the

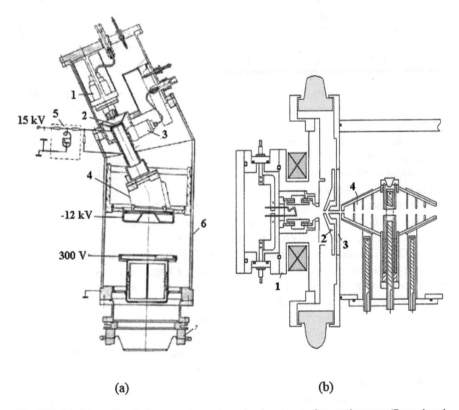

(a)   (b)

Fig. 3.7 (**a**) Schematic of charge-exchange negative ion source for accelerators. (Reproduced from Roslyakov [83] with permission of G. Roslyakov). (1) plasma generator, (2) extraction system with charge-exchange target, (3) gas valve, (4) bending magnet, (5) thyratron for protection extraction electrode, (6) source body, (7) high-voltage electrode of accelerator. (**b**) Charge-exchange source of negative ions for implanters. (Republished from N. White, Patent USA, 4980556, issued Dec. 25, (1990) with permission N. White). (1) ion source, (2) locking electrode, (3) grounded electrode, (4) magnesium target

charge-exchange target, and the negative ion beam is accelerated at the exit from the charge-exchange target. Despite the low H⁻ ion current density, the brightness of this beam is relatively high due to the very small emittance of the proton beam.

Donnally discovered in 1965 [86] the increased efficiency of H⁻ ion formation by charge exchange of low-energy protons in targets of alkali metal vapors, making it possible to substantially increase the generation efficiency of negative ions in charge-exchange ion sources [87]. When using alkali metals, it is important to minimize the release of the working substance beyond the target, since alkali metal vapor contaminates the system, initiates breakdown in high-voltage gaps, and disrupts normal operation of the source. In this case, it is sensible to use targets in the form of a directed jet of vapor, as first applied by Fogel and coworkers [72].

In pulsed sources, jet targets are used with condensation of the working substance on a cooled surface [88]; in continuous mode, jet targets with recirculation of the working substance are used. Some quite sophisticated designs of such targets have been developed recently [89, 90], including those with a large aperture [91]. These kinds of targets have been used to obtain H⁻ ions, He⁻ ions, and polarized H⁻ ions, in combination with single-aperture primary beam formation systems [37].

The use of such targets in combination with multi-aperture primary beam formation systems makes it possible to obtain intense beams of ions at low energy (systems with high perveance). In 1971, Osher and Hamilton developed a multi-aperture negative hydrogen ion source with reduced perveance of about 1.5 mA $V^{-3/2}$ (energy 1 keV, current 1 A) [92]. Such a high perveance is due to the large area of the emission surface (6.3 cm in diameter) and deceleration of the beam after extraction. Using this source in 1972, Osher formed H⁻ beam with an energy of 1.5 keV and a total current of 50 mA [93]. It was not possible to use the full potential of the source due to breakdown initiated by cesium removal.

As noted above, because of difficulties in obtaining high values of primary ion emission current density at low extraction voltage, it turns out to be better to use sodium rather than cesium in charge-exchange targets [83].

In 1973, by Dimov and coauthors, a beam of H⁻ ions with current 76 mA was obtained by charge exchange in a sodium vapor jet target of 7 keV protons [82]. When switching to deuterium, a D⁻ ion beam was obtained with current 104 mA at 8 keV ion energy and a He⁻ ion beam with current 12 mA [82, 84, 88].

In 1975, Semashko's group at the Kurchatov Atomic Energy Institute produced an H⁻ ion beam of current 150 mA [23]. Using a sodium jet target with sodium recirculation, beams of H^+, H_2^+, and H_3^+ ions at energy 5 keV and current 1 A, generated by a nonmagnetic source, were charge-exchanged.

The last of these developments aimed at creating a powerful injectors of H^0 atoms with an energy of hundreds keV for controlled thermonuclear fusion. With the development of a surface-plasma method for obtaining negative ions, the development of high-current charge-exchange sources has been discontinued.

The charge-exchange method is also used to produce heavy negative ions for ion implantation application [94]. A schematic of a charge-exchange ion source for this purpose is shown in Fig. 3.7b. Positive ions are formed in a source with a discharge in a magnetic field, and beam extraction is along the magnetic field. The extracted beam is passed through a magnesium charge-exchange target, and the necessary component of negative ions is selected by a magnetic analyzer, which is accelerated in a tandem accelerator. Beams of boron B⁻ ions with current 0.4 mA, phosphorus P⁻ at 1 mA, antimony Sb⁻ at 1 mA, and arsenic As⁻ at 1 mA have been produced.

In the proposed CW 10 mA charge-exchange ion source in BINP [95], the negative ion beam is produced via dissociation of a primary molecular ion beam and subsequent charge-exchange in a hydrogen gas target. The primary ion beam is formed with an energy two times higher, which corresponds to the maximum yield of negative ions in hydrogen gas target. The ion source has to produce a beam with a high molecular ion fraction. They dissociate in the gas target into half energy fragments. An advantage of this method is the operation of the plasma emitter in the ion source at low power and low gas pressure in the discharge compared with sources that produce beams with a high content of protons. In this case, continuous operation of the plasma source can be more reliable. An issue would be beam separation from a high flux of co-streaming positive ions and neutrals, which must be properly disposed onto appropriate beam dumps. In our case, the negative ion beam is deflected by a magnet and then accelerated to an energy of ~100 keV. Injection of the beam into the accelerator at high energy enables a reduction of beam losses at the initial stage of acceleration.

The cutaway view of the negative ion source is shown in Fig. 3.8. The ion source (1) has to provide a primary positive ion beam with a current of ~1 A and an energy of 30 keV. The beam passes through the gas target, diaphragm, and separation magnet. Inside the vacuum chamber (2) with differential pumping, the positive ion and neutral dumps are placed, as shown in Fig. 3.8. The one-gap accelerator (3) is applied to produce 105 keV energy beam. The ion source and vacuum chamber have to be placed on the high-voltage platform, which is biased up to −90 kV potential. This provides an acceleration for the 15 keV negative ion beam after the separation

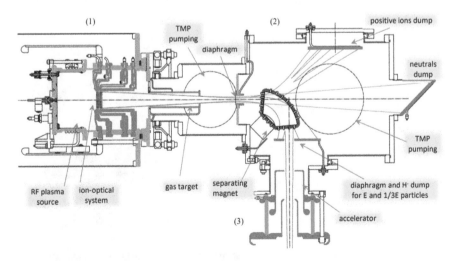

Fig. 3.8   Negative ion source design (top view): (1) ion source, (2) vacuum chamber, (3) accelerator

magnet to the full energy of 105 keV. The positive ion source consists of radio-frequency (RF) plasma source and a four-electrode ion-optical system. In the RF plasma source, there is a cylindrical alumina ceramic tube with a Faraday screen installed along the inner cylindrical surface to protect the ceramics from the thermal load of the plasma, sputtering, and metallization. The back plate facing the plasma is made of molybdenum. The copper cylinder and molybdenum disk have water cooling though feedthroughs placed on the rear flange of the plasma chamber. Hydrogen gas is puffed into the discharge chamber by using an electromagnetic valve. The ignition unit is placed on the rear wall of the plasma source. The four turns RF antenna is wound around the ceramic tube. It is made of 6 mm cooper tubing with heat shrinkable tubing insulation. The schematic of multi-aperture water cooled four-electrode ion-optical system a photograph of assembled extraction system are shown in Fig. 3.9. The electrodes have a spherical form with a radius of 0.5 m for ballistic focusing of the beam. They are made of chromium zirconium bronze, and they have 109 round apertures with a diameter of 4 mm arranged in a hexagonal pattern inside a diameter of 6 cm. The transparency of the electrodes is about 50%. Each electrode has a water-cooling channel surrounding the apertures. Water channels are inside the electrode holders and go out through the flanges compressed between alumina ceramic insulators. The gas target is formed in a copper tube with an internal diameter of 5 cm and a length of 19 cm. It has a circular cooling line at the output and a gas input in the middle part.

The diaphragm splits the vacuum vessel into two volumes for differential pumping. It is made of magnetic steel to avoid penetration of weak stray magnetic fields from the separation magnet through the diaphragm into the left side. This prevents the deflection of the ions in front of the diaphragm. The internal part of the diaphragm is made of copper.

The separating magnet is designed to deflect the negative ion beam by 90°. A general view of the separation magnet is shown in Fig. 3.8. The magnet consists of

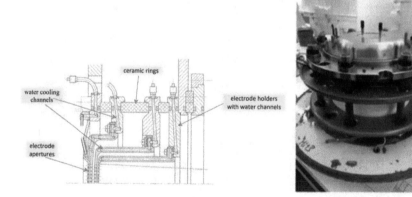

Fig. 3.9 Schematic of ion-optical extraction system (left) and assembled water cooled extraction system (right)

an open magnetic core with two poles, a set of permanent magnets to create the main magnetic field, overhead profiled poles that create a nonuniform magnetic field for radial focusing of the beam, and magnetic field correction coils. The correction coils are made of copper tubes through which water flows for cooling. The field in the central plane of the magnet is 0.175 T. The same magnet was used in the experiments for transportation and acceleration of the negative ion beam for the tandem accelerator. The water-cooled plate is installed after the separating magnet. It has a diaphragm for the negative ion beam to pass through with an energy of 15 keV. The negative ion species with full and one-third energy are dumped on the plate.

Ion sources have neutral dump, TMP pumping, diaphragm and positive ion dump.

3.5 Charge-Exchange Polarized Negative Ion Sources

The charge-exchange method has been used to obtain beams of polarized H^-, D^- ions and H^+, D^+ ions [96, 97]; a schematic of such a source is shown in Fig. 3.10. A beam of atoms is formed in an RF dissociator, and the polarized atoms are separated by a focusing sextupole filter with high-frequency transitions between atomic levels. Polarized H^o atoms are ionized by resonant charge exchange in a jet of dense deuterium plasma generated by an arc-discharge source, and the resulting ions extracted by a grid extraction system and selected by a magnet. In this way, polarized proton beams, with current 10 mA and polarization 85%, and polarized deuterons, with current 10 mA and polarization 85%, have been obtained.

Polarized negative ions can also be formed [99–101]. To this end, at the suggestion of Dudnikov at Novosibirsk, a jet of a dense plasma is converted into negative ions on a cesiated surface-plasma ionizer followed by resonant charge exchange of polarized atoms in a jet of negative ions. A schematic of this surface-plasma converter is shown in Fig. 3.11a. Up to 4 mA of polarized H^- was obtained with a polarization of 85% as shown in Fig. 3.11b.

Other methods of polarized H^- and D^- production by charge exchange of polarized D^o and H^o have been proposed [102, 103]. An ionizer is shown schematically in Fig. 3.12a. Its operation is basically the same as a normal magnetron surface-plasma source of negative ions, but with inverted geometry, i.e., the cathode is the outer and the anode is the inner of two concentric cylinders. H^- ions are produced on the Cs-coated molybdenum cathode with spherical concavity and are accelerated away from the cathode through small holes in the anode and toward the central region of the ionizer. Sufficient H^+ ions from the source plasma diffuse into the central region to provide space-charge neutralization of the H^- and D^- ions. Polarized D^o atoms pass axially through the central region and are ionized. The short length of this ionizer compared to other techniques provides the advantage large acceptance for the polarized atomic beam. The spherical geometrical focusing of the negative ions produced increases the gas efficiency and energy efficiency of the ionizer. A proposed high polarization negative ion source with surface-plasma

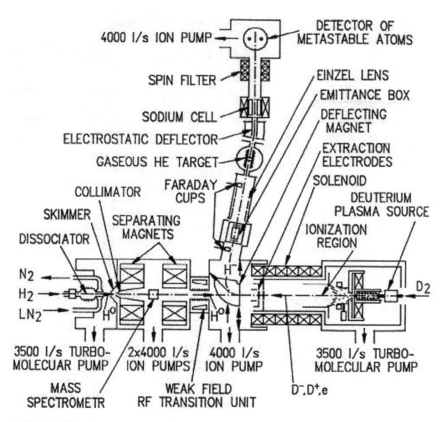

Fig. 3.10 Diagram of a source of polarized hydrogen ions with deuterium ionizer. (Reproduced from Belov et al. [98] with permission of Elsevier Publishing)

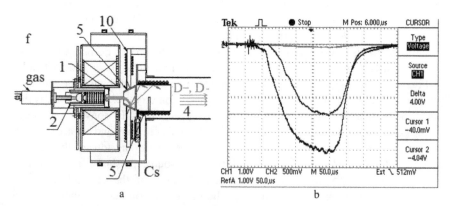

Fig. 3.11 (a) Generation of a cold D⁻ plasma jet for resonant charge-exchange negative ionization of polarized hydrogen atoms (INR version). (b) Oscillograms of a polarized H⁻ ion current of up to 4 mA. (1) arc-discharge channel, (2) triggering, (4) negative ion beam, (5) converted H+->H⁰ and H⁰->H-, (10) magnet compensator. (Reproduced from Belov et al. [98] with permission of Elsevier Publishing)

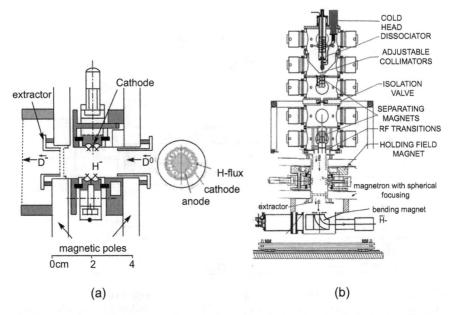

Fig. 3.12 (**a**) Resonance charge-exchange ionizer with spherical focusing for ionization of polarized H° or D°. (**b**) High polarization negative ion source with surface-plasma ionizer with multispherical focusing. (Reproduced from Dudnikov and Dudnikov [103])

ionizer with spherical focusing is shown in Fig. 3.12b. It consists of a source of polarized atoms, as above, with a multispherical focusing surface-plasma ionizer, extraction grids, and bending magnet for separation of polarized D⁻ and non-polarized H⁻ ions.

Similar method of polarized negative ion was tested in [104] with using two-dimensional cylindrical focusing magnetron with ~10 times large gas flow, strongly scattering the flow of polarized particles.

The charge-exchange method has been used in the OPPIS polarized proton source [105–107] at Brookhaven, a schematic of which is shown in Fig. 3.13. A H° atom injector with geometric focusing and high beam current density with neutralization in a hydrogen target is used. Atoms are introduced into a magnetic field and ionized in a helium target. They then pick up a polarized electron in a rubidium target with optical pumping, pass through the transition zone (Sona transition), and are charge-exchanged into negative ions in a sodium jet target. Up to 0.5 mA of H⁻ ions are formed in 0.5 ms pulses with a polarization of up to 86%, which is enough to fill, using charge-exchange injection, the AGS booster up to the space-charge limit and to ensure the operation of RHIC with full intensity. For the development of these sources, Belov and Zelensky received the Veksler Award in 2006.

It has been proposed [109, 110] to use the charge exchange of He⁺ ions into metastable He⁻ ions to produce ion beams that are polarized in nuclear spin. A schematic of the polarized ³He⁻ ion source is shown in Fig. 3.13. An arc-discharge ion source can be used for the generation of high-brightness beams of He⁺ with an intensity of up to 3 A at an energy of ~10–15 keV. A pulsed Xe gas target can be used for space-charge compensation and for metastable He* production. The

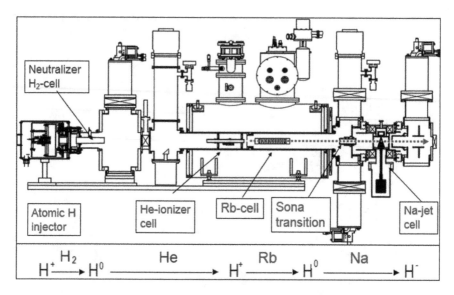

Fig. 3.13 Schematic of the OPPIS polarized H⁻ ion source with optical pumping of a rubidium target and charge exchange of polarized atoms into H⁻ ions in a sodium jet. (Reproduced from Zelenski et al. [108] with permission from AIP Publishing)

second potassium vapor jet target can be used for He⁻ production. The ³He⁻ ions are produced in double collisions (³He⁺ + Cs → ³He*, ³He* + Cs → ³He⁻) of 25 keV ³He⁺ in a cesium charge-exchange process.

A He⁺ beam of more than 3 A intensity is required to produce a 10 mA He⁻ beam, as the conversion efficiency of the double-charge-exchange process from He⁺ ions is less than several percent. In addition, the He⁺ beam is required to converge at the center of the charge-exchange cell to form the He⁻ beam emitted from a narrow region in the cell.

Ions with momentum components 1/2 and 3/2 should be autoionized (up to 95%), leaving only ³He⁻ ions with components |5/2, 5/2>. Then, using RF to induce a transition of one of the components to the zero state, one can produce a ³He⁻ beam with a nuclear polarization close to ~95%. A schematic of the proposed experiment using BNL equipment to measure the He⁻ beam production is shown in Fig. 3.14. A high-brightness He⁺ ion beam (7) with an intensity of up to 3 A and an energy of ~10–15 keV is generated by an arc-discharge plasma source (1) and formed by a multigrid focusing extraction system (2). A pulsed Xe gas target (3) is used for space-charge compensation, and short-lived He⁻ ions can eject electrons during their flight in the decay channel (6) with solenoid and RF transition, producing a polarized ³He⁻ beam (10) as shown in the figure. To prevent intra-beam stripping, the He⁻ beam is separated from the intense He⁺ and He⁰ beams by a bending magnet (5). For metastable He* production, a vapor jet target (4) (K, Rb, or Cs) can be used for He⁺ to He⁻ (9) conversion.

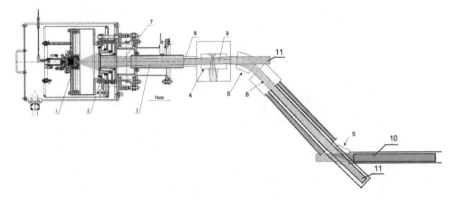

Fig. 3.14 Schematic of ^{3}He$^-$ ion source. (Reproduced from Dudnikov et al. [109]). (1) arc-discharge He$^+$ source, (2) extraction system, (3) space-charge compensation, (4) Cs (or Rb or K) target, (5) bending magnet, (6) decay channel with solenoid and RF transition and, (7) He$^+$ beam, (8) space-charge compensated beam, (9) He$^-$ beam, (10) polarized ^{3}He$^-$ beam, (11) ^{3}He neutral beam

3.6 Cold Muonium Negative Ion Production

Ultraslow muons up to now have been generated by resonant ionization of thermal muonium atoms (Mu) formed at the surface of a hot tungsten foil placed at the end of an intense "surface" muon beam line. In order to efficiently ionize Mu near the tungsten surface, a resonant ionization scheme via the 1s-2p unbound transition has been used. Low emittance muon beams have been discussed in several scientific reports [111–114]. A complex laser system has been used to efficiently ionize Mu atoms near a tungsten or aerogel surface [111, 112].

Another possibility for efficient ionization of muonium atoms is resonant charge exchange with slow ions [115]. The cross section for this process is very high. The cross section for resonant charge exchange with protons σ_8, shown in Fig. 3.3, is 5×10^{-15} cm^2 at 10 eV energy, while the cross section for resonant charge exchange with negative hydrogen ions σ_7, also shown in Fig. 3.3, reaches 10^{-14} cm^2 at 10 eV energy.

These processes have been successfully used for highly efficient ionization of polarized hydrogen and deuterium atoms [96, 97, 99, 100]. For the generation of highly dense plasma, an arc-discharge plasma source with diaphragmed channel, developed at BINP [79, 80], was used. A schematic of this plasma source is shown in Fig. 3.15. It consists of a pulsed gas valve 1, trigger electrode 2, cathode 3, barrier diaphragm 4, washer channel 5, and anode 6. An arc discharge with current up to 300 A is established in the washer channel, activated by trigger electrode 2. A dense plasma flux is formed in the anode channel. The muonium cooling target 7 accepts the muons flux and produces cold muonium atoms. The muonium atoms leaving the muonium cooling target are ionized by resonant charge with a dense proton flux, extracted by an extractor system and separated from the protons by a magnetic analyzer.

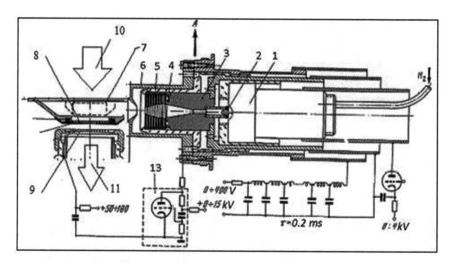

Fig. 3.15 Schematic of hydrogen plasma source and muonium ionization system. (Reproduced from Dudnikov and Dudnikov [115] with permission of AIP Publishing). (1) pulsed gas valve, (2) trigger electrode, (3) cathode, (4) barrier diaphragm, (5) washer channel, (6) anode, (7) muonium cooling target, (8) plasma grid, (9) extraction grid, (10) flux of hot muons, (11) flux of accelerated cold muons

We can estimate the efficiency of positive muonium production. Fast muons (10) are directed to the cooling target (7), and cold muonium atoms then exit the cooling target in the gap between target and extraction plasma grid (8). The flux of protons from the arc-discharge plasma source can be efficiently converted to a flux of negative hydrogen ions in the surface-plasma ionizer shown in Fig. 3.9. A plasma jet is generated by a discharge in a diaphragmed channel (4–6), and a magnetic field is applied for focusing the jet. The plasma jet flux density can be up to 10 A/cm², with corresponding flux density $j = 6 \times 10^{19}$ particles/cm². The proton energy in the jet is $W \sim 5$ eV; the cross section for resonant charge exchange at an energy of 5 eV is 6×10^{-15} cm²; the velocity of 100 meV muonium atoms is ~1.3×10^{6} cm/s, and in 10^{-6} s muonium, atoms can travel ~1.3 cm. The gap between the cooling target and the plasma grid should be ~1 cm. Thus, the ionization rate $P \sim 6 \times 10^{19} \times 6 \times 10^{-15}$ $= 3.6 \times 10^{5}$ s^{-1}, and the probability of ionization in 10^{-6} s is ~36%. The ionized muons can be accelerated by a voltage applied between grids (8) and (9). By variating the plasma density, it is possible to increase the probability of ionization yet further. The muons can be separated from protons after extraction by magnetic analysis.

This flux of negative ions can be used for resonant charge exchange with muonium atoms for efficient production of negative muonium ions, in the same way as polarized negative ion have been produced [99]. The efficiency of polarized negative ion production was up to 12%.

Another possibility for polarized atom ionization has been proposed [116]; a schematic of the device is shown in Fig. 3.16. Plasma sources for intense negative ion production rely on the efficient conversion of positive plasma ions to negative

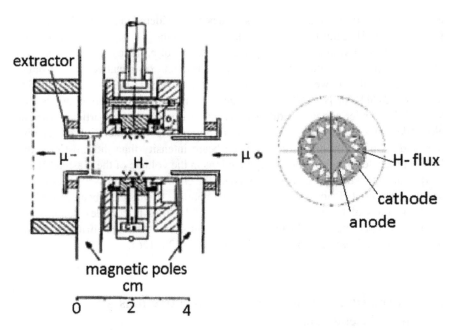

Fig. 3.16 Surface-plasma source-based multispherical ionizer for resonant charge-exchange ionization of muonium atoms to extract negative muonium ions. (Reproduced from Dudnikov and Dudnikov [115] with permission of AIP Publishing)

ions on surfaces with reduced work function [117, 118]. The emission is enhanced by using cesium to lower the work function of the emission surface, a well-known and often-used technique. Shaping the emission surface concave, known as geometric focusing or self-extraction [109, 119], results in the natural focusing of the negative ions, so increasing the negative ion current density. This increase in negative ion emission even on cesiated surfaces has been demonstrated [120].

The concave spherical emitter electrodes in a surface-plasma source can be arranged such that the generated negative hydrogen ions are focused onto the apertures leading away from the discharge into a charge-exchange region. The ionizer combines a surface-plasma source with a short charge-exchange region into which unpolarized H⁻ ions are injected radially. The muonium atoms are injected on-axis into one end of this charge-exchange region with the extraction grid at the opposite end. The concept is illustrated in Fig. 3.14.

3.7 Negative Ion Beam Formation from Gaseous Plasmas

By analogy with plasma sources of positive ions, negative ion sources, in which negative ions of the desired kind are formed by the interaction of particles within the volume of the gas-discharge plasma, also exist. The regularities in the formation of negative ions in a gas-discharge plasma, the corresponding elementary processes, and the optimization of plasma sources have been considered in a number of reviews [3, 37, 121–124]. Until recently (1971), it was thought that negative ions are supplied to the gas-discharge plasma by the interaction of particles in the plasma volume. Therefore, all sources with direct extraction of negative ions from a gas-discharge plasma were referred to as plasma sources [123, 124]. Investigation of the production of H⁻ ions in discharges of a mixture of hydrogen and cesium showed that, under certain conditions, the interaction of plasma particles with the electrode surfaces can provide significantly higher rates of generation of negative ions and significantly higher of extracted beam intensity than the most effective known processes of formation of negative ions in the volume of the plasma [2].

Recently, new data have been obtained on the elementary processes of formation and destruction of H⁻ ions within the plasma volume, studies have been carried out on the kinetics of formation and destruction of H⁻ ions in complex-composition plasmas, and parameters of beams obtained from sources that are attributed to volume production have been significantly improved [125, 126].

3.8   Formation and Destruction of Negative Ions in a Gaseous Plasma

The most important elementary processes of formation and destruction of H⁻ ions in a hydrogen gas-discharge plasma are shown in Figs. 3.1, 3.2, and 3.3, where the dependencies of the relevant reaction cross sections on energy and velocity of the particles are given. Although the ionization of a neutral particle to form a positive ion requires an energy not less than the ionization energy $I \sim 10\text{–}20$ eV and when a free electron is captured to form a negative ion, the electron affinity energy $S \sim 1$ eV is freed up, nevertheless the energy cost for formation of an H⁻ ion in a plasma source is much greater than the cost for formation of a positive ion. This is due to the small value of the cross section for the negative ion formation in the plasma and the high probability of their destruction, which inhibit the emission of negative ions from the plasma.

Under typical plasma source conditions, the capture of free electrons to the electron affinity level of hydrogen atoms does not play an important role, because of difficulties with utilization of excess energy in elementary processes. The cross section for radiative capture of free electrons with spontaneous emission of excess energy is very small, $\sigma_r = 6 \times 10^{-23}$ cm² at an electron energy of ~1 eV [11], and the capture rate of electrons in triple collisions with transfer of excess energy to the third particle at a density of $10^{15}\text{–}10^{16}$ cm⁻³ should only be equal to the rate for radiation capture.

However, for certain special conditions, in the presence of an intense flux of photons providing induced free-bound transitions and in a plasma with high atomic hydrogen density and low electron temperature, these processes can provide a high negative ion generation rate. H⁻ ions are formed more efficiently in the gas-discharge plasma volume due to collisions of electrons with molecules and molecular ions.

The formation of negative ions in the interaction of electrons with ground-state hydrogen molecules has been well studied, but more work remains. The features of this process have been reviewed [42]. Data on the dependence of the cross section σ_1 for this process on energy are shown in Fig. 3.1. As a result of electron capture by the H_2 molecule, the autoionization state of the H_2^- ion is formed, which decays through various channels, including the formation of an H^- ion. H^- ions begin to appear at an electron energy W_e greater than 3.7 eV—the difference between the 4.47 eV molecular dissociation energy and the electron affinity of 0.75 eV. This process has a strong isotopic effect. At electron energy $W_e = 4$ eV, the cross section σ_1 reaches 1.7×10^{-19} cm^2 for H_2 molecules. For HD molecules, σ_1 is 10^{-22} cm^2, and for D_2 molecules, it is 8×10^{-24} cm^2. The threshold for H^- ions formation is much smaller than the threshold appearance of H_2^+ (15 eV), but the cross section for all electron energies is much less than the maximum cross section for the formation of H_2^+, reaching 1.5×10^{-16} cm^2. For electron energy $W_e = 3.7$–13.5 eV, resonances have amplitudes up to 1.3×10^{-20} cm^2 through the H_2^- decay channel $H_2 + e \rightarrow H_2^- \rightarrow H^- + H$. The electron energy excess above the threshold is transformed into kinetic energy of the reaction products. The resonance at 14 eV energy with an amplitude of 2×10^{-20} cm^2 corresponds to the reaction $H_2 + e \rightarrow H_2^- \rightarrow H^- + H^*$. According to relative measurements [127], in the electron energy range $W_e = 18$–40 eV, the cross section for the formation of H^- ions increases by a factor of seven. In a number of papers, it was assumed that this cross section increases monotonically, reaching a value of 1–2×10^{-19} cm^2 at an energy of $W_e = 200$ eV.

The cross sections for the production of negative deuterium ions are much smaller. With a much greater probability, the H_2 ion decays into an electron and a molecule that has passed to a neighboring vibrationally excited state. The cross section for this process, σ_3 in Fig. 3.1, reaches 5.5×10^{-17} cm^2 at an electron energy of $W_e = 2$ eV. The cross section for the dissociation of molecules from the ground state is also much larger than the cross section σ_1. At a maximum electron energy of 15 eV, the dissociation cross section is 9×10^{-17} cm^2 [128]. Because of the relatively high direct dissociation threshold of 8.8 eV, in a plasma with a high concentration of low-energy electrons, the role of stepwise dissociation through intermediate vibrationally excited states should be significant, and consequently, the population of vibrationally excited states should be high, and their role should be high in the formation of H^-.

In a 1974 review of negative ion sources, Abroyan and coworkers [37] assumed that negative ions are formed upon dissociative attachment, not only to hydrogen molecules in the ground state but also to vibrationally excited molecules. The authors of [129, 130] tried to explain a number of features of a glow discharge in hydrogen mixtures with alkali metals by the formation of H^- ions in the interaction of electrons with vibrationally excited hydrogen H_2 molecules. The vibrational excitation should reduce the threshold for the appearance of H^- ions, and on the whole, increase the generation rate of H^- ions by slow electrons. Indeed, the cross section for the formation of H^- dissociative attachment ions upon collision of electrons with vibrationally and rotationally excited molecules increases significantly upon excitation to high levels, as shown in Fig. 3.17. Molecular excitation can be efficiently

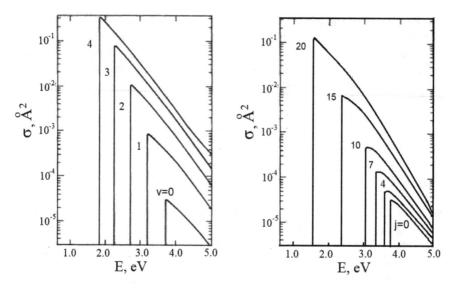

Fig. 3.17 Cross sections for the formation of H⁻ ions in collisions with vibrationally (ν) and rotationally (j) excited H_2 molecules. The cross sections increase with increasing excitation levels. (Reproduced from Wadehra and Bardsley [62], with the permission of APS Publishing)

produced by high-energy electrons as shown in Fig. 3.18. The dissociative attachment cross sections for rotationless ($j = 0$) vibrational states of H_2 from $v = 10$–15. for these exothermic processes rise from the threshold as $e^{1/2}$, which is consistent with the Wigner law for capture of p-wave electrons. Note also that the peak cross section decreases slightly with increasing vibrational quantum number v, in contrast to the drastic enhancement of the peak threshold cross section with the initial vibrational motion for $v < 9$ [126].

Figure 3.19 shows the increase in cross sections for dissociative capture upon excitation to high vibrational levels for hydrogen and deuterium. When excited to levels with $\nu = 5$, the cross sections increase by a factor of 10^5. The authors of a number of papers [131, 132] have considered the possibility of effective formation of H⁻ ions in the interaction of electrons with molecular hydrogen ions. Figure 3.1 shows the calculated value [133, 134] of the cross section σ_2 for the formation of H⁻ ions, as a function of electron energy, in the reaction $H_2^+ + e \rightarrow H^- + H^+$. For the electron energy of $W_e = 2$–4 eV, the cross section $\sigma_2 \sim 1.3 \times 10^{-17}$ cm² $\gg \sigma_1$.

The formation of H⁻ ions in the interaction of crossed beam of electrons and H_2^+ ions has been investigated experimentally [135, 136]. The dependence obtained, σ_2^* in Fig. 3.1, is explained by the estimated population of highly vibrationally excited states of the ion beams used, obtained from a H_2^+ plasma source. The dependence corresponds to an almost uniform filling of the first five vibrationally excited states. The excitation of vibrational states substantially reduces the threshold for formation of H⁻ ions from H_2^+ ions. Furthermore, the channel for dissociation of H_2^+ ions contains many paths not leading to the formation of H⁻ ions. For example, Fig. 3.1 shows the cross section σ_4 for the dissociation of H_2^+ into H and H* atoms when

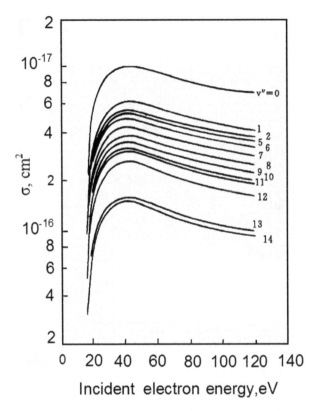

Fig. 3.18 Cross sections for vibrational excitation of H_2 molecules by electrons. (Reproduced from Hiskes et al. [64])

interacting with an electron, which exceeds $\sigma_2{}^*$ by almost a factor of 500 [137]. The cross section for dissociation of $H_2{}^+$ into an atom and a proton is 10^{-15} cm^2 at an electron energy of $W_e = 10$ eV [128].

No matter how small the cross sections for the processes considered for the formation of H$^-$ ions, an arbitrarily high H$^-$ generation rate in the plasma can be obtained by an increase in the density of molecules, molecular ions, and electrons. However, in this case, the destruction processes of H$^-$ ions in the plasma also intensify, so that the rate of growth of their density in the plasma decreases, and only ions generated in increasingly thin near-surface-plasma layers will be included in the H$^-$ ion flux of interest to us. In practice, the intensity is limited by the limiting gas density, which destroys negative ions at the source outlet.

The two important processes for the destruction of H$^-$ ions in a hydrogen plasma are H$^-$ + e → H + e + e and H$^-$ + H$^+$ → H + H*. The variation of these two cross sections (σ_5 and σ_6, respectively) on electron energy (for σ_5) and on the total particle energy in the center-of-mass system (for σ_6) is shown in Figs. 3.2 and 3.3. Of these two processes, the most important role is played by the destruction of H$^-$ ions by the electron impact. Several studies have received considerable attention [138–140].

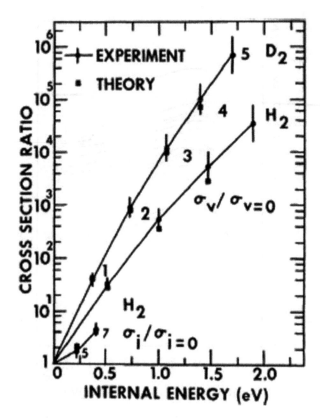

Fig. 3.19 Increasing cross section ratio $\sigma_v/\sigma_{v=0}$ for H$^-$ negative ion formation by vibrational excitation of H$_2$ and D$_2$ molecules and by rotational excitation $\sigma_v/\sigma_{j=0}$. (Reproduced from Allan and Wong [63], with the permission of APS Publishing)

The most reliable results were obtained by an intersecting beam method. The threshold for this reaction is 0.75 eV; at an electron energy of $W_e = 10$–20 eV, there is a maximum with complex structure, reaching 3.5–4 × 10^{-15} cm^2.

The dependence of the ion-ion recombination cross section for H$^-$ ions with protons on the total ion energy in the center-of-mass system (σ_6 in Fig. 3.3) was obtained in [141, 142] and refined in [143], where data obtained in earlier work are also reported. At an H$^-$ ion energy in the laboratory system of $W < 30$ eV, the cross section changes approximately in inverse proportion to the H$^-$ ion velocity in the laboratory frame, so that it can be approximated by the expression $\sigma_6 \sim 1.4 \times 10^{-7}$ cm^{3}s^{-1}/v$^-$, where v$^-$ is the H$^-$ velocity. At H$^-$ ion energy of ~0.05–0.5 keV, a plateau with complex structure oscillating about $\sigma_6 \sim 2 \times 10^{-14}$ cm^2 is observed. At an energy greater than 2 keV, the cross section σ_6 rapidly decreases.

Resonant charge exchange of H$^-$ ions with hydrogen atoms (σ_7 in Fig. 3.3) does not cause irreversible death of H$^-$ ions; however, the conversion of a fast H$^-$ ion into a slow ion strongly decreases the probability of its escape from the plasma. This process ensures effective cooling of H$^-$ ions by converting fast H$^-$ ions with a considerable momentum spread into H^- ions with the temperature of the colder dissociated hydrogen. Resonant charge exchange has been reported [144, 145]. For comparison, the cross section σ_7 for resonant charge exchange of a proton with a hydrogen atom is shown in Fig. 3.3.

The only irreversible destruction of H^- ions in interaction with atomic hydrogen is related to high energies. The cross section for this process (σ_9 in Fig. 3.3) is $\sim 10^{-15}$ cm^2 at an energy of 10^3–10^4 eV [146]. The data of [147] about this process are apparently greatly inflated.

The cross section for the destruction of H^- ions in interaction with hydrogen molecules (σ_{10} in Fig. 3.3) increases from 10^{-16} cm^2 at an energy of several eV to 10^{-15} cm^2 at an energy of hundreds eV [148–151]. This process, σ_{10}, as for the process σ_9, determines the destruction of H^- ions during beam formation, acceleration, and transport.

Figure 3.3 also shows the cross section σ_{11} for charge exchange of protons on molecular hydrogen; this process determines proton loss in proton sources.

Available information on the destruction of H^- ions by interaction with molecular ions (σ_{12}) is inconsistent. A number of estimates [146] indicate the comparability of the cross sections σ_{12} and σ_6 at low energy (apparently $\sigma_{12} \sim 5 \times 10^{-14}$ cm^2 at an energy of 1 eV). It seems that the data [152] on the cross section for this process are greatly underestimated. In general, ion-ion recombination processes are characterized by a weak dependence on the type of positive ions [153], since an electron from a negative ion is captured to levels with large quantum numbers.

3.9 Beam Formation from Negative Ions Generated in the Plasma Volume

The production of H^- ions in the plasma volume has been considered in a number of reviews [5–7, 13–15, 32, 36, 37, 41, 42, 45, 46, 154]. In earlier work, it was thought that H^- ions are formed through the interaction of electrons with molecules in the ground state. The resulting calculated H^- density was very low compared to that measured in experiments. The formation of H^- is more realistically due to interaction with vibrationally and rotationally excited molecules. In this case, one can have a low electron temperature in the negative ion generation region. The H^- generation rate is proportional to the density n_g of gas molecules in the excited state and the electron density n_e near the emission aperture. After extraction, H^- ions can be destroyed in collisions with molecules (σ_{10} in Fig. 3.3). Taking into account this destruction, the H^- flux density should decrease exponentially. The thickness of the molecular target is proportional to the gas density in the gas-discharge chamber and to the diameter of the emission aperture or the width of the slit. This relationship can be represented by

$$J = A n_g n_e \exp{-\left(B\sigma_{-10}\, dn_g\right)} \tag{3.8.1}$$

where J is the beam current density, A is a numerical coefficient, n_g and n_e are the densities of gas molecules and of electrons, B is a numerical coefficient, σ_{-10} is the cross section for the process $H^- + H_2 \rightarrow H + H_2 + e$ (Fig. 3.3), and d is the aperture diameter. The dependence of J on n_g increases approximately linearly with n_g for small n_g and decays exponentially for large n_g with a maximum at $n_g = 1/\sigma_{-10}Bd$ (or $B\sigma_{-10}dn_g = 1$). The maximum current density $J_{max} = An_e/e\sigma_{-10}Bd\exp(1)$ is inversely proportional to d, and the extracted current $I = J\pi d^2/4 = \pi dAn_e/e\sigma_{-10}B$ is proportional to d. For this reason, the emission current density can be large for a small aperture, but cannot be larger than $J \sim 10$ mA/cm² with an aperture of diameter $d \sim 1$ cm or greater. The accelerated current is a factor of e smaller than that emitted after the emission aperture, since the main part $(1 - e^{-1})$ is stripped after the emission aperture. The experimental dependence of the intensity on the gas supply is shown in Fig. 3.20. The straight line extrapolates the intensity without stripping. The difference between this line and the experimental curve is the intensity of the neutrals generated by stripping on the outgoing gas. This stripping generates a large stream of fast neutrals, which affect the subsequent acceleration of H^- ions.

Vibrationally excited molecules must be supplied to the H^- ion generation region from the discharge with high electron temperature (the concept of a tandem source with magnetic filter). The region of generation of vibrationally excited molecules is separated from the region of generation of ions by a magnetic filter with transverse magnetic field, which returns fast electrons. The concentration of vibrationally excited molecules can be several percent.

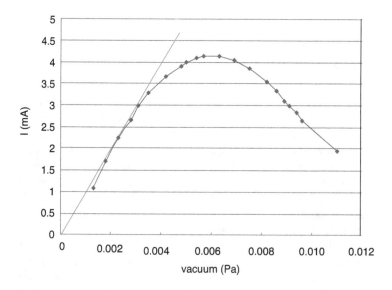

Fig. 3.20 Experimental dependence of the intensity of H^- ion beam on gas flow rate (vacuum in the chamber). Straight line—extrapolated intensity without destruction of negative ions in the escaping gas

3.10   Plasma Volume Sources of Negative Ions

The formation of negative ions within the plasma volume has been known for a long time [155], but for many years, it was not possible to achieve effective generation of H^- ion beams in this way. Naturally, in the first attempts to extract H^- beams from gas-discharge plasma, plasma sources were used that had been well-developed for positive ions. It should be emphasized that because of the complexity of the processes that accompany the production of negative ion beams, theoretical analysis of these processes for a long time had almost no effect on the optimization of sources. However, following improvements based on empirical observations, significant progress has been made.

In the studies of Fite in 1953 [71], only 10^{-7} A of H^- ions could be obtained from the low-current discharge. Then, Moak and coworkers in 1959 [156], using a more intense discharge in a standard duoplasmatron with extraction of H^- ions along the discharge axis, obtained a beam with a current of 11 μA, comparable to beams from charge-exchange sources. In the standard discharge mode as used for proton generation, an H^- ion beam with current 0.5 μA was extracted, but with increased hydrogen supply, the intensity of the H^- beam increased by a factor of 20. At the same time, the flux density of H^- ions through an emission hole 0.1 mm in diameter reached 0.1 A/cm^2. Together with the negative ions, a stream of accompanying electrons with current 5 mA was formed and accelerated by the extraction voltage. Beginning with these first experiments, minimizing the flow of accompanying electrons has become one of the main problems in the direct extraction of negative ions from a gas-discharge plasma. By improving the configuration of the gas-discharge region, intensifying the discharge, increasing the hydrogen density, and most importantly increasing the area of the emission aperture from axially symmetric duoplasmatrons, it was possible to produce H^- beams with current up to 70 μA [157], but the co-extrated electron flow exceeded this flow of H^- ions by a factor of $(1–2) \times 10^3$.

Using a duoplasmatron with axial extraction, Demirkhanov and coworkers [132] produced an H^- ion beam with a current of 200 μA. Investigations of the ion energy spectra revealed that the ion generation regions are localized before layers with increased voltage drop. In particular, considerable generation was observed in the cathode layer. In this work, the assumption was made that in H^- ion generation the interaction of slow electrons with molecular ions plays a dominant role. Further intensification of the generation of H^- ions in duoplasmatrons with axial extraction was limited by excessive increase in the flux of accompanying electrons, since the large-area extraction aperture serves as a drain for both primary and secondary (electron-generated) plasma because of penetration of the extractor electrode electric field.

For high-quality H^- ion beam sources for charge-exchange accelerators, duoplasmatrons with a shifted emission hole were used. In 1968, Collins and coworkers [158] and Lawrence and coworkers [159] found that the ratio of the H^- ion current to the accompanying electron current increases substantially, if the emission hole is displaced from the axis of the discharge column toward the periphery. Under optimal conditions, these sources produced H⁻ ion beams with a current of 80–100 µA and an accompanying electron current of 2–4 mA. The electron energy spread was 3–6 eV. The loss of electrons from the discharge column, which is constricted by the magnetic field, is significantly lowered at the peripheral region, whereas the motion of H⁻ ions is little affected by the magnetic field. In the following years, various duoplasmatron modifications with displaced emission hole were developed for charge-exchange accelerators [160, 161]. With small emission aperture dimensions, it is possible to obtain beams of H⁻ ions from these sources with comparatively high brightness. A normalized emittance of 0.05 π mm.mrad is characteristic of beams with a current of 50–70 µA. The substantial reduction in the flow of accompanying electrons provided an opportunity for further improvements in the generation of H⁻ ions using duoplasmatrons with displaced emission hole. However, in steady-state operational mode, it was not possible to form beams of over a milliamp, basically because of discharge power limitation to about 1 kW and also a hydrogen pressure limitation. Dubarry and Gantherin [162] increased the area of the emission aperture and obtained a beam of 400 µA. The H⁻ ion beam current density in the sources under consideration was limited to 20–50 mA/cm². By increasing the discharge power to 10^4 W in pulsed mode, Abroyan and coworkers [163] increased the H⁻ ion beam current from these kinds of sources up to 2.2 mA.

A higher usage efficiency of the generated H⁻ ion flux is realized in the tube-shaped duoplasmatron developed and studied in detail by Golubev [164, 165]. In this source, the main tubular discharge column with fast electrons is formed by an annular thermocathode, and a collimating rod, passing through a hole in the annular cathode along the center of a hole in an intermediate electrode until the intermediate electrode, is cut off. The collimating rod limits the arrival of fast electrons to the axis of the system along the magnetic field. H⁻ ions are extracted through an emission hole on the axis from the region with an increased concentration of H⁻ ions shaded from fast electrons by the collimating rod. In the first experiments with this source, in pulsed mode, H⁻ ion beams were obtained with a current of up to 7 mA and an emission current density of up to 0.14 A/cm². Such a beam was accompanied by a stream of electrons with a current of 300–500 mA. The normalized emittance of the ion beam at a current of 4 mA was estimated as 0.1 π cm mrad. Subsequently, the intensity of the beam was increased to 14 mA by increasing the hydrogen pressure up to 1.6 torr. For discharges in duoplasmatrons with a high degree of magnetic constriction of the plasma, instabilities that cause significant fluctuations in the parameters of the gas-discharge plasma are characteristic. Because of this, the parameters of the generated beams strongly deteriorate. In a modification of the tube-discharge source as described in [166], the annular cathode is immersed in a magnetic field formed by a ferromagnetic electrode, and the end of the molybdenum collimating rod is 3–6 mm away from the emission hole. The electron flux is reduced by the spherical surface of the thermal cathode and the geometry of the magnetic field. High discharge stability was noted. In 0.1 msec duration pulsed mode, H⁻ ion beams with an intensity of 15–17 mA at an emission density of about 200 mA/cm² and co-extracted electron current of not more than 100 mA were obtained. The authors indicate a very low value of 2–10×10^{-3} torr for the optimum hydrogen pressure in the discharge chamber. Hydrogen was fed into the discharge through the collimating rod, so that the local hydrogen density in the generation region could be much greater. An increased yield of H^- ions upon increasing the discharge voltage to 300–500 V, due to a decrease in cathode heating, was considered as an argument in favor of the formation of H^- ions in the plasma volume by interaction of electrons with molecules. The resulting intensity of the H^- ion beam was 15–17 mA and the emission current density 0.2 A/cm^2, and could be provided by the formation of H^- ions due to direct volumetric processes. But the analysis carried out above shows that this requires a high density of hydrogen, exceeding the actual value by a factor of 10^2–10^3. At the same time, favorable conditions for surface-plasma generation of H^- ions are realized in the source under consideration. The coefficient of sputtering and reflection of H^- ions from high-work-function surfaces is very small, 10^{-3}–10^{-4}. But by sputtering the oxide cathodes used in these sources, the surfaces of the collimating rod and the intermediate electrode must be covered with barium and barium oxide films, which reduces their work function to 2 eV. Under these conditions, the formation of H^- ions due to the capture of electrons from electrodes with reduced work function to the electron affinity levels of the sputtered and reflected particles can be quite effective. In such sources, the yield of H^- ions is maximal at the potential of the collimating rod, which is close to floating potential that is negative with respect to the anode. Intense bombardment of the collimating rod by accelerated ions and fast atoms produced by reflection from the electrodes can ensure the generation of intense H^- ion beams. The proximity of the end of the collimating rod to the emission aperture and the lower density and electron temperature between collimating rod and the emission aperture favor the production of fast H^- ions through surface-plasma generation.

An extensive program of research and development of duoplasmatrons with tubular discharge was carried out at the Brookhaven National Laboratory [167, 168]. In the first experiments, H^- ion beams were obtained with currents up to 5 mA [168] from a source close in construction to that described in Ref. [165], and it was possible to increase the current up to 15 mA [169] by supplying cesium. Details of studies of H^- ion generation in a hydrogen discharge and in a discharge with added cesium are described in [167]. From a duoplasmatron with hydrogen discharge, H^- ion beams with current up to 11 mA were obtained, with an extraction aperture diameter of 2.45 mm, a 200 A discharge current, a flux of accompanying electrons 1.6 A, and a normalized beam emittance of 1 mm mrad. From a discharge with added cesium, the H^- ion beam current was up to 60 mA, with an emission current density up to 1.3 A/cm^2 and an electron current of 1.9 A. The normalized emittance of the beam at a current of 40 mA was estimated to be 0.23 cm mrad. An investigation of the energy spectrum of H^- ions from discharges with added cesium revealed that a significant fraction of H^- ions is accelerated by the voltage between the collimating rod and the anode [169]. There is no doubt that in discharges with cesium, the surface-plasma mechanism for the formation of negative ions dominates. The authors explain the formation of H^- ions in discharges without cesium by volumetric processes, but also in this source that generates plasma in a discharge with a cold oxide cathode the work function of the electrodes would be small because of sputtering of the oxide cathode and the efficiency of surface-plasma negative ion generation relatively high, although less than in discharges with cesium. To identify the specific mechanism of H⁻ ion generation in these sources, experiments are necessary with the exclusion of impurities that reduce the work function. The optimum hydrogen pressure in the source plasma chamber is 0.1 torr. There is a very weak dependence of H⁻ ion beam intensity on hydrogen pressure for beams extracted from a discharge without cesium, whereas the dependence on hydrogen density must be significant for H⁻ ions formed in the volume.

Comparatively intense beams of H⁻ ions, suitable for practical applications, could be obtained from sources using a discharge in a uniform magnetic field as widely used to obtain positive ions. In 1963, Ehlers and coworkers [170], using a Penning discharge with cold cathodes adjacent to an anode emission slit adapted to extract positive ions across the magnetic field, produced a beam of negative deuterium ions with a current of 0.5 mA at a discharge voltage of 3 kV by increasing the gas supply. By separating the main discharge column from the anode wall to the emission slit, the beam current was increased up to 0.75 mA with a lower gas flow and a much smaller flux of accompanying electrons. After replacing the cold cathode with heated cathodes, a beam with a current of 1.2 mA was obtained from a direct current discharge without a reflexing electrode and up to 2 mA from a discharge with reflexing electrode. From a similar source with optimized geometry, H⁻ ion beams were obtained with current up to 5 mA [171]. With a fixed supply of hydrogen, the intensity of the H⁻ ion beam increases to saturation when the discharge current is increased.

Figure 3.21 shows a photo of the Ehlers source in the high-voltage terminal of the Van de Graaf accelerator at the Institute of Nuclear Physics, Novosibirsk, and a schematic of this source. This negative ion source with heated LaB_6 cathode can produce up to 8 mA of pulsed H⁻ beam (1 ms, 2 Hz).

With increasing hydrogen supply, saturation occurs at high discharge current and high H⁻ion beam intensity. Saturation of the intensity of the beam with respect to hydrogen density in the steady-state regime cannot be achieved due to the limited pumping rate, but experiments with pulsed hydrogen supply [172, 173] have shown that an increase in intensity with increasing hydrogen supply is replaced by a rapid decrease in the beam current at high gas density. Shifting of the main discharge column with fast electrons from the anode wall with emission slit made it possible to reduce the flux of accompanying electrons from 200–300 mA per 1 mA of H⁻ ion current to 10–30 mA. But the remaining flow of electrons greatly complicates work with the source. Thus, intercepting this electron flow with special electrodes is one of the most difficult problems to be solved when developing sources of this type.

Other sources of this type were developed and investigated in many laboratories [170–175]. The available information on the dependences of ion beam current on the characteristics of the gas-discharge plasma are in satisfactory agreement with ideas on the plasma generation of H⁻ ions, as described above. Vibrationally excited hydrogen molecules are formed in the main discharge column, which enter the H⁻ ion generation region with cold plasma adjacent to the emission slit.

Fig. 3.21 Upper: photograph of the Ehlers source on the high-voltage terminal of the Van de Graaf accelerator at the Institute of Nuclear Physics, Novosibirsk. Lower: schematic of the source (Reproduced from Dimov [81] with the permission of G. Dimov). (1) thermocathode, (2) cathode current leads, (3) radiators, (4) cathode insulator tube, (5) silicone rubber seal, (6) permanent magnet, (7) buffer body, (8) gas valve, (9) valve cooling, (10) extractor, (11) gas tube, (12) source body, (13) anode, (14) anode insert, (15) electron collector, (16) quadrupole lens, corrector, (17) quadrupole magnetic shielding, (18) magnetic shielding of accelerator tube, (19) rubber membrane

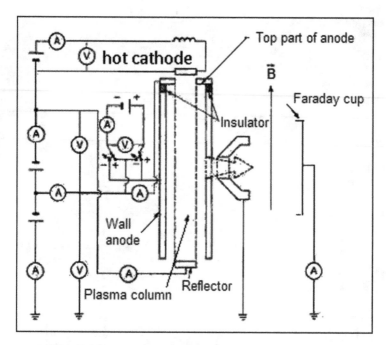

Fig. 3.22 Schematic of source with isolated plasma electrode. (Reproduced from Jimbo et al. [174] with permission of Elsevier Publishing)

A source with isolated plasma electrode is shown in Fig. 3.22. Ehlers and coworkers had changed the plasma electrode voltage [174]. The diameter of the reflexing electron column is 4.8 mm, and the diameter of the plasma electrode is 9.6 mm. The yield of negative ions increased in a magnetic field of 5 kG. Under normal conditions, when the plasma electrode potential is equal to the anode potential, the negative ion current was 4.6 mA and the electron current was 20 mA. When the plasma electrode potential was reduced to −6 V, the H⁻ ion current increased to 9.7 mA, and the electron current to 100 mA. When the potential was increased to 5 V, the ion emission decreased to 3 mA, and the electron current to 15 mA. With negative plasma electrode potential, the positive ion current was a factor of two less than the negative ion current. With a positive plasma electrode potential, the proton current was twice the negative ion current. The proton content was up to 90%, with 4% H_2^+. The maximum positive ion current was 4.1 mA, half the H⁻ ion current.

The Dudnikov-type Penning source [176] has been operated successfully with LaB_6 cathodes in a cesium-free discharge. It is found that the extracted H⁻ current density is comparable to that of the cesium-mode operation and the H⁻ current density of 350 mA/cm² has been obtained for an arc current of 55 A (in the Dudnikov-type Penning source with cesiation, the emission current density was up to 2 A/cm² [177]). The H⁻ yield is closely related to the source geometry and the applied magnetic field. Experimental results demonstrate that the majority of the H⁻ ions extracted are formed by volume processes in this type of source operation [178].

The plasma of a Penning-type discharge forms two regions: the central hot plasma region, containing fast electrons oscillating between the hot cathode and the anticathode (reflector), and the cold plasma region between the hot plasma column and the inner cylindrical wall of the source (Fig. 3.22). The ionization and the vibrational excitation of molecules occur in the central region. Outside this central hot plasma column, the electron temperature decreases in the radial direction. H$^-$ ions are formed by dissociative attachment to vibrationally excited molecules in the relatively cold surrounding region, close to the emission slit. Goretsky et al. [179, 180] reported the extraction H$^-$ ion current of 40 mA, with the current density of 80 mA/cm^2.

A source using a reversed magnetron discharge [181] is shown in Fig. 3.23. The plasma is generated by a discharge in a reversed magnetron and diffuses to the emitter. It was possible to extract up to 50 mA of H$^-$ ion current, and an electron current of 200 mA, from this source at a discharge current of 140 A.

A successful negative ion source was developed at TRIUMF [182]. A schematic of the source is shown in Fig. 3.24. A discharge with heated cathode is ignited in a chamber with a multipole magnetic cusp wall. The plasma diffuses to the plasma electrode through a magnetic filter. By supplying a regulated voltage to the plasma electrode relative to the plasma, the current of accompanying electrons can be reduced. Magnets are inserted into the extraction electrode, which deflects the accompanying electrons. A magnetic corrector is placed on the ground electrode. Using an emission hole of 13 mm in diameter, it is possible to extract from this source up to 15 mA of H$^-$ ions at a discharge power of 4 kW, a cathode power of 1.5 kW, and an accompanying electron current of 60 mA. Cesium was first added in

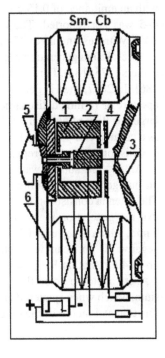

Fig. 3.23 Source with reversed magnetron discharge. (Reproduced from Litvinov and Savchenko [181]). (1) cathode, (2) anode-emitter, (3) plasma electrode, (4) diaphragm, (5) gas valve, (6) permanent magnet

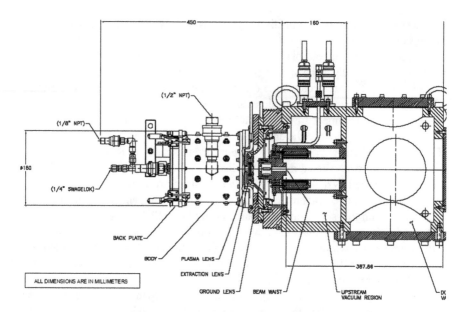

Fig. 3.24 The TRIUMF negative ion source. (Reproduced from Jayamanna et al. [183])

1998, and it now runs well in either mode. After cesiation, an emittance drop from $\varepsilon_{n90\%} = 0.2$ to 0.13 π mm mrad at 28 keV and a current of 20 mA was reported. The maximum beam current increased with cesiation by a factor of 1.25 to 25 mA dc with an emittance of 0.15 π mm mrad. A drop in the already very low e/H⁻ ratio of 4–5 was not reported. The extracted electrons are dumped into an extractor at 8 kV. With cesium, up to 25 mA of H⁻ ions were produced from these sources in a surface-plasma generation regime. These sources were duplicated in many research centers [175, 184]. Further improvements to the source have been described [185, 186]. In 2017, the H⁻ beam current was increased up to 60 mA with an emission aperture of 14 mm, a discharge power of ~15 kW, a cathode current of 800 A, and a large gas flow of 66 sccm. This ion source was licensed by D-Pace and distributed as a commercial product [187].

This ion source was repeated in [188].

An H⁻ ion source was developed at the Lawrence Berkley Laboratory with plasma generation by a high-frequency discharge using an internal antenna [189]. The antenna consists of a copper tube covered with an insulating enamel. The negative ion generation region is a collar with a diameter of 15 mm and a height of 9 mm, which is separated from the plasma generation region by a magnetic filter. A negative H⁻ ion beam was obtained from a 5.6 mm emission hole with a current of up to 40 mA at an RF power of 50 kW and an accompanying electron current of 1 A. It should be noted that the enamel on the antenna contains up to 17% potassium, which when sputtered and accumulated on the collar reduces the surface work function and increases the efficiency of surface-plasma negative ion generation.

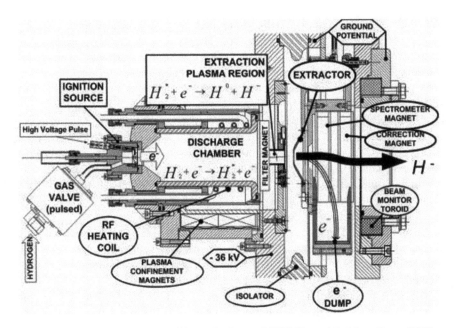

Fig. 3.25   Diagram of the RF source of negative ions at DESY. (Reproduced from Peters [190])

At DESY, Germany, this source was improved by replacing the internal antenna with an external one [190]. This source is shown schematically in Fig. 3.25. The high-frequency discharge is ignited in an $Al_2 O_3$ chamber by an RF antenna located outside the chamber. The production of up to 80 mA of H^- ions from an emission aperture of 6 mm at an RF discharge power of up to 80 kW and an accompanying electron current of 2.5 A has been reported. This source uses parts made of processible MACOR ceramic, which has a high potassium content. The transfer of potassium to internal surfaces intensifies the surface-plasma generation of negative ions. A long-term negative ion current of 40 mA with an emittance of 0.46 mm mrad has been reported. This source was not used for further negative ion acceleration, and for this reason, the reported beam intensity is questionable.

In a review article by Tarvainen and Peng [191], the authors state "RF ion sources relying on volume production of H^- are typically operated in continuous mode. Pulsed operation allows replenishing the cesium coverage between the discharge pulses which effectively inundates the benefits of the volume production. Thus, pulsed high-current RF H^- sources are rarely found. There are however, two well-documented exceptions to this trend. The HERA RF volume source developed at DESY has routinely delivered an H^- current of 40 mA at 8 Hz with short pulses using an RF power of 20 kW at 2 MHz. The highest reported H^- current extracted from the HERA source is 80 mA in 200 µs pulses. In long pulse operation, i.e. 3 ms pulses, the source has delivered 30–40 mA of H^-, with approximately 30% current droop over the pulse. The design of the source features an external antenna wound around a ceramic (Al_2O_3) plasma chamber and a collar structure extruding inwards from the plasma electrode. The HERA ion source design was later adopted to CERN as a prototype of the Linac4 H⁻ ion source. The source did not, however, meet the performance goals of the Linac4 project as the electrons, co-extracted with the H⁻ and dumped onto the extraction electrode at 45 kV, caused severe structural damage. Later an improved design of the ion source demonstrated H⁻ beam current of 20 mA with an e/ H⁻ ratio of 50–60 in volume production mode with 30 kW of RF power. Further developments at CERN have lead into a redesign of the HERA ion source concept and resulting in 30 mA of H⁻ current (e/ H⁻ ratio of 20) at 45 keV produced with 30 kW of RF power in 500 μs pulses. However, the performance goal of the Linac4 project is 80 mA of H⁻ and it is considered unlikely to achieve this level with a volume source. Therefore the recent research activities at CERN have been focused on cesiated operation of the RF ion source causing the volume source development at CERN to become stagnant."

As remarked by Peters [192], "The antennas used by LBL, SSC and DESY were all coated by the same manufacturer." An analysis of the coating found a high percentage of K and Na (Table 3.2). Different contents of potassium might have contributed to reported differences in source performance.

The HERA external antenna source has parts made from MACOR machinable ceramic, which has a high concentration of potassium. This potassium can be liberated by the plasma and transported to the internal surfaces of the source.

To work efficiently, the CERN source requires a long activation time to liberate alkali impurities. Lettry has remarked [193], the reconditioning time needed after 2 months storage to reach 20 mA H⁻ is 25 days. Conditioning time needed to reach 20 mA of H⁻ is 30 days. 1/2 day stop to replace the Al_2O_3 filter magnet housing by molybdenum with restart time to reach 20 mA H⁻: 3.3 days. Ratio $e/H⁻$ is decreased during conditioning from 100 to 60. This fact clearly confirms the surface-plasma operation of the CERN RF source without cesiation.

In the J-PARC negative ion source [194], a surface-plasma mechanism of H⁻ formation was discovered without Cs injection.

The HERA RF source and CERN RF source with high beam current work in surface-plasma production mode with low ionization impurities from the ion source parts. This volume source is a TRIUMF negative ion source with an emission current density below 10 mA/cm².

Peking University reported the production of intense H⁻ ion beams (up to 45 mA in pulsed mode at 2 kW, and up to 29 mA in cw mode at 1 kW, claimed efficiency

Table 3.2 Results of analysis of the antenna coating

Elements	Percentage
Si	46.9
Ti	29.6
K	15.2
Al	5.7
Na	2.6

~29 mA/kW) from a source with a microwave discharge [195–197]. But in these publications, only 1.2 mA of H⁻ is directly measured after a magnetic analyzer, with a current to the collector up to the analyzer of 40 mA, and large currents are obtained by extrapolation through the relationship of electrons and ions after the bending magnet. In particular, the ratio of ion current to electron current is found to be 4 (incredibly large for a source with a longitudinal magnetic field). By adjusting the voltage on the suppression electrode, it is possible to ensure the focusing of the negative ion beam and good passage of the beam to the collector after the magnetic analyzer (which is accomplished by formation of positive ions). The electron beam can be deflected by a magnet near the extractor and does not pass to the collector after the analyzer. The current predicted at the collector after the analyzer in this case is erroneous, not legitimate.

Figure 3.26 shows the current signal to the direct collector FC1, and the current to the collector FC2, after the magnetic analyzer when tuning to the negative ions peak and to the peak of electrons, from the PKU microwave source. It is seen from this figure that the shape of the total current on FC1 reproduces the shape of the FC2 current when tuned to the electron peak and differs strongly from the FC2 negative ion signal, which should not be for a H⁻/e ratio of 4.

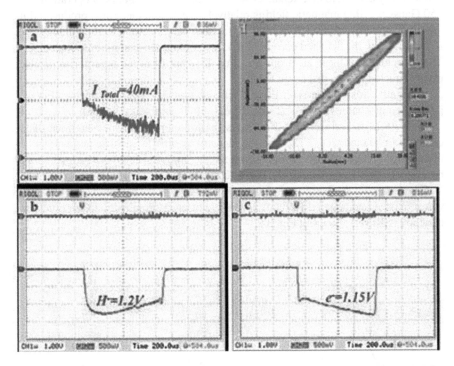

Fig. 3.26 Peking University microwave H⁻ source data: oscillograms of the total current to the collector FC1, the current to the collector FC2 when tuned to the peak of negative ions and to the peak of the electrons. (Reproduced from Zhang [196])

A more realistic characterization of this ECR negative ion source has been presented [198]. A practical 2.45-GHz microwave-driven Cs-free H⁻ source was developed, based on the experimental H⁻ source at Peking University (PKU). Several structural improvements were implemented to meet the practical requirements of the Xi'an Proton Application Facility (XiPaf). Firstly, the plasma chamber size was optimized to enhance the plasma intensity and stability. Secondly, the filter magnetic field and electron deflecting magnetic field were enhanced to reduce the co-extracted electrons. Thirdly, a new two-electrode extraction system with a larger electrode gap and enhanced water-cooling ability to diminish spark and sputter during beam extraction were applied. Finally, a direct H⁻ current measuring method was adopted by the arrangement of a new pair of bending magnets before the Faraday cup (FC) to remove residual electrons. With these improvements, electron cyclotron resonance (ECR) magnetic field optimization experiments and operational parameter variation experiments were carried out on the H⁻ ion source, and a maximum 8.5 mA H⁻ beam was extracted at 50 kV at a microwave power of 5.4 kW (efficiency 1.5 mA/kW, standard for plasma negative ion sources, not 29 mA/kW as was claimed before) with a time structure of 100 Hz/0.3 ms. The root-mean-square (RMS) emittance of the beam was 0.25 π·mm·mrad. These improved H⁻ source and extraction system were maintenance-free for more than 200 hours of operation. The ~50 Gauss bending magnetic field is relatively week, and electrons, produced during destruction of negative ions after extraction, can penetrate to the Faraday cup. For this reason, the detected current of 8.5 mA could include some electron current. The trajectories of these electrons are very different from the trajectories of electrons extracted from a discharge in He.

A negative ion source with ECR discharge was developed at Saclay (France) [199]. At a discharge power of 2 kW, 3.5 mA of H⁻ ions was obtained.

In 1979, M. Bacal developed a laser stripping method for measurement of H⁻ density in plasmas [200]. The high H⁻ density in diffuse gas discharge was confirmed using this method, which supports the importance of vibrationally and rotationally excited molecules in H⁻ production.

Cavity ringdown absorption spectroscopy (CRD) for measurement of H⁻ density was proposed and adopted [201]. A schematic of cavity ringdown absorption spectroscopy (CRD) is shown in Fig. 3.27.

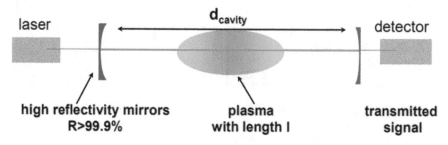

Fig. 3.27 Schematic of cavity ringdown absorption spectroscopy (CRD)

Pumping of an optical cavity by a (tunable) laser source measurement of signal decay after laser source is switched off. The measurement of laser light attenuation trapped in a high-finesse optical cavity transfers the absorption signal from wavelength into time dependence empty cavity with decay time τ additional absorption τ to τ',

$$I(t) = I_o e - t / \tau; \tau = d / c(1-R); \ n = d\left(1/\tau' - 1/\tau\right)/ I d \sigma c.$$

The recent review of volume negative ion production is presented in [202]. This topical review gathers the last updates concerning cesium-free negative ion sources. Hence, beyond the frame of this course, this topical review addresses both the theoretical and experimental work performed during these last few years and complexities represented by the conception of a negative ion source ranging from the creation of negative ions to their neutralization.

To test the ability to make suitable quality beams of heavy negative ions [203], an experiment was conducted at Lawrence Berkeley National Laboratory using chlorine in an RF-driven ion source. Without introducing any cesium (which is required to enhance the negative ion production in hydrogen ion sources), a negative chlorine current density of 45 mA/cm² was obtained under the same conditions that gave a 57 mA/cm² of positive chlorine, suggesting the presence of nearly as many negative ions as positive ions in the plasma near the extraction plane. The negative ion spectrum was 99.5% atomic chlorine ions, with only 0.5% molecular chlorine and essentially no impurities. Although this experiment did not incorporate the type of electron suppression technology that is used in negative hydrogen beam extraction, the ratio of the co-extracted electrons to Cl⁻ was as low as 7 to 1, many times lower than the ratio of their mobilities, suggesting that few electrons are present in the near-extractor plasma. This, along with the near-equivalence of the positive and negative ion currents, suggests that the plasma in this region was mostly an ion-ion plasma. The negative chlorine current density was relatively insensitive to pressure and scaled linearly with RF power. If this linear scaling continues to hold at higher RF powers, it should permit the current densities of 100 mA/cm², sufficient for present heavy ion fusion injector concepts. The effective ion temperatures of the positive and negative ions appeared to be similar and relatively low for a plasma source.

Ion generation in H⁻ discharges consumes ~0.2–1 MeV energy per ion, but because of the large flux of accompanying electrons, the energy cost of the H⁻ ion must be larger than this, as compared with the energy price of positive ions, 1–5 keV/ion.

The gas economy [204] of plasma sources is at a 10⁻³ level. Despite the low intensity, comparatively large emittance, and such high cost for H⁻ ion generation, the use of plasma volume sources of H⁻ is in some cases more effective than for proton sources. Plasma volume sources have been used in accelerator technology [155] and in plasma experiments, but their low efficiency and limited intensity do not allow their use to produce beams of high-energy neutral atoms as required for controlled fusion application.

3.11 Thermionic Production of Negative Ion Beams

With the limited possibilities of prior methods of generating negative ions, it seemed tempting to produce negative ion beams by thermionic emission from surfaces having a low work function. The thermionic emission of negative ions was known a long time ago as evaporating of oxygen negative ions from oxide cathode sputtered luminophore in oscillograph tubes.

The possibility of using thermionic emission for the production of negative ions had been pointed out in principle a number of times [205–207]. The stability of the emission surface and the low-energy spread of the ions formed are prerequisites for obtaining high-brightness beams, even given the low ion flux emissivity. The effectiveness of thermionic emission is characterized by the degree of thermal ionization α^+, equal to the ratio of the ion flux leaving the surface to the neutral particle flux simultaneously leaving the surface due to thermal desorption. Under the simplest conditions, the degree of thermal surface ionization is determined by the well-known Saha-Langmuir relation

$$\alpha^+ = A^+ \exp\left(\varphi - V\right)/T; \quad \alpha^- = A^- \exp\left(S - \varphi\right)/T$$

for positive and negative ions, respectively. In these relations, $A^+ = Q^+/Q$ and $A^- = Q^-/Q_0$ are the ratios of the statistical weights of positive Q^+ and negative Q^- ions to the statistical weight of neutral particles Q (for atoms with one valence electron $Q = 2$, $Q+ = 1$, $Q^- = 1$); φ and T are the work function and surface temperature, respectively; V is the ionization energy; and S is the electron affinity. More complex situations have been considered in a number of monographs [204, 206]. To effectively convert the working substance into positive or negative ions, it is necessary to ensure the ratio $(\varphi - V) > T > 0$ or $(S - \varphi) > T$, respectively. Note that due to the increase in electric field strength at the surface, the degree of surface ionization for $S < \varphi$ can be dramatically increased; however, the local spread of transverse energy can greatly increase also.

After significant effort, it was possible to create highly efficient thermionic sources of positive cesium ions ($V = 3.89$ eV) with the working substance supplied to the ionizing tungsten surface ($\varphi = 4.5$ eV) through pores in the emitter [208]. In such sources, it is possible to convert to ions up to 90–99% of the supplied cesium at an emissive current density up to 10^{-1}–10^{-2} A/cm^2, limited by the electric field of the space charge of ions at the ultimate extraction voltage of 30 kV. Many results of the technology for obtaining positive ion beams can be used for the production of negative ion beams.

The possibility of obtaining negative ions is primarily determined by the possibility of providing a sufficiently low work function. Current effective electron emitters that are resistant to poisoning at elevated temperatures, such as lanthanum hexaboride, impregnated cathodes, L-cathodes (cathodes with barium diffusing through pores to the electron emission surface), and the like, have a work function $\varphi = 2.6$ eV, which is less than the electron affinity for a number of elements (for halogens $S = 3.1$–3.7 eV). There is an experience of obtaining relatively intense beams of negative iodine ions

$S = 3.1$ eV due to thermal ionization on lanthanum hexaboride [152, 209–212]. The difficulties of creating surfaces with stable low work function complicate the use of the thermionic method for obtaining negative ions with smaller electron affinity. According to the literature, the work function of oxide cathode can be reduced to 1 eV and even to 0.8 eV [213]. The best samples of oxygen-cesium photocathodes have a work function of 0.7 eV [214]. The work function of tungsten by adsorption of oxygen and cesium can be lowered to 0.75 eV [205]. Recently, emitters with negative electron affinity have been created, in which the energy levels of the conduction band in the volume of the crystal are higher than the energy level of the electron at rest outside the crystal [215, 216]. It seems that providing a high degree of ionization of thermally desorbed hydrogen ($S = 0.754$ eV) is solvable. The preservation of a low work function at the elevated temperature necessary to ensure a high rate of thermal desorption of particles is a complex problem. It can be hoped that by continuous regeneration of the surface structure, this problem will be solved.

Only atomic negative hydrogen ions are stable, whereas under normal conditions, hydrogen exists as molecules of hydrogen or other hydrogen-containing compounds. Dissociation of molecules can occur during adsorption, but at a not very high surface temperature, the dissociation rate can limit the ion emission. A possible reason for limiting the secondary emission of H⁻ ions in work reported by Bender and coworkers [217] may be an insufficient rate of atomization of molecular hydrogen adsorbed from the gas phase. In this connection, there is a need for a dissociator of hydrogen supplied (gas discharge, high-temperature furnace, etc.). Dissociation can be facilitated by the use of hydrogen-containing substances with low dissociation energy.

Mass spectrometric observations of electron-stimulated and thermionic emission of H⁻ ions from a heated molybdenum surface in the presence of cesium vapor have been reported [218]. The observed formation efficiency of H⁻ ions and D⁻ was low, but significantly higher than the formation efficiency of H⁻ ions on pure molybdenum.

The available data confirm the possibility of using thermal surface ionization to produce negative ions, including hydrogen isotope ions.

Targeted studies of the complex of problems under discussion have practically not been carried out. It is not ruled out that the resolution of the technical problems described above will allow the production of high-quality intense beams of negative ions by the thermionic method. Providing conditions for the formation of H⁻ ions by thermal surface ionization in surface-plasma sources is one of the reserves for increasing the efficiency of the surface-plasma method for obtaining of negative ions.

3.12   Secondary Emission (Sputtering) Production of Negative Ion Beams

The possibility of formation of negative ions by bombarding a surface with fast particles originated in 1931, when Woodcock discovered this process experimentally [219]. These processes can be characterized by the following values: the sputtering and reflection coefficients $K_s = I_s/I_1$ and $K_r = I_r/I_1$, the coefficient of negative ionization of the fluxes of sputtered and reflected particles $\beta^-_s = \Gamma^-_s/I_s$, $\beta^-_r = \Gamma^-_r/I_r$, and the coefficient of secondary emission (sputtering) of negative ions and the conversion of particles to negative ions upon reflection $K^-_s = \Gamma^-_s/I_1$, $K^-_r = \Gamma^-_r/I_1$. In these relations, I_1 is the flux of bombarding particles, I_s and I_r are the total fluxes of sputtered and reflected particles, and Γ^-_s and Γ^-_r are the fluxes of sputtered and reflected converted negative ions. It is useful to introduce the total ionization and secondary emission factors: $\beta^- = \Gamma^-/I$, $K^- = \Gamma^-/I_1$, where $\Gamma^- = \Gamma^-_s + \Gamma^-_r$, $I = I_s + I_r$.

Subsequent studies of these processes were conducted under more controlled conditions. The bombarded surface was placed in a container with controlled vacuum conditions. The surface was cleaned, and its state and its properties controlled. Beams of bombarding particles were obtained from a separate source, from which portions isolated by monochromators were fed to the target, and more sophisticated recording equipment was used. A large number of studies were carried out under these conditions, which served as the basis for presently existing ideas about these processes. The results obtained have been systematized in a number of reviews [220–223], monographs [224–227], and conference proceedings [228–231]. A bibliography on this topic has been systematized [232].

These results indicate that when surfaces with work function $\varphi \sim 3\text{--}5$ eV are bombarded with beams of energy $W \sim 10^2\text{--}10^6$ eV, the maximum secondary emission coefficient for negative hydrogen ions is $K^- \sim 10^{-3}\text{--}10^{-4}$, the β^- coefficient is $\sim 10^{-2}\text{--}10^{-3}$, and $K^- = K^-_s + K^-_r \sim 10^{-3}\text{--}10^{-4}$. When metals surfaces with work function $\varphi = 5.4$ eV are bombarded with atomic ion beams of elements with large electron affinity (halogens), the conversion rate can reach tens of percent, with $\beta^- \sim 1$ [233]. The energy spectra of secondary H^- ions formed when bombarding a tungsten surface with H^- and H_2^+ ions with energies of order 1 keV have been investigated [234]. A low-energy peak associated with the knockout of H^- ions from the adsorbate was observed at surface temperatures below 1168 K. The maximum of this peak is located at an energy of 2–4 eV, and the half-width at half-maximum (HWHM) is about 3 eV; toward high energies, this peak extends to 26 eV with an average height of 5–19% of the maximum amplitude. The high-energy peak associated with the reflection of hydrogen particles in the form of H^- ions does not depend on the surface temperature. It has a linearly increasing part stretching up to 200 eV; then, a plateau follows up to an energy of 60% of the incident particle energy; this peak ends with an almost linear decay to the bombarding particle energy. The conversion rate $K^-_r \sim 10^{-4}$. In measuring the secondary emission of negative adsorbate ions (H^-, O^-, etc.), the working substance was supplied to the surface from residual gas. The experimental dependencies of the secondary current of these ions on the pressure of the residual gas at the surface have maxima.

In the bombardment of metallic surfaces by argon ions with an energy of 13 keV and a current density of 3 mA/cm^2, the flux of secondary H^- ions had a weak peak at a residual gas pressure of 10^{-5} torr [235]. This dependence is explained by most authors by the destruction of negative ions by residual gas, which should be recognized as untenable. The cross sections for the destruction of negative ions are not so great [18, 236], and thus their flow to the collector decreases markedly at a pressure $\sim 10^{-4}$ torr.

The dependence of the steady-state secondary H$^-$ ion current obtained by bombarding a copper surface with 2 keV cesium ions was obtained [237]. For constant hydrogen pressure near the surface, one could expect saturation of this dependence. In fact, after reaching a maximum at a bombardment current density of 30 µA/cm^2, the current of secondary hydrogen ions decreases and at a current density of more than 60 µA/cm^2 becomes almost an order of magnitude smaller.

An important step in the development of the secondary emission method for obtaining negative ions was the discovery by Ayukhanov [238] and Kron [239] of a significant increase in the K$^-$ coefficient when alkali metals were deposited on the surface. It has been shown [220, 240, 241] that the secondary emission of negative ions increases with decreasing surface work function. On the basis of these studies, the first secondary emission sources of negative heavy ions were developed, as described by Mueller and Hortig [242–244] and Ayukhanov and Chernenko [245]. In these sources, ions are emitted from a plane surface and accelerated normal to the surface by an ion-optical system. The working substance is contained in a near-surface layer. Cesium vapor is supplied to the emissive surface. The bombardment is by positive ions of inert gas or alkali metal. The intensity of the beam obtained is in the range of ~10^{-3} − 10 µA. For Muller and Hortig's source, the H$^-$ ion current was of order 0.01 µA [242–244].

Secondary emission sources with a geometry proposed by Middleton have been developed successfully; these sources were called "sputtering sources" by the authors [246–249]. In these sources, negative ions are emitted from the inner surface of a cone bombarded by a cesium ion beam and extracted through a hole in the cone. An alkali metal is deposited on the emitting surface by means of vapor deposition or as a beam. Typically, the working substance is contained in the cone material; in some cases, it can be fed to the emitting surface as a stream or gas jet.

For negative hydrogen ion production, titanium cones saturated with hydrogen are also used. Bombardment of the surface by positive inert gas ions or by cesium is mainly along the cone axis. The negative heavy ion intensity from the Middleton source exceeds 100 µA. The maximum H$^-$ and O$^-$ ion current is 25 µA. Sputtering sources have found a wide application in tandem accelerators. The construction of various sputtering sources has been described in a book by Middleton [250].

One variants of Middleton's spray source is shown in Fig. 3.28. Cesium is supplied through a tube from a container to a hot tungsten spherical surface cesium ionizer. Cesium ions are extracted by a high voltage and are focused into a small spot on the surface of the sputter target. Sputtered negative ions are extracted by the same voltage and are formed into a beam.

Description of some sputtering negative ions are presented in review of G. Alton [251].

The sputtering negative ion source was developed in BINP [252].

Schematic of this ion source is presented in Fig. 3.29. A step motor 11 rotate, an isolation shaft 12, thru Wilson seal 10 a drum with samples for sputtering 1. A thermal surface cesium ionizer 2 ionize a cesium flux, delivered cesium furnace 7, 8 with valve 9. The cesium ions accelerated and focused by voltage applied between ionizer 2 and dram with samples 1. A sputtered negative ions 4 extracted by voltage between ionizer 2 and extraction electrode 3 and focused by electrostatic lens 5.

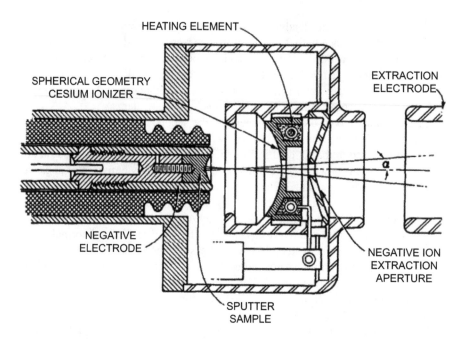

Fig. 3.28 A variant of a negative ion sputtering source. (Reproduced from Middleton [249])

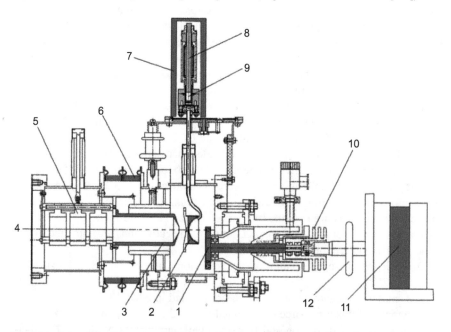

Fig. 3.29 Schematic of cesium sputtering ion source of BINP for AMS. The second-generation sputtering negative ion source, operated at BINP AMS. 1 the rotating turret holding target samples, 2 thermal ionizer, 3 extractor electrode, 4 beamline, 5 electrostatic lens, 6 HV isolator, 7 external heater of the cesium evaporation oven, 8 mechanism for cracking cesium ampules, 9 cesium ampules container, 10 air-cooling radiator, 11 stepper motor, 12 electric isolator.

Commercial versions of sputtering sources are available from the National Electrostatic Corporation (NEC, pelletron.com) [253].

Quantitative measurements were initiated [217] of the secondary emission of H^- ions from surfaces coated with alkali metal films and the behavior of the secondary emission of H^- ions, connected with the supply of hydrogen to the surface from the gas phase. In this work, the secondary emission of H^- ions from tungsten and tantalum surfaces covered by a cesium film, upon exposure to H_2 gas, was studied, with bombardment by cesium ions with an energy up to 5 keV and a current density up to 15 mA/cm^2. The cesium and hydrogen fluxes were regulated. Along with the measurement of the secondary emission coefficient, its behavior over time was studied. In the steady-state regime, the secondary emission coefficient K^- for H^- ions from the tungsten surface did not exceed 5–7%. At the same time, an increase in the hydrogen pressure at the tungsten surface from 10^{-5} to 10^{-4} torr increased the H^- ion yield only by a factor of 3, and a further increase in the hydrogen density had a very little effect on the K^- value. A significant increase in the K^- coefficient was observed in the first 1–100 s after the bombarding beam was switched on. The peak value of K^- and the duration of increased H^- ion emission depended on the exposure time of the surface without bombardment and with the supply of cesium and hydrogen. For an exposure time of less than 0.5 s, an increased yield of H^- ions was not observed. When hydrogen was supplied to the bombarded tungsten surface from the rear through pores in the tungsten, it was possible to increase the steady-state K^-_s coefficient to 23%. Thus, the H^- ion output depends on the temperature, and the maximum temperature was 400 °C.

In the following experiment, described by Bender and coworkers [217, 254], a peripheral annular part of a rotating tungsten disk was used as the bombarded emitter of H^- ions. A cesium vapor stream, a hydrogen stream, and a cesium ion beam were used to serially bombard through 120° sections of the rotating disk. Between the adsorption of hydrogen and the cesium bombardment, there was an adjustable delay time of order in the period of rotation. Stationary and peak (after bombardment) H^- ion yield depended on the rotational speed of the disk. At a speed of 2–10 rpm, the steady-state and peak values of K^- were equal, reaching a maximum value of 80% over a wide range of bombarding ion flux density. These authors reported [217, 254] that for effective sputtering of hydrogen in the form of H^- ions, a long exposure of the cesium surface to the flow of molecular hydrogen is necessary. Obviously, H^- ions are effectively sputtered by the beam from the surface only from certain adsorption states, and for the surface hydrogen to transition into these adsorption states the hydrogen particles need to be on the surface for about 0.5 s (or more). It follows from the data obtained in this work that in the transition regime after prolonged exposure in hydrogen with enhanced efficiency, 10^{16}–10^{17} H^- ions/cm^2 are emitted, which is much greater than a one-monolayer adsorbed coating. It is not excluded that during activation, a complex multilayer coating may be formed, possibly containing Cs H, and in a transient mode, secondary H^- ions could be desorbed with high efficiency from this coating. The decrease in the yield of H^- at high exposures is due to the competing adsorption of other, more active substances from residual gases. The reason for the increase in the steady-state emission of

H⁻ ions when hydrogen is supplied to the surface through pores in the heated emitter could be increased hydrogen dissociation, which facilitates adsorption to H⁻ states favorable for emission. In [217, 254], the secondary emission sources of H⁻ ions were essentially developed with beam intensities of order 2.5 mA.

A series of experiments, designed to study kinetic emission of H⁻ ions during bombardment of tungsten and tantalum emitters by accelerated Cs + ions, was conducted by BINP researchers in 1975–1977 [217, 254]. To reduce the emitter surface work function, cesium ions were fed continuously either by ion beam or through emitter pores. The schematic of an experiment on the activation of the emitter surface by hydrogen plasma is presented in Fig. 3.30. A rotating disk was utilized as the emitter of H⁻ ions. The site of interest on the disk was successively exposed to the source of cesium vapors, then to the plasma gun to introduce hydrogen onto the surface, and thereafter to the source of cesium ions with an energy of 1–2 keV. Hot tungsten plates were used for cesium surface ionization. Positive voltage (1–2 kV) with respect to the rotating disk was applied to the cesium feeder (2 in Fig. 3.26) to accelerate Cs ions toward the emitter. The same voltage was used to extract from the emitter surface H⁻ ions formed via sputtering of hydrogen introduced into the emitter; the ions were analyzed following the separation in the magnetic field [254]. In experiments on negative ion desorption from a hydrogen plasma-activated MoCs emitter, the secondary emission coefficients of hydrogen negative ions were $Y = 1.2–1.45$ [252], i.e., 2–3 times larger than those observed earlier in studies on the sputtering of the coat formed by cesium and hydrogen deposition from a gas

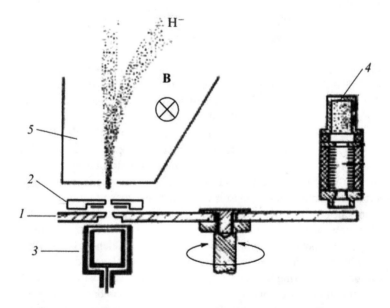

Fig. 3.30 Schematic of experiment to study the emission of H⁻ ions upon the activation of the emitter surface by hydrogen plasma: (1) rotating emitter, (2) cesium ion and atom feed unit, (3) cesium ion flux meter, (4) plasma gun, and (5) magnet poles

over the surface [249]. The data obtained provided a basis for the creation of a continuously operating sputter type source of H⁻ ions that generated negative ions by virtue of secondary ion emission [252]. Experiments conducted with this device confirmed the high efficiency of surface negative ion generation under conditions of electrode surface bombardment by H Cs plasma.

The secondary emission process involves the production of negative ions when fast positive ions are passed through solid-state films [52, 72, 255, 256]. Protons transmitted through the film, similarly as for sputtered and reflected particles, capture electrons upon exiting the film surface. Processes during the passage of protons through the thickness of the film do not play a noticeable role in the electron capture. A significant effect of the properties of the film exit surface on the yield of H⁻ ions was reported [199]. Even in the first experiments with films, it was possible to obtain a significant yield of H⁻ ions, up to 10–20% of the proton current. By deposition of alkali metal films on the exit surface, this yield can apparently be significantly increased. Unfortunately, solid-state films are transparent only for particles with a sufficiently high energy of tens of keV or greater. Due to destruction of the film, the primary particle flux density is limited to a low level of ~10 μA/cm^2.

The negative ion formation in the cesium sputter ion sources occurs on the surface of a cathode containing the ionized material. The cathode is covered by a thin layer of cesium (Cs), which lowers the work function of the surface enhancing the negative ion formation. Recently, JYFL-ACCLAB has revealed that the photo-assisted production of negative ions can be provoked by lasers at various wavelengths with the photon energy exceeding a certain threshold, which questioned the resonant ion pair production hypothesis. Furthermore, the laser-assisted production of negative ions of oxygen (O-) as well as aluminum (Al-) was observed with the off-resonance diode lasers [257]. This observation opens the door for practical applications of photo-assisted negative ion production also for other negative ion species, not just those with their electron affinity states in resonance with the excited states of neutral Cs. In presentation is present that the beam current enhancement does not depend on the resonant ion pair production, it depends on the applied laser power, ion source conditions, and the extracted beam current can be enhanced by a factor > 9.

3.13 Summary

The destructibility of negative ions, which is the basis of their effective use in charge-exchange technology, at the same time makes it difficult to obtain intense negative ion beams. Their production is further complicated by the low probability of formation of negative ions in known processes, and beam formation is complicated by a significant removal of the working substance and the intense flux of accompanying electrons. All this makes the problem of obtaining beams of negative ions much more complicated than obtaining beams of positive ions.

The high efficiency of using beams of negative ions stimulated the development of methods for their production. In 1967, the intensity of negative ion beams was limited to 10^{-3} A. However, the intensity and brightness of beams obtained by the available methods was much inferior to the corresponding characteristics of positive ion beams and did not allow full realization of the advantages of charge-exchange technologies.

Investigations of charge-exchange, plasma volume, secondary emission, and thermionic methods for producing negative ion beams have prepared the basis for further progress and, in particular, have served as the basis for developing the more effective surface-plasma method for producing beams of negative ions.

References

1. V. Dudnikov, *Negative Ion Sources* (NSU, Novosibirsk, 2018). В. Дудников, *Источники отрицательных ионов*. НГУ, Новосибирск, 2018
2. V. Dudnikov, *Development and Applications of Negative Ion Sources* (Springer, Switzerland AG, 2019)
3. M.D. Gabovich, *Physics and Technology of Plasma Ion Sources* (Atomizdat, Moscow, 1972). М.Д. Габович, *Физика и техника Плазменных источников ионов*, Москва, Атомиздат, 1972
4. H. Zhang, *Ion Sources* (Springer, Switzerland AG, 1999)
5. I. Brown (ed.), *The Physics and Technology of Ion Sources* (Wiley-VCH Verlag GmbH & Co. KGaA, New York, 2004)
6. A.T. Forrester, *Large Ion Beams* (Wiley, New York, 1988)
7. B. Wolf, *Handbook of Ion Sources* (CRC Press, Boca Raton, 1995)
8. H.J. Hopman, R.M.A. Heeren, Negative ion source technology, in *Plasma Technology*, ed. by M. Capitelli, C. Gorse, (Springer, Boston, MA, 1992). https://doi.org/10.1007/978-1-4615-3400-6_13
9. J. Thomson, Phyl. Mag. **21**, 225 (1911)
10. M. Born, *Atomic Physics* (M. Mir, 1965). М. Борн, *"Атомная физика"*, М. Мир (1965)
11. T. Bete, E. Solpiter, *Quantum Mechanics of Atoms with One and Two Electrons* (Fismatgis, Moscow, 1960). Т. Бете, Э. Солпитер, *Квантовая механика атомов с одним и двумя электронами*, Москва, Физматгиз, (1960)
12. B.M. Smirnov, *Atomic Collisions and Elemental Process in Plasma* (Atomizdat, Moscow, 1968). Б.М. Смирнов, *Атомные столкновения и элементарные процессы в плазме*, Москва, Атомиздат, 1968
13. B.M. Smirnov, *Ions and Excited Atoms in Plasma* (Atomizdat, Moscow, 1974). Б.М. Смирнов, *Ионы и возбуждённые атомы в плазме*, Москва, Атомиздат, 1974
14. H.S.W. Messy, *Negative Ions* (Cambridge University Press, 1976). Г. Месси, *Отрицательные ионы*, Москва, Мир, Moscow, 1979
15. I. Mac Daniel, *Collision Processes in Ionized Gases* (Mir, Moscow, 1967). https://doi.org/10.1063/1.555524. И. Мак-Даниель, *Процессы столкновений в ионизованных газах*, Москва, Мир, 1967
16. H. Hotop, Binding energies in atomic negative ions. J. Phys. Chem. Ref. Data **4**, 539 (1975)
17. B.M. Smirnov, *Negative Ions* (Mc Graw-Hill, New York, 1976)
18. S.K. Allison, Rev. Mod Phys **30**, 1137 (1958)
19. N.V. Fedorenko, ZTF **40**, 2481 (1970).; Н.В. Федоренко, ЖТФ, 40, 2481 (1970). N. V. Fedorenko, ZTF, 40, 2481 (1970)
20. A.R. Tawara, Charge changing processes in hydrogen beams. Rev. Mod. Phys. **45**, 178 (1973)

21. T. Andersen, Atomic negative ions: Structure, dynamics and collisions. Phys. Rep. **394**, 157–313 (2004). www.elsevier.com/locate/physrep

22. M.D. Gabovich, N.V. Pleshivtsev, N.N. Semashko, *Beams of Ions and Atoms for Controlled Nuclear Fusion and Technologies* (Energoatomizdat, Moscow, 1986) М. Д. Габович, Н.В. Плешивцев, Н.Н. Семашко, *"Пучки ионов и атомов для управляемого термоядерного синтеза и технологических целей"*, Москва, Энергоатом издат, 1986

23. N.N. Semashko, Investigation and Development of Injectors of fast ions and atoms of Hydrogen for stationary Magnetic trap, Dissertation for Doctor of Fis-Mat nauk, IAE im. Kurchatova, Moscow (1974). Н.Н. Семашко, *"Исследование и создание инжекторов быстрых ионов и атомов водорода для стационарных магнитных ловушек"*, Диссертация на соискание учёной степени доктора физ.-мат. наук, ИАЭ им. И.В. Курчатова, Москва, 1974

24. G.I. Dimov, V.G. Dudnikov, Cross sections for stripping of-1-MeV negative hydrogen ions in certain gases. Zhur. Tekh. Fiz **36**, 1239 (1966).; Sov. Phys. -Tech. Phys. 11, 919 (1967); Г. Димов, В. Дудников, «Измерение сечений обдирки ионов H- с энергией 1 МэВ в некоторых газах", ЖТФ, 36, 1239 (1966)

25. N.S. Buchelnikova, UFN, 65 (1957); Н.С. Бучельникова, УФН, 65, (1958)

26. Y.M. Fogel, UFN **71**, 243 (1960).;.Я.М. Фогель,УФН,71, 243, (1960)

27. V.A. Oparin, R.N. Il'in, I.T. Serenko, E.S. Solov'ev, ZETF **66**, 2008 (1974).;.В.А. Опарин, Р.Н. Ильин, И.Т. Серенко,Е.С. Соловьев, ЖЭТФ, 66, 2008 (1974)

28. H.J. Kaiser, E. Heinicke, H. Baumann, K. Bethege, Proc. Symps. On ion source and formation ion beam, BNL 50310, 237, Brookhaven, 1971. Nucl. Instrum. Methods **95**, 389 (1971)

29. G.D. Alton, Negative ion formation and processes and sources, in *Electrostatic Accelerators: Fundamentals and Applications*, ed. by R. Hellborg, (Springer, Heidelberg), pp. 223–273

30. A. Valter (ed.), *Electrostatic Accelerators of Charged Particles* (Gosatomizdat, Moscow, 1971). *"Электростатические ускорители заряженных частиц"*, под редакцией А.К. Валтера, Москва, Госатомиздат, 1963

31. P.H. Rose, Nucl.Instrum. Methods **28**, 146 (1964)

32. K.W. Ehlers, Design considerations for high-intensity negative ion sources. IEEE Trans. Nucl. Sci. **12**, 3 (1965)

33. C.H. Coldie, Nucl. Instrum. Methods **28**, 139 (1964)

34. R.P. Bastide, N.B. Brooks, A.B. Wittkover, P.H. Rose, K.H. Purser, IEEE Trans. Nucl. Sci. **NS-12**(3), 775 (1965)

35. K.H. Purser, IEEE Trans. Nucl. Sci. **NS-20**(3), 136 (1973)

36. K. Prelec, T. Sluyters, Rev. Sci. Instrum. **44**, 1452 (1973)

37. M.A. Abroyan, V.P. Golubev, V.L. Komarov, *Negative Ion Sources* (NIIEFA, Review OD-4, Leningrad, 1976) М.А. Аброян, В.П. Голубев, В.Л. Комаров, Г.В. Чемякин, *Источники отрицательных ионов*, НИИЭФА, Обзор ОД-4, Ленинград, (1976)

38. D.C. Faircloth., Ion sources for high-power hadron accelerators, http://arxiv.org/ftp/arxiv/papers/1302/1302.3745.pdf

39. M. Stockli., Volume and surface-enhanced ion sources, http://cas.web.cern.ch/cas/Slovakia-2012/Lectures/Stockli.pdf

40. M. Stockli, T. Nakagava, Ion injectors for high-intensity accelerators. Rev. Accel. Sci. Technol. **6**, 197–219, World Sientific Publishing Company (2013)

41. J. Sherman, G. Rouleau, New developments with H- sources, in *Seventeenth International Conference on the Application of Accelerators in Research and Industry (CAAR12002) Denton*, Texas, November 12–16, (2002)

42. M. Bacal, Negative hydrogen ion production in fusion dedicated ion sources. Chem. Phys. **398**, 3–6 (2012)

43. Y. Belchenko, V. Dudnikov, Surface negative ion production in ion sources, in *Production and Application of Light Negative Ions, 4rd European Workshop*, ed. by W. Graham, (Belfast, 1991), pp. 47–66

44. Y. Belchenko, Surface negative ion production in ion sources. Rev. Sci. Instrum. **64**, 1385 (1993)

45. M. Bacal, M. Sasao, M. Wada, Negative ion sources. J. Appl. Phys. **129**, 221101 (2021). https://doi.org/10.1063/5.0049289
46. M. Bacal, M. Wada, Negative hydrogen ion production mechanisms. Appl. Phys. Rev. **2**, 021305 (2015)
47. Proc. of the 15th International Conference on Ion Sources, 9-13 September 2013, Chiba, Japan, Rev. Sci. Instrum. **85**(2I), part II (2014)
48. Proc. of the 14th Internat. Conf. on Ion Sources, 11-16 September 2011, Giadian Naxos, Italy, Rev. Sci. Instrum. **83**(2), Part 2 (2012)
49. Proc. of the 13th Internat. Conf. on Ion Sources, 20-25 September 2009, Gatlinburg, TN, USA, Rev. Sci. Instrum. **81**(2), part II (2010)
50. AIP 1655: Fourth International Symposium on Negative Ions, Beams and Sources (NIBS 2014) Edited by Werner Kraus, Paul McNeely, published (2015)
51. Proceedings, 3rd International Symposium on Negative Ions, Beams and Sources (NIBS 2012): Jyväskylä, Finland, September 3–7, 2012, Olli Tarvainen (ed.) , Taneli Kalvas (ed.) 2012AIP Conf.Proc. 1515 (2012)
52. AIP CP 1097; Negative Ions, Beams and Sources, 1ˢᵗ International Symposium, Edited by E. Surrey and A. Simonin
53. AIP CP 925, 11ᵗʰ International Symposium on the Production and Neutralization of Negative Ions and Beams, Santa Fe, NM, USA, 13–15 September 2006. Edited by M. Stockli
54. AIP CP 763, 10ᵗʰ International Symposium on the Production and Neutralization of Negative Ions and Beams, Kiev, Ukraine, 14–17 September 2004. Edited by. J. Sherman and Yu. Belchenko
55. AIP CP 639, Ninth International Symposium on the Production and Neutralization of Negative Ions and Beams, Gif-sur-Yvette, France, 30–31 May 2002. Edited by M. Stockli
56. Proc. Second Symp. Production and Neutralization of Negative Hydrogen Ions and Beams, Brookhaven, 1980 (BNL, Upton, NY, 1980), BNL- 51304, edited. By Th. Sluyters
57. Proc. Symp. Production and Neutralization of Negative Hydrogen Ions and Beams, Brookhaven, 1977 (BNL, Upton, NY, 1977), BNL- 50727, edited. By Th. Sluyters and C. Prelec
58. C. Schmidt, "Production and neutralization of negative ions and beams", (Report on the 5th International Symposium, Upton, NY, USA, 30 October – 3 November 1989)
59. AIP CP158, Production and neutralization of negative ions and beams (Report on the 4th International Symposium, Upton, NY, USA, 1986) edited by. J. Alessi
60. H.J. Hopman, R.M.A. Heeren, Negative ion source technology. Plasma Technol, 185–201 (1992)
61. V.G. Dudnikov, Surface plasma method of negative ion production, Dissertation for doctor of Fis-Mat. nauk, INP SBAS, Novosibirsk, 1976
62. M. Wadehra, J.N. Bardsley, Phys. Rev. Lett. **41**, 1795 (1978)
63. Allan, S.F. Wong, Phys. Rev. Lett. **41**, 1791 (1978)
64. J.R. Hiskes, M. Bacal, G.W. Hamilton, *LLL UCID-18031* (Lawrence Livermore Laboratory, Livermore, CA, 1979)
65. W.G. Graham, Proc. 2nd International Symposium on Production and Neutralization of Negative Hydrogen Ions and Beams, (BNL, Brookhaven, Upton, 1980) BNL Report-51304, p. 126
66. H. Tawara, A. Rassek, Rev. Mod. Phys. **45**, 178 (1973)
67. A.S. Schlachter, Production and Neutralization of Negative Ions and Beams, in *3rd International Symposium*, (AIP, New York, 1984). AIP Conference Proceedings No.111), p. 300
68. V. Dudnikov, Surface plasma source with anode layer plasma accelerator. Rev. Sci. Instrum. **83**, 02A713 (2012)
69. L.W. Alvarez, Rev. Sci. Instrum. **22**, 705 (1951)
70. S.K. Allison, J. Cuevas, M. Garsia, Minos. Phys. Rev. **120**, 1266 (1960)
71. W.L. Fite, Phys. Rev. **89**, 411 (1953)

72. Y.M. Fogel, L.I. Krupnik, V.A. Ankudinov, ZTF **26**, 1208 (1956). Я.М. Фогель, Л.И. Крупник, В.А. Анкудинов, ЖТФ, 26, 1208 (1956)

73. J.A. Philips, J.L. Tuck, Rev. Sci. Instrum. **27**, 97 (1956)

74. J.A. Weinman, I.K. Cameron, Rev. Sci. Instrum. **27**, 288 (1956)

75. L.E. Collins, A.C. Riviera, Nucl. Instrum. Methods **4**, 121 (1959)

76. G. Dimov, B. Sukhina, PTE **1**, 16 (1968). Г.И. Димов, Б.Н. Сухина, ПТЭ, 1, 16 (1968)

77. G. Gornicyn, B. Dyachkov, V. Zinenko, PTE **5**, 74 (1969). Г.А. Горницин, Б.А. Дьячков, В.И. Зиненко, ПТЭ, 5, 74, (1969)

78. N.B. Brooks, P.H. Rose, A.B. Wittkover, R.P. Bastiede, Nucl. Instrum. Methods **28**, 315 (1964)

79. G. Dimov, Y. Kononenko, O. Savchenko, V. Shamovsky, Г.И. Димов, Ю.Г. Кононенко, О.Я. Савченко, В.Г. Шамовский, ЖТФ, 38, 997 (1968). ZTF **38**, 997 (1968)

80. G. Dimov, O. Savchenko, ZTF **38**, 2002 (1968). Г.И. Димов, О.Я. Савченко, ЖТФ, 38, 2002 (1968)

81. G.I. Dimov, Charge exchange Injection in accelerators and storage rings, Dissertation for Doctor of Fis-Mat. Nauk, INP SBAS, Novosibirsk, 1968

82. G. Dimov, G. Roslyakov, Г.И. Димов, Г.В.Росляков, ПТЭ,1, 29 (1974). PTE **1**, 29 (1974)

83. G. Roslyakov, Pulsed chare exchange source of negative ions, Dissertation for candidate Fis-Math nauk, Novosibirsk, 1974. Г.В. Росляков, "Импульсный перезарядный источник отрицательных ионов", диссертация на соискание учёной степени кандидата физ.-мат. Наук, Новосибирск, 1974

84. G. Dimov, G. Roslyakov, PTE **2**, 33 (1974). Г.И. Димов, Г.В. Росляков, ПТЭ, 2, 33 (1974)

85. J.A. Fasolo, IEEE Trans. Nucl. Sci. **NS-22**(3), 1165 (1975)

86. B.L. Donnally, Patent USA 3424904, filed 3/5/65, issued 28/1/65

87. B.L. Donnally, Proc. III Internat. Symp. Polarization Phenomena, Madison, 295 (1970)

88. G. Dimov, G. Roslyakov, PTE **3**, 31 (1974). Г.И. Димов, Г.В. Росляков, ПТЭ, 3, 31 (1974)

89. B. Dyachkov, V. Zinenko, At. Energy **24**, 18 (1962). Б,А. Дьячков, В.И. Зиненко, Атомная энергия, 24, 18 (1962)

90. H.R. Hiddleston, J.A. Fasolo, D.C. Manette, et al., 1976 Linear Accelerator Conf., Chalk River, Rep. AELC-5677, p. 387 (1976)

91. B. Dyachkov, A. Krylov, V. Kuznetsov, preprint IAE-2523, Moscow (1975). Б.А. Дьячков, А.И. Крылов, В.В. Кузнецов, препринт ИАЭ-2523, Москва (1975)

92. J.E. Osher, G.W. Hamilton, Proc. of the 2nd Internat. Conf. on Ion Sources, Vienna, 1972, p. 157, (1972)

93. J.E. Osher, F.J. Gordon, G.W. Hamilton, Proc. of the 2nd Internat. Conf. on Ion Sources, Vienna, 1972, p. 876, (1972)

94. N. White, Patent USA, 4980556, issued Dec. 25, (1990)

95. I. Shikhovtsev, V. Amirov, K. Anikeeva, et al., A 10 mA, steady-state, charge exchange negative ion beam source. AIP Conf. Proc. **2373**, 040001-1–040001-8 (2022). https://doi.org/10.1063/5.0057483

96. A.S. Belov, S.K. Esin, S.A. Kubalov, et al., Pis'na JETF **42**, 319 (1985)

97. A.S. Belov, S.K. Esin, S.A. Kubalov, et al., Nucl. Instrum. Methods Phys. Res. **A255**, 442 (1987)

98. A.S. Belov, V.G. Dudnikov, Y.I. Kusik, et al., A source of polarized negative hydrogen ions with deuterium plasma ionizer. Nucl. Instrum. Methods Phys. Res., Sect. A, P. **333**(2–3), 256 (1993)

99. A.S. Belov, V.G. Dudnikov, V.E. Kuzik, et al., Nucl. Instrum. Methods Phys. Res. **A333**, 256 (1993)

100. A.S. Belov, V.G. Dudnikov, S.K. Esin, et al., Rev. Sci. Instrum. **67**, 1293 (1996)

101. V Dudnikov, MAC Cummings, RP Johnson, Polarized deuteron negative ion source for nuclear physics applications, IPAC 2018 (2018)

102. V. Dudnikov, A. Dudnikov, Polarized negative ion source with multiply spherically focusing surface plasma ionizer. AIP Conf. Proc. **2052**, 050019 (2018)

103. V. Dudnikov, A. Dudnikov, Polarized Negative Ion Source with Multiply Sphericaly Focusing Surface Plasma Ionizer, ArXiv, 1808.06001, (2018)
104. A.I. Hershcovitch, J.G. Alessi, A.E. Kponou, Gas limitations on the performance of a new polarized negative ion source. Nucl. Inst. Methods Phys. Res. A **345**, 411–416 (1994)
105. A.N. Zelenski, S.A. Kohanovskii, V.M. Lobashev, V.G. Polushkin, Pis'ma JETF **42**, 5 (1985)
106. A. Zelenski, Rev. Sci. Instrum. **81**, 02B308 (2010)
107. A. Zelenski, V. Davydenko, A. Ivanov, et al., Phys. Part. Nucl. **45**, 308 (2014)
108. A. Zelenski, G. Atoian, D. Raparia, The RHIC polarized H$^-$ ion source. Rev. Sci. Instrum. **87**, 02B705 (2016). https://doi.org/10.1063/1.4932392
109. V. Dudnikov, V. Morozov, A. Dudnikov, High intensity source of he negative ions, IPAC 2015
110. V. Dudnikov, V. Morozov, A. Dudnikov, Polarized ^{3}He$^-$ ion source with hyperfine state selection. AIP Conf. Proc. **1655**, 070006 (2015)
111. M. Yoshida , F. Naito, S. Artikova, Y. Kondo, Hayashizaki, Y. Iwashita, K. Torikai, Re-acceleration of ultra cold muon in j-parc mlf, WEPWA018 Proceedings of IPAC2015, Richmond, VA, USA, (2015)
112. Y. Miyake, Y. Ikedo, K. Shimomura, P. Strasser, N. Kawamura, K. Nishiyama, A. Koda, H. Fujimori, S. Makimura, J. Nakamura, T. Nagatomo, R. Kadono, E. Torikai, M. Iwasaki, S. Wada, N. Saito, K. Okamura, K. Yokoyama, T. Ito, W. Higemoto, Hyperfine Interact. **216**, 79 (2013)
113. M. Bogomilov et al., The MICE Collaboration, JINST 7, P05009, 2012 Proceedings of IPAC2016, Busan, Korea TUPMY002 03
114. S. Artikova, F. Naito, M. Yoshida, PASJ2013 Proc. (Nagoya, Japan, 2013), SAP045
115. V. Dudnikov, A. Dudnikov, Ultracold muonium negative ion production. AIP Conf. Proc. **2052**, 060001, ArXiv, 1806.03331 (2018). https://doi.org/10.1063/1.5083774
116. V. Dudnikov et al., Polarized deuteron negative ion source for nuclear physics applications, IPAC 2018, Vancouver (2018)
117. V. Dudnikov, Method of negative ion obtaining, Patent cccp, 411542, 10/III. 1972.; http://www.findpatent.ru/patent/41/411542.html. В.Г. Дудников, "Способ получения отрицательных ионов", Авторское свидетельство, М. Кл.Н 01 J 3/0,4, 411542, заявлено 10/III,1972,
118. Yu Belchenko, G. Dimov, V. Dudnikov, "Physical principles of surface plasma source operation", *Symposium on the Production and Neutralization of Negative Hydrogen Ions and Beams, Brookhaven, 1977* (Brookhaven National Laboratory (BNL), Upton, NY, 1977), pp. 79–96; Ю. Бельченко, Г. Димов, В. Дудников, «Физические основы поверхностно плазменного метода получения пучков отрицательных ионов», препринт ИЯФ 77–56, Новосибирк 1977. http://irbiscorp.spsl.nsc.ru/fulltext/prepr/1977/p1977_56.pdf; Yu Belchenko, G. Dimov, V. Dudnikov, "Physical principles of surface plasma source method of negative ion production", Preprint IYaF 77–56, Novosibirsk 1977
119. V. Dudnikov, Y. Belchenko, J. Phys. **40**, 477 (1979)
120. Y.I. Belchenko, V. Dudnikov, et al., Ion sources at the Novosibirsk Institute of Nuclear Physics. Rev. Sci. Instrum. **61**, 378 (1990)
121. G.I. Dimov, I.Y. Timoshin, V.V. Demidov, V.G. Dudnikov, PTE **4**, 30 (1976) Г. И. Димов, И, Я. Тимошин, В. В. Демидов, В. Г. Дудников, ПТЭ, 4, 30 (1967)
122. K.W. Ehlers, Nucl. Instrum. Methods **32**, 309 (1965)
123. Th. Sluyters, K. Prelec, IX Internat. Conf. on H. E. Accel., Polo Alto, (1974)
124. M. Bacal, A. Hatayama, J. Peters, IEEE Trans. Plasma Sci. **33**, 1845 (2005)
125. F. Taccogna et al., Latest experimental and theoretical advances in the production of negative ions in cesium-free plasmas. Eur. Phys. J. D **75**, 227 (2021). https://doi.org/10.1140/epjd/s10053-021-00228-y
126. Y. Xu, I.I. Fabrikant, Dissociative electron attachment rates for H$_2$ and its isotopes. Appl. Phys. Lett. **78**(17), 2598 (2001)
127. V. Khvostenko, V. Dukelsky, JTEF **33**, 851 (1957) В.И. Хвостенко и В.М. Дукельский, ЖЭТФ, 33, 851, (1957)

128. S.M. Corrigan, J. Chem. Phys. **43**, 4381 (1965)

129. V. Kuchinsky, V. Mishakov, A. Tibilov, A. Shukhtin, V Allunion conference on physics of atomic collision, Abstracts, Uzhgorod, 189 (1972). В. Кучинский, В. Мишаков, А. Тибилов, А. Шухтин, Всесоюзная конференция по физике атомных и электронных столкновений, тезисы докладов, Ужгород, стр. 189(1972)

130. V. Kuchinsky, V. Mishakov, A. Tibilov, A. Shukhtin, Opt. Spectrosc. **39**, 1043 (1965) В. Кучинский, В. Мишаков, А. Тибилов, А. Шухтин, Оптика и спектроскопия, 39, 1043, (1965)

131. J.L. Magee, M. Barton, J. Am. Chem. Sos. **72**, 1965 (1950)

132. F. Demirkhanov, Y. Kursanov, N. Blagoveschensky, ZTF **40**, 1911 (1970) Ф. А. Демирханов, Ю. В. Курсанов, Н. Ф. Лазарев, У. М. Благовещенский, ЖТФ, 40, 1911, (1970)

133. V. Dubrovsky, V. Ob'edkov, Theor. Exp. Chem. **2**, 715 (1966). В. Г. Дубровский. В. Д. Объедков, Теоретическая и экспериментальная химия, 2, 715 (1966)

134. V.G. Dubrovsky, V. D. Ob'edkov, R. K. Janev, in "V Internat. Conf. on Pysics of electronic and Atom. Collision" Leningrad, Abstracts, p. 342 (1967)

135. B. Peart, K.T. Dolder, J. Phys. B, Atoms and Mol. Phys. **8**, 1570 (1975)

136. B. Peart, K.N. Dolder, Comments Atom., Mol. Phys. **1**(4), 97 (1976)

137. V.G. Dubrovsky, V.D. Obedkov, Theor. Exp. Chem. **2**, 715 (1966) В. Г. Дубровский, В. Д. Объедков, Теоретическая и экспериментальная химия, 2, 715, (1966)

138. D.F. Dance, M.F. Harrison, R.D. Rundel, Proc. Roy. Soc., A **299**, 525 (1967)

139. B. Peeart, D.S. Walton, K.T. Dolder, J. Phys., B **3**, 13646 (1970)

140. D.S. Walton, B. Peart, K.T. Dolder, J. Phys., B **4**, 1343 (1971)

141. K.D. Rundel, K.L. Aitken, M.F. Harrison, J. Phys., B **2**, 954 (1969)

142. J. Moseley, W. Abert, J.R. Peterson, Phys. Rev. Lett. **24**, 435 (1970)

143. B. Peart, R. Grey, K.T. Dolder, J. Phys., B **9**, L369 (1976)

144. A. Dolgarno, M.R.C. McDowell, Proc. Phys. Soc. **A74**, 457 (1959)

145. D.G. Hummer, R.F. Stebbings, W.L. Fite, L.M. Branscomb, Phys. Rev. **119**, 668 (1960)

146. D.D. Bates, R.L.F. Boyd, Proc. Phys. Soc. Lond. **A69**, 910 (1956)

147. J. M. Sautter, Ph. D. thesis (University de Grenoble), 1968

148. J.B. Hasted, R.A. Smith, Proc. Roy. Soc., A **235**, 349 (1956)

149. J.F. Williams, Phys. Rev. **154**, 9 (1967)

150. T. Jorgensen, C.E. Kuyatt, W.W. Lang, D.C. Lorents, C.A. Sautter, Phys. Rev. **140A**, 1481 (1965)

151. R. Smythe, J.W. Toevs, Phys. Rev. **139A**, 15 (1965)

152. Yu.I. Geller, G.I. Dimov, G.V. Roslyakov, Preprint INP 40–71, Novosibirsk (1971). Ю. И. Геллер, Г. И. Димов, Г. В. Росляков, Препринт ИЯФ 40–71, Новосибирск (1971)

153. B.M. Smirnov, *Assymptotic Methods in Theory of Atomic Collisions* (M. Atomizdat, 1973). Б. М. Смирнов, *Ассимптотические методы в теории атомных столкновений*, **М., Атомиздат, 1973**

154. D.F. Hunt, F.W. Crow, Electron capture negative ion chemical ionization mass spectrometry. Anal. Chem. **50**(13), 1781 (1978)

155. W.H. Bennet, P.F. Darby, Phys. Rev. **49**, 97 (1936)

156. C.D. Moak, H.E. Banta, J.H. Thuraton, J.W. Johnson, R.F. King, Rev. Sci. Instrum. **30**, 694 (1959)

157. A.W. Wittkover, R.P. Bastide, N.B. Beooks, P.H. Rose, Phys. Lett. **3**, 336 (1963)

158. L.E. Kollins, R.H. Gobbett, Nucl. Instrum. Methods **35**, 282 (1965)

159. G.P. Lawrence, R.K. Beawchamp, J.L. Mckbben, Nucl. Instrum. Methods **32**, 357 (1965)

160. A.V. Almazov, Y.M. Khirnyi, C.G. Kochemasova, PTE **6**, 36 (1966). А. В. Алмазов, Ю. М. Хирный, Л. Н. Кочемасова, ПТЭ, 6, 36 (1966)

161. V.P. Golubev, V.L. Komarov, S.C. Tsepakin, Premiere conf. instrument sur les source d'Ions. Sacle-France **129** (1969)

162. M. Dubarry, G. Gantherin, in Premiere Conf. Int. Sur les sources d'Ions. (I.N.S.T.N., Saclay,) p. 133 (1969)

163. M.A. Abroyan, G.A. Nalivaiko, ZTF, 876 (1972). М. А. Аброян, Г. А. Наливайко, С. Г. Цепакин, ЖТФ, 876 (1972)

164. V.P. Golubev, G.A. Nalivaiko, S.G. Tsepakin, in Proc. Proton Linear Accelerator Conf., (Los Alamos Rept N. LA-5115,) p. 356 (1972)

165. V.P. Golubev, "Production of intense beams of hydrogen negative ions from ion source of duo-plasmatron type", Aothoreferat of dissertation for candidate of physics-math. Since, NIIEFA Leningrad (1974). В. П. Голубев, "Получение интенсивных пучков отрицателтных ионов водорода из источников дуоплазматронного типа", Автореферат кандидатской диссертации, НИИЭФА, Ленинград, (1974)

166. V.P. Golubev, A.V. Morozov, A. Nalivaiko, Proc. IV Allunion workshop on charged particle accelerators, 1, 319 (1975). В. П. Голубев, А. В. Морозов, Г. А. Наливайко, Труды четвертого Всесоюзного совещания по ускорителям заряженных частиц, Москва, Наука, , т.1, стр. 319 (1975)

167. T. Sluyters, K. Prelec, Nucl. Instrum. Methods 113, 299 (1973)

168. M. Kaboyashi, K.K. Prelec, T. Sluyter, Rev. Sci. Instrum. 47, 1425 (1976)

169. K. Weisemann, K. Prelec, T. Slyuters, BNL AGS report. J. Appl. Phys. 48, 2668 (1977)

170. K.W. Ehlers, B.F. Gavin, E.L. Hubbard, Nucl. Instrum. Methods 22, 87 (1963)

171. K.W. Ehlers, Nucler Instrum. Methods 32, 309 (1965)

172. I.Y. Timoshin, *Pulsed high current source of hydrogen negative ions for electrostatic accelerators* (INP SBAS, Novosibirsk, 1966). И. Я. Тимошин, "Импульсный сильноточный источник отрицательных ионов водорода для электростатических ускороителей", кандидатская диссертация, ИЯФ СОРАН, Новосибирск 1966

173. G.I. Dimov, I.Y. Timoshin, O.Y. Savchenko, Y.G. Kononenko, V.G. Shamovsky, *Allunion Conference for Chared Particle Accelerators* (Nauka, Moscow, 1968) Г. И. Димов, И. Я. Тимошин, О. Я. Савченко, Ю. Г. Кононенко, В. Г. Шамовский, Всесоюзное совещание по ускорителям заряженных частиц, Москва, Наука, (1968)

174. K. Jimbo, K. Ehlers, K. Leung, R. Pyle, Nucl. Inst. Methods A 248, 282 (1986)

175. S.V. Grigorenko, Development and research of system of external injection of hydrogen neative ion for cyclotrons, Autoreferat of dissertation foe candidate of fis-math nauk, Sankt Petersburg (2011) Григоренко Сергей Викторович, "Разработка и исследование систем внешней инжекции отрицательных ионов водорода для циклотронов", автореферат диссертации на соискание ученой степени кандидата физ.-мат. наук, Санкт-Петербург (2011)

176. V.G. Dudnikov, Surface plasma source of Penning geometry, IV USSR National Conference on Particle Accelerators, M. Nauka, v.1, 323 (1974); В. Дудников, «Источник отрицательных ионов с Пеннинговской геометрией», Труды Всесоюзного совещания по ускорителям заряженных частиц, М. Наука, т. 1, 323 (1974)

177. G.I. Dimov, G.E. Derevyankin, V.G. Dudnikov, A 100 mA negative hydrogen-ion source for accelerators. IEEE Trans. Nucl. Sci. 24(3), 1545 (1977)

178. K.N. Leung et al., Operation of a Dudnikov type Penning source with LaB$_6$ cathodes*, Conference: 4. international symposium on the production and neutralization of negative ions and beams, Upton, NY, USA, (1986); Y. D. Jones, R. P. Copeland, M. A. Parman, E. A. Baca, "Xenon Mixing In A Cesium-Free Dudnikov-Type Hydrogen Ion Source", *Proceedings Volume 1061, Microwave and Particle Beam Sources and Directed Energy Concepts* (1989) https://doi.org/10.1117/12.951842

179. V.P. Goretsky, A.V. Ryabtsev, I.A. Soloshenko, et al., Kinetic processes in the negative hydrogen ion sources under high power discharge. AIP Conf. Proc. 287, 201 (1992)

180. M. Golovinsky, V.P. Goretsky, A.V. Ryabtsev, et al., Cesium influence on H− ion emission from a source with reflection-type discharge. AIP Conf. Proc. 287, 430 (1992)

181. P. Litvinov, I. Savchenko, A new plasma H⁻ source, in *Proceedings of the 7th International Symposium on the Production and Neutralization of Negative Ions and Beams*, AIP Conf. Proc No.380, 272 (1995)

182. T. Kuo et al., Rev. Sci. Instrum. **67**, 1314 (1996)

183. K. Jayamanna, I. Aguilar, I. Bylinskii et al., A 20 mA H⁻ ion source with accel-accel-decel extraction system at TRIUMF, report TUPPT022, Proceedings of Cyclotrons 2013, Vancouver, BC, Canada (2013)

184. H. Etoh, Y. Aoki, H. Mitsubori, et al., Progress in the development of an H⁻ ion source for cyclotrons. AIP Conf. Proc. **1655**, 030014 (2015)

185. T. Kuo et al., Further development for the TRIUMF H⁻ D⁻ multicusp source. Rev. Sci. Instrum. **69**(2), 959–961 (1998)

186. K. Jayamanna, I. Aguilar, I. Bylinskii et al., A 20 mA H- ion source with accel-accel-decel extractionsystem at TRIUMF, report TUPPT022, Proceedings of Cyclotrons 2013, Vancouver, BC, Canada (2013). K. Jayamanna, F. Ames, I. Bylinskii, M. Lovera, B. Minato, A 60 mA DC H⁻ multi cusp ion source developed at TRIUMF. Nucl. Instrum. Methods Phys. Res., **A 895**, 150 (2018)

187. http://www.d-pace.com/?e=141

188. H. Etoh et al., Development of a 20 mA negative hydrogen ion source for cyclotrons. AIP Conf. Proc. **1869**, 030050 (2017). https://doi.org/10.1063/1.4995770

189. K.W. Ehlers, K.N. Leung, Rev. Sci. Instrum. **53**, 1423 (1982)

190. J. Peters, Review of negative hydrogen ion sources high brightness/high current, in *2002 Proceedings of 21st Linear Accelerator Conference* (LINAC 2002), Gyeongju, Korea, p. 42 (www.jacow.org)

191. O. Tarvainen, S.X. Peng, Radiofrequency and 2.45 GHz electron cyclotron resonance H⁻ volume production ion sources. New J. Phys. **18**, 105008 (2016)

192. J. Peters, Review of negative hydrogen ion sources high brightness/high current, in *2002 Proceedings of 21st Linear Accelerator Conference (LINAC 2002)* (Gyeongju, Korea) p42 (www.jacow.org)

193. J. Lettry et al., CERN's Linac4 ion sources status. AIP Conf. Proc. **1655**, 030005 (2015)

194. A. Ueno, H. Oguri, K. Ikegami, Y. Namekawa, K. Ohkosh, Interesting experimental results in Japan Proton Accelerator Research Complex H− ion-source development. Rev. Sci. Instrum. **81**, 02A720 (2010)

195. H.T. Ren, S.X. Peng et al., *of IPAC13*, MOPFI034, 360–362, Shanghai, China, (2013)

196. T. Zhang, S.X. Peng, H.T. Ren, et al., *Proceeding of LINAC2014* (THPP115, Geneva, Switzerland, 2014)

197. S.X. Peng, H.T. Ren, Y. Xu et al., CW/Pulsed H⁻ ion beam generation with PKU Cs-free 2.45 GHz microwave driven ion source, AIP CP1655, 070005 (2015)

198. C. Wimmer, L. Schiesko, U. Fantz, Investigation of the boundary layer during the transition from volume to surface dominated H− production at the BATMAN test facility. Rev. Sci. Instrum. **87**, 02B310 (2016). https://doi.org/10.1063/1.4932985

199. R. Gobin, O. Delferriere, R. Ferdinand, et al., Developmentof an H⁻ ion source based on the electron cyclotron resonance plasma generator at CEA/Saclay. Rev. Sci. Instrum. **75**, 1741 (2004)

200. M. Bacal, G.W. Hamilton, H- and D- production in plasmas. Phys. Rev. Lett. **42**, 1538 (1979)

201. C. Wimmer, L. Schiesko, U. Fantz, Investigation of the boundary layer during the transition from volume to surface dominated H− production at the BATMAN test facility

202. F. Taccogna, S. Bechu, A. Aanesland, Latest experimental and theoretical advances in the production of negative ions in cesium-free plasmas. Eur. Phys. J. D **75**, 227 (2021). https://doi.org/10.1140/epjd/s10053-021-00228-y

203. L.R. Grisham, S.K. Hahto, S.T. Hahto, J.W. Kwan, K.N. Leung, Experimental evaluation of a negative ion source for a heavy ion fusion negative ion driver

204. M. Dehnel, T. Stewart, An industrial cyclotron ion source & injection system, Triumph inj19P18CIS

205. Л.Н. Добрецов, М.В. Гомоюнова, *Эмиссионная электроника*, М. Наука, 1966. L. N. Dobretsov, M. V. Gomoyunova, Emission electronic, M. Nauka, 1966

206. Э.Я. Занберг, Н.Н. Ионов, *Поверхностная ионизация*, М. Наука, 1967. E. Ya. Zanberg, Surface Ionization, M. Nauka, 1967

207. E.Y. Zandberg, N.I. Ionov, Surface ionization. Usp. Fiz. nauk **67**, 581 (1959) Э. Я. Зандберг, Н. И. Ионов, Поверхностная ионизация, УФН, 1959, том 67, номер 4, 581–623 (1959)

208. T. Kuskevich, Rev. Sci. Instrum. **6**, 16 (1966) Кускевич, Томпсон, Приборы для научных исследований, 6, 16, (1966)

209. E. Stahlinger, Proc. Sympos. on Ion Source and Form. Ion Beams, P. 47, BNL 50310 (1977)

210. K. Wallace, Pat. USA, kl. 250-43, Non3336475, published 08.1. (1967)

211. G.I. Dimov, G.V. Roslyakov, ZTF **42**, 1186 (1972) Г. И. Димов, Г. В. Росляков, ЖТФ, 42, 1186 (1972)

212. J. Pelletier, C. Pomot, J. Cocagne, Negative surface ionization: Intense halogen-ion source. J. Appl. Phys. **50**, 4517 (1979). https://doi.org/10.1063/1.326558

213. V. Fomenko, I. Podchernyaeva, *Emission and Adsorbcion Property of Substances and Materials* (M. Atomisdat, 1975) В. А. Фоменко, И. А. Подчерняева, *Эмиссионные и адсорбционные свойства веществ и материалов*, М. Атомиздат, (1975)

214. A. Sommer, *Photoelectron Materials* (M. Energiya, 1973) А. Соммер, *Фотоэмиссионные материалы*, М., Энергия, 1973

215. D.F. Williams, J.J. Tietjen, Proc. IEEE **59**, 1489 (1971)

216. N.N. Petrov, ZTF **41**, 2473 (1971) Н. Н. Петров, ЖТФ, 41, 2473, (1971)

217. E. Bender, G. Dimov, M. Kishinevskii, Proc. 1 allunion seminar for secondary ion-ion emission, Kharkov, 1975 (deponir VINITI N 2783–75 dep) p. 119–131, preprint INP 75–9, Novosibirsk (1975) Е. Д. Бендер, Г. И. Димов, М. Е. Кишиневский, Труды 1 всесоюзного семинара по вторичной ионно-ионной эмиссии, Харьков, 1975 (Рук. Деп. ВИНИТИ No 2783–75 деп.) Стр. 119–131. Препринт ИЯФ-75-9, Новосибирск, (1975)

218. A.H. Ayukhanov, E.O. Turmashev, T. Bakhramov, Interraction of atomic particles with solid, 1, 176, Naukova Dumka, Kiev (1974). А. Х, Аюханов, Э. О. Турмашев, Т. Бахрамов, *Взаимодействие атомных частиц с твердым телом*, ч. 1, стр. 176, Наукова Думка, Киев, (1974)

219. K. Woodcock, The emission of negative ions under the bombardment of positive ions. Phys. Rev. **38**, 1696 (1931)

220. Y.M. Fogel, UFN **91**, 75 (1967) Я. М. Фогель, Успехи физических наук, 91, 75 (1967)

221. R. Castaing, J.F. Henneguin, Adv. Mass Spectrom. **5**, 419–425 (1971)

222. R.J. MacDonald, Adv. Phys. **19**, 457 (1970)

223. V.I. Vexler, Electronic technique, ser. 1, electronic SVCh, 12, 123 (1972). В. И. Векслер, Электронная техника, сер. 1 Электроника СВЧ в. 12, 123 (1972)

224. H. Messy, E. Barkhop, *Electron and Ion Collisions* (Inostrannaya literature, M, 1958) Г. Месси, Е. Бархоп, *Электронные и ионные столкновения*, Иностранная литература, М., 1958

225. U.A. Arifov, *Interraction of Atomic Particles with Surface of Solids* (M. Nauka, 1968) У. А. Арифов, *Взаимодействие атомных частиц с поверхностью твердого тела*, М., Наука, 1968

226. V.T. Cherepin, M.A. Vasil'ev, *Secondary Ion-Ion Emission of Metals and Alloys* (Kiev, 1975) В. Т. Черепин, М. А. Васильев, *Вторичная ионно-ионная эмиссия металлов и сплавов*, Киев, 1975

227. G. Carter, J.S. Colligan, *Ion Bombardment of Solid*, vol 6 (Heineman, London, 1968)

228. Second Allunion symposium of ineration of atomic particles with a solid, colletion of reports, Moscow, 1972. Второй Всесоюзный симпозиум Взаимодействие атомных частиц с твердым телом, сборник докладов, Москва, 1972

229. 15 Allunion conference on emission electronics, short review of reports, 1, 2, Kiev 1972. 15 Всесоюзная конференция по эмиссионной электронике, краткое содержание докладов, т. 1, т. 2, Киев, 1972

230. Interraction of atomic particles with a solid, 1 and 2, Kiev 1974. *Взаимодействие атомных частиц с твердым телом*, т. 1, т. 2, Киев, Наукова думка, 1974

231. Interraction of atomic particles with a solid, v. 1, 2 , 3, 4 Allunion conference, Kharkov, 1974. *Взаимодействие атомных частиц с твердым телом*, т. 1, т. 2, т. 3, 4 Всесоюзная конференция, Харьков 1974

232. Secondary ion-ion emission, Systematic catalog of literature 1947–1974, Kharkov 1975. *Вторичная ионно-ионная эмиссия*, Систематический указатель литературы 1947–1974, Харьков 1975

233. E.Y. Zandberg, ZTF **8**, 1387 (1955) Э. Я. Зандберг, ЖТФ, 25, в. 8, 1387, (1955)

234. L.P. Levine, H.W. Berry, Phys. Rev. **118**, N1, 158 (1961)

235. M. Pedrix, Palettos, R. Gotte, C. Guiland, Nucl. Instrum. Methods **56**, 23 (1967)

236. N.V. Fedorenko, ZTF, 2481 (1970) Н. В. Федоренко, ЖТФ, 2481, (1970)

237. I.K. Abdullaeva, A.K. Aukhanov, *Ion Bombardment-Method Investigation of Surface* (FAN, Tashkent, 1975) И. К. Аблуллаева, А. Х. Аюханов, *Ионная бомбардировка-метод исследования поверхности*, Ташкент, ФАН, 1975

238. U.A. Arifov, A.H. Ayukhanov, Isvestiya AN Us. CCR. seriya Fis-mat. nauk **6**, 34 (1961) У.А. Арифов, А.Х. Аюханов, Изв. АН Уз. ССР, серия физ-мат. наук, 6, 34, (1961)

239. V.E. Kron, J. Appl. Phys. **34**, 3523 (1962)

240. M.K. Abdullaeva, A.Kh. Aukhanov, U.B. Shamsiev, Second, allunion symposium on interaction of atomic particles with surface, collection of reports, Moscow, p. 169, 1972. М. К. Аблулаева, А. Х. Аюханов, У. Б. Шамсиев, Второй Всесоюзный симпозиум по заимодействию атомных частиц с твердым телом, сборник докладов, Москва, 1972, стр. 179

241. A.K. Aukhanov, Y.A. Golubev, V.N. Chernenko, *Interaction of Atomic Particles with Solid* (Naukova dumka, Kiev, 1974) А. Х. Аюханов, Ю. А. Голубев, В. Н. Черненко, *Взаимодкйствие атомных частиц с твердым телом*, 3 Всесоюзная Конференция по взаимодействию атомных частиц с твердым телом, Киев, Наукова думка, 1974, стр. 116

242. M. Muller, G. Hortig, IEEE Trans. Nucl. Sci. **NS-16, N3**, 38 (1969)

243. M. Muller, G. Hortig, Premiere Conference Internat. Sur les Source d'Ions, 18–20 Juin, INSTN, Saclay-France, p. 159 (1969)

244. G. Hortig, P. Mokler, M. Muller, Z. Physics **210**, 312 (1968)

245. A.K. Aukhanov, V.N. Chernenko, PTO **2**, 150 (1972) А. Х, Аюханов, В. Н. Черненко, ПТЭ, 2, 150, (1972)

246. International Conf. on Heavy Ion Sources, IEEE Nucl Sci., NS-23,N2, (1976)

247. R. Middleton, C.N. Adams, Nucl. Instrum. Methods **118**, 329 (1974)

248. G.V. Chemyakin, A.G. Troshikhin, Interaction of atomic particles with solid, 4 allunin conference, Kharkov, v. 3, 110 (1974). Г. В. Чемякин, А. Г. Трошихин, *Взаимодействие атомных частиц с твердым телом*, 4 Всесоюзная конференция, Харьков, т. 3, стр. 110 (1974)

249. R. Middleton, A review of sputter negative ion sources. IEEE Trans. Nucl. Sci. **NS-23**(2), Apit (1976)

250. R. Middleton, A Negative-Ion Cookbook, Department of Physics, University of Pennsylvania, Philadelphia, PA 19104, October 1989 (Revised February 1990)

251. G.D. Alton, Ion sources for accelerators in materials research. Nucl. Instrum. Methods Phys. Res. **B73**, 221–288 (1993)

252. N. Alinovsky et al., RUPAC 2008 (2008)

253. http://www.pelletron.com/negion.htm

254. E.D. Bender, M.E. Kishinevskii, I.I. Morozov, Preprint No. 77–47 (Novosibirsk: Budker Institute of Nuclear Physics, SB USSR Acad. Sci., 1977)

255. J.A. Pillips, Phys. Rev. **91**(2), 455 (1953)

256. Y.M. Fogel et al., JETF **28**, 711 (1955) Я. М. Фогель и др. ЖЭТФ, 28, 711 (1955)

257. A. Hossain et al., "Laser-assisted negative ion production in cesium sputter ion source", NIBS 2022, Padova, Italy, 2022

Chapter 4
Surface Plasma Production of Negative Ions

Abstract Topics covered in this chapter include: Early experiments on negative ion production in cesiated discharges; studies of negative ion emission from hydrogen plasma with added cesium; energy spectra of H⁻ ions in surface plasma sources; advanced design options for surface plasma sources; emissive properties of electrodes in surface plasma source discharges; plasma parameters and negative ion destruction in the plasma; cesium in surface plasma sources; physical basis of the surface plasma method of negative ion production; regularities in the formation of reflected, sputtered, and evaporated particles; electron capture to the electron affinity levels for sputtered, reflected, and evaporated particles; implementation of surface plasma production of negative ion beams.

4.1 Early Experiments on Negative Ion Production in Cesiated Discharges

In order to obtain a plasma with a more favorable electron energy distribution and to create greater steady-state electric fields in the plasma, Dimov, at Novosibirsk, suggested in 1964 using a gaseous magnetron discharge for negative ion production. The first studies of plasma sources of H⁻ ions with magnetron gas discharge cells were performed by Dimov and Ustyuzhaninov in the period from 1965 to 1967. A $40 \times 40 \times 100$-cm³ vacuum chamber, pumped by a diffusion pump with liquid nitrogen trap, was used, and two copper coils with internal diameter of 6 cm, external diameter of 25 cm, and thickness of 6 cm were installed. The spacing between the centers of the coils was 10 cm. When powered, the coils produced a magnetic field of up to 0.3 Tesla for up to 1.5 s in the on-axis central region. Cylindrical axially symmetric diodes were mounted on the system axis. The inner diameters of the stainless steel anodes used from were 10–15 mm. Cold cathodes in the form of a rod or incandescent cathode spirals wound on a rod were located along the anode axis. The discharge region was limited by ceramic washers near the chamber ends. Current leads were attached to the ends of the cathode, passing through the central

© The Author(s), under exclusive license to Springer Nature Switzerland AG 2023

V. Dudnikov, *Development and Applications of Negative Ion Sources*, Springer
Series on Atomic, Optical, and Plasma Physics 125,
https://doi.org/10.1007/978-3-031-28408-3_4

apertures of the ceramic washers, through which voltage was applied to the discharge gap and heater current for the thermal cathode. Hydrogen was fed into the plasma volume within the magnetron by an electromagnetic-pulsed valve outside of the main magnetic field. Rectangular 1 msec pulses of amplitude up to 4 kV to ignite and support the plasma discharge were fed to the cathode-anode gap by a thyratron from a pulse-forming line of impedance 25 Ω through a ballast resistor to limit the discharge current. The magnetron anode was grounded. Negative ions were extracted from the plasma through an emission slit with dimensions 1×15 mm^2 oriented along the magnetic field, with ion extraction across the magnetic field as in all plasma sources of traditional design. A d.c. extraction voltage of up to 19 kV was applied to the extraction electrode. A negative ion collector, held at a slightly lower potential, was located in a container held at the extraction electrode potential. The signal from the collector was transmitted to the oscilloscope through a high-voltage isolation capacitor. Electrodes of various designs were provided to intercept the flow of accompanying electrons. Various discharge regimes in magnetrons with heated tungsten cathode and cold cathodes of various materials were tested in this setup. The voltage drop in discharges with cold cathodes was 500–600 V at discharge current up to 5–7 A and decreased to 80 V for current greater than 10 A; in discharges with tungsten thermal cathodes, the voltage was 150 V. In the regimes investigated, negative ion beams were extracted with current up to 1 mA. However, because of instability in the discharge ignition and burning, poor collection of accompanying electrons, and frequent breakdown, it was not possible to achieve stable operation of this device at that time.

Studies of the emission of negative hydrogen ions from discharges in crossed fields in the installation described above were resumed at the end of 1967. The main focus was on cold cathode discharges. When working with magnetron cells of previous design, it was not possible to obtain stably reproducible discharge modes. The discharge transferred into breakdown over the washer surfaces due to metal deposition on insulator surfaces. Stable operation of high-current magnetron arc discharges was achieved using glass insulators with projections shielding part of the insulator surfaces from the deposition of cathode metal. A schematic of the gas discharge cell used is shown in Fig. 4.1a.

Hydrogen gas was puffed in by a pulsed electromagnetic gas valve 13 and fed through tube 12 to a series of small diameter holes tangential to the inner surface of anode 7 so that a swirling gas flow formed in the cell. The profile of the emission slit, parallel to the axis of the magnetron, was supposed to contribute to a downward deflection of the outgoing gas flow from cell 14, and the negative ion beam 2 should be deflected up by the main magnetic field.

In this way we hoped to reduce the effective thickness of the gas layer destroying the H$^-$ ions extracted from the gas discharge cell. In this design, it was possible to eliminate the effect of the insulator surfaces 10 on discharge ignition. A high-current discharge in a cylindrical magnetron with anode inner diameter of 7–8 mm and cathode diameter of 1 mm was established in a magnetic field of 0.04 Tesla. The voltage drop across the discharge increased linearly from 40 to 80 V when the magnetic field was increased from 0.04 to 0.1 Tesla. At a field of 0.1 Tesla, the voltage

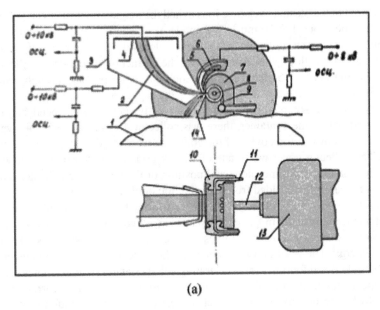

(a)

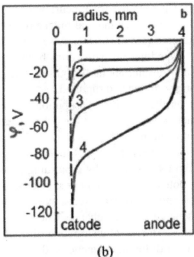

(b)

Fig. 4.1 (**a**) Schematic of magnetron gas discharge setup. (Reproduced from Dudnikov [1]). (1) Electromagnetic coil, (2) ion beam, (3) equipotential container, (4) ion collector, (5) electron flow, (6) extractor with scallop-collecting electrons, (7) magnetron anode, (8) magnetron cathode, (9, 12) channels for hydrogen gas, (10) glass insulators, (11) cathode current supply, (13) pulsed gas valve, (14) hydrogen flow through the emission slit. (**b**) Radial potential distribution for different magnetic fields [2]: (1) $B = 0$, (2) $B = 0.037$ Tesla, (3) $B = 0.085$ Tesla, (4) $B = 0.3$ Tesla

drop abruptly decreased to 55 V and then increased linearly to 110 V for a field 0.3 Tesla. The characteristic drop at 0.1 Tesla is apparently connected with a change in the conditions for the collection of positive ions on the cathode, accelerated by the electric field in the plasma. Negative ions exited the magnetron through an emission slit with dimensions 1×10 mm². The anode of magnetron 7 was grounded. The extraction voltage was applied to the extraction electrode. In the electron drift direction, a plate 6 "scallop" was assembled on the extraction electrode to collect the accompanying electron 5 along the magnetic field. After passing through a slit of size 1.5×13 mm² in the extraction electrode, the negative ion beam entered into the inner space of equipotential container 3, at a positive potential independently controlled, and within the container, there was collector 4 to receive the negative ion beam. The removable top covers of the container were fitted with screens of dense mesh. The collector cover was made of a thick grid with a phosphor layer which served to visualize the dimensions and position of the beam. The collector and its cables were carefully shielded from parasitic ion or electron flows.

Thus it was possible to obtain relatively well-formed beams of negative ions with collector current up to 7.5 mA from discharges in cylindrical magnetrons. The beam dimensions as visualized on the luminescent mesh cover were about 10×20 mm² at 10 keV energy. The position of the beam on the collector was in good agreement with that calculated for different accelerating voltages and different magnetic fields. The dependency of the negative ion collector current on discharge current and on source hydrogen density had a form characteristic for the generation of negative ions in the plasma volume. The source extractor current exceeded the ion current by a factor of 30–50. Lifetime tests of cylindrical magnetron discharge sources have shown that at a pulse repetition rate of 0.5 Hz, pulse duration of 1 msec, and discharge current of 130 A, providing a negative ions beam current of 5–7 mA, the cathode of the source can withstand about 10^5 pulses. The cathode contacts with glass insulators were strongly eroded and ruptured.

To obtain information on the potential distribution in the plasma, special end insulators were made which were mounted in the cylindrical magnetron instead of glass insulators. The partitioned insulator consisted of 17 thin-walled nickel cylinders, the diameters of which increased from 2.4 to 7.8 mm, separated by thin insulating layers, screened from the discharge by protruding parts of the cylinders. An oscilloscope recorded the potentials acquired in the discharge of 1, 5, 9, 13, and 17 cylinders connected to the anode through a resistance of 10^7 Ohms. The external characteristics of the discharge did not change when the glass insulators were replaced with sectioned ones. The radial distribution of floating potential acquired by the cylinders are shown for different magnetic field strengths in Fig. 4.1b. As the magnetic field is increased, the electric field intensity in the plasma column and the pre-anode potential jump increase substantially. The dependencies of the potential distribution and the total voltage drop on the hydrogen flow and discharge current were weak.

In traditional designs of ion sources with extraction across the magnetic field, the emission slit is oriented along the magnetic field. It is known that there is a significant difference in the coefficients of plasma diffusion across the magnetic field and along the field, causing a rapid decrease in the plasma density as it diffuses across the magnetic field, with metal electrodes at either end of the plasma column that limit the plasma along the magnetic field. The characteristic length of the exponential density decay is proportional to the distance between the electrodes. This implies the natural proposition that a decrease in the extent of the emission slit perpendicular to the magnetic field should greatly reduce the flux of magnetized electrons from the plasma through the emission slit and significantly weaken the flux of unmagnetized negative ions. Thus there was hope for the possibility of a significant decrease in the flux of accompanying electrons by orientation of the long narrow emission slit across the magnetic field. To test this assumption, studies were begun of a flat magnetron with cathode in the form of a flat plate with dimensions $1 \times 10 \times 10$ mm^3 surrounding the oval anode with flat side walls and shaped insulators at the ends.

The planar magnetron shown in Fig. 4.2a had an emission aperture of size 0.5×10 mm^2 oriented perpendicular to the magnetic field, and the planar magnetron shown in Fig. 4.2b had an emission aperture of 5×10 mm^2, separated by a jalousie, oriented perpendicular to the magnetic field with five narrow emission slits. The voltage drop in planar magnetron discharges depends linearly on the magnetic field, increasing from 45 to 230 V with magnetic field increasing from 0.02 to 0.3 Tesla. For discharge current in the range 2–150 A and fixed hydrogen density, the voltage drop in discharges in both planar and cylindrical magnetrons did not depend on the cathode material. Oscillograms of the discharge voltage always showed high-frequency fluctuations of considerable amplitude that were not noticeable on the discharge current oscillograms. There were traces of cathode erosion in the form of shallow transverse grooves uniformly covering the cathode surfaces (traces of running cathode spots). These features allow us to conclude that an arc discharge with cold emission was established in the gas discharge cells studied and that the increase in voltage drop across the discharge with increasing magnetic field strength is due to a decrease in the mobility of magnetized electrons.

It was not possible to extract an H$^-$ ion beam from planar magnetron discharges at that time. In the electrode configurations used, and with the existing magnetic

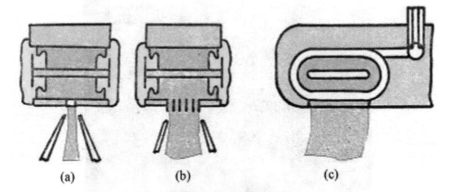

Fig. 4.2 Plain magnetron (planotron) gas discharge cell. (**a**) One-slit planotron, cross-section along the magnetic field, (**b**) multislit version, (**c**) cross-section perpendicular magnetic field. (Reproduced from Dudnikov [1])

field configurations, discharges leading to breakdown in the extraction gap were ignited.

In the 1969–1970 period, new attempts were made to increase the yield of negative ions from the plasma. In these attempts, hopes for increased yield were based on the use of hydrogen-containing molecules with lower dissociation energy and ionic-bonded molecules for working materials, for which the threshold for negative ion formation is smaller and the production cross-section is larger than for hydrogen molecules.

The test stand for source research was restructured. A new vacuum chamber with internal dimensions $24 \times 26 \times 12$ cm^3 was connected to the old vacuum chamber as shown in Fig. 4.3. Poles from Armco iron of diameter 50 mm were introduced into this box through side walls and were connected externally by a yoke with copper coil to establish a magnetic field. Most subsequent experiments were carried out

Fig. 4.3 Modernized vacuum chamber for testing negative ion sources

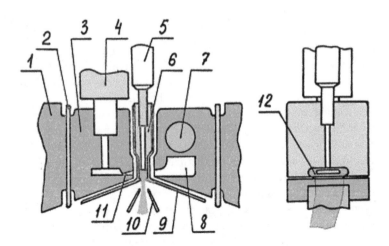

Fig. 4.4 Schematic of a Penning discharge source in a gas jet. (Reproduced from Dudnikov [1]). (1) Electromagnet pole, (2) insulating gaskets, (3) pole tips, (4) gas supply valve, (5) heated cesium container, (6) insulated anode, (7) liquid nitrogen cavity, (8) jet receiver, (9) screen anode, (10) extractor, (11) flat supersonic nozzle, (12) anode window

using this magnetic system. Gas discharge chambers were installed between the poles of this electromagnet.

An attempt was made to form a plasma with a large gas density gradient between the gas discharge plasma and the vacuum and with only a small thickness of neutral gas layer beyond the boundary of the dense plasma. For this purpose, a source comprising a high-current Penning discharge in a supersonic gas jet was fabricated. The design of this source is shown in Fig. 4.4.

Pole tip 3 served as the Penning discharge cathodes and were insulated with dielectric spacer 2 from grounded magnet pole 1. Anode 6 was installed with a rectangular window 12 of dimensions 2×10 mm^2 in the gap between the pole pieces. A flat supersonic nozzle 11 with an 0.1×8 mm^2 critical section and an 0.5×8 mm^2 output section was located in one pole piece. The working gas was supplied to the nozzle by the pulsed valve 4. The gas jet formed by the nozzle passed through a window in anode 12 and entered a cavity 8 of the second pole piece cooled by liquid nitrogen 7. The pole tips in the discharge region were reinforced with tantalum foil. Particles could escape from the discharge through the anode window to the extraction electrode 10 through an emission slit of size 0.5×8 mm^2 oriented across the magnetic field. Anode screen 9 prevented gas from flowing through the gaps between the cathode and the anode in the extraction gap. For a diborane B$_2$H$_6$ discharge, the discharge voltage drop was 450 V. When cesium was released from the heated container 5 that was connected to the discharge space, the voltage on the discharge fell to 120 V. In this device, very-high-current glow discharges were obtained, which later ensured the effective generation of negative ions in surface plasma sources. It was not possible to extract and form beams of negative ions from these discharges at that time. When applying voltage to the extraction gap, a discharge was already ignited at hundreds of volts, preventing further increase of the extraction voltage. As it became clear later, in this case, as in the extraction of negative ions from arc discharges in planar magnetrons, discharge ignition in the extraction gap was favored by the curvature of the magnetic field lines in the direction of the extraction electrode; because of this, favorable conditions were created for the accumulation and multiplication of electrons in the extraction gap.

In 1970, a number of attempts were made to extract negative ion beams from discharges in magnetrons and from hollow cathode discharges of various designs, with polyatomic hydrogen-containing substances and ionically bond substances used as feed material, such as diborane, thermal decomposition products of lithium hydride, and decomposition products of cesium borohydride. In all these cases, no convincing results were obtained.

At the same time, we tried to extract negative ions through the plasma layer that should have been formed by secondary negative-ion emission on the cathode bombarded by fast ions in a hydrogen and cesium discharge.

A plasma device was installed between the poles of the electromagnet as shown in Fig. 4.5. A flat cathode was surrounded by an oval anode 2 with flat lateral walls. The discharge space was limited by the side-shields of cathode 4, preventing electrons from leaving to anode 2 along magnetic field lines from poles 1. This time, the insulators 8 were located out of the plasma region. Cesium borohydride was placed in a thin-walled container 3 heated by direct current. The decomposition products of the cesium borohydride were introduced into the plasma chamber through holes in the cathode central plate and a tube connecting the cavity in the central plate of the

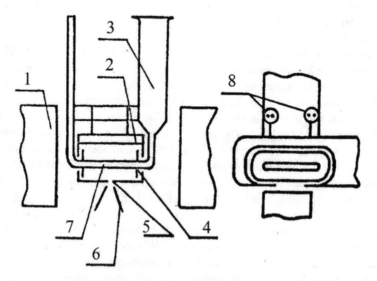

Fig. 4.5 Schematic of a gas discharge source of planotron configuration. (Reproduced from Dudnikov [1]). (1) magnet poles, (2) anode of the plasma cell, (3) heated container with working substance, (4) side cathode shield, (5) emission gap, (6) extractor, (7) cathode, (8) cathode fixation insulators

cathode with the container. Voltage to ignite and support the discharge was applied across the cathode-anode gap. In this system it was possible to establish quasistationary discharges, but due to difficulties in regulating the supply of the working substance, poor sealing of the gas discharge cell, and other design flaws, it was not possible to investigate negative ion emission from this system.

In early 1971, we again turned to the study of the emission of negative ions from hydrogen plasma discharges in crossed fields. Taking into account the acquired "negative" experience gained after a number of trial experiments, the design of a new source was brought to the form depicted schematically in Fig. 4.6. The body of plasma cell 3 was fixed on insulators 2 of Plexiglas in the gap between the pole pieces 1. The plates of extraction electrode 14 were welded to special protrusions from the pole pieces 1, creating a magnetic field in the high-voltage gap (but not a Penning trap configuration). A pair of pole pieces with the source was installed between the grounded poles of the electromagnet. A plasma cell with planotron configuration is formed by the cathode, consisting of the central plate of cathode 10 and cathode side shields 11, and a cathode-enclosing anode formed by parts of the plasma cell body 3 and anode insert 5. A cathode made of a 0.2-mm-thick tantalum foil was attached to tantalum current leads 12 passing through the wall of anode insert 5 and insulated from it by ceramic tubes 13. The volume of the plasma cell was minimized as much as possible. Gaps between the cathode and the anode, in which a discharge should not burn, were reduced to 1 mm. Hydrogen was supplied to the plasma cell through a short channel by a pulsed electromagnetic valve 4. Emission slit 7 with dimensions 0.5×10 mm^2 oriented across the magnetic field was cut in the thin-walled body of the plasma cell. From the discharge region,

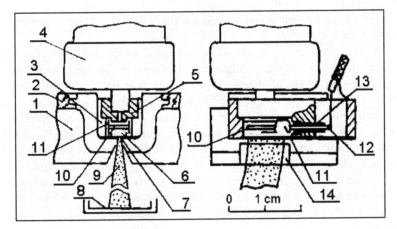

Fig. 4.6 Schematic of planotron discharge configuration for producing a negative ion beam. Left cross-section along magnetic field, right cross-section perpendicular to magnetic field. (Reproduced from Dudnikov [1]). (1) Electromagnet pole tips, (2) high-voltage insulators, (3) body of plasma chamber, (4) pulsed gas inlet valve, (5) anode insert, (6) screen of emission slit, (7) emission slit, (8) collector, (9) negative ion beam, (10) central cathode plate, (11) cathode side shields, (12) cathode holders, (13) cathode insulators, (14) extractor plates

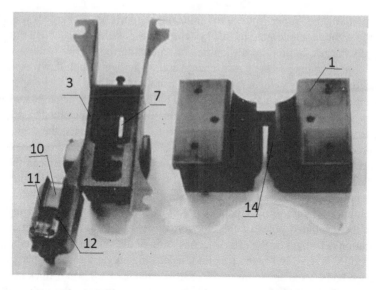

Fig. 4.7 Photograph of the first planotron. (1) Electromagnet pole tips, (3) body of plasma chamber, (7) emission slit, (10) central cathode plate, (11) cathode side shields, (12) cathode holders, (14) extractor plates. (Reproduced from Dudnikov [1])

particles could pass to the emission slit through the gap between the anode projections 6 shielding the emission slit from the dense, high-current plasma.

The old power supply systems were used for pulsed hydrogen gas injection, plasma ignition and support of the discharge, and ion extraction. A beam collector 8 was installed to monitor the beam current 9. The body of the plasma cell was held at the negative-polarity extraction voltage, and the collector was held at low voltage.

A photograph of the first planotron is shown in Fig. 4.7.

After some initial testing, a mode of operation was established in which a pulsed discharge with 1 msec rectangular current pulses burned stably at a discharge current up to $I_d \sim 30$ A. The discharge voltage drop was about $U_d \sim 400$ V when the discharge current, hydrogen supply, and magnetic field strength were varied widely. Under optimum conditions, a negative ion beam with collector current up to $I^- \sim 4.5$ mA was obtained from these sources with an extraction voltage $U_0 \sim 20$ kV. In this case, the total current drawn in the extraction circuit was no greater than $I_0 \sim 15$ mA. The parasitic current due to accompanying electrons and all other processes exceeded the negative ion beam current by only a factor of two, whereas in all previous negative ion sources, the electron current exceeded the ion beam current by a factor of tens or hundreds. Thus, hopes for a decrease in the flux of accompanying electrons by decreasing the length of the emission gap along the magnetic field received some strong reinforcement.

Steady work on the source again aroused interest in non-traditional methods of increasing the generation of negative ions. First of all, we wanted to test the effect of cesium vapor on the formation of H$^-$ ions in the source discharge. The first experiments of this kind were performed on July 1, 1971.

A tablet made from a pressed mixture of cesium chromate with titanium and containing 1 mg of released cesium was attached to the side wall of the anode insert, facing the discharge. It was assumed that cesium would be released when the tablet was heated by the plasma. In this experiment, the gap between the anode projections, through which the plasma particles passed to the emission slit, measured 0.7×10 mm^2. At approximately the same discharge parameters, a beam of H$^-$ ions with current $I^- = 1.5$ mA was registered at the collector. After a while, the collector current at the end of the 1 msec pulse began to increase as shown in Fig. 4.8. The current pulse acquired an almost triangular shape: the collector current increased during the pulse from $I^- = 1.5$ mA to $I^- = 4$ mA. After 20 min of source operation at a repetition rate of 3 Hz, the increase on the trailing edge disappeared, and the oscillogram acquired an ordinary rectangular shape with an amplitude $I^- = 1$ mA. The assumption was made that the increase of collector current was associated with the release of cesium into the discharge.

In the plasma discharges investigated, the cathodes are heated much more strongly than the anode. Therefore, in subsequent experiments, the cesium tablets were located on the central plate of the cathode, covered with a thick nickel grid. At the same time, the gap between the anode projections through which the plasma passed to the emission slit was expanded to 2.5 mm. After these changes, the negative ion

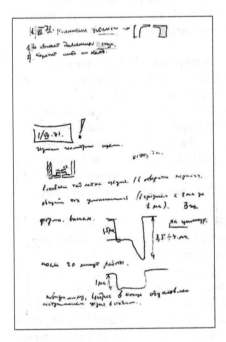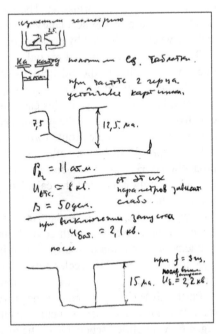

Fig. 4.8 Pages from the laboratory log book for July 1, 1971, describing the evolution of intensity of the negative ion beam when cesium is added to the discharge. (Reproduced from Dudnikov [1])

current to the collector increased to 12.5 mA as shown in Fig. 4.8, and after optimizing the gas supply and discharge current, the negative ion collector current increased to 15 mA. Investigation of the beam trajectories by means of a movable probe and a luminescent screen convinced us that electrons do not enter the collector—using hydrogen as the feed gas, the collector current is mainly due to a hydrogen H^- ion beam. Heavy ions are also formed and extracted from the hydrogen discharge; they are separated from the H^- ion beam by the source magnetic field. Observation of the discharge parameters showed that the emission of H^- ions increases with decreasing voltage on the discharge from $U_d = 400$–500 V to $U_d = 150$–100 V.

Thus, following these pioneering experiments, it was found that the addition of cesium to the gas discharge cell intensifies the emission of H^- ions from the discharge [1, 3–6]. By reducing the length of the emission aperture along the magnetic field, it is possible to improve the separation of the negative ion flux from the accompanying electron flux.

4.2   Studies of Negative Ion Emission from Hydrogen Plasma with Added Cesium

The addition of cesium gave a clearly manifested increase in the emission of H^- ions from the discharge. Thus the question arose about the mechanism for the intensification of negative ion formation. The most probable explanation was the already-familiar increase in the secondary emission of negative ions from the planotron cathode when bombarded by positive ions.

The available data did not exclude other mechanisms for the increased H^- production. According to the calculations of Smirnov [7, 8] in some cases, negative ions should be effectively formed by charge capture by slow atoms of weakly bound electrons in excited atoms. In this case, the $Cs^* + H^0 \rightarrow Cs^+ + H^-$ process could be effective. In the volume of the plasma and on electrode surfaces, ionically bonded cesium hydride molecules could form, which in the plasma volume could dissociate with high probability upon collision with electrons to form H^- and Cs^+ ions. The production of H^- beams with current up to 15 mA from a 0.5×10 mm^2 emission slit did not strongly contradict the calculations for obtaining H^- ion beams from the plasma volume.

However, this possible mechanism for the production of H^- beams with current 15 mA from a discharge with cold electrodes provided a solution to the original problem—the creation of sources satisfying the requirements for charge-exchange injection of protons into accelerators and storage rings. In this connection, attempts were made to empirically optimize H^- ion beam generation in these sources.

Further research on planotron negative ion sources was carried out using the same test-stand. The main structural elements of the source variant are shown in Fig. 4.6. A simplified schematic of the overall system and the diagnostic equipment for monitoring the parameters that characterize the source operation is shown in Fig. 4.9.

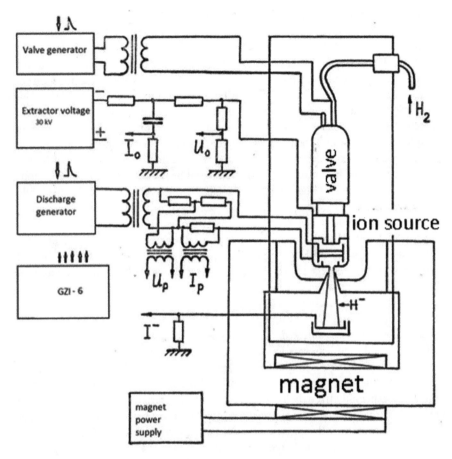

Fig. 4.9 Schematic of experimental setup for the operation of a surface plasma source (SPS) and measurement of the process characteristics. (Reproduced from Dudnikov [1])

The magnetic field strength B was controlled by the electromagnet current. The discharge was initiated by voltage pulses with amplitude up to 3 kV and duration 1 msec applied, through a thyratron switch, across the anode-cathode gap via an isolation transformer fed by a pulse-forming line of impedance 25 ohms. The discharge voltage U_p was monitored with a resistive voltage divider, and the discharge current I_p by a low-resistance shunt. These signals were transmitted from the high-voltage part of the circuit to oscilloscopes through isolation transformers. An extraction voltage U was applied to the body of the gas discharge chamber. The total current drawn in the extractor circuit, I_o, was monitored by means of the voltage across a resistor included in the high-voltage capacitor circuitry. Hydrogen from a gas cylinder was fed to a pulsed electromagnetic valve, which is at high voltage, along a copper tube with insulating section. Current pulses to operate the pulsed valve were transmitted from a special circuit through an isolation transformer. The hydrogen density n_g in the plasma chamber before discharge ignition was controlled by varying the gas cylinder pressure, varying the amplitude of the valve-opening current pulses, and measuring the delay between initiation of valve-opening and triggering the discharge. The pulsed supply of hydrogen Q was estimated from the chamber pressure change, and the shape of the gas pulses was determined from signals from miniature ionization gauges. The negative ion beam current I was monitored by a movable collector, and the current density distribution $J^-(r)$ by small collectors located behind apertures in the main collector, and by a luminescent phosphor screen. At high beam intensity, the current density distribution could be visually assessed from the thermal glow of the collector. The H$^-$ ion beam is spatially separated by the source magnetic field from heavy negative ions and D$^-$ ions formed when deuterium was supplied, so that the intensities of these beams could be measured separately. The pulsed devices were controlled by a GZI-6 six-channel pulse delay generator.

Without cesium supply, various types of discharges were explored, from arcs with voltage U_d = 40–50 V to high-voltage discharges with U_d = 1–2 kV, and of these only glow discharges with U_d = 400–600 V provided efficient generation of H$^-$ ions. With the addition of cesium, the variety of possible discharge operational modes became completely immeasurable.

In this connection, first of all, we had to deal with elucidation of the conditions ensuring stable burning of the necessary high-current glow discharge.

It was possible to suppress arc discharges by making all the cathode parts and their fastenings in the plasma chamber from low vapor pressure refractory metal such as molybdenum, tungsten, and tantalum, thoroughly cleaning their surfaces of impurities, especially from dielectric inclusions and abrasives, and careful shielding the contacts to these parts with cathode insulators. For metal surfaces with higher vapor pressure and in the presence of contaminants, a constricted arc discharge forms in the vapor of cathode metal or contaminants. The purity of the hydrogen used is important. The switching to an arc mode is facilitated by significant atmospheric impurities. The purity of technical hydrogen is adequate for the stable burning of high-current glow discharges, but it is necessary to pump the gas system to vacuum after filling it with hydrogen and to use a sufficiently high hydrogen pressure in the pulsed gas valve. The surfaces are quickly cleaned of minor amounts of polluting impurities that appear on the cathode during its processing and installation, by conditioning arc discharges that gradually transform into a glow discharge.

High-current glow discharges with a voltage U_d that is almost independent of the discharge current I_p are established under conditions that ensure the efficient use of the energy of fast electrons accelerated in the near-cathode potential drop to form plasma ions. To realize these conditions, the size of the region of electron oscillations perpendicular to the magnetic field should exceed the Larmor diameter for electrons of energy eU_d, and the hydrogen density should be sufficient for a noticeable perturbation of the motion of emitted electrons, so as to ensure irreversible capture of electrons in the oscillation region. If the strength of the magnetic field or the density of hydrogen is insufficient to meet these conditions, and if parasitic arc discharges are adequately suppressed, low-current glow discharges with limited discharge current and voltage close to the amplitude of the pulses are ignited.

To eliminate parasitic discharges, special electrode configurations for the plasma cells were designed, creating conditions for confinement of electrons only in the main gas discharge gap, and eliminating electron oscillations in the remaining gaps. Discharges in hydrogen with added cesium burned more stably if the cesium chromate/titanium tablets were placed not under a nickel grid as in the first experiments, but in a cavity in the central plate of the cathode, made in the form of a rectangular enclosure made of 0.2–0.3-mm tape or in side shields of the same design, and with a Penning electrode configuration. Glow discharges with low voltage drop, $U_d = 100$ V, and providing the largest H$^-$ ion flux were steadily ignited only at an elevated plasma chamber temperature. To ensure the efficiency of the source at elevated temperatures, it was necessary to switch to high-voltage ceramic insulators, to upgrade the pulsed gas valve supplying hydrogen [9], and to provide more rigid mounting of all elements.

The fast-pulsed gas valve is an important component of the ion source. A schematic of the fast-pulsed gas valve is shown in Fig. 4.10.

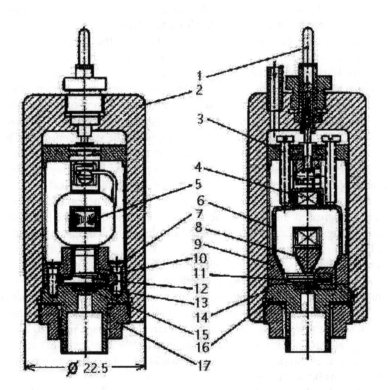

Fig. 4.10 A schematic of the pulsed gas valve. Two cross-sections. (1) contact, (2) body, (3) lock screw, (4) coil, (5) magnetic conductor, (6) grid, (7) screws, (8) insert, (9) ring, (10) elastic disk, (11) armature, (12) support, (13) saddle, (14) gasket, (15) measuring orifice, (16) base, (17) screw

The entire mechanism of the valve is contained in the Armco steel body 2 which is filled with the gas being fed. Copper gasket 14, clamped between base 16 and body of screw 17, ensures that the valve will be vacuum-tight. The gas flows through measuring orifice 15. Armature 11, made of Armco steel, acts as the valve of the valve. In the "closed" state the gas is kept from reaching the measuring orifice by the armature whose polished surface is pre against the "Viton" saddle 13 and held into the base with four elastic disks 10 (vacuum rubber 9024 "A"). The elastic disks are attached to a ring 9, which is fastened to the base with two screws 7. The gas flows into the feeding unit when the armature is forced from the saddle by the pulsed magnetic field generated in the gap between the armature and magnetic conductor 5 by an impulse in the current flowing through coil 4. The magnetic conductor is made of an alloy of CrWPt steel (thickness 0.1–0.3 mm). The electromagnet is attached to the ring with two locking screws 3. The size of the working gap (about 0.2 mm) is adjusted with shims between the magnetic conductor and the ring. Support 12, made of Viton, prevents contact between the armature and the magnetic conductor. Copper grid 6 and insert 8, decreasing current dispersion, help to concentrate the magnetic field in the working gap. The activating current impulse, which flows to the coil through electrical contact 1, is discharged from capacitor C, into the induction coil of the valve's electromagnet. The capacitor is switched from charge to discharge by thyristor switches T1 and T2, respectively, and diode D. To decrease the amount of energy required and increase the frequency, resonance charge transfer and recharging of the capacitor through inductors (L1) and (L2), respectively, is used. When the electromagnet is activated by a current impulse with a period of 300 us and an amplitude of 400 ampere-turns, a gas burst is formed which has a duration of 300 us and a rise time of 100 us. The amount of gas admitted per impulse depends on the diameter of the measuring orifice, the gas pressure on the valve, and the amplitude of the activating impulse. The valve constructed in this way works at frequencies up to 10^3 Hz. In tests the valve was operated at a frequency of ~700 Hz for a long time. After 10^9 cycles (corresponding to 3000 h of continuous operation at a frequency of 100 Hz), its characteristics had not changed. The seal normally operates at temperatures under 150 °C. If necessary, the temperature may be increased. In this case the rubber disks must be replaced with steel or tungsten springs. A photograph of assembled/disassembled valves is shown in Fig. 4.11.

These modifications made it possible to increase the pulse repetition rate up to 10 Hz, accelerated the conditioning of electrodes and the transition to a low-voltage glow discharge, and ensured stable operation of the source at a discharge current up to $I_d = 100$ A. The electrical strength of the high-voltage insulators and gaps was increased, so that it was possible to raise the extraction voltage up to 20 kV. After these improvements, an H⁻ ion beam of current up to $I^- = 25$ mA was extracted from the plasma with electrode configuration shown in Fig. 4.5, using an emission slit with dimensions 0.5×10 mm^2.

Subject to the above conditions, a range of variants of the gas discharge cell electrodes were produced and tested, differing in configuration, geometric dimensions, and materials. The best results were obtained with molybdenum cathodes.

Fig. 4.11 A photograph of assembled/disassembled valves

Fig. 4.12 A source variant
with optimized
configuration of the gas
discharge cell.
(Reproduced from
Dudnikov [1])

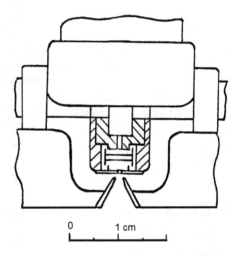

The configuration, geometric dimensions, and material of the electrodes of the beam formation system were also varied.

The highest-current H⁻ ion beams were produced using gas discharge cells with a planotron electrode configuration, which are close to those shown schematically in Fig. 4.12.

By early 1972, H⁻ ion beams in discharges with cesium were obtained from such sources with intensity up to $I^- = 75$ mA for emission slit dimensions 0.5×10 mm², with $I_{ext} = 0.2$ A; up to $I^- = 170$ mA for emission slit dimensions 1×10 mm², with

$I_{ext} = 0.5$ A; and up to $I^- = 230$ mA for emission slit dimensions 1.5×10 mm^2, with $I_{ext} = 0.8$ A (I_{ext} is the total current drawn from the extractor power supply, including H$^-$ ion beam current, accompanying electron current and back-accelerated positive ion current).

These beam currents and current density up to 1.5 A/cm^2 significantly exceeded all previous achievements in the production of H$^-$ ion beams. A proportional increase in beam intensity with increasing emission slit area established preconditions for the full satisfaction of accelerator technology needs for H$^-$ ion beams and secured applications requiring yet higher intensity beams.

The question arose as to the mechanisms of such efficient generation of negative ion beams.

To obtain information about the location and mechanism of the formation of negative ions, special experiments were performed on these sources.

For a more detailed explanation of the role of cesium, an exact replica of the source was made as in Fig. 4.10 from which H$^-$ ion beams were produced with an emission current density of up to 1.5 A/cm^2 and an intensity of up to 230 mA for extractor slit dimensions 1.5×10 mm^2, when cesium was injected.

At this time, special precaution were taken, even precluding accidental contamination of the cesium source, and investigations were carried out of the emission H$^-$ ions from the hydrogen plasma discharge of the planotron and of the Penning electrode geometry formed from the planotron by removing the central cathode plate. In these cells, discharges were ignited with almost identical characteristics, completely analogous to discharges in the first versions of the source without cesium feed. Oscillograms of the discharge current and discharge voltage, Fig. 4.13, display the current-voltage characteristic as obtained by powering the discharge with half-sinusoidal voltage pulses. For this measurement, the inductances of the pulse-forming line were short-circuited, and the voltage pulse was formed by the parasitic inductance of the isolation transformer. The H$^-$ ion current monitored by the beam collector and the total extractor current were similar in shape to the discharge current pulse.

Fig. 4.13 Oscillograms of discharge voltage and discharge current in a planotron without cesium. The horizontal scale is 0.2 ms/div. (Reproduced from Dudnikov [1])

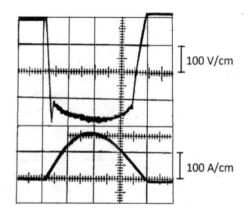

100 V/cm

100 A/cm

In plasma cells with a 1-mm gap between the central cathode plate and the edges of the anode protrusions, high-current glow discharges were ignited at a magnetic field $B = 0.12$ Tesla, and the estimated initial hydrogen gas density $n_g = 10^{16}$ mol/cm^3. As the discharge current was increased, the discharge voltage increased slightly, but decreased somewhat with decreasing hydrogen supply and magnetic field strength.

The H$^-$ ion current was practically independent of magnetic field over the operating range of its variation. The dependencies of the negative ion beam current (collector current I_H^-) extracted from planotron discharges on the extraction voltage U_o, on the discharge current I_p, and on the pulsed hydrogen supply Q, are shown in Figs. 4.14, 4.15, and 4.16. Dependencies of the total current in the extraction circuit I_{ext} from the extraction voltage U_o, discharge current I_d, hydrogen supply Q, and magnetic field are shown in Figs. 4.14, 4.15, and 4.17. The characteristic distributions of the H$^-$ ion beam current density as measured at a collector located 10 cm from the extractor gap are shown in Fig. 4.18.

Beams of heavy negative ions are well separated from the H$^-$ ion beam by the magnetic field of the source. Under optimal conditions, the heavy ion beam current decreased to 10% of the H$^-$ ion beam current, but under certain conditions, especially with tantalum cathodes, their intensities could be comparable.

By variation of emission gap dimensions in discharges of pure hydrogen, H$^-$ ion beams with emission current density up to $J^- = 200$–270 mA/cm^2 were obtained. The emission current density decreased somewhat as the width of the emission slit was increased.

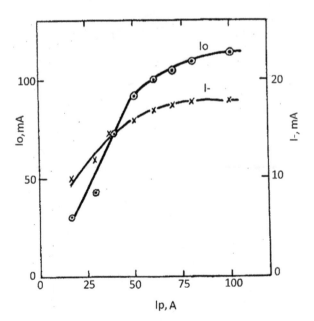

Fig. 4.14 Negative ion current I^- and total extractor I_o as a function of discharge current I_p. (Reproduced from Dudnikov [1])

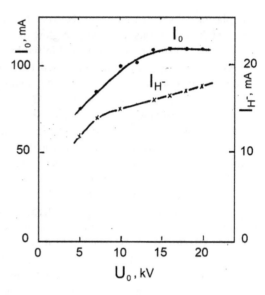

Fig. 4.15 Dependence of the negative ion beam current I^- and the total extractor current I_0 on the extraction voltage U_0. (Reproduced from Dudnikov [1])

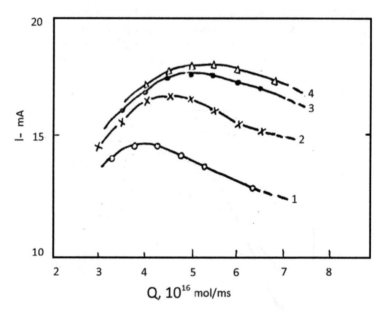

Fig. 4.16 Dependence of the negative ion beam current I^- at different discharge currents on the hydrogen gas supply Q. (Reproduced from Dudnikov [1]). (1) $I_d = 35$ A, (2) $I_d = 50$ A, (3) $I_d = 75$ A, (4) $I_d = 100$ A

The total extractor current I_0 increased very rapidly with increasing extraction gap width δ and decreasing wall thickness h of the extraction slit, and it was not possible to investigate the emission of H$^-$ ions at high-discharge currents from a

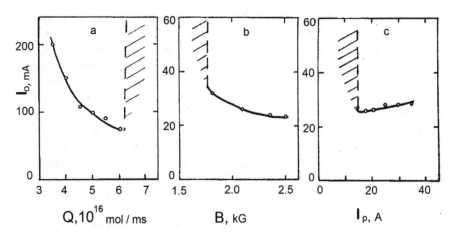

Fig. 4.17 Dependence of the total extractor current I_o on the pulsed hydrogen supply Q, on the magnetic field strength B, and on the discharge current I_d. (Reproduced from Dudnikov [1])

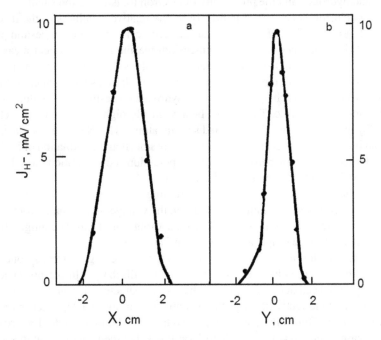

Fig. 4.18 H⁻ ion beam current density distribution at the collector, (**a**) parallel to the magnetic field, and (**b**) perpendicular to the magnetic field. (Reproduced from Dudnikov [1])

source with $\delta = 1.5$ mm. H⁻ ion beam current up to 22 mA was extracted with an emission slit of dimensions 1×10 mm², whereas H⁻ beams with current up to $I^- = 170$ mA were obtained from an identical source by adding cesium.

On the whole, the results obtained did not contradict the idea of the generation of H⁻ ions in the volume of the hydrogen plasma. Since the possibility of optimizing conditions for the formation of H⁻ ions in plasma sources of traditional design, and forcing parameters even in pulsed mode are limited by the very large coextracted electron beam current, an increase in ion current by a factor of several seemed reasonable without invoking new assumptions about negative ion generation mechanisms.

However, the results of studies of H⁻ ion production in hydrogen-cesium discharges were completely incompatible with this framework, primarily because of the high density of the H⁻ ion flux generated, which was many times greater than the most optimistic estimates of the generation of H⁻ ions in the plasma volume.

Estimates of the cross-sections for electron capture by hydrogen atoms from excited cesium atoms, using the results of Refs. [7, 8], implied that the probability of this process cannot significantly exceed the probability of charge-exchange in unexcited cesium, since quasi-crossing states that Cs* + H and Cs⁺ + H⁻ occur at distances too large, for which the intersection of the wave functions of excited cesium and hydrogen and the probability of electron transfer are too small.

This conclusion is confirmed indirectly by the relative smallness of the inverse process H⁻ + Cs⁺ → Cs* + H, observed spectroscopically in hydrogen-cesium plasmas [10, 11]. Experience with hydrogen-cesium plasma and subsequent measurements of cesium ejection from gas discharge cells show that for conditions optimal for the generation of H⁻ ions, the cesium density in the plasma is small and it is highly ionized, and thus charge exchange of hydrogen on cesium in the volume cannot play an important role. There must be a relatively high probability of electron capture by hydrogen atoms from excited sodium atoms (H + Na* → H⁻ + Na⁺), for which an inverse process is probable [12]. It is possible that the formation of H⁻ ions in a hydrogen-potassium plasma with high potassium concentration, which was studied [13], is really determined by this process.

The discharge voltage decreases with increasing cesium feed into the discharge, and for sufficiently high-cesium concentration, the voltage oscillograms from high-current glow discharges in plasma cells with a variety of electrode configurations acquire a characteristic shape, as illustrated in Fig. 4.19.

From oscillogram of the discharge voltage U_p and discharge current I_p, one can determine the current-voltage characteristics of the discharge. In the presence of cesium, a pronounced dependence of the discharge characteristics on the prehistory is observed, and indirectly (through the cesium concentration) a dependence on the discharge current. Typically, the discharge voltage for the first hundred microseconds decreases from $U_d = 150$–200 V to $U_d = 100$ V, but then continues at this level at least up to tens of milliseconds. Most effective ions are generated in these regimes. By decreasing the discharge voltage, the Larmor radius of emitted electrons is less than the gap between the cathode central plate and the anode at lower magnetic field strength. With minimum value of $d = 1$–1.3 mm for this design, discharges with cesium are ignited for field strength $B > 0.05$ Tesla. Due to the large scattering cross-sections, the minimum values of initial hydrogen density $n_g \sim 3 \times 10^{15}$ mol/cm³ are also smaller.

Fig. 4.19 Oscillograms of discharge voltage U_p, discharge current I_p, H⁻ ion current I^-, and total current in the discharge gap I_o. (Reproduced from Dudnikov [1])

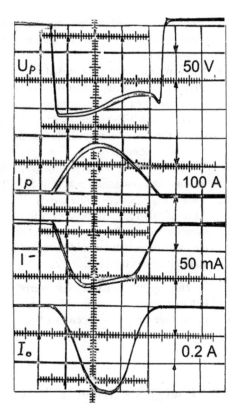

The dependencies of H⁻ ion current on collector current I^- and of the total extractor current I_o on the extraction voltage U_o, discharge current I_p, and magnetic field B in sources with cesium were similar to the corresponding dependencies for sources without cesium, but the ratio I^-/I_p of collector current to discharge current for the same emission slit dimensions and the limiting values of the H⁻ ion beam intensity upon addition of cesium was appreciably greater, as was the ratio I^-/I_o. The ratio of discharge power to beam intensity decreased even more. As a rule, H⁻ ion beams were extracted with maximum current from discharges with cesium with minimal hydrogen supply but still sufficient for stable burning of a high-current glow discharge. The beam current decreased comparatively sharply with the increase of the hydrogen supply above the minimum value, whereas for sources without cesium, fairly smooth maxima were observed in the dependence of I^- on Q.

When deuterium was injected in place of hydrogen, D⁻ ion beams were extracted with approximately the same intensities.

A number of observed effects indicated a significant role of cathode processes in the formation of negative ions:

(a) As the gap between the cathode central plate and the emission slit increases, the maximum H⁻ ion current clearly decreases, although it remains relatively large.

From sources with a Penning electrode configuration, with the same dimensions of the side shields and the same discharge currents, the ion current of H^- was about half that.

(b) Mass spectrometry of extracted negative ion beams showed that the heavy negative ion flux consists mainly of O^- ions. When using cesium fluoride, instead of a mixture of cesium chromate and titanium, a significant amount of F^- ions was registered. The intensity of the O^- ion flux strongly depended on the material of the cathode central plate and was minimal for a molybdenum cathode melted in vacuum.

(c) The extraction of negative ions from discharges with the addition of lithium, sodium, and potassium was tested. In this case, hydrides of the corresponding metals were placed in the cavity of the cathode central plate. The vapors of these metals reduced the glow discharge voltage to U_d = 200–250 V, and the intensity of the H^- beams under comparable conditions was greater than from the discharges without additives, but for K and Na 2–3 times less than with the addition of cesium [14].

The entirety of discharge characteristics in gas discharge cells with planotron and Penning electrode configurations provided hope that the greater part of the discharge voltage drop is concentrated in a narrow near-cathode layer, as in high-current Penning discharges [15].

4.3   Energy Spectra of H^- Ions in Surface Plasma Sources

We wanted to concentrate on ions formed at the cathode and to carry out energy analyses of the extracted negative ion beams.

A schematic of the experimental setup is shown in Fig. 4.20. A negative ion source 2 with pole pieces 4 was installed in a small chamber between the poles of the electromagnet so that the H^- ion beam passed horizontally into a large vacuum chamber. The heavy ion beam was separated from the H^- ion beam in the source magnetic field. Part of the H^- beam passes through the entrance slit to the analyzer 10 at an angle of 45° to the plates. Ions deflected through 90° by the electric field between the plates fall on the detector (Faraday cup, or the first dynode of electron multiplier 11), through the analyzer output slit. To increase the speed and sensitivity, the signals from the Faraday cup and the electron multiplier were transmitted by means of cathode follower circuitry.

When studying the energy spectra, H^- ions were extracted through emission slits 7 of width δ = 0.2–0.5 mm and length l = 1–10 mm. The power supplies provided long-term stability of ~1 V at a level of 10^4 V of the extraction voltage U_o and the voltage U_k across the analyzer plates. For accurate measurement of these voltages, a digital voltmeter with precision dividers was used. Rapid changes in these voltages were transmitted to oscilloscopes through capacitive dividers. The input and output slits of the analyzer plates had a width of 0.05 mm and had sharpened edges;

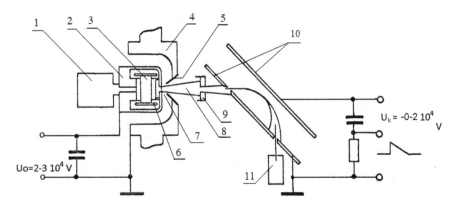

Fig. 4.20 Scheme of registration of the energy spectrum of H⁻ ions from the planotron and the Penning discharge. (Reproduced from Dudnikov [1]). (1) Gas valve, (2) source body-anode, (3) central cathode plate, (5) extractor, (6) side cathode plates, (7) emission slit, (8) negative ion beam, (9) collector, (10) energy analyzer, (11) collector with secondary electron multiplier

the distance between the slits was 20 cm. These parameters correspond to a calculated energy resolution of 5 eV for the beam energy used, 10^4 eV. To quickly view the spectrum on an oscilloscope, a sawtooth sweep was added to the DC voltage across the analyzer plates. It was possible to record the beam energy spectrum, at selected times throughout the discharge pulse, in a time of 10^{-4} s. The source operated at a repetition rate of 0.1–10 Hz with full reproducibility from pulse to pulse.

Simultaneous high-accuracy scanning of the beam-accelerating voltage U_0 and the voltage across the analyzer plates U_k, corresponding to tuning for a sharp maximum in the energy spectrum over a wide range of variation of these voltages, made it possible to reduce the error in determining the absolute energies of the detected ions to $\delta W < 20$ eV at a beam energy $W \sim 10^4$ eV 16]. The energy scale for fast viewing of the energy spectra was calibrated by the shift of the spectra for small changes in the beam extraction voltage.

Examples of the H⁻ ion beam energy spectra obtained from discharges in a planotron with the addition of cesium are shown in Fig. 4.21, which shows a selection of oscillograms. In the energy spectra of planotron beams, two peaks are observed, separated by a dip. The energy width of the first peak at lower energy decreases with increasing hydrogen supply, when the intensity of discharge noise is reduced. From discharges without recorded fluctuations, obtained in sources without cesium addition, and with Penning electrode configurations with cesium addition, H⁻ ion beams with a small energy width of the first peak, comparable with the resolving power of the analyzer, were obtained.

Multiple careful measurements of the energy corresponding to the maximum of the first peak showed that when a negative accelerating voltage U_0 was applied to the body of the gas discharge chamber, then within the accuracy of measurement $W_0 = eU_0 \pm \sim 20$ eV. When an accelerating voltage was applied to the planotron cathode, then a similar relationship was obtained for the energy W_c corresponding to the

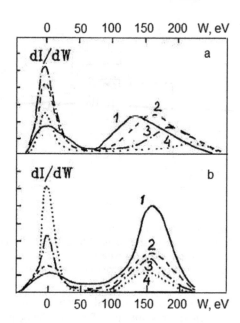

Fig. 4.21 Energy spectra of H⁻ ion beams extracted from discharges in cells with a planotron electrode configuration; discharge with fluctuations. (Reproduced from Dudnikov [1]). (**a**) Different discharge voltages U_d: (1) 120 V, (2) 150 V, (3) 160 V, (4) 210 V. (**b**) Different hydrogen feeds Q: (1) 10^{16}, (2) 1.2×10^{16}, (3) 1.7×10^{16}, (4) 2.2×10^{16} mols/ms. (Note that zero energy on the spectrum x-axis corresponds to the extraction energy eU_o)

second peak, but in this case, because of the larger peak width, the uncertainty in the energy determination was greater.

The distance between the maxima, $\Delta W = W_c - W_o$, was close to or slightly greater than eU_d. This can clearly be seen in the series of oscillograms shown in Fig. 4.19a, for which the discharge voltage was varied by changing the cesium flow.

These results suggest that ions of the first peak (with $<W> = W_o$) are formed in the anodic region and in the nearly equipotential volume formed by the anode projections near the emission slit, and exit through the emission slit with low velocity, while ions of the second peak are formed at the plasma-bombarded cathode by electron capture from the cathode at the electron affinity levels of sputtered and reflected particles, and accelerated to the emission slit by the total discharge voltage [16, 17]. The energy width of the second peak is comparable to the distance between the maxima because of discharge voltage fluctuations and also because of the large initial velocity of emitted H⁻ ions resulting from momentum exchange between particles bombarding the cathode and condensed phase particles on the cathode and adsorbate material.

In beams obtained from discharges with cesium in a planotron with a small gap between the cathode and anode, and with minimum hydrogen supply Q_{min}, almost all the beam ions are concentrated in the second peak near energy W_c. With increasing hydrogen supply, the intensity of the ion flux at $<W> = W_c$ decreases, while the ion flux at $<W> = W_o$ increases (Fig. 4.19). These data suggest that under the conditions investigated, the main component of slow ions in the anode region near the emission gap is formed by resonant charge exchange of hydrogen atoms with fast primary H⁻ ions formed at the cathode by interaction of plasma particles with the reduced work function surface.

Fig. 4.22 Energy
spectrum of H⁻ ions
emitted from a Penning
discharge. (Reproduced
from Dudnikov [1])

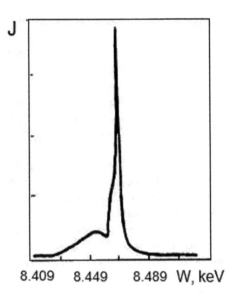

8.409   8.449   8.489 W, keV

   The energy spectrum of ion beams extracted from a Penning discharge is shown
in Fig. 4.22, where only ions of the first peak with $<W> = W_0$ are seen.
   In beams from discharges without cesium, the intensity of the second cathodic
peak was lower than the intensity of the first anodic peak by a factor of 10^3 even for
minimum hydrogen flow, and the distance between the peaks is less than
$eU_d = 400–500$ eV. Even a slight increase in hydrogen feed for discharges without
cesium, a second high-energy peak was not seen.
   In discharges with cesium, apparently only a small fraction of first peak ions is
formed by the usual plasma processes in the plasma volume. Under certain conditions, an appreciable contribution to the ion flux with energy $<W> = W_0$ (the first
peak) can be made by H⁻ ions from the anode walls, from anodic production, from
the emission slit bombarded by fast atoms formed by reflection of ions from the
cathode in the form of atoms, and from the destruction and charge-exchange with
fast negative ions in the plasma. The work function of these surfaces should decrease
due to cesium adsorption, and a potential distribution that is favorable for the emission of H⁻ ions should be created on these surfaces by ambipolar diffusion of the
plasma. It is possible that this process provides significant emission of H⁻ ions from
discharges in cells with large distances between cathode surfaces and the emission
slit, in which only a very small fraction of ions can come from the cathode to the
emission aperture without destruction.
   Ions of the first peak from discharges without noise reveal a complex energy
distribution pattern (Fig. 4.20). It is possible that some of the broadening is due to
resonant charge exchange in the extraction electric field. With the high field strength
in the extraction gap, the effective extent of the region with high charge-transfer
probability is small, so that the spectral broadening due to this effect is small despite
the high density of atomic hydrogen in the extraction gap. The appearance of a

small peak with energy lower than the energy of the main peak at 19–20 eV is due, apparently, to the charge exchange sequence H⁻ → H° → H⁻ in the extraction field. Similar transformations of the proton energy spectrum have been observed [18] in investigations of fast hydrogen atom diagnostic injectors.

For small separation between the central cathode plate and the emission slit (in these embodiments the gap d between the central cathode plate and the anode projections could be reduced to 1 mm) and low magnetic field strength, the spectra are almost identical throughout the beam cross-section. With an increase in the gap d to 3 mm and an increased magnetic field strength, spatial separation of ion groups with different energies was observed: the cathode group ions are strongly deflected by the magnetic field even when moving from cathode to emission slit.

From a helium discharge with added cesium, appreciable fluxes of H⁻ ions were detected, the intensity of which increased substantially when hydrogen was injected into the gas discharge cell between discharge pulses. Energy analysis of these fluxes revealed that in this case H⁻, ions are formed, only at the cathode, by electron capture by desorbed atomic hydrogen.

Other registration of energy spectrum of emitted H⁻ was presented in Ref. [19]. A schematic diagram of the experimental arrangement is shown in Fig. 4.23. The device is a cylindrical ion source of the bucket type with a diameter of 20 cm and a length of 18 cm. The stainless steel chamber of this bucket source is surrounded externally by ten columns of cobalt-samarium magnets ($B \sim 3.6$ kG) to form a line-cusp configuration for confinement of primary electrons and plasma. The end flange of the chamber is also equipped with magnets. These magnets, mounted in four rows, are connected to the ones on the cylindrical wall. The extraction side of the chamber is closed by the plasma electrode. The extraction slit in this plasma

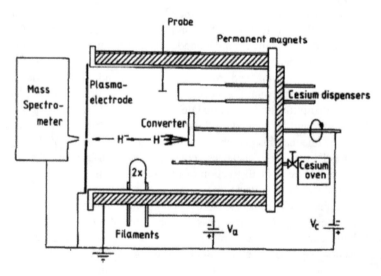

Fig. 4.23 Schematic diagram of the multicusp negative ion source. The source is not drawn to scale. The diameter of the source is 20 cm; its length is 18 cm

electrode has an area of 0.5×11 mm^2. The actual pressure in the source is not measured. However, in normal operation, the pressure indicated by a gauge located downstream of the mass spectrometer is 1×10^{-4} Torr. Calculations, based on the conductance of the holes in the plasma electrode, indicate that the pressure in the source is ~1.5×10^{-3} Torr. The base pressure of the vacuum system is 1×10^{-6} Torr. During operation a continuous flow of hydrogen is maintained. A steady-state hydrogen plasma is generated by electrons emitted from two tungsten filaments (diameter 0.5 mm). These filaments are biased at -70 V with respect to the source chamber wall (anode). Normal operation included an arc current of 4 A, a hydrogen pressure of 1.5×10^{-3} Torr, and a converter voltage of -200 V. Langmuir probe measurements indicate a plasma density of 3×10^{10}cm^{-3} and an electron temperature of 0.7 eV. They further show that the plasma potential is ~2.5 V positive with respect to the anode potential. At a discharge current of 4 A and a pressure of the order of 1×10^{-3} Torr the species ratio in this source was measured to be $H^+:H_2^+:H_3^+ = 4:35:61$. A circular water-cooled converter is inserted into the source chamber through a high-voltage insulator mounted on the end flange. The converter consists of an oxygen-free copper plate with a diameter of 5.7 cm and a thickness of 1 cm. On this copper plate, a polycrystalline molybdenum foil (thickness 0.25 mm) is brazed with Au-Ni brazing material. A tungsten polycrystal, a tungsten monocrystal (110), and a molybdenum monocrystal (110) are brazed directly onto the copper.

The spectra for the contaminated surfaces correspond with the H$^-$ peaks in Fig. 4.24. With respect to the measurement resulting in Fig. 4.24, the sensitivity along the energy scale is increased by a factor of 5 and along the current scale by a factor of 10. The discharge conditions for the energy spectra taken on contaminated surfaces are the same as in Fig. 4.25. For the case of clean surfaces, the arc current was reduced to 3 A. The surfaces are cleaned by sputtering with Ar through holes in the molybdenum foil. The dimensions of the crystals are $16 \times 19 \times 1$ mm^3. The crystals were mechanically polished to a grain size of 1 pm. Subsequently, the crystals are electrolytically polished. The converter can be rotated about its center. The extraction slit in the plasma electrode is situated off center. Only one type of converter material at a time is seen by the viewing system formed by the extraction slit and the entrance slit of the mass spectrometer. The latter is situated 3.8 mm behind the extraction slit. The distance from the converter to the extraction slit is approximately 9 cm. Typically the converter is biased at -200 V with respect to the source chamber wall.

Cesium is introduced into the discharge either via SAES Getters dispensers or via a molybdenum pipe connected to a cesium oven. The dispensers or the hole in the molybdenum pipe are placed such that a large part of the cesium vapor is directed toward the converter surface. In normal operation, with cesium added to the discharge, the Langmuir probe measurements indicate that the plasma density is 7×10^9 cm^{-3}. The electron temperature is 0.8 eV. The mass spectrometer, consisting of an electromagnet and a Faraday cup, has an entrance slit width of 0.12 mm. The maximum mass number that can be detected is about 100. The half-opening angle of the system formed by the extraction and the entrance slit is $2.8°$. The mass

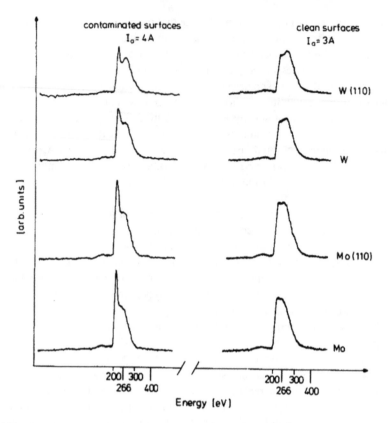

Fig. 4.24 Energy spectra of the H⁻ ions leaving the four converter surfaces for bot~ contaminated and clean surface conditions

spectrometer is operated at the same potential as the source chamber wall and the plasma electrode. For negative ions with the same mass-charge ratio, the mass spectrometer acts as an energy analyzer.

The source was first operated without cesium, with an arc voltage of 70 V, an arc current of 4 A, and a source pressure of ~1.5×10^{-3} Torr. The converter voltage was −200 V and the converter current was about 80 mA.

Figure 4.26 shows the measured H⁻ yield as a function of the energy of the H⁻ ions for the four converter surfaces. These H⁻ peaks are measured with a higher sensitivity. Both the H⁻ energy spectra from contaminated and clean surfaces are shown. The discharge conditions are the same as mentioned above. The energy of the detected H⁻ ions ranges roughly from eV_c where V_c is the converter voltage, to $2eV_c$. The energy spectra are mainly composed of two peaks, indicating two different populations of H⁻ ions. The high-energy part of the peak is explained by reflection (backscattering) of the hydrogen ions from the converter. Plasma ions (H^+, H_2^+, and H_3^+) are accelerated across the sheath with the molecular ions H_2^+ and H_3^+ dissociating at the converter surface. Upon leaving the surface, some of the hydrogen

Fig. 4.25 Energy spectra of the H⁻ ions leaving the four converter surfaces. The discharge current is 4 A; the discharge voltage is 70 V. The converter voltage is −200 V. The converter current is approximately 100 mA. The estimated pressure in the discharge chamber is 1.5 × 10⁻³ Torr. Cesium is added to the discharge by means of a cesium dispenser. The detection sensitivity is decreased by a factor of 500 with respect to the measurements without cesium, e.g., Fig. 4.24

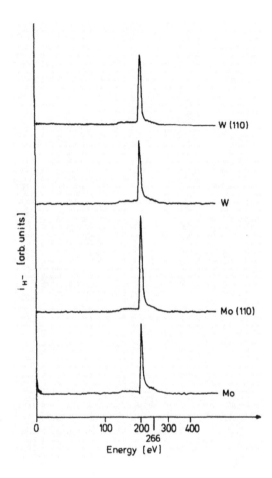

particles become negatively charged. These H⁻ ions are then accelerated back into the plasma across the sheath. Thus, at a converter voltage Vc, the detected H⁻ ions can have energies at 2eVc in the case of incident H⁺ ions, 1.5eVc in the case of H_2^+ ions, and 1.33eVc in the case of H_3^+ ions. Consequently, due to this species effect, the high-energy part of the H⁻ energy spectrum should consist of three superimposed groups. Combined with the energy dependence of H⁻ formation on backscattering, the peak resulting from H_3^+ reflection is the highest. The H⁺ reflection peak is not visible. The low-energy part of the H⁻ energy spectrum shows a peak at an energy slightly higher than eVc. This peak is due to ion-impact desorption of negative hydrogen ions from the surfaces. Comparison of the peak height of the low- and high-energy H⁻ peaks of the four contaminated converter surfaces (Fig. 4.26) show that they are different for tungsten and molybdenum. The low-energy desorption peak is 40% smaller in the case of tungsten compared to molybdenum. The high-energy backscattering peak is 25% smaller in the case of tungsten. The differences between the energy spectra of the polycrystals and the monocrystals are negligible.

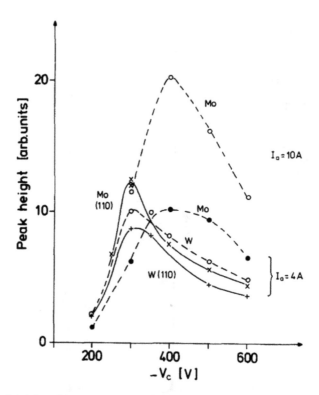

Fig. 4.26 Peak heights of the energy spectra of the H⁻ ions as a function of the converter voltage for an approximately constant cesium pressure. The discharge parameters are the same as in Fig. 4.25. One curve is shown with an arc current of 10 A

The right-hand side of Fig. 4.24 shows the energy spectra of the H⁻ ions leaving clean converter surfaces. The spectra are taken directly after cleaning of the surfaces by adding argon or xenon to the hydrogen discharge. The discharge parameters are the same as those for the energy spectra shown in the left-hand side of Fig. 4.24, except for the arc current, which is reduced to 3 A. During this cleaning procedure, the ion-impact desorption peak of the H⁻ energy spectrum is observed to increase. After a while, typically 5 min (depending on the amount of noble gas introduced), the desorption peak decreases again. It returns rapidly after the argon inlet is stopped. In about 10 min, the H⁻ energy spectra again look like those shown in the left-hand side of Fig. 4.24. However, refractory metals contain small concentrations of alkalis (mainly sodium) as an impurity. Small fractions of a monolayer (0.01) will have measurable influences on the H⁻ production on "clean" surfaces. Differences in peak height can be attributed to the influence of these dilute alkali impurities [20].

After the measurements with pure hydrogen, cesium was introduced into the discharge. Figure 4.25 shows energy spectra of H⁻ ions leaving the four converter surfaces in the presence of cesium. The arc voltage is 70 V, the arc current is 4 A, and the hydrogen pressure ~1.5 × 10^{-3} Torr. The converter voltage is −200 V with respect to the anode, and the converter current is approximately 100 mA. With cesium added to the discharge, the H⁻ peak height increased by a factor of 1000 compared to the case without cesium. Cesium adsorption on tungsten or molybdenum surfaces lowers the work function of the surface and hence increases the negative ionization probability. It can be seen from Fig. 4.25 that for that particular cesium pressure in the source, the monocrystalline (110) molybdenum converter surface yields a higher H⁻ production than the other surfaces. The H⁻ energy spectra in Fig. 4.25 also show that the H⁻ ions are mainly desorbed (sputtered) from the surfaces. This is in agreement with the interpretation of their experimental data by Bel'chenko et al. [4]. The backscattered fraction is small, less than 5%, compared to the desorbed fraction. The desorbed H⁻ fraction probably is high in the presence of cesium on the surface because of the large amount of hydrogen bound to the converter by the cesium.

Figure 4.26 shows the heights of the H⁻ peaks as a function of the converter voltage for one particular cesium inlet. The optimum H⁻ yield is found at a converter voltage of about −300 V, except for the polycrystalline molybdenum surface, where the optimum is about −400 V, independent of the arc current. It is observed that these optima shift to higher (absolute) converter voltages when the amount of cesium added to the discharge increases. The highest optimum voltage we measured was −600 V. This is in contradiction with earlier measurements where an optimum converter voltage of between −150 and −200 V was measured.

An explanation can be found in the enhanced sputtering yield when the incident Cs + ion energy is increased. When the cesium pressure rises, more cesium is adsorbed. Therefore, to maintain an optimum coverage of a partial monolayer, the sputtering rate must also increase, which is done by raising the converter voltage. Thus maximum H⁻ yield is expected when the converter voltage at the particular value of the cesium pressure is such that the cesium sputtering and the cesium adsorption dynamically establish a coverage of about half a monolayer.

The experiments considered have demonstrated unambiguously the dominant role of the interaction of the gas discharge plasma flow with electrode surfaces with reduced work function in the generation of intense negative ion fluxes.

The planotron electrode configuration has proved to be well suited for expression of the surface plasma negative ions generation mechanism. The following features characterize the process.

(a) Intense plasma particle fluxes provide effective hydrogen adsorption to the cathode central plate and efficient generation of a stream of fast sputtered and reflected hydrogen particles.

(b) Cesium is locked near the negative electrode by the discharge. Desorbed cesium ionizes and returns to the negative electrode in the form of ions. A number of features contribute to the preservation of the low work function surface necessary for the effective formation of negative ions, even with intense bombardment of the cathode by fast particles.

(c) The ability to ignite the discharge with reflexing electrons in the magnetic field
    at low gas density and with small gap between the cathode central plate and the
    emission slit and the rapid acceleration of formed negative ions to the emission
    gap in a narrow space-charge-compensated layer of near-cathode voltage drop
    contribute to the efficient transport of generated negative ions to the beam form-
    ing system.
(d) The orientation of the emission slit perpendicular to the magnetic field effi-
    ciently filters the negative ion flux from the electron flux and allows extraction
    and formation of H⁻ ion beams with intensity many times greater than previ-
    ously achievable.

A schematic of the processes involved in the surface-plasma method of negative
ion production is shown in Fig. 4.27. The working gas and cesium are supplied to
the gas discharge. The cathode is bombed by ions and atoms of working gas and
cesium. The streams of desorbed and reflected negative ions escape from the cathode, accelerated by the cathode drop. The negative ion flux is partially attenuated
when moving through the plasma. Fast negative ions are resonantly charge-exchange
with dissociated gas.

Determination of the dominant mechanism of negative ion formation has guided
the direction of further improvement of these sources. In subsequent source design

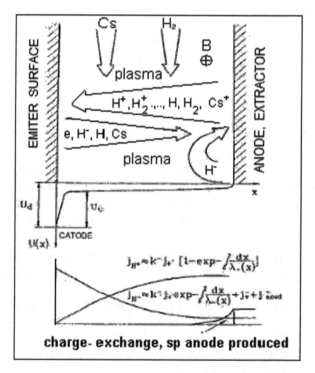

Fig. 4.27 Schematic of surface plasma processes involved in the production negative ions.
(Reproduced from Dudnikov [1])

versions, which could now reasonably be called surface plasma sources (SPS), measures were taken to optimize the conditions for surface plasma formation of negative ions and for more efficient transport of the formed ions to the extraction system.

Although in previous source designs, it was possible to generate stable negative H⁻ ion beams with intensity up to hundreds of milliamperes with an emission current density up to 1.5 A/cm², after the dominant role of the surface plasma negative ion formation mechanism became apparent, the shortcomings of these earlier sources also became apparent. Thin-walled, almost completely thermally insulated cathodes were easily heated by the discharge, and allowed the cesium to be rapidly released from tablets, consisting of a cesium chromate/titanium mixture, placed in internal cavities in the cathodes. At the same time, because of the low thermal inertia of the cathodes, the temperature regime of cesium adsorption changed very strongly during each discharge pulse, so that optimal cesium concentration, ensuring maximum electron capture efficiency at electron affinity levels of the sputtered and reflected particles, were most likely not established for any appreciable time. Low rigidity of cathode mounting did not allow reduction of the gap between the cathode central plate and the anode projections to less than 1–1.3 mm. When moving through such a plasma layer, a significant fraction of the ions formed would be destroyed.

4.4 Advanced Design Options for Surface Plasma Sources

In accordance with these considerations, in the following embodiments of SPS design, the central cathode plate was made from a massive molybdenum bar. A more rigid fixation of the cathode provided better heat dissipation and allowed reliably use of small gaps as small as 0.5 mm between the cathode central plate and the anode projections.

The source was located between the poles of the magnet with improved high-voltage insulation, the dimensions were reduced, the electrodes of the beam formation system were more carefully made, and a new small electromagnetic-pulsed valve was used [9]. A schematic drawing of such a source is shown in Fig. 4.28.

In the manufacture of the first source, precautions were taken to avoid accidental contamination of its parts by cesium. First, the extracted H⁻ ions were investigated with the source operating without cesium and with a massive cathode of size $4 \times 6 \times 14$ mm³ and a small gap of 0.5 mm between the cathode and the anode protrusions about the extraction slit. A decrease in the size of the source along the magnetic field made it possible to reduce the gap between the magnetic poles, so that the magnetic field strength was sufficient to ignite the discharge in pure hydrogen at $d = 0.5$ mm.

The discharge characteristics were quite similar to those previously obtained. For a discharge without cesium in this source, an H⁻ ion beam with current up to 15 mA was extracted through a 0.4×5-mm² emission slit; the corresponding emission

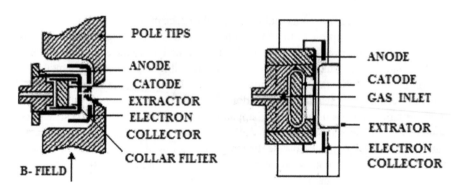

Fig. 4.28 Surface plasma source with massive cathode. (Reproduced from Dudnikov [1])

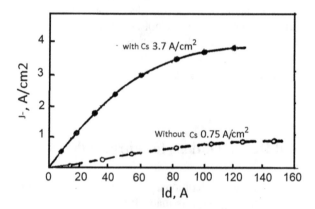

Fig. 4.29 Dependence of emission current density on discharge current for planotrons with and without cesium. (Reproduced from Dudnikov [1])

current density $J^- = 0.75$ A/cm^2 was significantly greater than the $J^- = 270$ mA/cm^2 previously produced The dependence of J^- on I_p is shown in Fig. 4.29.

In this case, the emission current density is significantly greater than the calculated capabilities for volume production of negative ion formation in a hydrogen plasma. With increasing gap, the maximum intensity of the extracted beam sharply decreased. Apparently, when heated to high temperature, alkali metals were extracted from the source components, which catalyzed the surface-plasma formation of negative ions.

Subsequent experiments investigated H$^-$ ion emission from discharges with the addition of cesium. Cesium was introduced to a source with cathode central plate dimensions $4 \times 6 \times 14$ mm^3, as before in the form of tablets of a mixture of cesium chromate with titanium placed in a cavity in the cathode central plate under spot-welded cathode side shields.

The cathode was strongly heated when the discharge was ignited without cesium. But after release of cesium and reducing the discharge voltage from $U_d = 400$ V to $U_d = 100$ V, the discharge power expended at the cathode decreases and the cathode

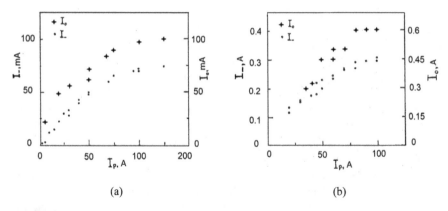

(a)   (b)

Fig. 4.30 H⁻ ion beam collector current I^- and total current in the extraction gap I_o as a function of discharge current I_p. Extraction slit dimensions (**a**) 0.4×5 mm² and (**b**) 1×10 mm². (Reproduced from Dudnikov [1])

heating decreases. In typical operation, these sources worked quite stably. To obtain beams with maximum intensity, careful optimization of the gas supply and the cesium release were required. The dependence of the H⁻ ion beam collector current I^- and the total current in the extraction gap I_o on discharge current I_p under optimized conditions, for a source with cathode dimensions $4 \times 6 \times 14$ mm³, central cathode plate area 2.5 cm², and emission slits dimensions 0.4×5 mm² and 1×10 mm², are shown in Fig. 4.30.

For the initial linear part of the beam current dependence on discharge current, $I^-(I_p)$, the emission current density reaches 10–12% of the discharge current density. In the saturation region of this dependence for discharge current greater than 100 A, the emission current density reached was 3.7 A/cm², as shown in Fig. 4.30.

The H⁻ ion beam intensity increases approximately in proportion to the area of the emission slit. The current of accompanying electrons is 13% of the negative ion current for slit area 0.4×5 mm² and is equal to the negative ion current for slit area 1×10 mm². At the achieved values of emission current density, space-charge effects become very significant. To extract a beam of H⁻ ions with current 0.3 A through an emission slit of area 1×10 mm², it was necessary to increase the extraction voltage to 30 kV.

An H⁻ ion beam of current 0.2 A for slit dimensions 1×10 mm² was obtained at a discharge current of 50 A and a discharge voltage of 100 V. These parameters correspond to the energy price of the H⁻ ion in this, source $P = 25$ keV/ion at an emission density of 2 A/cm². Taking into account the fact that in this case the area of the emission aperture is 25 times smaller than the area of the central plate of the cathode, and the intensity of the extracted beam increases in proportion to the area of the emission aperture, these results establish a precondition for bringing the H⁻ ion generation efficiency in the SPS to 1 keV/ion, comparable with the efficiency of the best positive ion sources.

For thermonuclear research, stationary and quasi-stationary beams of H⁻ ions are needed. With this application in mind, the production of H⁻ ion beams from these sources was tested at large pulse duration. In this case, the voltage across the anode-cathode gap was fed from the 50-Hz electric power network through a step-up transformer switched to the source by a thyristor switch. In this case, bell-shaped pulses of discharge current with duration up to 8 msec and amplitude of up to 50 A were obtained. In this regime, the dependence of the H⁻ ion beam current on discharge current remained practically unchanged. For long pulses, the optimal cesium feeding regime was even more stable than for short pulses.

Emission of H⁻ ions from discharges with Penning electrode configuration when the cathode central plate was removed was investigated. The cathodes of the Penning cell were side shields. Upon removing the central cathode plate, an anode window was formed of size 6×16 mm^2. Cesium tablets were located in foil boxes on the side shields of the cathodes. A more rigid cathode attachment allowed to study the emission of ions in forced modes. In commonly used discharges with electrons reflexing in the magnetic field, the discharge voltage and current fluctuate at frequencies in the range 10^5–10^7 Hz with amplitude up to tens of percent of the steady value. With decreased magnetic field and increased gas density, the level of fluctuations decreases, and for some values of these parameters disappears, passing sometimes through a stage of coherent oscillations. The greater the magnetic field strength, the greater the gas density needed to stabilize the fluctuations. The noise level and its dependence on magnetic field and gas density are affected by the geometric dimensions and configuration of the electrodes, discharge current, average discharge voltage, cesium concentration, and many other factors.

In planotron gas discharge cells with small anode-cathode gaps and small transverse size of the reflexing region, a high magnetic field is required to ignite the discharge, and discharges without noise are obtained at such high gas density that negative ions formed at the cathode do not reach the emission slit without destruction.

At minimum gas density, the beam current and current density are maximum, but fluctuations in the discharge greatly increase the transverse energy spread and the emittance, and greatly reduce the brightness.

In the resulting cells with Penning electrode configuration, because of the large width of the window, high-current glow discharges were ignited at low magnetic field strengths, $B > 0.03$ T. Thanks to this, discharges without noise were obtained with a low hydrogen density, which ensures completely normal functioning of the source. Without cesium, discharges without noise were obtained without any effort. When cesium was supplied to produce discharges without noise, some optimization was necessary. Without cesium the secondary electron ion emission coefficient is low, and current transverse to the magnetic field is carried by plasma electrons. With cesium the secondary electron ion emission coefficient is high, and current transverse to the magnetic field is carried by cathode electrons.

Under these conditions, H⁻ ion beams were obtained from Penning cells with the same discharge current, at beam current approximately three times less than from planotron cells. But beams from discharges without noise were formed much better, had smaller divergence, and higher collector current density, despite the lower intensity. The H⁻ ion current increased proportionally to the discharge current up to 150 A, so that beams with current up to 150 mA were extracted through a 1×10-mm² emission slit, which is a factor of two less than the maximum current from a planotron cell. When the magnetic field was decreased, the current in the extraction gap increased, but due to the thickening of the walls of the emission slit, it was reduced.

Assuming that only part of the cathode working area makes an effective contribution to the generation of H⁻ ions, we decided to investigate the dependence of ion emission on the width of the anode window. In this case, cesium tablets were placed in a thin-walled container affixed to the upper anodic protuberance, and served as an element of the anode window. With this design it was possible to ensure the normal release of cesium and the stable burning of high-current glow discharges. With the existing magnetic field strength B up to 0.3 Tesla, high-current glow discharges were ignited steadily with a window width greater than 2.7 mm. Reducing the field or width of the window, fluctuations in the flow of cesium caused transitions of the high-current glow discharge into a high-voltage discharge with limited current (sometimes several times per pulse).

Discharges without noise were produced in a cell with a 3×16-mm² window at a field strength $B \sim 0.08$–0.12 Tesla and an acceptable hydrogen supply. The ratio of beam current to discharge current was only a factor of 1.5 less than for a planotron cell in its optimal regime. At a discharge current of 150 A, stable, well-formed H⁻ ion beams with current up to 0.3 A were extracted through the 1×10-mm² emission slit. The discharge current density at the cathode in this case was significantly higher than the current density at the cathode of a planotron cell.

These results established prerequisites for obtaining intense high-quality beams with high brightness. Since brightness is the most important characteristic for sources used in accelerator technology, we subsequently decided to use such gas discharge cell configurations in SPSs for accelerators.

Thus two directions were distinguished with slightly different tasks:

- Production of intense beams at maximum intensity with acceptable ion-optical characteristics
- Production of high-quality beams with high brightness at suitable intensity and efficiency

To confirm the veracity of the identified patterns, a central plate of dimensions $4 \times 6 \times 35$ mm³ and cathode of area 5 cm² was placed in a source with large emission slit in the same housing. A schematic drawing and photograph of this source is shown in Fig. 4.31 [21–23]. Cesium tablets were located in the cavity of the cathode central plate and in a removable upper anode protrusion.

Discharges with the same characteristics were stable supported in this substantially enlarged cell. The emissive properties of this source, with emission slit dimensions 0.9×30 mm², are shown in Fig. 4.32. Under optimum conditions, H⁻ ion beams with current up to 0.88 A, for a discharge current of 450 A, were extracted in a perfectly reproducible way. Thus the possibility of obtaining beams with yet higher intensity seemed to be adequately supported [24].

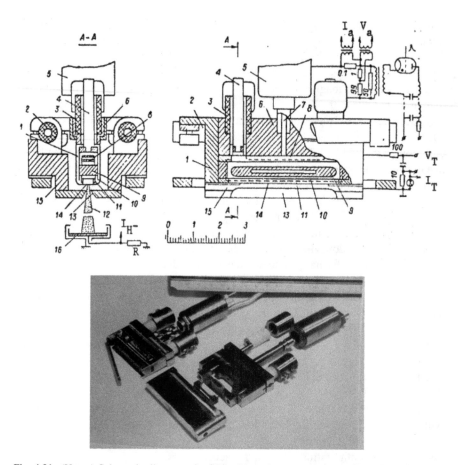

Fig. 4.31 (Upper) Schematic diagram of a SPS with a planotron plasma cell configuration and (lower) photograph of the partially disassembled source. (Reproduced from Dudnikov [1]). (1) Source body plasma cell, (2) high-voltage insulators, (3) cathode insulators, (4) cathode holders, (5) pulsed hydrogen supply valve, (6) anode insert, (7) channel for hydrogen supply, (8) detachable anode protrusion with cavity for cesium, (9) central cathode plate, (10) cathode side shields, (11) anode protrusion walls, (12) H⁻ ion beam, (13) extractor plates, (14) emission slit, (15) magnet pole tips

It turned out to be much harder to study the emission of H⁻ ions from sources with a larger emission slit width. As the width of a single-emission slit increases, the flow of accompanying electrons increases greatly. In the first studies of H⁻ ion emission from discharges in crossed fields, we attempted to extract ions through sectioned emission slits made in the form of mesh strips, as is effectively done to form beams in charge-exchange sources, but as a rule in these cases, the emission density is smaller than in single slits.

As the width of the emission slit increases, with a single-aperture extractor that is not sectioned, the effective extension of the extraction gap increases, so that a substantial increase in extraction voltage is necessary to obtain the previous H⁻ ion

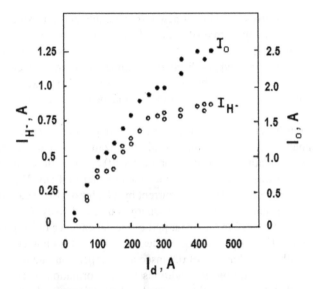

Fig. 4.32 H⁻ ion beam collector current I^- and total extractor current I_o as a function of discharge current I_d. (Reproduced from Dudnikov [1])

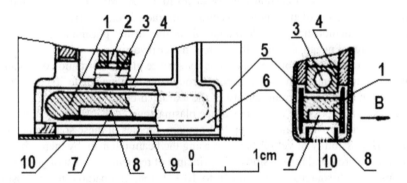

Fig. 4.33 Schematic diagram of an SPS with sectioned emission slit. (Reproduced from Dudnikov [1]). (1) Central cathode plate, (2) channel for hydrogen injection, (3) cesium container in the anode protuberance, (4 and 5) body of the gas discharge chamber, (6) cathode side shields, (7) cavity for cesium pellets, (8) expander in anode protuberance, (9 and 10) sectioned emission slit

emission current density. The emission aperture of the previous source was blocked by a grid of strips so that the extent along the magnetic field of the slits so formed was small. The main experiments were carried out with emission slit dimensions 2.8×10 mm², sectioned by a jalousie (a shutter made of a row of slats) into four identical emission slits. The cross-section of the strips was 0.1×0.5 mm². The molybdenum strips were fixed in the mandrel at one end and could expand freely without deformation when heated. The design of this device is shown schematically in Fig. 4.33.

This source was investigated on a new test-stand adapted for extraction voltage up to 70 kV. Without any source structural changes at all, and after improving the high-voltage insulators, carefully screening insulator contacts with electrodes, carefully cleaning the housing and gas discharge chamber and extraction electrodes, and with a minimum gap between the emission port wall and the extractor electrode edges of 3 mm, we were able to extract H^- ion beams at extractor voltage up to 60 kV.

Stable H^- ion beams with intensity up to 0.8 A per cm of emission slit were obtained comparatively easily. With careful optimization, beams with an intensity of about 1 A were obtained. At a collector located 10 cm from the emission gap, the beam cross-section was close to circular with a diameter of 4 cm. The total current in the extractor gap exceeded the beam current by a factor of two. Again in this case, the ratio of ion current density J^- to discharge current density J_p at the cathode reached a value $J^-/J_p = 0.12$ in the linear part of the dependence of ion current H^- on discharge current. The strips used quite steadily survived prolonged operation in forced modes with pulse duration of 0.8 msec and a repetition rate up to 3 Hz.

The empirical optimization of conditions for the formation of H^- ions in discharges with added cesium and the accumulated experience with such discharges made it possible to increase the intensity of H^- ion beams up to hundreds of milliamperes at an emission current density of up to 1.5 A/cm^2. Subsequent studies, particularly a study of the energy spectra of extracted ions carried out in 1972, convincingly demonstrated the dominant role of the surface plasma mechanism for the formation of negative ions in the sources under investigation.

Focused optimization of the conditions for surface plasma formation of negative ions and improvement in ion transport through the plasma made it possible to increase the emission current density of the extracted beams up to 3.7 A/cm^2. On this basis, it was possible to create surface plasma sources that stably generate H^- ion beams with a pulsed intensity up to 1 ampere. Studies of these sources demonstrated the possibility of yet further increasing the beam intensity by increasing the emission slit area, and the possibility-in-principle of reducing the cost of generating H^- ions in such sources at least down to 10 keV/ion, comparable to the cost of energy for obtaining positive ions in the best sources [25–33].

The realization of effective H^- ion generation in discharges without fluctuations in the plasma parameters has created the prerequisites for construction of sources of high-quality H^- ion beams with high brightness.

The resulting formation efficiency of negative ions and the produced beam intensity are determined by the efficiency of formation and acceleration of bombarding ions in the plasma, the likely emission of negative ions from the surface of electrodes when the surface is bombarded with plasma particles, the probability of destruction of negative ions during their transport through the plasma to the forming system, and the efficiency of this system. In turn, the emission properties of electrode surfaces are strongly influenced by processes accompanying the interaction of the plasma with the electrodes, and the properties of the plasma depend on the emission properties of the electrodes.

4.5   Emissive Properties of Electrodes in Surface Plasma Source Discharges

From general ideas about mechanisms for supporting high-current glow discharges with reflexing electron in a magnetic field, it follows that for fixed cold-cathode secondary electron emission coefficient γ (due to bombarding ions), electrons emitted into the plasma from the cathode and accelerated to energy eU_d by the cathode voltage drop, on average, must create in the plasma $1/\gamma$ ions that fall to the cathode. Faced with the dependence of γ on bombarding ion energy and the dependence of the ion generation rate on electron energy, a definite value of the near-cathode potential drop U_d is established. The potential distribution in the plasma must ensure that electrons are collected by the anode across the magnetic field, consistent with the loss of ions to the cathode and anode. When electron diffusion is limited by the magnetic field, electron transport mechanisms must be dominated by the electric field. The potential distribution created must prevent ions from leaving to the anode and increase the efficiency of ion collection by negative electrodes. For small transverse dimensions of the gas discharge cell, the potential drop in the plasma must be small in comparison with the cathode drop, concentrated in a thin layer near the cathode. The thickness of the layer depends on the ion current density at the cathode, the magnitude of the near-cathode potential drop, and the negative ion current density and ion mass, in accordance with the Child-Langmuir law. If the cathode is separated from the anode by a gap with insulated ends with a high magnetic field, the voltage drop in the plasma can exceed the cathode drop. Thus the discharge burning voltage should be determined to a large extent by the emission properties of the cathode. A significant decrease in discharge voltage when cesium is added indicates a significant increase in the secondary electron emission coefficient. The plasma density could be determined from the cathode current density, but to estimate it, information on the secondary emission coefficient of electrons from the cathode in SPS discharges is also needed.

Unfortunately, information on the emission properties of electrodes in discharges with reflexing electrons, even with a supply of pure gas, is fragmentary and inconsistent [4, 15, 34, 35].

Information on the effect of adsorption of alkali metals on secondary emission by ion bombardment is quite extensive [36], but data on secondary emission that we are interested in bombardment by hydrogen particles of energy 10^2 eV are very vague [37, 38]. The possible effect of the adsorption of cesium on the emission properties of a molybdenum surface is illustrated by data shown in Fig. 4.34 [26]. The authors of this work studied the dependence of the electron secondary emission coefficient γ and the negative ion emission coefficient K^- on cesium surface concentration θ, upon bombardment of a cesiated molybdenum surface of by beams of Ar^+ argon ions. The increase in γ and K^- is characteristic for adsorption of cesium. Note the presence of a maximum near the minimum value of work function.

Naturally, when the discharge is ignited, the emission characteristics change radically, but the information given can serve as a guide.

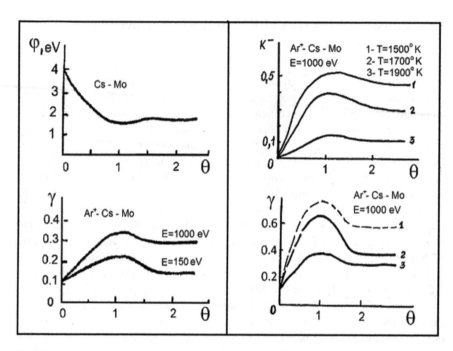

Fig. 4.34 Variation with surface concentration θ of cesium on molybdenum (in units of the optimal concentration) of (upper left) φ, work function of the cesiated surface; (lower left) γ, coefficient of electron emission by ion bombardment, at two different ion energies; (upper right) K^-, negative ion emission coefficient for several heat treatment regimes; and (lower right) γ, coefficient of electron emission by ion bombardment for several heat treatment regimes. (Reproduced from Ortykov et al. [36])

Because of the strong dependence of the discharge voltage U_d on the emission properties of the electrodes, the discharge voltage can be used as a parameter characterizing the electrode emission properties. Since discharges with practically identical characteristics are obtained in cells with different electrode configurations, providing conditions for the efficient use of fast electron energy for ionization in the plasma, it can be assumed that the same discharge voltages are obtained in different cells with similar electrode emission properties.

The conditions for negative H^- ion generation are homogeneous over the entire surface of the central plate of the planotron cathode. For an extracted beam emission current density J^-, power $J_d U_d$ (where J_d is the discharge current density at the cathode and U_d is the discharge voltage drop) is expended on ion generation in the discharge region per unit area. The energy cost P^- for the formation of an H^- ion in the source is determined by the ratio of the product of discharge current density J_d and discharge voltage U_d to the emission current density J^-:

$$P^- = J_d U_d / J^- = U_d / \kappa.$$

Fig. 4.35 $\kappa = J^-/J_d$ (ratio of H$^-$ extracted ion beam current density J^- to discharge current density J_d at the cathode) and K^- (estimated coefficient of secondary emission of H$^-$ ions from the planotron cathode) as a function of discharge voltage U_d, for cells with planotron geometry. (Reproduced from Dudnikov [1])

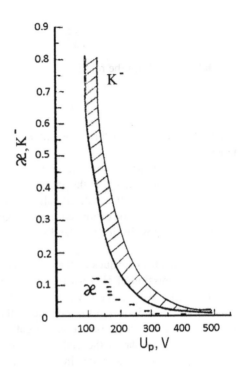

The values of $\kappa = J^-/J_d$, determined from the linear part of the $I^-(I_d)$ dependence and corresponding to different planotron discharge voltages under optimized conditions, are shown in Fig. 4.35. As the cesium feed increases, κ increases from 0.01 at $U_d = 500$ V to 0.1–0.12 at $U_d = 100$ V, and simultaneously P^- decreases from 50 keV/ion to 1 keV/ion. In most of the implemented SPS variants, the area S_e of the emission slit is smaller than the area S_c of the central cathode plate, so only a small fraction of the generated H$^-$ ion flux is extracted, and the energy cost for obtaining an H$^-$ ion in the extracted beam turns out to be increased by a factor of S_c/S_e. A number of SPS designs (with closure of electron drift in one plane [2, 39], without closed electron drift [40], and with geometric focusing) should allow almost complete recovery of the entire H$^-$ ion flux reaching the emission surface of the beam formation system, so that a decrease in the energy costs to $P^- \sim 1$ keV/ion is also possible when ions are generated by a discharges with cold electrodes.

The ratio $P^- = U_d/\kappa$ takes into account the energy expenditure for the emission of electrons P_e, the cost of generating positive ions P^+ and their acceleration to bombardment energy E^+, the probability of emitting H$^-$ ions in a single bombardment event K^-, and destruction of H$^-$ ions during transport to the emitting surface of the beam formation system and during beam formation.

For plasma densities that are not too high, the probability w^- of the passage of H$^-$ ions through the system without destruction can be made close to unity, so that a general expression for the energy cost P^- of generating H$^-$ ions in the extracted beam can be written:

$$P^- = \left(P^+ + E^+ \right) S_c / \left(K^- w^- S_e \right)$$

which has as its limit the relationship

$$P^-_{\min} = \left(P^+_{\min} + E^+ \right) / K^-,$$

which depends only on the values of K^- and the energy of the bombarding particles E^+.

For small gap width between the planotron cathode central plate and the emission slit, the main fraction of the extracted ion flux consists of ions emitted from the cathode. In this case, the realized values of negative ion secondary emission coefficient $K^- = J^-/J^+$ can be estimated from the known values $\kappa = J^-/J_d$, if we determine the fraction of positive ions in the discharge current at the cathode using $J^+/J_d = 1/(1 + \gamma + K^-)$.

First, estimates of values of the ratio J^+/J_p were obtained early in our investigation of the emission of H$^-$ ions from the considered discharges. The positive ion current density at the cathode was estimated from their emission through a slit with dimensions 0.3×5 mm^2 in the cathode of the reversed planotron. In this case, the walls of the molybdenum box served as cathode, and a plate-like anode was located in the center, similar to the cathode of the direct planotron. Discharges in the reversed planotron were quite similar to discharges in the cells of other configurations, but at that time in the reversed planotron, it was not possible to sustain a stable discharge with current $I_d > 30$ A because of switching to arc mode. From this cell, positive ion bunches with an emission current density up to 7 A/cm^2 were extracted. With increased cesium release from tablets placed in the anode cavity, the ratio J_d/J^+ increased from values close to unity at $U_d = 500$ V to 2–3 at $U_d = 150$ V.

Later [41] we attempted to estimate the secondary emission coefficient of electrons from the cathode γ from the ratio of power released at the cathode W_c to power released at the anode W_a of a cell with Penning configuration and with electrodes cooled by an air flow as used in SPS accelerator embodiments. A schematic of the plasma cell is shown in Fig. 4.36.

We measured the average power W_c carried away by cooling air from the cathode, and from the anode W_a. The source was operated in a pulsed mode with pulse duration 200 μsec and repetition rate 10–50 Hz. The release of cesium was regulated by varying the average discharge power. The air temperature at the outlet of the cooling system was monitored by thermocouples. To calibrate the cooling channel, heat-insulated calibrated electric heaters were installed. The air flow was controlled by the differential pressure on the cooling line. The obtained ratios $\alpha = W_c/W_a$ for different discharge voltages are shown in Fig. 4.37.

As the cesium release increases, the ratio $\alpha = W_c/W_a$ decreases from $\alpha = 5.5$–6 at $U_d = 400$ V to $\alpha = 2.5$ at $U_d = 100$ V. The current I^+ of positive ions to the cathode dissipates a power at the cathode of $U_d I^+$, and due to neutralization of ions, radiation, excited particles, etc., there is additional power PI^+ proportional to the ion current, but not dependent on their generation efficiency in the plasma at different discharge

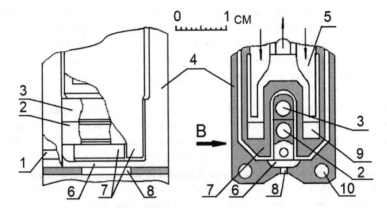

Fig. 4.36 Diagram of a gas-cooled plasma cell with Penning electrode configuration. (Reproduced from Dudnikov [1]). (1) Hydrogen supply channel, (2) cesium container, (3) anode bridge cooling channel, (4) plasma chamber body, (5) cathode-cooling channel, (6) expander, (7) cathode, (8) emission gap, (9) magnetic inserts, (10) cooling channels for anode wall with emission slit

Fig. 4.37   Ratio $\alpha = W_c/W_a$ of average power input to the cathode (W_c) to the average power input to the anode (W_a) as a function of discharge voltage of the gas discharge cell. (Reproduced from Dudnikov [1])

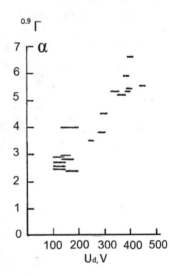

voltages. Electrons emitted from the cathode introduce the power $I_e U_d = U_d \gamma I^+$ into the plasma, from which a part PI^+ is returned to the cathode, and the remaining $W_a = (\gamma U_d - P)I^+$ falls on the anode. Under these assumptions, the effective coefficient of secondary electron emission Γ is expressed in terms of the quantities introduced by the relation

$$\gamma = 1/\alpha + (1 + 1/\alpha)P/U_d.$$

Unfortunately, this ratio includes the quantity P, information about which is very uncertain. It can be assumed that at high-discharge voltages, the relative roles of particle reflection from the cathode, radiation from excited and fast atoms, and ion neutralization in the heating of cathodes are relatively small [42]. In this case $U_d = 400$–500 V and the values $\gamma = 0.2$–0.3 should correspond. At low-discharge voltage, the uncertainty of the coefficient P is transferred to the obtained values of γ. Assuming $P \sim U_d$ we obtain an estimate for the secondary emission coefficient $\gamma \sim 3$ for $U_d = 100$ V.

Direct measurements of the ratio of discharge current density to positive ion current density to the cathode were performed in a gas discharge cell with Penning electrode configuration, a schematic circuit of which is shown in Fig. 4.38 [43].

Special measures made it possible to eliminate parasitic discharges outside the anode window. The positive ion flux accelerated in the cathode potential drop and emerging through a narrow slit in the cathode was recorded. The slits used in the molybdenum cathode were of width 0.01–0.05 mm with sharp edges. The dimensions of the slits were measured by a microscope and by diffraction of laser radiation. A grid with mesh size 0.2 mm and transparency 80% was placed between the slit and the collector, on which an adjustable retarding potential was applied. The volt-ampere characteristics of the collector revealed a nearly monochromatic flow of positive ions with energy less than eU_d for 20–30 eV and with an energy spread ~10–30 eV. The ratio of discharge current to collector ion current was monitored oscillographically when the discharge was powered by half-sinusoidal voltage pulses. Measurements were carried out at discharge currents for which the cathode

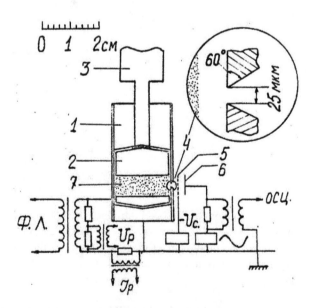

Fig. 4.38 Schematic of the setup for determining the ratio of the Penning cell cathode current density J_p to the positive ion current density at the cathode J^+. (Reproduced from Dudnikov [1]). (1) Gas discharge cell body with molybdenum cathodes, (2) anode, (3) valve-supplying hydrogen, (4) narrow slit in cathode, (5) analyzer grid, (6) collector, (7) anode window (7×12 mm^2) with gas discharge plasma

Fig. 4.39 Ratio of discharge current density at the cathode J_d to positive ion current density at the cathode J^+ in a Penning cell at different discharge voltages U_d. (Reproduced from Dudnikov [1])

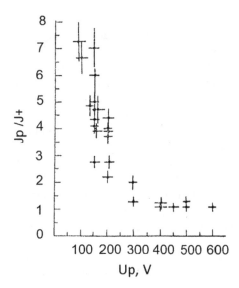

layer width is greater than the gap width. At high-discharge current, the ratio of ion current to discharge current sharply decreased, apparently due to curvature of the plasma boundary and defocusing of the ion flux entering the gap. The value of J_d/J^+ for different discharge voltages, recalculated taking into account the slit size and the discharge current distribution, is shown in Fig. 4.39. This time containers of metallic cesium were placed in the anode cavity, covered from the discharge plasma. With increasing cesium release, the ratio of current densities increased from close to unity for $U_d = 600$–300 V up to $J_d/J^+ = (1 + \gamma + K^-) = 7 - 8$ for $U_d = 100$ V.

Uncertainty in the results of individual measurements is due to discharge voltage changes during the pulse, voltage measurement uncertainty, measurement uncertainty of the current and gap size, and some uncertainty in the current density distribution. The scatter of the results of individual measurement cycles characterizes the irreproducibility of the experimental conditions with changes in the magnetic field, the supply of hydrogen, the adsorption of gases, changes in the discharge current, and so on.

According to Ref. [44], the photoemission of electrons increases greatly with decreasing work function of the oxidized tungsten because of cesium adsorption near the minimum of the work function. Because of this, the contribution of photoemission to the effective secondary emission coefficient can become significant when cesium is adsorbed, whereas under normal conditions the contribution of the photoelectric effect is small [45]. The contribution of the photoelectric effect to electron emission in discharges with cesium has been studied [46]. With decreasing work function due to adsorption of cesium, the probability of photoemission increases by a factor of 2.5–3.5, while for a thick cesium film, the probability of photoemission decreases to 60–70% of the initial value.

If the emission properties of the electrodes in cells with planotron and Penning electrode configurations differ only slightly for identical discharge voltages, it follows from these data that with an increasing cesium concentration at the electrodes, the coefficient of secondary emission of H^- ions should increase from $K^- < 0.01$ for $U_d = 500$ V to $K^- \sim 0.6$–0.8 for $U_d = 100$ V (Fig. 4.35).

From the J_d/J^+ values, it is possible to estimate the energy expenditure in the plasma on the formation of positive ions incident on the cathode, $P^+ = eU_d(J_d/J^+ - 1)$. From data shown in Fig. 4.39, it follows that as the discharge voltage decreases, the energy cost increases from $P^+ = 0.1$ keV/ion at $U_d = 300$ V to $P^+ = 0.5$ keV/ion at $U_d = 100$ eV. Such a strong increase in energy cost is due to a change in the plasma electron energy spectrum. In the relatively small volume V of discharge cells, a significant current of fast electrons is injected from the cathodes and accelerated by the cathode potential drop to an energy $W_e = eU_d$. Apparently the retention time of fast electrons in the reflexing region is comparable with the time for them to slow down to the Maxwellian electron energy of several electron volts. Owing to these features, the density of fast electrons $n_{eo} \sim I_e\tau/V$ (where I_e is the electron current, τ is the slowing down time, and V is the discharge cell volume) for small neutral particle density can be comparable to the plasma electron density n_e, and the electron energy distribution function must decrease slightly to energies $W_e \sim eU_d$. It is the fast electrons that should play the main role in the generation of positive ions. The ionization cross-section for molecular hydrogen reaches a maximum at an electron energy of 80 eV, whereas the excitation cross-section reaches the same value at 20 eV electron energy. Electrons with energy $W_e \sim 300$ eV can expend about half of the energy for ionization, whereas electrons with energy near the ionization threshold of 20–30 eV expend their energy on excitation and transmission to Maxwellian electrons due to collisions and collective processes. This inhibition becomes more and more significant, so that as the discharge voltage decreases, an ever smaller fraction of the power entering the discharge is consumed by ionization. It is possible that part of the increase in energy cost is due to the escape to the anode of electrons with relatively high energy, and the escape to the anode of a portion of the positive ions formed. In discharges with cold electrodes used both for plasma generation and as emitters of negative ions, it is difficult to simultaneously optimize the emission properties, the energies of bombarding particles, and the conditions for generation of bombarding particles.

In SPSs with separated functions, with independent optimization of conditions for implementation of these processes, one can count on the conservation of energy costs at a level $P^+_{min} = 0.1$ keV/ion for $K^- \sim 0.8$, so it seems possible to reduce the cost of H^- ion formation to $P^-_{min} = (P^+_{min} + E^+)/K^- = 0.25$ keV/ion.

4.6   Plasma Parameters and Negative Ion Destruction in the Plasma

Data on the emission properties of electrodes allow us to make more definite conclusions about the plasma parameters and the effect of H^- ion destruction during transport from the emitting electrode to the beam formation system. Information about measured values of the ratio $J_d/J^+ = (1 + \gamma + K^-)$ at different discharge voltages

(Fig. 4.39) allows one to estimate the positive ion current density at the cathode under different SPS operating conditions. For judgments about the plasma density, information is needed on the flow velocity of the ions v^+ at the boundary between the quasineutral plasma and the cathode layer.

In a quasineutral plasma, near the electrodes at negative potential, electric fields should be created that convert the total plasma pressure into an ion momentum flux falling on the electrode [23, 47–49]. Without a magnetic field and when ions are collected along the magnetic field, these fields are localized in the near-electrode region with dimensions of order the mean free path. When current flows through the plasma across the magnetic field, an electric field of considerable strength may exist in the plasma column, and ions can be accelerated to the negative electrode as in plasma accelerators with closed electron drift (Hall accelerators), etc., with extended acceleration zone [50]. Note that it is advantageous in the SPS to provide the greatest possible intensity of bombarding energetic particle flux at possibly lower plasma density in the space between the emitting electrode and the beam formation system.

Unfortunately, the available information about streaming ion velocity v^+ (energy of directional motion $W^+ = \frac{1}{2}Mv^{+2}$) in discharges is quite vague. At high fast electron density in the plasma, the total charged particle pressure, which determines the directed ion velocity, must be high and at a low Maxwellian electron temperature as recorded by probes. If we assume that a significant fraction of the energy spread of ions emerging through a narrow gap in the Penning cell cathode is due to their acceleration in a quasineutral plasma, an estimate is obtained for the energy of the ion directed motion at the boundaries of the near-electrode layer $W^+ \sim 10$ eV.

If the ions experience almost no collisions with neutral particles when moving toward a negative electrode, the plasma density at a discharge current density J_p is determined by the relation:

$$n_e = J^+ / v^+ = J_d / v^+ \left(1 + \gamma + K^-\right),$$

and the effective thickness d of the plasma layer is given by

$$\langle n_e d \rangle = J_d d / v^+ \left(1 + \gamma + K^-\right).$$

The cross-section for resonant charge exchange of protons on atomic hydrogen, which determines the transfer of ion momentum to neutral particles, is $\sigma_g = 5 \times 10^{-15}$ cm^2 for $W^+ = 1$–10 eV. If the total thickness of the plasma layer is $\langle n_e d \rangle > 1/\sigma_g = 2 \times 10^{14}$cm^{-2}, the interaction becomes significant, and the rate of escape of ions from the generation region will be determined by diffusion, and not by free acceleration in electric fields. In this regard, with an increase of $\langle n_e d \rangle$ above the critical value of 2×10^{14} cm^{-2} due to an increase in gas density n_g or gap width d, the effective thickness of the plasma layer can increase very rapidly, and thus the destruction of H$^-$ ions during transport through the plasma must depend on $\langle n_g d \rangle$ very strongly.

The destruction of H⁻ ions during their motion through plasma and gas limits the intensity of the beams obtained from plasma sources. In SPSs, the role of these processes is significantly weakened by the rapid passage of ions through the dense plasma layer. In the SPS, primary positive ions are drawn onto the surface of the emitting electrode by a voltage applied between the plasma and the emitting electrode. With the same voltage, the H⁻ ions emitted from the surface are accelerated to the beam formation system up to an energy $W = eU_d$. Under such conditions, the spatial charge of negative ions should not reduce the efficiency of their emission from the electrode to current densities comparable to the primary ion current density, i.e., to $K^- \sim 1$. To reduce negative ion destruction during transport to the beam formation system and during beam formation, the thickness of the plasma and gas layer in their path should be as small as possible. Primary ions should be generated with a small gap d between the emitter and the beam formation system and at a minimum hydrogen density. The use of a planotron plasma cell configuration proved successful in this sense. In this case, the decrease in the gap between the central cathode plate by the emitter and the emission slit is limited to a level $d = 0.5$ mm by the electron Larmor radius for energy $W_e = eU_d \sim 100$ eV. Because of the good magnetic confinement of electrons in the reflex region, high-current discharges ignite at an initial hydrogen pressure $p_2 = 0.1$ Torr even with a cold cathode.

Let us estimate the achievable density of the H⁻ ion flux and their destruction in the plasma. The ion flux H⁻ of current density J^- obtained from the surface of the emitting electrode bombarded by a stream of ions with density J^+ is given by

$$J^-(0) = K^- J^+.$$

Due to negative ion destruction in the plasma and in the beam formation region, the current density of H⁻ ions decreases to

$$
\begin{aligned}
J^- &= K^- J^+ \exp\left\{ -\int n_e \left(\sigma_5 v_e / v^- + \sigma_6 \right) dx - \int \left[n_1 (\sigma_7 + \sigma_9) + n_2 \sigma_{10} \right] dx \right\} \\
&= K^- J^+ \exp\left\{ -\left\langle n_e \left(\sigma_{5eff} + \sigma_6 \right) d \right\rangle - \left\langle n_1 (\sigma_7 + \sigma_9) + n_2 \sigma_{10} \right\rangle D \right\} \\
&= K^- J^+ \exp - \int dx / l(x) \\
&= K^- J^+ w^-.
\end{aligned}
$$

In this expression, σ_5 is the H⁻ ion destruction cross-section by electron impact, σ_6 is the recombination cross-section for H⁻ ions with protons (we suggest that the recombination cross-section with H_2^+ ions is close), n_e is the electron density equal to the total density of H⁺ and H_2^+ ions minus the density of negative ions, n_1 is the density of hydrogen atoms, n_2 is the density of H_2 molecules, σ_7 is the resonant charge-exchange cross-section for H⁻ ions on H, σ_9 is the cross section for the destruction of H⁻ ions by atomic hydrogen, σ_{10} is the destruction cross-section on molecular hydrogen, d is the effective thickness of the plasma layer, D⁻ is the effective thickness of the gas layer, and v^- is the velocity of H⁻ ions with energy $W^- = 100$ eV. Cross-sections for destruction of H⁻ are shown in Figs. 3.2 and 3.3.

The reaction rate for destruction of H$^-$ ions by electron impact $<\sigma_5\, v_e>$ in a plasma with a Maxwellian velocity distribution reaches a maximum $<\sigma_5 v_e> = 8 \times 10^{-7}$cm^3/s at an electron temperature $T_e \sim 15$ eV. This value corresponds to the effective cross-section for the destruction of H$^-$ ions with energy 100 eV by electrons $<\sigma_{5ef}> = 5 \times 10^{-14}$cm^2. At an effective value of $T_e \sim 3$ eV, this cross-section is comparable to $<\sigma_9> = 1 - 2 \times 10^{-14}$cm^2. Under the considered conditions, $\sigma_7 = 7 \times 10^{-15}$cm^2, $\sigma_9 = 1 - 2 \times 10^{-15}$cm^2, and $\sigma_{10} = 10^{-15}$cm^2. Due to the processes σ_9 and σ_{10}, the H$^-$ ions are destroyed mainly at the beam formation stage. The effective thickness of the gas layer D can significantly exceed the thickness of the plasma layer due to gas escaping through the emission slit, especially for large emission aperture area.

The resulting ion current density can be expressed in terms of the total current density at the emitting electrode J_p:

$$J^- = K^- J_d \,/\left(1+\gamma+K^-\right)\exp\left\{-J_d\,/\left(1+\gamma+K^-\right)v^+ - \left\langle\sigma_{5ef}+\sigma_6\right\rangle d - \left\langle n_1\left(\sigma_7+\sigma_9\right)+n_2\sigma_{10}\right\rangle D\right\}.$$

At constant γ, K^-, T_e, n_1, and n_2, the dependence $J^-(J_d)$ should have a maximum at

$$J_d\,/\left(1+\gamma+K^-\right)\left\langle\sigma_{5ef}+\sigma_6\right\rangle d = n_e\left\langle\sigma_{5ef}+\sigma_6\right\rangle d = 1,$$

$$J^-_{\max} = K^- v^+ /\left\langle\sigma_{5ef}+\sigma_6\right\rangle \exp\left\{-1-\left\langle n_1\left(\sigma_7+\sigma_9\right)+n_2\sigma_1\right\rangle D\right\}.$$

Thus the maximum H$^-$ ion flux density should depend on the secondary emission coefficient of negative ions K^-, the effective escape velocity of the bombarding ions from the electrode v^+ (corresponding to directed energy W^+), the destructive properties of the plasma $<\sigma_{5ef} + \sigma_6>$, and the thickness of the plasma layer d. For $\sigma_{5ef} = 5 \times 10^{-14}$cm^2, $\sigma_6 = 2 \times 10^{-14}$cm^2, and $d = 1$ mm, the optimal plasma density $n_e = 10^{14}$ cm^{-3}, the optimum bombarding ion current density $J^+ = 25$ A/cm^2 for $W^+ = 1$ eV, and $J^+ = 80$ A/cm^2 for $W^+ = 10$ eV, for $(1 + \gamma + K^-) = 6$, the corresponding value is $J_d = 150$–500 A/cm^2. Without destroying the gas at $K^- = 0.7$, for the maximum H$^-$ ion current density, we obtain $J^-_{\max} = 6 - 20$ A/cm^2. In practice we produced $J^-_{\max}$ up to 8 A/cm^2.

In the SPS with an effective layer thickness of the plasma between the central cathode plate and the emission slit $d = 2$ mm, saturation of the H$^-$ ion emission density was observed at $J_p = 50$ A/cm^2 and $J^- = 3 - 4$ A/cm^2. Such values of emission density are practically limiting for the beam forming systems used, so that the beam intensity can be limited by the capabilities of the beam formation systems even before the intensity is limited by the destruction of H$^-$ ions during transport to the beam formation system. Other mechanisms also lead to intensity limitation: deterioration of electrode emission properties, reduction of the H$^-$ ion formation efficiency and increase of their destruction due to increased hydrogen dissociation, increased cesium desorption, etc. It is possible that the effective rate of escape of ions from the generation region is not as high as expected. A decisive influence on the limiting intensity can be caused by the destruction of H⁻ ions by neutral particles, both during transport to the beam formation system and within the beam formation region.

For a rough estimate of the effective thickness of the gas layer, it can be assumed that the density of the gas emerging through the emission gap of width δ and length l is approximated by the expression

$$n(x) = n(0)\delta / (\delta + x),$$

and the total thickness of the destructive target is

$$\langle nD \rangle = n(0) \int \delta dx / (\delta + x) = n(0)\delta \ln(1 + l / \delta).$$

For SPSs with single-aperture emission slits with $\delta = 1$ mm and $l = 10$ mm, the effective thickness of the gas layer yields an estimate of $D = 5$ mm, and for large-aperture formation systems, even larger values. The average cross-sections for the destruction of H⁻ ions by atomic and molecular hydrogen is $<\sigma> = 1 - 4 \times 10^{-15}$ cm², so that the destruction will be small only for $(n_1 + n_2) < 1/ <\sigma D> = 0.5 - 2 \times 10^{15}$ c m⁻³. In SPSs with planotron configuration without heated cathodes, the discharges are ignited at an initial H₂ pressure of $P_2 = 0.1$ Torr.

An experimentally derived diagram of planotron ignition and discharge support is shown in Fig. 4.40. The ignition and combustion region for a discharge with fluctuations is indicated by single hatching in B-P space (2D space of magnetic field B and hydrogen pressure P), and double hatching shows the region for discharge burning without fluctuations. In these measurements, the SPS was located in a pumped vacuum chamber, and the pressure was measured by vacuum gauges with a thermocouple and radioactive sensors. If such a hydrogen density were retained in the compartment during discharge combustion, then the initial flux of H⁻ ions from the emitting surface would be weaker by tens of times, and the dependence of ion current H⁻ on gas supply and the dimensions of the emission slit would be catastrophically sharp. In fact, under optimal conditions, this is not observed. The dependence of the intensity of the H⁻ ion beam on hydrogen supply shows that for these operating conditions, $<n_1(\sigma_7 + \sigma_9) + n_2\sigma_1 > D < 1$. The density of neutral particles is significantly reduced due to their heating and hydrogen displacement by the plasma.

Oscillograms showing that the gas is confined by the discharge plasma are shown in Fig. 4.41. The gas flow signal without discharge q_0 is more than five times greater than the gas flow signal with a q_p discharge. The gas is confined by the discharge and is released when the discharge is off.

The required values of gas density n_2 for discharge ignition and the values of magnetic field strength B are consistent with estimates based on the assumptions that the electron Larmor radius for energy $W_e = eU_d$ should be less than the width of the reflex region d and the gas density n_2 should be sufficient for perturbations of the motion of emitted electrons. In the presence of plasma, the irreversible capture of emitted electrons in the plasma is due to their scattering by plasma particles, so the

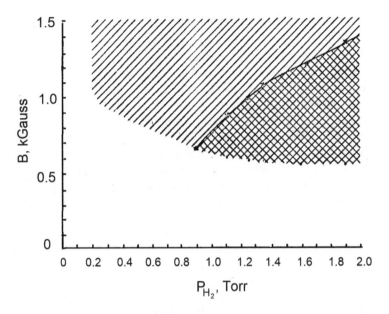

Fig. 4.40 Diagram of regions of discharge ignition and burning in a planotron with cesium, in the space of magnetic field and initial hydrogen pressure. Single hatching refers to the discharge burning region with fluctuations, and double shroud refers to the discharge burning region without fluctuations. (Reproduced from Dudnikov [1])

Fig. 4.41 Oscillograms of discharge current signal I_d, hydrogen flows from the SPS with a discharge q_p, and hydrogen flows from the SPS without a discharge q_o. δ is crosstalk. (Reproduced from Dudnikov [1])

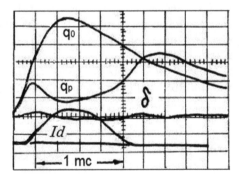

discharge can burn at even lower neutral particle density. According to rough estimates, up to 5% of hydrogen supplied to the flat-cathode planotron discharge cell is converted to a H⁻ ion beam.

It can be hoped that for SPSs with separated functions with geometric focusing during plasma generation in discharges with thermal cathodes, the initial hydrogen density in the gas discharge cell can be reduced to values used in positive ion sources, thereby significantly improving the gas efficiency of SPSs [51].

4.7   Cesium in Surface Plasma Sources

In the SPS, the decrease in work function of the electrodes, which has a decisive influence on the efficiency of formation of negative ions, is provided by adsorption of cesium fed into the gas discharge cell [1, 3, 5, 6].

Cesium is the heaviest of stable alkaline elements, with a minimum work function $\varphi = 2.14$ eV. Even lower work function is provided by metals and semiconductors with a partial coating of cesium. Cesium reacts with many elements. Table 4.1 shows cesium compounds with hydrogen and air components.

Pure cesium melts at 28.45°C and has a boiling point of 671°C.

The patterns of cesium adsorption-desorption and its release through the emission slit determine the generation of negative ions and the feasibility of practical applications of SPS. Data on the effect of cesium on the work function of refractory metals are systematized in reviews on thermionic transducers [38, 39]. The work function φ of cesium adsorbed on refractory metals decreases almost linearly from $\varphi = 4$–5 eV to a minimum $\varphi_{min} = 1.5 - 1.6$ eV when the cesium surface concentration is $N_{opt} = 2 - 4 \times 10^{14} cm^{-2}$ and then increases to 2.14 eV for $N > 5 - 6 \times 10^{14} cm^{-2}$. The work function of fluorinated or oxidized surfaces can decrease to $\varphi_{min} = 1.1 - 1.2$ eV for $N = 1 - 2 \times 10^{14} cm^{-2}$, and with adsorbed hydrogen up to 1.5 eV (Fig. 4.42) [52]. In this publication the adsorption of hydrogen and Cs alone as well as the coadsorption of Cs and hydrogen on W(110) were studied by metastable impact electron spectroscopy (MIES) and work function measurements. The main conclusions are as follows: (a) ionization of the Cs(6 s) level is not observed for Cs coverages $\theta_{Cs} \leqslant 0.4$ monolayer; (b) the formation of a H(1S)-metal bond occurs at a binding energy $W_B = 4.7$ eV; and (c) upon Cs and hydrogen coadsorption, an independent layer of hydrogen is formed between the substrate and the Cs adlayer. The charge density of the Cs adlayer is lowered by about 25% upon formation of the hydrogen intermediate layer. The work function of a thick layer of cesium with adsorbed hydrogen is considerably higher than that of a thick layer of cesium on pure tungsten.

To maintain the optimum surface concentration of cesium at the electrodes, it is necessary to compensate for the desorption of cesium from the surface. At low-electrode-temperature T, the cesium flux from the volume that compensates thermal desorption is sufficiently small. With increasing T, the rate of thermal desorption

Table 4.1 Cesium compounds with hydrogen and air components

Formula	Name	Color	Melting point
CsH	Cesium hydride	White	170°C
Cs_2O	Dicalcium oxide	Yellow or orange	490°C
CsO_2	Cesium superoxide	Yellow to orange	600/432°C
Cs_2O_2	Dicetium peroxide	Yellow	590°C
$CsNO_3$	Cesium nitrate	White	414°C
$CsCO_3$	Cesium carbonate	White	610°C
CsOH	Cesium hydroxide	Colorless	272°C

Fig. 4.42 Dependence of the work function of the surface of pure tungsten and tungsten with adsorbed hydrogen on the surface concentration of cesium. (Reproduced from Maus-Friedrichs et al. [52] with permission of Elsevier Publishing)

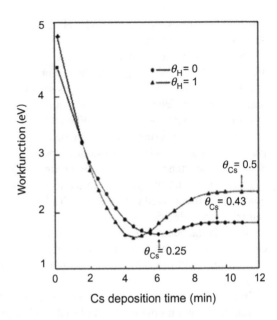

and the necessary compensating flow increase in proportion to $e^{-l/T}$, where l is the energy of desorption of cesium from the optimal coatings. For optimum cesium concentration on molybdenum, $l = 2$ eV; the heat of desorption of cesium from a pure molybdenum surface is $l = 3$ eV. The dependence of the energy of desorption for neutral Cs, l_o, and energy of desorption for Cs ions, l^+, on the surface concentration of cesium on tungsten is given in [53]. At $T = 600$ K, a cesium flux density $q = 3 \times 10^{14}$ cm^{-2} s^{-1} is needed to compensate for thermal desorption, provided at a volume Cs density of $N_v = 5 \times 10^{10}$ cm^{-3}, close to the density of saturated cesium vapor at $T_{cs} \sim T/2 = 300$ K. Note that when adsorbing cesium on oxidized surfaces, the heat of desorption is higher, and the surface concentration of the optimal coating is less, which should lead to a decrease in the equilibrium cesium flux density maintaining the optimal cesium concentration at the surface.

In the SPS, the cesium adsorption-desorption kinetics on electrodes bombarded by a fast particle flux from the discharge plasma are complicated by sputtering, pulsed surface overheating, ionization, and other factors. With a power density P/S_c per electrode in a time t from the beginning of the discharge pulse (where P is the power at the cathode and S_c is the cathode area), the temperature T of the electrode surface will increase by $\Delta T = 2P(t)^{1/2}/S_c^{1/2}(\pi\lambda c\rho)^{1/2}$, where λ is the thermal conductivity, ρ is the density, and c is the heat capacity. For molybdenum the coefficient degrees sec$^{1/2}$. At $P/S_c = 10^4$ W/cm^2, the temperature of the thick molybdenum planotron cathode increases by $\Delta T \sim 100$ K in $t \sim 10^{-3}$ s. For a cathode with $T \sim 10^3$ K, the thermal desorption flux from the optimal coating can increase by 5–10 times (a factor of $\exp(l_o\Delta T/T^2)$ due to such pulsed superheating.

During the discharge pulse, cesium is desorbed from the surface by sputtering of fast particles from the discharge. The cathode is bombarded by H$^+$, H$_2^+$, H$_3^+$, and Cs$^+$

ions and fast atoms and the anode by H^+, H_2^+, H_3^+, Cs^+, and H^- ions and fast atoms. A hydrogen particle with energy eU_d can transmit energy $W_{max} = 4M_H eU_d/M_{cs}$ to an adsorbed cesium atom. Under discharge conditions ($U_d = 150$–200 V, $M_H = 1$–3) $W_{max} = 4$–18 eV. Since $W_{max} > l_0$, the cesium atom can be knocked off the surface, but the coefficient of such sputtering is small due to the narrow range of impact parameters for which the energy pulse transmitted to the cesium particle is sufficient to overcome the potential barrier. When the fast particle flux density at the cathode is ~10^2 A/cm^2, the cesium lifetime on the cathode, determined by sputtering by fast particles, should be ~10^{-4} s. Cesium is sputtered more effectively by cesium ions. Several circumstances help maintain optimal cesium concentration on the cathode. In high-current discharges, cesium atoms desorbed from the cathode surface must quickly become ionized in the plasma discharge and should be quickly return to the cathode by the electric field.

To study the behavior of cesium in a gas discharge chamber with Penning geometry, the yield of ions and cesium atoms transmitted through the planotron emission slit and the yield of cesium ions through an aperture in the source cathode were explored.

In the SPS, various ways of supplying cesium to the discharge are used: from cesium titanium chromate tablets placed in discharge-heated electrodes of the gas discharge cell and from independently heated containers with cesium, cesium cartridges or cesium-bismuth alloys.

A photograph of cesium oven with Cs_2CrO_4 + Ti tablets is shown in Fig. 4.43.

To stabilize the release of cesium into the outlet tube of the container, a filter of molybdenum wires was used. The yield of cesium from the container with metallic cesium was measured as a function of container temperature.

When Cs_2CrO_4 + Ti tablets are heated, the cesium is released due to its reduction by titanium: $4Cs_2CrO_4 + 5Ti \rightarrow 8Cs + 5TiO + 2CrO_3$ [54]. One tablet weighs 10 mg and contains 1 mg of extractable cesium. The measurements showed that the cesium yield depends on the specific batch of tablets and on the duration of their use.

Cs oven with metallic cesium commonly used at IPP Garching is shown in Fig. 4.44.

The first information about the neutral cesium flux through the emission slit was obtained from the observation of the current in the extractor gap. Because of the high electric field, cesium atoms escaping through the emission slit effectively are ionized on the slightly heated surface of the extraction electrode. The resulting Cs^+ ions are accelerated to the gas discharge chamber by the extraction voltage, creating an additional current in the extraction gap, reinforced by secondary processes. Figure 4.45 shows an oscillogram of the current in the extraction gap. The current pulse after the discharge current pulse ends is associated with cesium escape through the emission slit after the discharge is terminated.

More detailed information on the neutral cesium flux was obtained using a surface-ionization detector of cesium atoms. The measurement schematic is shown in Fig. 4.46. An ionizing tungsten hotplate ($T = 1500$ K) heated by a tantalum spiral grid are set at an angle of 45–60° to the flow of atoms from the emission slit. The

Fig. 4.43 A photograph of cesium oven with Cs_2CrO_4 + Ti tablets. (37) Cesium oven body; (38) oven assembly; (39) heater; (40) thermal shield; (41) heart connector; (42) wire with connector; (43) plug with copper gasket; (44) press nut; (45) cesium pellets; (46) press form body; (47) press form piston; (48) press form bolt

ions produced by surface ionization are accelerated and formed into a beam by a 2 kV extraction voltage applied between the plate and the ionizer housing. The ion beam is analyzed by a magnetic mass spectrometer with $B < 0.25$ Tesla ($R = 15$ cm) and recorded by a Faraday cup collector. The efficiency of the cesium registration from the source was determined by the cesium flux from a heated container installed in place of the source. The measurements showed that the collector recorded 0.2% of the flux of cesium coming from the source. Information on the density of cesium ions on the cathode and anode regions was obtained by direct extraction of cesium ions from these regions by a 10 kV voltage, followed by analysis by a magnetic mass spectrometer. In the same way, measurements were made of the light ion H^+, H_2^+, and H_3^+ fluxes.

Figure 4.47 shows a typical oscillogram of the cesium ion current from the collector of the mass spectrometer, illustrating changes in the cesium ion flux from the source in time at a high (~1000 K) planotron cathode temperature, in conjunction

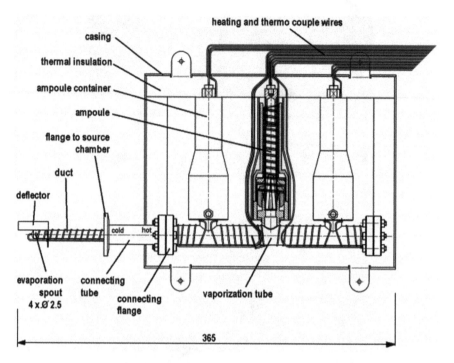

Fig. 4.44 Cs oven with metallic cesium commonly used at IPP Garching. (Original design: J. Steinberger, IPP)

with oscillograms of discharge current I_p and discharge voltage U_d. One can see that cesium atoms leave the source mainly after the end of the discharge pulse. Cesium release during the pulse is small, since cesium is highly ionized and the extraction voltage blocks the escape of cesium ions [55]. It is established that the delay time from the beginning of the signal relative to the end of the discharge pulse depends linearly on the distance between the source and the ionizer. The velocity of cesium atoms determined from this corresponds to a temperature of 10^3 K. Figure 4.47 shows the pulsed cesium ion signal from the detector at different discharge chamber temperature. The change in temperature was achieved by varying the discharge repetition rate at fixed discharge current. As the temperature of the source gas discharge chamber rises, the cesium impulse discharge increases quadratically as a result of an approximately linear increase in the amplitude and duration of the pulsed cesium release, determined at the $1/e$ level. The signal amplitude is almost independent of pulse duration at fixed average discharge power input. The repetition rate varies inversely with the discharge pulse duration. The amplitude and duration of the signal grow with hydrogen input. The exponential growth with anode temperature is shown in Figs. 4.47 and 4.48.

The quasistationary component of the flux of atoms in intervals between discharges depends only on the average power deposited in the discharge.

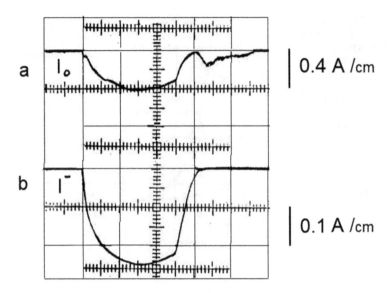

Fig. 4.45 Oscillograms of (**a**) total current I_0 at the extraction gap at elevated electrode temperature of the gas discharge cell and (**b**) H⁻ ion collector current. Horizontal scale is 200 μsec/div. (Reproduced from Dudnikov [1])

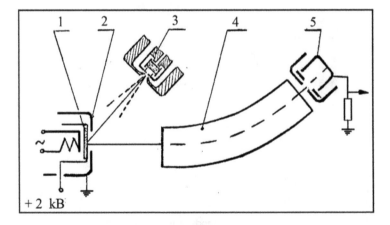

Fig. 4.46 Schematic of the setup for measuring the cesium atom flow escaping through the emission slit of the SPS. (Reproduced from Dudnikov [1]). (1) Ionizer, (2) accelerating electrode, (3) gas discharge source chamber, (4) analyzing magnet, (5) mass spectrometer collector

The cesium ion current density per cathode as a function of discharge current density is shown in Fig. 4.49. The cesium ion current density at the cathode of the Penning cell increases with discharge current and depends weakly on the temperature of the gas discharge chamber.

The main process of cesium desorption from the electrode surface during the discharge pulse is sputtering, as evidenced by the significant relative thermal

Fig. 4.47 Characteristic
oscillogram of cesium ion
current from mass
spectrometer collector,
illustrating the time
variation of cesium ion
flux from a planotron
source at high cathode
temperature (~1000 K)—
also showing oscillograms
of discharge current I_d and
discharge voltage U_d.
(Reproduced from
Dudnikov [1])

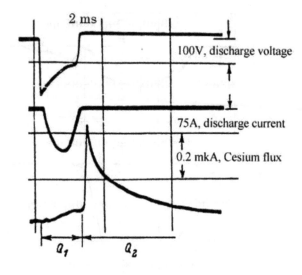

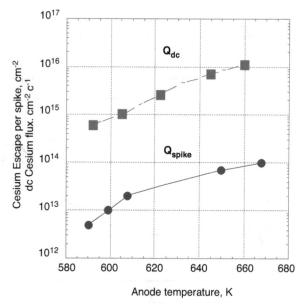

Fig. 4.48 Dependence of pulsed cesium atom flow from the SPS (Q_{spike}), and the quasistationary
atom flow (Q_{dc}) on the temperature of the body of the gas discharge chamber (Reproduced from
Dudnikov [1])

desorption of cesium current density at the cathode, and the growth of flux density
with discharge current density J_d. The fact that the cesium flux density through the
emission slit is less than the flux density per cathode by a factor ~500 indicates that
the cesium desorbed from the anode is rapidly ionized in the discharge plasma and
transferred to the cathode by the electric field, and after desorption from the cesium

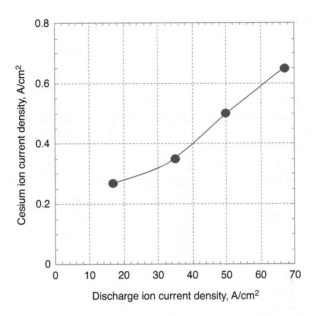

Fig. 4.49 Cesium ion current density per cathode as a function of discharge current density. (Reproduced from Dudnikov [1])

cathode returns to the cathode after ionization. Cesium ions accelerated in the near-cathode layer can sputter cesium from the surface, as a result of which the integrated cesium surface scattering coefficient increases. It is known that for a pure surface of molybdenum, the scattering coefficient of cesium ions of energy ~100–200 eV is $K_s = 0.6$. Estimates of the knockout coefficient K_s from a cesium coating with $\Theta = 0.3$–0.6 (fractional coverage) for the hard-sphere model give $K_s = 0.2$–0.3. As a result, for the integrated cesium scattering coefficient, we have $K_s = 0.8$–0.9, i.e., a Cs particle makes 5–10 oscillations near the cathode. At a fast hydrogen particle flux density of $J^+ = 10^{20}$ cm^{-2} s^{-1} ($I_d = 100$ A), the cesium ion flux density per cathode is $J_{cs} = J^+_H(1 - K_s)^{-1} = 10^{18}$cm^{-2}s^{-1}, where K_H is the coefficient of sputtering of cesium by hydrogen particles, which agrees with the measured value of J_{cs}.

Measurement of the flow of H$^+$, H$_2^+$, and H$_3^+$ light ions showed that the number of protons, i.e., the degree of hydrogen dissociation, increases rapidly with increasing discharge current. Therefore, at low-discharge current, when the molecular ions H$_2^+$ and H$_3^+$ predominate in the ion current to the cathode, the cesium ion flux density per cathode is higher because of the large cesium sputtering coefficient by molecular ions. The surface concentration of cesium at the cathode increases toward the end of the discharge pulse, which is experimentally confirmed by an increase in cesium ion flux density through a hole in the cathode and a decrease in the discharge voltage. The average lifetime of cesium at the cathode surface is $\tau \sim N/J_{cs} \sim 10^{-4}$s (where N is the cathode surface cesium concentration), which is much greater than the residence time in the plasma. The cesium atom ionization time is $\tau_i \sim 10^{-7}$ s at a cathode plasma density of 10^{14} cm^{-3}, and the average electrons energy is 10 eV; thus the total amount of cesium present in the plasma is small $N_v \sim j_{cs}\tau_i \sim 10^{11}$ particles/cm^2 (for cathode area $S_c = 2$ cm^2).

In the SPS, with cesium release due to electrode heating by the discharge, the supply of cesium to the cell is automatically boosted when the electrode temperature is increased, so that efficient generation of negative ions is ensured when the electrode temperature T varies over a wide range. Adjustable electrode cooling allows a significant reduction in cesium consumption, but one must also use an adjustable cesium supply. Measurements of the cesium flux through the emission slit showed that the SPS cesium flux during the discharge is very small, but increases sharply after the discharge terminates (Fig. 4.47). Removal of cesium during discharge is hindered by its rapid ionization in the discharge. The extraction voltage prevents the release of cesium ions through the emission slit. Because of the predominance of the electric field mechanism of electron transfer across the magnetic field, cesium ions must be efficiently transported to electrodes at negative potential. Cesium desorbed from these electrodes quickly ionizes and returns back. The sharp increase in cesium flux after the end of the discharge pulse is due to the free release of neutral cesium, which appears in the volume after the discontinuation of the discharge, and a gradual decrease in the flux of cesium adsorption on cooling electrodes. Reduction of sputtering and thermal desorption due to decreased bombarding particle energy contributes to a further increase in the cesium surface concentration. With a sufficient flow of cesium due to feedback, the discharge voltage decreases to $U_d = 100$ V, corresponding to a maximum secondary electron emission coefficient with minimum work function. The decrease in binding energy with increasing concentration limits the accumulation of excess cesium. From these considerations it follows that the concentration of cesium, somewhat greater than optimal, should be stable. It is hoped that when the electrodes are cooled, the optimal cesium concentration will also be retained in a high-current steady discharge.

In SPSs for accelerators [56], H$^-$ ion beams with current 0.1 A, pulse repetition rate 100 Hz, and pulse duration 0.25 msec are generated at an average cesium consumption rate of $<q> \sim 10^{-3}$g/h$\sim 10^{15}$ particles/s. In these sources, the average H$^-$ ion flux exceeds the average cesium atom flux by a factor of 10, and the ratio of intensities during the discharge burning exceeds 10^3–10^4. It turns out that H$^-$ ions are efficiently generated in the SPS below the vacuum condition requirements. In fact, for typical SPS conditions, cesium is accumulated on the walls of the gas discharge cell in the form of compounds—oxides, hydrides, etc.—so the cesium flux from the source in the interval between discharge pulses proves to be small at elevated gas discharge cell body temperatures. It is known that the bombardment of the surface causes an irreversible change in adsorption and emission properties. Bombardment by ion beams of alkali metals greatly increases the secondary emission coefficient. When metals are bombarded in an atmosphere of hydrogen and cesium by cesium ions, the secondary emission coefficient of H$^-$ ions, which is significant at the beginning, rapidly decreases to very small values. The slow rotation of the bombarded disk eliminates this effect, while with rapid rotation it remains.

Up to now, changes in the properties of surfaces under SPS conditions have been studied very poorly. The preservation of high H$^-$ ion formation efficiency upon surface bombardment by intense particle fluxes from the plasma is a non-trivial and very important concern. It is hoped that an in-depth study of this issue will help increase the H⁻ ion formation efficiency and increase the stability and durability of SPSs.

The best results so far have been obtained with ionizing molybdenum electrodes. Tungsten is preferred for a number of features, but numerous experiments have not confirmed this assumption. Perhaps, because of looser atomic packing, the reflection of hydrogen particles from tungsten is limited by deeper penetration into the lattice. With a molybdenum emitter, smaller fluxes of heavy ions, mainly O⁻ ions, are obtained. Because molybdenum atoms are lighter than cesium atoms, the reflection coefficient of accelerated cesium ions from molybdenum is much less than the reflection coefficient from other refractory metals. This effect contributes to the preservation of optimal cesium concentration on the molybdenum surface under intense bombardment, since cesium ionized in the discharge and accelerated to the cesium emitter is better captured by the molybdenum surface.

The escape of cesium through the emission aperture can be reduced by ionizing cesium between pulses by a weak discharge during excitation of cesium by resonant laser radiation [57]. Laser diagnostics of the cesium density in SPS discharges have been proposed [57] and developed [58]. A simulation of cesium redistribution in a large volume RF SPS has been presented in [59]. The strong increase of the co-extracted electron current suggests that a deterioration or vanishing of the Cs layer takes place during long pulse operation; hence, the maintainability of a constant low work function at the cesiated PG is thus the key parameter to assure good performance of SPS. At present, the work function of bulk Cs (2.1 eV) can be reached at the ion sources; however several factors contribute to the formation and maintenance of the Cs layer, affecting the resulting surface work function as well. During vacuum phases, the Cs high chemical reactivity together with the presence of residual gases in the source (due to the limited vacuum of 10^{-7}–10^{-6} mbar) leads to the contamination and degradation of the Cs layer, and hence to work functions higher than for bulk Cs (>2.7 eV). When a plasma is applied in front of the cesiated surface, the plasma exposure leads to a redistribution of Cs all over the source surfaces and to a cleaning of the Cs layer (thanks to the impinging plasma particles and/or to the UV and VUV plasma radiation), establishing the low work function of 2.1 eV and having positive effects on the ion source performance. However, the plasma also leads to a deterioration or removal of the cesium layer on the PG during long plasma pulses, modifying the work function of the plasma grid and decreasing the source performance. In order to maintain the Cs layer on the PG and assure in this way high source performance, fresh Cs must be continuously evaporated into the source. Typical neutral Cs densities are of the order of 10^{14}–10^{15} m^{-3}, but experiments at ELISE have shown that higher Cs densities are needed during long pulses in deuterium to counteract the increase of the co-extracted electron current. However, a high Cs content in the source is not recommended, since it significantly changes the plasma parameters and it also causes breakdowns between the grids due to the outflow. In particular in view of the DEMO reactor, the ion source requires high standards of reliability, availability, maintainability, and inspectability (RAMI requirements); hence, a better Cs management in order to maintain a sufficient Cs flux toward the surface minimizing the Cs consumption or otherwise a valid alternative to Cs evaporation is highly needed.

Lowering the work function by implantation of cesium ions into copper and molybdenum has been tested [60]. Cesium ions were implanted at an energy of 10–20 keV to a dose of 10^{18}–10^{19} cm^{-3}. The samples were heated to 900 °C. The estimated work function was $\varphi \sim 1.9$ eV. The secondary emission of negative ions from residual gas in such treatment increased by a factor of 7 to 8.

4.8   Measurement of Work Function in Ion Source Conditions

The experimental apparatus used for measurements of parameters important to clarify the fundamental processes leading negative ion surface production did not have large empty volume but stored movable components to measure the work function occupying the volume in front of the target. On the other hand, any measurement device cannot cover the area under exposure of plasma; the device blocks the plasma bombardment of the target area, and the device is irradiated by the plasma. Thus, it is common to measure the work function of an electrode in an ion source using photoelectric effect, as any delicate counter electrode is required for collecting produced photoelectrons [61].

Fundamental surface process data are usually collected in UHV environment with the samples heated to remove adsorbates prior to alkali metal deposition. On the other hand, vacuum environments of the system testing ion sources may contain substantial amount of residual gas. Fantz et al. [63] recognized that the work function of the plasma grid must be properly evaluated in an ion source operation condition and assembled a dedicated system to confirm correlations between Cs flux and the measured work function. The original system measured the photoelectric work function after turning off the plasma, while Gutser et al. [62] established an electronics control sequence to measure photoelectric current during less than 10 ms time interval, while the plasma was extinguished. Thus, the work function was measured in the environment right after the plasma ions were lost from the region in front of the target. Their measurement system composed of a RF plasma generator and the monochromatic light source system are schematically shown in Fig. 4.50. Line radiations from a mercury discharge lamp are selected with narrow band interference filters passing through the corresponding wavelengths. Quantum efficiencies are usually small, while optical components such as interference filters and vacuum windows limit the solid angle to collimate the incident light. Thus, the photoelectric current is usually amplitude modulated with an optical chopper to improve the signal-to-noise ratio by phase sensitive detection using a lock-in amplifier. The work functions are determined by fitting the measured quantum efficiency curve to Fowler's theoretical curve. Using the arranged experimental setup, Friedl

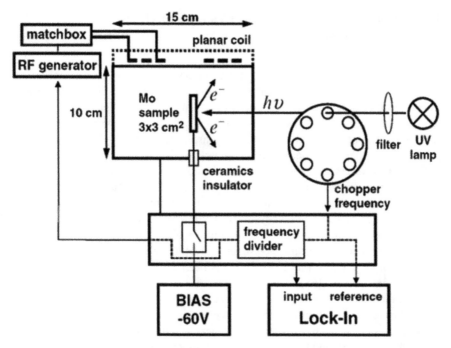

Fig. 4.50 An experimental arrangement to measure the work function of a target irradiated by a plasma. Monochromatic light at different wavelengths is produced by using a narrow band path interference filter. (Reproduced from Gutser et al. [62] with the permission of AIP Publishing)

and Fantz [64] investigated the work function of Cs-covered stainless steel in a 10^{-6} mbar pressure environment with a turbo-molecular-pumped chamber by monitoring the residual gas with a quadrupole mass analyzer. They found that the Cs-covered surface deactivated by residual gas can recover a low work function condition by heating up the target. They further measured the impact of D_2 plasma and H_2 plasma exposures to the Cs covered (stainless steel) sample after turning off the plasma to measure the photoelectric current from the sample [65]. They showed some lower minimum work function for a D_2 plasma than a H_2 plasma and concluded that adsorption of residual gas can increase the work function of the cesiated target, while hydrogen discharge removed adsorbed residual gas to recover a low work function surface. Cristofaro et al. [66] have recently demonstrated that the work function of the Cs-covered stainless surface can be as low as about 1.8 eV, which is the typical value of the work function minimum of the metal surface with the partial monolayer coverage of Cs [55]. This work function minimum condition was realized by hydrogen plasma exposure but was higher than the minimum work function of Cs-covered 304 stainless steel measured by using the thermionic method [67]. A cold target may adsorb more residual gas, and the temperature of the target can affect the minimum work function of a Cs-covered metal surface exposed to a plasma under residual gas contamination.

4.9   Cesium Management and Cesiated Surface Work Function

Although there are several methods for cesium detection in the vacuum phase, for example, the photoelectric current measurement (work function method), a surface ionization detector (Langmuir detector), a quartz microbalance, or a mass spectrometer, they cannot be easily used in combination with plasma operation and in the harsh environment of an high-power rf ion source operating at high voltage.

In addition diagnostic access to the source is also very limited. A diagnostic method which offers direct access to the ground state density of a particle in the vacuum and in the plasma phase is absorption spectroscopy utilizing a resonant transition. This well-known technique is being used in many applications using either white light absorption spectroscopy or (tunable diode) laser absorption spectroscopy; however the application to cesium in discharge chambers for cesium quantification is not well established so far. The white light absorption spectroscopy of the cesium resonance line at 852 nm has been already successfully applied to laboratory experiments [68].

The ground state of cesium has a $5p^6$ 6 s configuration resulting in a 6 $^2S_{1/2}$ state. For the first excited configuration $5p^6$ 6p, the fine structure results in two electronic states, the 6 $^2P_{1/2}$ and the 6 $^2P_{3/2}$ state. The resonance transitions correspond to the D1 line at 895.3 nm and the D2 line at 852.1 nm with transition probabilities for spontaneous emission (A_{ik}) of 2.87×10^7 s^{-1} and 3.276×10^7 s^{-1}, respectively. Due to the lower wavelength, which implies a better availability of light sources and sensitivity of detectors, and the higher transition probability the D2 line is been chosen for the absorption diagnostics. In addition, this 852.1-nm line is also used for optical emission spectroscopy. The nuclear spin $I = 7/2$ results in a hyperfine structure characterized by the total angular momentum quantum number of the atom F. Thus, the ground state 6 $^2S_{1/2}$ splits up into two hyperfine levels ($F = 3,4$) and the 6 $^2P_{3/2}$ state into four hyperfine levels ($F = 2,3,4,5$). According to the selection rules for optical transitions, $\Delta F = 0, \pm 1$, the cesium D2 line consists of six hyperfine lines. Because of the narrow energy splitting of the 6 $^2P_{3/2}$ hyperfine states ($\approx 8 \times 10^{-7}$ eV), the Doppler broadening of the individual lines reflects the energy splitting of the ground state ($\approx 4 \times 10^{-5}$ eV) only resulting in two peaks. An example is shown in Figure 4.50a for a gas temperature of 1000 K which is typical for low-pressure ion sources. Due to the strong overlap of the three individual lines ($\Delta\lambda_{FWHM} = 1.86$ pm) in each of the two lines, it is reasonable to use a Gaussian profile for each of the two lines ($\Delta\lambda_{FWHM} = 1.87$ pm left peak, $\Delta\lambda_{FWHM} = 1.91$ pm right peak) separated by 21.4 pm. In the case of white light absorption spectroscopy, the apparatus profile of the spectroscopic system determines the spectral resolution of the diagnostic system. The typical spectral resolutions of a 1-m spectrometer with a grating of 1800 grooves per mm is similar to the line separation resulting in a spectrum in which the two lines are barely separated. In the case of tunable laser absorption, the line profile of the laser determines the spectral resolution which is in most cases orders of magnitudes below the line separation. A comparison of the absorption signals

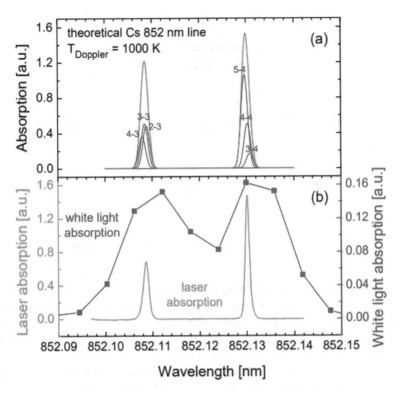

Fig. 4.51 (**a**) Calculated caesium resonance line (6 2 P3/2–6 2 S1/2 transition, 852 nm line) in which a Doppler profile at 1000 K is used for the individual hyperfine components. (**b**) Absorption signals measured with laser absorption and white light absorption

obtained with the white light laser absorption for quantification of cesium densities
4 absorption technique (spectral resolution $\Delta\lambda_{FWHM}$ = 15 pm) and a tunable diode
laser (line profile 0.01 pm) is shown in Fig. 4.51b. As expected the line separation is
only indicated in the white light absorption with a low number of data points,
whereas the laser absorption yields clearly two separated Doppler-broadened lines.

The complete experimental setup has been applied first to a laboratory experiment which provides a good access and can be equipped simultaneously with other
cesium diagnostics as well as with different cesium supplies. The experimental
setup for the laser absorption is shown in Fig. 4.52, showing also the line of sight
and the geometry for the cesium dispenser with the typical amount of 10 mg
mounted at the bottom plate.

The plasma is generated by inductively coupling (ICP) using a planar coil with a
Faraday screen at the top of the stainless steel cylinder (15 cm in diameter, 10 cm in
height). The RF generator has a maximum output power of 600 W (f = 27.12 MHz).
A matchbox is used to adjust the plasma impedance. As shown in Fig. 4.52, the
chamber is equipped with several diagnostic ports; the bottom plate houses several
feedthroughs and ports as well. Hydrogen plasmas can be generated in the pressure

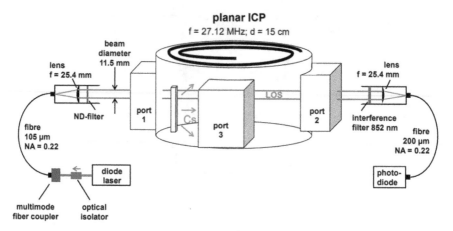

Fig. 4.52 Experimental setup for the laser absorption at the planar inductively coupled plasma (ICP)

range of 1–50 Pa. Mixtures with helium are often used to measure the electron temperature by optical emission spectroscopy and to influence the electron temperature and density actively. Typical parameters of hydrogen plasmas at 10 Pa are electron densities of 10^{16} m^{-3} and electron temperatures of 2 eV, measured by Langmuir probes and optical emission spectroscopy. The inductively coupled plasma source operates at similar vacuum conditions and plasma parameters as the ion sources. Similar cesium evaporation rates are used as well using either cesium dispenser from SAES [36] which contain Cs_2CrO_4 embedded in Al and Zr acting as oxidant or the standard cesium oven of the ion sources at IPP (liquid cesium reservoir) as utilized at the IPP test facilities. From the optical emission spectroscopy of the cesium line at 852 nm, it is known that the amount of cesium evaporated is in the similar density range. Alternatively to the laser absorption setup, the white light absorption spectroscopy can be installed at the same ports. For the latter, a stabilized high-pressure xenon arc lamp (OSRAM XBO 250 W) acts as light source, and a 1-m spectrometer with a CCD detector (apparatus profile (Gaussian) $\Delta\lambda_{FWHM} = 15$ pm) and fiber optics are used for the measurements. Typically five spectra are averaged allowing for a recording time of two data points per minute.

The response of the laser absorption during a dynamic change of the signal and thus the recording of time traces has been checked by a comparison with the signal of the surface ionization detector (SID). The SID measures the cesium ionized at a hot tungsten surface by applying a voltage between the outer spiral wire and the inner wire. The SID is mounted on the bottom plate in the line of sight of the cesium dispenser. The measurements were taken at a background vacuum pressure of 10^{-5} mbar. As shown in Fig. 4.53a, a very good agreement between the two methods is obtained keeping in mind that they measure at different locations with different detection geometries. At $t = 0$ the cesium dispenser has been switched on (after being stopped a certain time to achieve a certain dynamic range in cesium density), and both signals follow the abrupt cesium enhancement and the slow decay of the

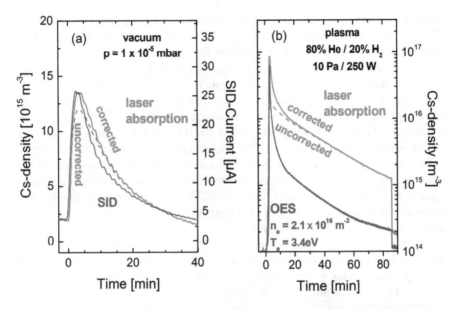

Fig. 4.53 Time traces of the cesium density determined by laser absorption compared to the signal obtained by the surface ionization detector (SID) in the vacuum phase (**a**) and by optical emission spectroscopy in the plasma phase (**b**)

signal. This is the typical behavior of the used SAES cesium dispenser evaporating at a constant heating current. Since the SID can only detect relative changes of the cesium flow, the scaling of the axes is set to match the maximum of the cesium density measured by the laser absorption spectroscopy. Line saturation effects have been taken into account being obviously relevant for reflecting the peak as the comparison of the corrected with the uncorrected signal demonstrates.

In case of plasma operation, the laser absorption signal is compared with the signal from the optical emission spectroscopy (OES) as SID measurements are not compatible with plasma operation. For simultaneous measurements, a line of sight perpendicular to one of the laser absorption but in the same plane is been chosen. Such differences in the geometry are of no relevance for the comparison because cesium is distributed homogenously in the chamber by the plasma. Since the spectroscopic system is intensity calibrated, cesium densities can be deduced from the line emission of Cs 852-nm line provided the electron density and temperature are known. Measurements of these parameters have been carried out with a Langmuir probe system in the same experiment at same discharge parameters, but without evaporation of cesium. Cesium however will change the plasma parameters depending on the amount of cesium in the discharge. Therefore it has to be kept in mind that the OES data of Fig. 4.53b are obtained using the n_e and T_e values of the pure H/He discharge. Due to the low ionization energy of cesium (3.89 eV), cesium is easily ionized in the plasma. The next ionization stage needs much higher energies (about 22 eV), which means the probability for doubly ionized cesium in low temperature plasmas is rather low. It has to be kept in mind that laser absorption and OES give access to the cesium density of the neutral cesium and that the values obtained in the plasma phase do not account for the ions and thus do not represent the total cesium amount in the chamber. Figure 4.52b compares the cesium densities from both diagnostic methods over a dynamic range of three orders of magnitude. As for Fig. 4.43a two curves are given for the laser absorption accounting for the line saturation effect (corrected) or not (uncorrected). The H_2/He plasma at 10 Pa pressure has been switched on first, and at $t = 0$ the cesium evaporation from a cesium dispenser has been started. The cesium density increases instantaneously followed by a decrease; a similar behavior as obtained in the vacuum case but the dynamic range is much higher in plasma operation. The signals are in reasonable agreement for the peak value (at $t \approx 0$), the OES densities falling more rapidly than the results from the laser absorption. As mentioned above, OES relies on the knowledge of electron density and temperature.

The automatic data acquisition and analysis together with the coupling of the trigger of BATMAN allows for routine measurements, before, during, and after the discharge. Such a time trace is shown in Fig. 4.54 for one of the discharges plotted in Fig. 4.53. In the vacuum phase, the cesium density is close to 10^{15} m^{-3}. As the RF power is switched on, a small peak or disturbance occurs until a more or less stable cesium density is been detected. The disturbance at the beginning of the discharge is caused by a change in the pressure due to a gas pulse which is commonly used at the IPP ion source to ensure plasma ignition.

During the discharge the density of neutral cesium atoms remains at a similar level as before which is quite surprising. At the plasma conditions of the ion source,

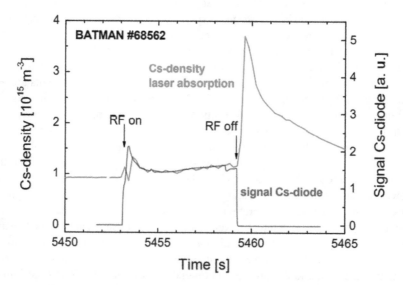

Fig. 4.54 Typical time traces of the cesium density determined by laser absorption 3 s before the discharge is turned on, during the discharge (rf on and rf off), and 5 s after the discharge (0.6 Pa, 60 kW, hydrogen)

i.e., electron densities in the range of 10^{17} m^{-3} and temperatures of 1 eV, above 90% of the cesium is ionized. This means, starting with the vacuum phase, a strong decrease of the signal is been expected. The nonappearance of the behavior led to the conclusion that a strong redistribution of cesium from the surfaces appears during the discharge. As also shown in Fig. 4.50, the OES signal is obtained by a simple photodiode equipped with a interference filter with central wavelength of the cesium 852-nm line. Since the photodiode is not calibrated, the scaling of its signal is set to fit the measured Cs density of the laser absorption spectroscopy. The comparison shows an excellent agreement in the temporal behavior. As stated from earlier OES measurements, the cesium density (neutrals) is mentioned to be in the range of 10^{15} m^{-3}, which of course depends on the cesium conditions but also on the knowledge of the exact plasma parameters, electron density, and temperature. As discussed for the results obtained in the laboratory experiments, the laser absorption is the more reliable method. After the rf power has been switched off, the cesium density increases rapidly which is attributed to the recombination of cesium ions into neutrals. Thus, the ionization degree of cesium might be deduced from the peak height. However, recombination takes place in a much shorter timescale than can be measured by the laser absorption, which means the actual height of the peak is, for the present time resolution given, point of speculation. Nevertheless, a lower limit for the ionization can be deduced from the measure peak height which is determined to be roughly 70%. The slow decay after the discharge reflects the diffusion time and sticking coefficient of the cesium particles toward and at the surfaces. For this example a decay time of ~2 s is obtained. In summary, it can be stated that these first examples demonstrated the applicability of laser absorption as a standard method for ion sources with first promising results giving already some new, quite surprising insight into the cesium dynamics. The method itself was for the first time established at cesium-seeded hydrogen ion sources and opens a window for more systematic investigations of the cesium dynamic which will be described in a following paper.

Cesium management in terms of optimizing consumption and understanding the redistribution dynamics during plasma operation with the two hydrogen isotopes is one of the most urgent and demanding tasks. An in situ and on-line work function measurement of the PG surface would be desirable for the optimization of the cesium management, but the installation of such a monitoring system is challenging in the harsh environment of ion source operation and is thus not established up to now. Therefore, dedicated studies are performed at the small-scale laboratory experiment ACCesS [69] in order to gain insights into the cesium work function dynamics under ion source conditions. At ACCesS, cesium is evaporated with the same cesium oven as used for the ion sources under vacuum conditions of $\sim 10^{-6}$ mbar onto a sample of the PG material (polycrystalline Mo layer on Cu). Hydrogen (or deuterium) plasmas are ignited via inductive RF coupling such that the plasma parameters are comparable to those close to the PG of the ion sources. The work function of the Mo surface is measured photoelectrically using an enhanced work function measurement setup, which is presented in detail elsewhere [64], and the neutral cesium density is continuously monitored by means of a TDLAS system along a line of sight parallel to and close to the sample surface. Figure 4.51 shows an example of the temporal behavior of the measured surface work function during a cesium-reconditioning campaign at ACCesS. The Mo surface was cesiated in a preceding campaign and left in vacuum for some days, leading to a degraded cesium layer with an initial work function of about 3 eV [65]. This work function is already substantially lower than the 4.3 eV measured after hydrogen plasma exposure without cesium. As soon as fresh cesium is evaporated in the vacuum phase at room temperature, the work function starts to decrease and reaches a value of about 2 eV [64] with typical neutral cesium densities in the range of 10^{14}–10^{15} m^{-3} [21, 66]. After the buildup of a fresh cesium multilayer under the given moderate vacuum conditions, H_2 gas is fed into the experiment. As can be measured with a residual gas analyzer attached to the chamber, the first H_2 gas inlet after an operational break of some days is accompanied by a substantial amount of impurity gases (mainly N_2, O_2, and H_2O), which have been accumulated in the gas line and are pumped out after some minutes. The release of the impurity gases leads to an instant drop of the measured neutral cesium density below the TDLAS detection limit of about 2×10^{13} m^{-3} and to an increase of the work function to slightly above 3 eV. This demonstrates the high sensitivity of the cesium evaporation and the vulnerability of the surface work function due to the high chemical reactivity of cesium. The deteriorated surface work function recovers, however, quickly by the application of H_2 plasma pulses and reaches a stable plateau of $\varphi \sim 1.75 \pm 0.10$ eV after a few 5 s plasma pulses. Hence, the plasma-surface interaction is an efficient way to condition the fresh cesium coating. The beneficial impact of the plasma pulses on the cesium work function explains the strong enhancement of the ion source performance during the conditioning phases, which are required after operational breaks for optimizing the extracted ion and co-extracted electron currents. Furthermore, the continuous monitoring of the neutral cesium density via TDLAS shows that a strong cesium redistribution is driven during the discharges and the cesium content during the vacuum phases gradually recovers. It reaches nearly the same level as before the first gas inlet after about half an hour in the present case (without changing the settings of the cesium oven). As can be seen in Fig. 4.55, the low surface work function can be maintained during prolonged vacuum phases and additionally kept after plasma exposure, since the cesium flux in the vacuum phases is sustained ($\sim 9 \times 10^{12}$ cm^{-2} s^{-1}). Moreover, the prolonged plasma phases up to several minutes lead to an enhanced cesium release and redistribution and to a steady improvement of the photoelectric quantum efficiency of the surface. When the cesium evaporation from the oven is stopped, however, the work function gradually increases in the vacuum phase. The degradation depends on the vacuum conditions and is on the order of 1 eV/h. By the application of continuous plasma pulses without active cesium evaporation, the low work function passivation is slowed down [39], depending on the cesium reservoirs in the source which are gradually depleted by the plasma recycling. These examples demonstrate the challenge to maintain a spatially and temporally stable low work function of the PG surface in the ion sources, especially during long-pulse operation, which critically determines the co-extracted electrons and thus the source performance limits. Since the temporal cesium dynamics are much

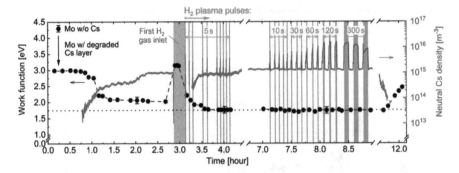

Fig. 4.55 Temporal development of the surface work function during a cesium re-conditioning scenario applied at ACCesS: a degraded cesiated Mo surface is re-cesiated and exposed to a sequence of hydrogen plasma pulses with variable length. The neutral cesium density is measured via TDLAS; the plasma-on time is indicated by the red shaded areas (10 Pa gas pressure and 250 W RF power applied)

more challenging in deuterium than in hydrogen [42, 69], stabilizing the PG surface work function seems to be mandatory for demonstrating the ITER requirements in deuterium. Consequently, efforts are needed to provide a sufficient cesium flux onto the PG, and one candidate would be the already mentioned cesium evaporation in the direct vicinity of the PG [43].

Negative hydrogen ion sources for NBI systems at fusion devices rely on the surface conversion of hydrogen atoms and positive ions to negative hydrogen ions. In these sources the surface work function is decreased by adsorption of cesium (work function of 2.1 eV), enhancing consequently the negative ion yield. However, the performance of the ion source decreases during plasma pulses up to 1 h, suggesting a deterioration of the work function. Fundamental investigations are performed in a laboratory experiment in order to study the impact of the plasma on the work function of a freshly cesiated stainless steel surface. A work function of 2.1 eV is achieved in the first 10 s of plasma, while further plasma exposure leads to the removal of Cs from the surface and to the change of the work function: a value of around 1.8–1.9 eV is measured after 10–15 min of plasma exposure (Fig. 4.56), and then the work function increases, approaching the work function of the substrate ($\geq$ 4.2 eV) after 5 h. The Cs removal must be counteracted by continuous Cs evaporation, and investigations performed varying the Cs flux toward the surface have shown that a Cs flux of at least 1.5×10^{16} m^{-2} s^{-1} is required to maintain a work function of 2.1 eV during long plasma exposure at the laboratory experiment [70].

Cesiated surface of molybdenum with low work function was produced in [71]. Cesium evaporation on molybdenum surface in "ion source vacuum," $10^{-5} - 10^{-6}$ Tor was used for production surfaces with low work function. Photoelectronic measurements were used for work function monitoring. In Fig. 4.57 variations of work function during cesium deposition are presented. Work function $\varphi \sim 1.2$ eV was produced.

Evolution of work function during exposure by residual vacuum and processing by VUV and gas discharge plasma is shown in Fig. 4.58a.

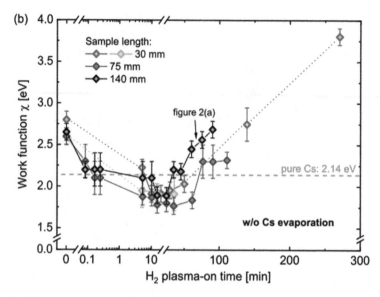

Fig. 4.56 Work function of freshly cesiated stainless steel samples of different lengths plotted against plasma-on time during hydrogen-pulsed plasma operation at 10 Pa and 250 W without further Cs evaporation. Dotted lines indicate missing data and are given as a guide to the eye

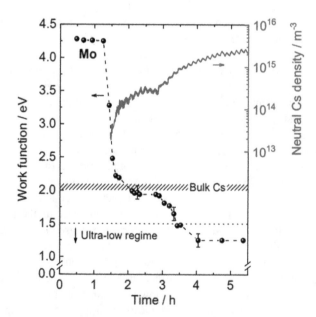

Fig. 4.57 Variations of work function during cesium deposition and neutral cesium density in equilibrium with deposited cesium

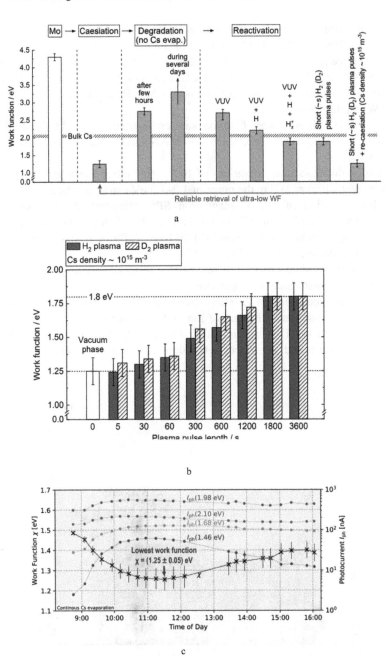

Fig. 4.58 (**a**) Evolution of work function during exposure by residual vacuum and processing by VUV and gas discharge plasma. (**b**) Work function of cesiated surfaces upon H_2 (D_2) plasma exposure. Durability of ultralow work function layer. (**c**) Demonstration of ultralow work function production in BATMAN surface plasma source

Work function of cesiated surfaces upon H_2 (D_2) plasma exposure and the durability of ultralow work function layer are presented in Fig. 4.58b.

The surface processing by plasma can restore a minimal work function $\varphi \sim 1.25$ eV. Stable WF of 1.8 ± 0.1 eV for Cs densities $\gtrsim 3 \times 10^{14}$ m^{-3} (Cs fluxes $\gtrsim 2 \times 10^{16}$ m^{-2}s^{-1}) during plasma phase can be reached. Reached WF independent of initial WF in the vacuum phase.

Production of ultralow work function was demonstrated in BATMAN SPS [72]. Demonstration of low work function in BATMAN surface plasma source is shown in Fig. 4.58c.

Cesiation under vacuum conditions of $10^{-6} - 10^{-5}$ mbar: Reliable formation of Cs layers with ultralow WF in the range of 1.2–1.3 eV can be reached.

During operation of DESY, magnetron SPS consumption of cesium was decreased from 10 mg/hour to 0.7 mg/day. In SNS RF SPS, the cesium consumption was decreased to 30 mg for 4 months [73]. In J-PARC SPS continues consumption of cesium was decreased to 35.8 Pg/hour [74].

These examples shows that accurate cesiation is possible with minimal cesium consumption not disturbing SPS operation.

Negative ion generation in Cs alternative materials → Cs-free ion source was investigated. Measurement of negative ion yields of various promising materials with low work function (lab experiment) is shown in Fig. 4.59. H$^-$ and D$^-$ negative ion generation was tested on surfaces of tantalum, tungsten, diamonds, boron-doped

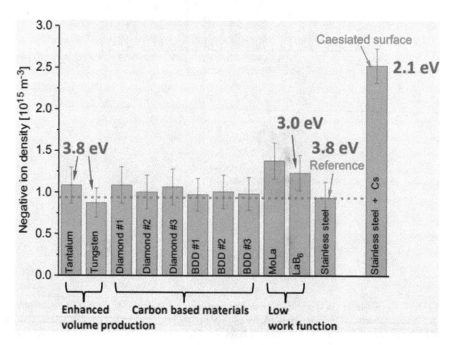

Fig. 4.59 Measurement of negative ion yields of various promising materials with low work function (lab experiment)

diamonds, MoLa, LaB$_6$, and stainless steel, but H$^-$ yield was 2.5 time larger for cesiated stainless steel with work function 2.1 eV. On the surface with lower work function as optimally cesiated molybdenum φ = 1.2 eV, the yield should be much higher.

Testing of electride ceramic (e 12CaO 7Al2O3 (4e-)) with work function ~!2.4 eV also do not hope for replacing cesiated surfaces with ultralow work function in surface plasma sources [75].

4.10   Physical Basis of the Surface Plasma Method of Negative Ion Production

From the results described in the previous sections, it follows that when cesium is added to gas discharge cells, the surface-plasma mechanism for generating negative ions becomes much more intense and dominates. If the right conditions are met, it is possible to ensure a high efficiency of negative ion generation by interaction of plasma particles with the electrode surfaces and to maintain a high level up to significant bombarding particle flux intensity, sufficient to produce intense beams of negative ions with emission current density up to several amperes/cm^2.

In this section we will consider a set of processes ensuring the production of intense beams of negative ions from the SPS, focusing primarily on the physical basis of the surface-plasma method for producing negative ion beams. This should serve as a foundation for the engineering phase of SPS development.

As already reported, the elementary processes of negative ion formation in the interaction of atomic particles with the surface of a solid were discovered experimentally near 90 years ago. The observation in 1961 of an increase in negative ion emission from a bombarded surface with decreased work function [76, 77] served as the basis for the development of secondary emission (sputtering) sources of negative ions, in which negative ion beams are obtained by bombarding a reduced work function surface by beams of positive ions. Some results of the investigation of these processes have been considered in Sect. 3.11.

When bombarding a surface with particle beams, the negative ion emission is limited to a relatively low level, up to tens or hundreds of microamps in most secondary emission sources. But by special activation of the surface, the intensity of H$^-$ ion beam produced by sputtering the adsorbate was increased to 2.5 mA with an emission density of about 10 mA/c [78, 79].

The production of negative ion beams of much greater intensity and current density by the SPS approach is possible due to the following circumstances.

Bombardment of emitting electrodes by an intense flux of low-energy plasma particles provides efficient generation of intense flows of sputtered and reflected particles of the working substance. A radical decrease in the work function of the emitting electrodes due to adsorption of cesium provides a high probability of electron capture from the electrodes to the electron affinity levels of the sputtered and reflected particles. In the SPS, it was possible to efficiently transport intense negative ion fluxes from the emitting electrodes, through the plasma, to the beam formation system. It was possible to ensure an effective separation of the negative ion flux from the flow of accompanying electrons. Realization of the necessary conditions in sources with a planotron plasma cell configuration made it possible to obtain beams of H⁻ ions with intensity of order 1 A at a current density up to 3–4 A/cm².

Beams of negative ions are formed in the SPS by a sequence of interrelated processes that can be conditionally divided into a number of elementary processes, as follows:

1. The formation of positive ions and atoms by ionization of the working substance in a gas discharge plasma.
2. These ions are accelerated in the plasma or in the near-electrode layer and fall on the emitting electrodes in the form of accelerated ions or fast neutral particles formed from accelerated ions by neutralization in the plasma or on electrodes.
3. As a result of momentum exchange between the bombarding particles and condensed phase particles, particles of the substance of the electrodes and adsorbate form a stream of working substance leaving the surface as sputtered and reflected particles.
4. As a result of electron exchange between the electrodes and the escaped particles, some of the sputtered and reflected particles leave the near-surface potential barrier in the form of free negative ions.
5. The flow of negative ions is accelerated and transported through the plasma and gas to the extraction gap of the beam formation system.
6. Negative ions that fall within the electric field of the beam formation system are accelerated and a negative ion beam is produced.

The totality of these processes should be organized in such a way as to ensure maximum efficiency of obtaining beams of negative ions of maximum intensity or maximum brightness. At the same time, naturally we have to make compromises. Information about many features of the processes under consideration is still uncertain, but in our understanding of a number of important points, significant progress has been made. The results of experiments, the results of investigations of SPS processes, and theoretical and experimental studies of the processes of sputtering, reflection, and electron exchange make it possible to consciously choose directions for the optimization of the totality of the processes under consideration and to estimate their cumulative efficiencies.

4.11   Regularities in the Formation of Reflected, Sputtered, and Evaporated Particles

The effectiveness of negative ion formation is largely determined by processes 3 and 4. In general, the formation of negative ions in single-interaction events between bombarding particles and the solid surface can be described as follows. On a surface area characterized by a number of parameters X_c, including the work function of the surface, its temperature T, the density and structure of particles in the solid, their mass and binding energy, the concentration of adsorbate particles, their mass, binding energy, and so on. A stream of bombarding particles with full intensity I_1 is characterized by the distribution function $f_1(X_1)$ in the set of parameters X_1, including the masses of these particles, electron shell configurations, ionization potentials, electron affinities, energies and directions of motion, charge states, etc. As a result of processes 3 and 4, a stream of sputtered particles with total intensity I_s leaves the surface, characterized by the distribution function $f_s(X_s)$ over a set of parameters X_s, including the energies and directions of motion of the sputtered particles, their mass, binding energy with the surface, electron affinities, charge states, etc. At the same time, a flux of reflected particles with intensity I_r and distribution function $f_r(X_r)$ is generated, as well as a flow of thermally desorbed particles with intensity I_T and distribution function $f_T (X_T)$. The change in the state of the surface (the values of the parameters X_c) due to bombardment and accompanying processes can greatly change the efficiency of the processes under consideration.

Some simplification is provided by the weak dependence of momentum exchange processes on the initial and final charge states of the interacting particles due to the rapid establishment of charge equilibrium for energies of interest to us. In this approximation, the distribution functions over the charge states of the sputtered particles f_s and the reflected particles f_r can be related to the parameters X_c and the properties of the bombarding flux through the differential scattering and reflection coefficients k_s and k_r:

$$f_s\left(X_s\right)=\int k_s\left[X_c,X_1,X_s\right]f_1\left(X_1\right)dX_1,$$
$$f_r\left(X_r\right)=\int k_r\left[X_c,X_1,X_r\right]f_1\left(X_1\right)dX_1.$$

In less detail, sputtering and reflection (scattering) characterize the integral sputtering coefficient $K_s = I_s/I_1$ and reflection coefficient $K_r = I_r/I_1$.

There is quite a lot of experimental data on the reflection of hydrogen particles with initial energy greater than 1 keV [80–85]. For reflection of particles with lower energy, we must use the results of numerical simulation [86, 87], which can later be substantially detailed, by extrapolating the experimental data on the scattering of particles with high energies and scattering of alkali metal particles [88, 89]. It follows from the available data that if the pair interaction potentials are successfully chosen, multiple scattering of bombarding particles on free target particles, taking energy losses into account, can adequately simulate the reflection of particles from the surface [52]. Integral reflection coefficients increase with decreasing initial energy due to a decrease in the depth of penetration into the target and reduction of energy loss. For the same reasons, K_r is larger for targets of particles of larger mass and with denser packing, as well as for deviation of the direction of motion of primary particles from the normal. Channeling can occur in single crystals. According to the calculated data [87], up to 30% of hydrogen atoms incident at the normal with energy 500 eV should be reflected from a copper target, and with a decrease in energy to 50 eV, the probability of reflection should increase to 60%. According to the data of [88], the coefficient of reflection from molybdenum at low energy is 10–20% greater than that from copper, and from carbon is 3–4 times less than that of copper. The reflection coefficient increases to $K_r > 0.8$–0.9 when bombarded at an angle to the normal $\theta > 60$–$80°$. The maximum energy of reflected particles is determined by the laws of pair collisions and at large target particles masses is close to the initial energy [50]. The energy distribution function of reflected particles has a maximum at high energy, so only a small fraction of reflected particles have low energy. When bombarded with particles of energies less than 10 eV, the reflection can be limited by the near-surface potential barrier, so that dependence on the final charge state can manifest itself [292]. Molecular particles dissociate into atomic particles upon collision with the surface.

The reflection coefficient should decrease due to hydrogen adsorption, but the sputtering coefficient from the adsorbate K_s should increase. Unfortunately data on the sputtering of hydrogen from the adsorbate is very scarce. In principle, sputtering due to a sequence of pair collisions can be investigated by numerical simulation. Because of the difficulties in recording slow hydrogen atoms, reliable data on the sputtering of hydrogen from adsorbate have not yet been found. The data on the sputtering of alkali metals from adsorbate on refractory metals [292] show that the sputtering coefficients from adsorbate can be high, $K_s \sim 0.3$–0.6, even for very small concentrations of these particles in the condensed phase.

According to these data, the probability of sputtering from the adsorbate increases with decreasing bombarding particle energy to dozens of eV. The distribution function of sputtered alkali metal particles has a maximum at an energy of several eV and is substantially reduced only at an energy of tens of eV. Both the maximum energy of the sputtered adsorbed particles and the position of the maxima are proportional to the binding energy of the target particles, but they are weakly dependent on the energy and type of bombarding particles. The fraction of slow particles in these spectra is low [292].

A high probability of desorption of alkali metals can be provided by triggering nonequilibrium oscillations of near-surface target particles by impact by bombarding particles. Due to the significant difference between the masses of hydrogen and target particles, the dominance of this mechanism in the sputtering of hydrogen is questionable, but there are no data excluding it. It may be that a decrease in bombarding particle energy favors both the reflection and sputtering of hydrogen from the adsorbate. In this regard, the bombardment of molecular particles can be more effective by many times. There is no doubt that at high hydrogen concentration in the adsorbate the intensity of the sputtered flux at epithermal velocities will be high, and the role of sputtering in the formation of H⁻ ions is significant. The data [78] can serve as a guide, wherein the sputtering coefficient, K^-, for hydrogen H⁻ ions when the adsorbate is bombarded with cesium ions of energy 1–3 keV, has a value of 0.8. High values of the hydrogen sputtering coefficients from adsorbate on cooled surfaces have been recorded [90], $K \sim 10^2$, but there are no experimental data on the energy of the sputtered particles.

The intensity I_T of the flux of thermally desorbed particles is proportional to the surface concentration of particle N_i in the adsorbate and depends on the surface temperature T and the binding energy l_i of the particles with the surface in accordance with the Frenkel formula:

$$I_T = \sum N_i \tau_i^{-1} \exp(-l_i / T),$$

where τ_i is a characteristic time of order the oscillation period of adatoms.

The binding energy l depends on many factors. At low hydrogen concentration on the surface of refractory metals, $l \sim 1$ eV, including in the presence of cesium, but at high hydrogen concentration the binding energy decreases greatly.

The evaporated particles have a Maxwellian velocity distribution.

Evaporation, sputtering, and reflection also determine the kinetics of the adsorption-desorption of cesium used in SPSs to reduce the work function of the electrodes.

4.12 Electron Capture to the Electron Affinity Levels for Sputtered, Reflected, and Evaporated Particles

The total flux of sputtered, reflected, and evaporated particles of interest to us contains fluxes of negative ions I_s^-, I_r^-, and I_t^- formed by these processes. The effectiveness of formation of negative ions by electron exchange between particles and a solid (process 4) is characterized by the corresponding ionization coefficients β:

$$\beta_s^- = I_s^- / I_s, \quad \beta_r^- = I_r^- / I_r, \quad \beta_t^- = I_t^- / I_t, \quad \beta^- = \left(I_s^- + I_r^- + I_t^-\right) / \left(I_s + I_r + I_t\right) = I^- / I,$$

and the total efficiency of formation of negative ions in processes 3 and 4 is characterized by the emission factors for negative ions due to these processes:

$$K_s = I_s^- / I_1, \quad K_r^- = I_r^- / I_1, \quad K_t^- = I_t^- / I_1, \quad K^- = \left(I_s^- + I_r^- + I_t^-\right) / I_1 = I^- / I_1$$

where I_1 is the flux of bombarding particles.

The coefficient of ionization of the fluxes due to thermal desorption can be estimated from the well-known Saha-Langmuir relationship. The degree of ionization α is given by

$$\alpha_t^- = \beta_t^- / \left(1 - \beta_t^-\right) = A^- \exp(S - \varphi) / T,$$

where A^- is the ratio of the statistical weight of a negative ion Q^- to the statistical weight of a neutral particle Q_0 (for atoms with one valence electron $Q_0 = 2$, $Q^- = 1$) and T is the surface temperature. The effective formation of an H^- ion through thermal desorption is possible only at very low work function φ, comparable to the electron affinity $S = 0.75$ eV. In the presence of a strong electric field E at the surface, the barrier height for charged particles decreases to $\Delta\varphi = e(eE)^{1/2}$ ($\Delta\varphi = 0.38$ eV at $E \sim 10^6$ V/cm). In a plasma with a flux to the negative electrode of H^+ ions with current density 100 A/cm^2, the average electric field strength at the electrode is about 10^5 V/cm for a potential difference of 100 V. The corresponding value of $\Delta\varphi = 0.1$ eV can strongly influence the formation of H^- ions at really attainable values of $\varphi \sim 1$ eV. For conditions of conventional plasma sources ($\varphi = 4$ eV, $T = 0.05$ eV), the ionization coefficient for evaporated particles is negligible $\beta_t^- = 10^{-30}$.

In the Arnot experiments [91, 92] on bombarding metal surfaces with hydrogen ions of energy 100 eV, the emission of sputtered and reflected H^- ions was detected with a total emission coefficient of $K^- \sim 10^{-4}$. Subsequent studies confirmed that the emission of slow H^- ions is due to adsorbate sputtering, which is removed from the surface at $T \sim 1000$ K, and the emission of fast H^- ions is independent of temperature and due to the reflection of bombarding particles. In these experiments the value $K^- = 10^{-4}$ was recorded, from which was estimated $\beta^- = 10^{-3}$–10^{-4} follows.

It was found in [18, 19] that the secondary emission of negative ions increases with decreasing work function. From the data available for such conditions, it was estimated that $\beta^- \sim 10^{-2}$–10^{-3}, many orders of magnitude greater than that estimated from the Saha-Langmuir equation, but still insufficient for the efficient generation of H^- ions.

For a long time, it was not possible to explain the substantial excess of the coefficient of negative ionization of the sputtered and reflected particle fluxes over the coefficient of ionization of the thermally desorbed particle flux, although this issue has been discussed since the time of its discovery by Arnot in 1936. Some of the numerous attempts to explain this phenomenon are given in monographs [91, 92]. The problem is to identify the mechanism moving electrons from the filled portion of the conduction band of the metal to the electron affinity level for a free particle with high efficiency, situated above the Fermi level $\Delta E = (\varphi - S) > > T > 0$. The probability of such a transition due to the thermal energy of the electron system of the solid is exponentially small. One of the first suggestions was that such a transition is realized by energy released during the Auger neutralization of bombarding ions. However, both the estimates and the experimental data show that this method of utilizing the neutralization energy is unlikely—the emission of negative ions depends on the charge state of the bombarding particles only very weakly.

For a long time, the only theoretical reference point for judgments about the possible degree of ionization of the fast component of the flux of sputtered and reflected particles was the Dobretsov formula [93]:

$$\alpha^-_{max} = \beta^-_{max} / \left(1 - \beta^-_{max}\right) = A \exp - \left(S\left(x_{kr}\right) - \varphi\right) / T,$$

where $S(x_{kr})$ is the electron affinity of the particle at the critical charge-exchange distance x_{kr} from the surface at which electronic exchange between the solid and the particle ends. The near-surface potential barrier for negative ions l^- is above the barrier for neutral particles l_0 by an amount $\Delta l = (\varphi - S)$. In this connection, the Dobretsov formula for a Maxwellian distribution of removed particle velocity goes over to the Saha-Langmuir relation. However, to express $S(x_{kr})$ and x_{kr} through the convenient characteristics of a solid and particles failed for a long time. Attempts have been made to determine experimentally by the positive ionization of reflected alkali metal ions [93]. The assumptions used did not give us any hope of obtaining significant ionization coefficients for streams of particles with small electron affinity $S < 1$ eV for attainable values of the work function $\varphi \sim 1.5$ eV.

The ideas explaining the high ionization efficiency of the particle flux moving away from the surface with increased velocities were developed almost simultaneously by Kishinevskii [94, 95], Kishinevsky [96, 97], and Janev [98, 99]. These representations are based on the analysis of the electronic exchange between a solid and a particle at small distance from the surface. The electronic exchange between a solid body and particles interacting with the solid was considered earlier in connection with the potential emission of electrons [4, 93, 100] when bombarding the surface with ions and excited particles, as well as in connection with neutralization of positive ions moving away from the surface [88, 101].

These earlier studies did not take into account the displacement of the levels associated with the particles. The shifting of affinity levels for particles approaching the surface due to the interaction of electrons with the metal has a decisive influence on the result of the interaction of particles with the surface. In the works mentioned, the shift of electron affinity level relative to the Fermi level of the metal was approximated by the action of an image potential. When approaching the surface, the level of the electron affinity acceptor levels is transformed to a zone, and the center of this region is lowered relative to the electron affinity level of free particles S by an amount $\Delta S = e^2/4x$, and the valence electron level of particles with low ionization energy V is transformed into a donor-level zone whose center is raised with respect to V by $\Delta V = e^2/4x$. Thus the dependence of the electron affinity on the distance between the particle and the surface x is approximated by the expression

$$S(x) = S + e^2 / 4x.$$

If the difference between the work function φ and the electron affinity S is not very large, then for particle distances $x < x_0 = e^2/4(\varphi - S)$, the electron affinity levels fall below the solid body Fermi level, so that the probability of resonant filling of this level by electrons from the solid becomes close to unity, in accordance with the Fermi thermodynamic equilibrium distribution. When a particle is removed from the region $x < x_0$, some particles leave in the form of negative ions. In the region $x > x_0$, energy of $\Delta l = \varphi - S$ is expended to remove the ion from the surface; much energy is required than to remove a neutral particle. Due to this, the level of the electron in the free negative ion turns out to be above the Fermi level of the solid by $\varphi - S$. The electron is thrown to the level of electron affinity due to the kinetic energy of the particle being removed. An analogous situation is realized in charge exchange in atomic collisions, when the bound electron is re-captured to a level with a lower binding energy. The probability of such a process is significant if the quasi-crossing of terms is realized due to the Coulomb interaction. The missing energy—the resonance defect in this case—is also borrowed from the kinetic energy of translational motion of the colliding particles. Thus the capture of electrons from electrodes to the electron affinity levels of sputtered and, especially, reflected particles can be regarded as a kind of charge exchange on a solid body, which advantageously differs from charge exchange of atomic particles by a huge geometrical cross-section. There is the possibility of providing low (~1 eV)-binding-energy electrons in the donor target, instead of the minimum 3.89 eV for atomic particles, if we ignore excited states, by the presence of a quasicontinuous zone of levels instead of a single level as for a free particle. The proposed model is applicable for interactions with the surface of relatively slow particles that move away from the surface with velocities v, smaller than the electron velocity at the Fermi level. For such particles, a thermodynamic equilibrium distribution of electrons is realized for the totality of energy levels of the solid and particles in the region $x < x_0$, and the approximation of the barrier by the potential of the polarization image forces is possible, so we can use the "resting-particle" approximation [102] for calculating electron transition probabilities between a solid and a particle.

Kishinevsky [59, 60] used the probability w for the transition of an electron between a particle and a solid the expression

$$w = A \exp{-ax}$$

as used earlier in semi-phenomenological theories of potential electron emission [102]. For a differential degree of negative ionization of a flow of monoenergetic particles, an expression similar to the Dobretsov formula is

$$\alpha^-(v,\varphi,S) = A - \exp\left[\left(S(x_{kr}) - \varphi\right)/T\right].$$

But in this case, the critical charge exchange distances x_{kr} and $S(x_{kr})$ are expressed through the surface work function φ, the electron affinity of the free particle S, the initial particle escape velocity from the surface v, and the coefficients A and a, which depend on the electron barrier structure between the solid and the particle.

The dependence of the rate of electron exchange between a solid and the electron affinity level of the H^- ion and the differential ionization coefficients of hydrogen particles moving away from the surface were calculated by Kishinevskii [57, 58]. It was assumed that the particles that have received enough energy due to scattering or by sputtering that initiates its exit from the surface leave the region $x < x_0$ as H^- ions with velocity v, sufficient to remove ions in the form $H^-(\frac{1}{2}Mv^2 > (\varphi - S))$.

The near-surface potential barrier for electrons was approximated by the image potential up to the bottom of the conduction band U_0:

$$U(\xi) = -e/4\xi, \text{ for } \xi > e/4U_0,$$

and

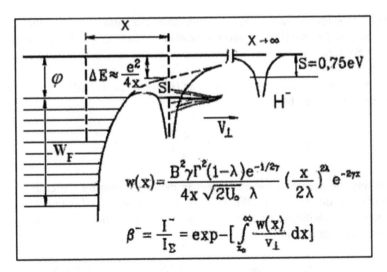

Fig. 4.60 Approximation of the potential for the electron transition between solid levels and the electron affinity of a particle. (Reproduced from Dudnikov [1])

$$U(\xi) = -U_o, \text{ for } \xi < e/4U_o.$$

(The shape of the potential for $\xi < x_o$ is unimportant for the above consideration, and for $\xi > x_o$ this approximation seems to be adequate.) The interaction potential energy of an electron with a particle, which ensures the presence of an electron affinity level, is approximated by a potential of zero width (i.e., a delta function) [103]. A schematic potential distribution is given in Fig. 4.60.

The differential negative ionization coefficient $\beta^-(v,S,\varphi)$ for particles having, at $x = x_o$, a velocity v upon removal from the surface is expressed in terms of the neutralization velocity of an H^- ion located at a distance x from the surface, $w(x)$, as

$$\beta^-(v,S,\varphi) = \exp - \left\{ \int (w(x)/v) dx \right\},$$

which determines the probability of particles leaving the surface without neutralization.

To determine the neutralization probability $w(x)$, we use a modification of the zero-radius potential method [103] used by Demkov and Drukarev to determine the probability of destruction of an H^- ion in a homogeneous electric field. The problem reduces to determining the imaginary part of the eigenvalue of the energy $\kappa^2/2$ of the three-dimensional Schrödinger equation, which has the following form in the atomic system of units:

$$\Delta\psi + \left(-\kappa^2/2 + 1/(2\xi) - 1/(2x)\right)\psi = \delta(r - R),$$
$$4\pi\partial\psi / \partial(r - R)\big|_{r=R} = -\gamma,$$

where r and R are the radius vectors of electron and particles, $\gamma^2/2$ is the electron affinity level of the free particle, and x is the distance from the particle to the surface.

The wave function ψ should decrease as $\xi \to \infty$ and have the form of an outgoing wave for $\xi < 1/(4\,U)_0$.

For the probability of neutralization of an H^- ion, located at a distance x from the surface, we obtain the following expression:

$$w(R) = B^2 \gamma \Gamma^2 (1-\lambda) e^{-\frac{1}{2\gamma}} \left(\frac{R}{2\lambda}\right)^{2\lambda} e^{-2\gamma R} / 4R (2U_0)^{1/2} \lambda$$

Multiply the square of the coefficient in the expression for the electron wave function in the free H^- ion $\psi = B^2 (2\gamma)^{1/2} e^{-r^2/(4\pi)^{1/2}} r$, where B^2 takes into account the finite radius of additional electron interaction with the particle. The expression for the differential ionization coefficient becomes

$$\beta^- = \exp -\propto \int_{R^0}^{\infty} w(R) \frac{dR}{v_\perp(R)} = \exp -\frac{B^2 \Gamma^2 (1-\lambda)}{2v_\perp R_0} \sqrt{\frac{\varphi}{U_0}} e^{-\frac{1}{2\gamma}} \left(\frac{R_0}{2\lambda}\right)^{2\lambda} e^{-2\gamma R_0}$$

where $R_0 = \tfrac{1}{4}(\varphi - S)$, $\lambda = \tfrac{1}{4}(2\varphi)^{1/2}$.

The calculated values of the differential coefficient of negative ionization of hydrogen, $\beta^-(v,\varphi)$, which has an electron affinity $S = 0.75$ eV ($\gamma = 0.236, B = 1.68$), as a function of the escape velocity from the surface v at $x = x_0$ and the work function φ are shown in Fig. 4.61.

For a realistic work function $\varphi > 1.7$ eV, an interpolation is suitable:

$$\beta^- = 0.12(v - v_0)/(\varphi - S),$$

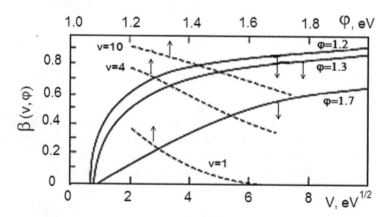

Fig. 4.61 Calculated dependences of the differential coefficient of negative ionization of hydrogen β^- on the escape velocity v for fixed work function φ, and on the work function φ at fixed escape velocity from the surface v (dotted line). (Reproduced from Dudnikov [1])

where $v_0 = (\varphi - S)^{1/2}$ is in $(eV)^{1/2}$ and $(\varphi - S)$ is in eV.

Cesiated surface of molybdenum with ultralow work function was produced in [69] after heating up to ~1000 °C. Cesium evaporation on molybdenum surface in "ion source vacuum," 10^{-5}–10^{-6} Tor, was used for production surfaces with low work function. Photoelectronic measurements were used for work function monitoring. Presented in Fig. 4.57 are the variations of work function during cesium deposition. Work function $\varphi \sim 1.2$ eV was produced. So calculation of ionization probability for work function $\varphi \sim 1.2$–1.3 eV realistic in condition of surface plasma sources with cathode heating up to ~1000 °C.

In a similar formulation, Janev [61, 63] has solved this problem by a WKB method; however, due to an error in the expression for the image potential, the probability of neutralization turned out to be greatly overestimated.

Kishinevskii considered qualitatively the features of the formation of H$^-$ ions on surfaces of the work function which is lowered by the adsorption of cesium. With coatings close to optimal, the formation of negative ions should occur on homogeneous surfaces with a corresponding work functions φ.

Naturally such a simplified analysis does not reflect all the features of the formation of H$^-$ ions under the most complicated conditions realized on surfaces in real SPS conditions. However, it provides an adequate estimate of the probability of formation of negative ions during sputtering and reflection.

Note that cesium atoms at optimum surface concentration are located on the surface of refractory metals rather sparsely: the average distance between atoms is 7 Å. Thus the close interaction of incoming atoms with cesium particles is unlikely.

Hiskes [104] considered the formation of H$^-$ ions under assumptions close to the work of Kishinevskii and produced similar results.

In the scheme considered, the slowing of ions by image fields, which reduce the already small coefficient of ionization of slow particles but does not affect the ionization of fast particles, was not taken into account. Particles having kinetic energy removal $\frac{1}{2}Mv^2 < (\varphi - S)$ at $x = x_0$ cannot escape over the potential barrier near the surface in the form of H$^-$ ions, or they are neutralized and are converted into free neutral particles or remain in the adsorbed state. The total H$^-$ ion flux from the surface, I^-, depends on the distributions $f_r(v,x_0)$ and $f_s(v,x_0)$ of reflected and sputtered particles at a rate of removal from the surface determined by the exchange of momentum between bombarding particles and condensed phase particles:

$$I^- = \int b^- (v, S, j) \left[f_s (v, x_o) + f_r(v, x_o) \right] dv.$$

The ionization coefficient β^- is determined by the expression:

$$\beta^- = I^- / I,$$

where $I = \int [f_r(v) + f_s(v)] dv.$

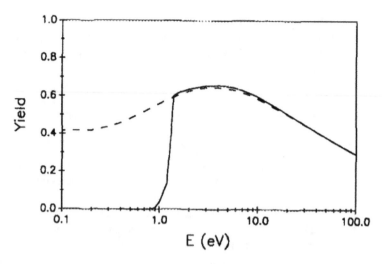

Fig. 4.62 H⁻ yield as a function of incident energy on a cesium-covered tungsten surface. $\varphi = 1.45$ eV. Solid line: variable velocity; dotted line: constant velocity

In turn, the distribution functions of the sputtered and reflected particles are determined by the regularities of the sputtering and reflection processes considered in the previous section.

The calculation of negative ionization probabilities on cesiated tungsten surface with work function $\varphi = 1.45$ eV was presented in [105].

H- yield as a function of incident energy on a cesiated tungsten surface ($\varphi = 1.45$ eV) is shown in Fig. 4.62: (solid line) variable velocity with retarding negative ions by imaging force.

A computer-calculated probability of negative ionization of hydrogen particles on metal surfaces with low work function was presented in [106].

The reason for the high ionization efficiency of the particle flux moving away from the surface with increased velocities bases on analysis of the resonant charge transfer (RCT) between a solid and a particle at small distance from the surface [69, 73]. Near the metal surface, the affinity level of a negative ion shifts downward due to the electrostatic interaction of electron with the metal. The shift of electron affinity level relative to the Fermi level of the metal can be approximated by an image potential. When approaching the surface, the level of the electron affinity acceptor is broadened, and the center of this band is lowered relative to the electron affinity level of free particles S by an amount $\Delta S = -1/4z$ (for convenience the atomic system of units is used, where $m_e = e = \hbar = 1$; e.g., 1 a.u. of distance is equal to 0.53 Å. The energies are given in electron-volts, relative to the vacuum level ($E_v = 0$)). Since the resonant negative ionization occurs for the ion-surface distances where the affinity level is below the Fermi level of the metal, the Fermi level of the metal (or the surface work function) has a decisive influence on the ionization probability. The common picture is that negative ionization probability increases exponentially with the decrease of the converter surface work function.

We consider one-electron process of negative ionization of hydrogen neutral atom by means of resonant charge transfer (RCT). The fraction of positive hydrogen ions is negligibly small, because H$^+$ is completely neutralized near the surface by means of Auger neutralization and RCT [106, 107]. Since the typical ion energies in surface NIS are 10–100 eV, the adiabatic approximation is applied. In this case the RCT is calculated by means of rate equation [108]. To find the population of the atomic state, we integrate the rate equation along the trajectory of ion motion [108, 109]:

$$\frac{dP^-}{dt} = -\Gamma_{\text{loss}}(z) \cdot P^- + \Gamma_{\text{capture}}(z) \cdot \left(1 - P^-\right) \tag{4.1}$$

where z is the normal distance to the surface, $\Gamma_{\text{loss}}(z) = g_{\text{loss}} \cdot \Gamma(z) \cdot F_{\text{loss}}$, $\Gamma_{\text{capture}}(z) = g_{\text{capture}} \cdot \Gamma(z) \cdot F_{\text{capture}}$, $\Gamma(z)$ is the RCT ion level width, and $F_{\text{loss}}(z)$ and $F_{\text{capture}}(z)$ are the electron loss and capture weights, respectively. The statistical factors are usually taken as $g_{\text{loss}} = 1$ and $g_{\text{capture}} = 0.5$ to account for the electron spin [108, 110].

The procedure of $\Gamma(z)$ calculation as a function of the ion-surface distance is described elsewhere [111, 112]. In brief, we perform a series of auxiliary calculations of H$^-$ decay in front of the metal surface for a fixed ion-surface distance. The hydrogen anion is considered as the hydrogen-like atom, consisting of a single *active electron* and a neutral atomic core. We use the three-dimensional realization of the wave packet propagation method [113, 114], which considers a direct study of the evolution of the active electron wave packet in the compound potential created by the surface and the projectile. Thus, we numerically solve the time-dependent Schrodinger equation (TDSE) with known initial conditions [115–118]:

$$i\frac{d\psi(r,t)}{dt} = \left(-\frac{\Delta}{2} + U(r,t)\right)\psi(r,t)$$
$$\psi(r,0) = \psi_0(r) \tag{4.2}$$

where $U(r, t) = V_{\text{e-ion}}(r, t) + V_{\text{e-surf}}(r) + \Delta V_{\text{e-surf}}(r, t)$ is the time-dependent potential felt by the active electron. Here $V_{\text{e-ion}}(r, t)$ describes the electron interaction with the projectile, $V_{\text{e-surf}}(r)$ describes the electron interaction with the metal surface, and $\Delta V_{\text{e-surf}}(r, t)$ describes the interaction with the polarization charge induced by the ion (ion-induced potential [119]). Note that $U(r, t)$ incorporates ion movement, since $V_{\text{e-ion}}(r, t)$ implicitly depends on ion position. The analytical potentials are available for the hydrogen anion [120] and some metal surfaces [121, 122]. Since the above set doesn't include analytical potential for the cesiated W surface, therefore, we use $\Gamma(z)$ from Ref. [108] for the cesiated Mo.

The TDSE numerical solution provides the time evolution of the system's wave packet $\psi(r,t)$. The projection of the current wave function on the initial state $\psi_0(r)$ gives the survival amplitude of the wave packet in the initial state:

$$A(t) = \psi_0(\boldsymbol{r}) | \psi(\boldsymbol{r},t), \tag{4.3}$$

that is a complex function. The square modulus of the survival amplitude gives the survival probability of the initial state, i.e., the probability that the H⁻ state is completely occupied:

$$P^-(t) = |A(t)|^2 \tag{4.4}$$

The ion level width $\Gamma(z)$, which characterizes RCT rate, is calculated from the following approximation:

$$P^-(t) = \exp(-\Gamma t). \tag{4.5}$$

Of note, in most cases, the so-called exponential decay takes place [108]. Hence, $\Gamma(z)$ can be easily calculated from linear approximation of $\log(P(t))$.

Finally, to find the negative ionization probability P^-, we integrate the rate Eq. (4.1) along the outgoing ion trajectory for distances 3 a.u. $< z <$ 20 a.u. In most of presented calculations (except Fig. 4.43), the ion is supposed to leave the surface with constant velocity, i.e., interaction with the image charge is not accounted.

Two main parameters, which characterize RCT and determine the ionization probability, are (i) negative ion affinity level $E_a(z)$ and (ii) level width $\Gamma(z)$. The procedure of $\Gamma(z)$ calculation is described above. The traditional approach for the affinity level is that ion level near the surface is shifted down on the value of image potential $E_a(z) = E(\infty) - 1/4z$.

In this section we compare our calculations to the experimental data and estimate possible NIS efficiency enhancement due to the parallel velocity effect. Most of the experimental data on negative ionization was obtained in LEIS experiments.

The main factor, which determines the negative ionization probability, is the difference between the Fermi level of the metal and affinity level of the negative ion. Of note, the Fermi level of the metal is related to its work function. Figure 4.63 shows the influence of work function on negative ionization probability. The model calculations were done with parameters from Refs. [94–108]; the parameter E_0 was taken as −2.0 eV. One can see that the negative ionization probability depends heavily on the work function. Of note, for small ion energies (~10 eV), the negative ionization probability is ~20% larger than in Ref. [94], but close to Ref. [108].

Figure 4.64 shows the normal energy/velocity dependencies of hydrogen negative ionization probability for the cesiated Mo and W(110) converter surfaces. The work function for the cesiated Mo surface is 1.5 eV [123], 1.45 eV for W(110) [108], and the parameter $E_0 = -1.47$ eV for both calculations. One can see that our calculation results quantitatively correspond to the experimental data [123]. This indicates that the appropriate theoretical model and numerical method were selected.

The theoretical negative ionization probability for Figs. 4.63 and 4.64 was calculated for the normal ion exit, i.e., with the zero parallel velocity component. Hence, the same $P^-(v_{\text{norm}})$ dependence can be obtained by means of the Anderson-Newns

Fig. 4.63 Influence of surface work function on negative ionization probability. The figure shows the negative ionization probability of hydrogen as function of the initial normal exit energy/velocity for different work functions

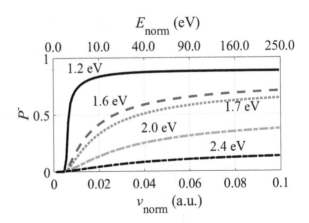

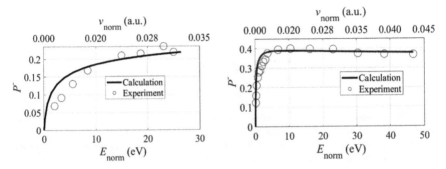

Fig. 4.64 Negative ionization probability of hydrogen on different converter surfaces as function of the exit normal energy/velocity. The calculations were done according to the model presented in Sect. 4.2. Upper panel: the cesiated Mo ($\varphi = 1.5$ eV) [123]; lower panel: the cesiated $W(110)$ ($\varphi = 1.45$ eV) [108]. The incident angle of the primary H+ beam is 10^0 from the surface and the exit angle varies in the range 0–20^0

model or the traditional rate equation. The key advantage of our model is that it can take into account the component of ion velocity, parallel to the surface. Figure 4.65 shows the H– formation probability on the cesiated $W(110)$ converter surface as function of the exit normal velocity during grazing scattering of H+ beam. Of note, different energies of primary beam correspond to the different values of hydrogen's parallel velocity in the diapason 0.14–0.25 a.u. The variation on the exit normal velocity is reached by the variation of the exit angle. The work function for the cesiated $W(110)$ surface is 1.45 eV [121], and the parameter $E_0 = -1.85$ eV. One can see that the calculation results quantitatively describe the experimental data. It should be mentioned that calculations in the original article [122] reproduce the experimental data slightly better. However in in [123], the entire dependence of the affinity level on the ion-surface distance $E(z)$ was approximated to fit the experimental data, while in our model, there is the only approximation parameter $-E_0$.

The negative ionization probability was calculated for hydrogen, deuterium tritium atoms, and muonium atoms escaping surfaces with different surface work

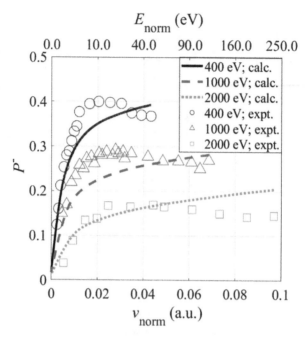

Fig. 4.65 Negative ionization probability of hydrogen on cesiated W(110) converter surface ($\varphi = 1.45$ eV) as function of exit normal velocity/energy for different ion beam energies (see legend). The experimental data is taken from Ref. [124]. The incident angle of the primary H^+ beam is 10^0 from the surface and the exit angle varis in the range $0\text{--}20^0$

functions with different speeds [124]. The integral dependence of the negative ionization for hydrogen, deuterium tritium atoms, and muonium atoms on the work function of the metal surface is shown in Fig. 4.63a. The integral probability was calculated taking into account the energy distribution function of the outgoing particles shown in Fig. 4.66b. This is a distribution function of atoms desorbed by 4 eV ions.

Surface production of muonium negative ions is discussed in [125].

Particles with larger mass have a lower velocity at the same energy and have an elevated probability for loss of extra electron.

The general pattern is that the probability of negative ionization decreases with increasing work function. This is in good agreement with the known experimental data and theoretical results on the study of the electron exchange of atomic particles with metal surfaces.

As cited in Ref. [126], the formation of negative hydrogen and deuterium ions in negative ion sources based on surface conversion is strictly depending on the work function of the converter surface. In order to study the isotopic difference in the negative ion surface formation between hydrogen and deuterium plasmas, simultaneous and absolute measurements of negative ion density and work function are performed at a well-diagnosed ICP plasma. The work function of the surface is

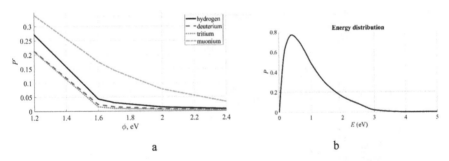

Fig. 4.66 (**a**) The integral dependence of the negative ionization for hydrogen, deuterium tritium atoms, and muonium atoms on the work function of the metal surface. The integral probability was calculated taking into account the energy distribution function of the outgoing particles shown in Fig. 4.66b

evaluated by means of the photoelectric effect, while the H$^-$ and D$^-$ densities are measured by cavity ring-down spectroscopy at 2-cm distance from the sample. The possibility of controlled and reproducible investigations allows to study the transition from negative ion volume formation to surface formation at low surface work functions. The negative ion density n_i shows a steep increase for work functions below 2.7 eV, while above this value, the main contribution to the negative ion density is due to volume formation. For a work function of 2.1 eV (bulk Cs), the H$^-$ density increases by at least a factor of 2.8 with respect to a non-cesiated surface, arising from densities of around 1×10^{12} cm^{-3} of the surface produced negative ions. By applying a deuterium plasma, no significant isotopic effect occurs regarding the surface-negative ion formation: the D$^-$ density measured for a work function of 2.1 eV increases by a factor of 2.5 with respect to a non-cesiated surface, achieving densities of the surface produced negative ions comparable to the hydrogen case. Since neutralization of the positive ions at the surface is not isotope dependent and since no isotope effect has been found in the surface conversion of the neutralized particles into negative ions, no isotopic difference is expected also for the conversion of atoms. It can thus be excluded that the isotope dependence of the co-extracted electrons observed in the ion source is due to an isotopic difference in the negative ion formation.

It follows from the above results that the most important reserve for increasing the efficiency of formation of negative ions is a decrease in the work function of the emitting surface. For $\varphi = 1.5$–1.6 eV, due to adsorption of cesium on refractory metals, we can calculate $\beta^- = 0.5$–0.6 for particles $\langle v \rangle = 5$ eV$^{1/2}$, and for $\varphi = 1.2$ eV, which is achievable with cesium adsorption in the presence of electronegative additives, this can be brought to $\beta^- = 0.8$–0.9. A further decrease in the emitting surface work function to $\varphi < 1$ eV would ensure the effective formation of ions from slow fragments with $\langle v \rangle = 1$ eV$^{1/2}$. Thus the theoretical concepts of reflection and sputtering and electron exchange between particles and the solid provide some predictions about the properties of the process, agreeing with the results of experimental studies of these processes in SPSs. Publication of the surface plasma mechanism of

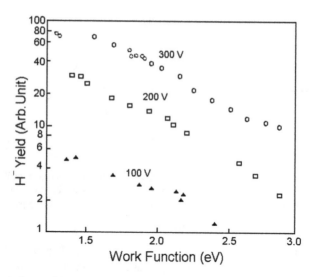

Fig. 4.67 Dependence of the yield of reflected negative ions on work function, for several different bombarding particle energies. (Reproduced from Wada et al. [127] with permission of AIP Publishing)

negative ion formation initiated studies of these processes in many laboratories around the world, and investigations [127] of the reflection of hydrogen particles in the form of negative ions from surfaces coated with alkali metals for different energies of bombarding particles. The review of methods for determination of surface work function in cesiated SPS were presented in Ref. [61].

The dependence on work function of the fraction of reflected negative ions at different bombarding particle energies is shown in Fig. 4.67. The fraction of reflected negative ions depends on the work function exponentially, as follows from the curves shown in Fig. 4.61.

Later, Janev repeated the calculations of β^- [129] and obtained results close to [58]. The β^- calculations were repeated in [130, 131], and formula

$$\beta^-(v) = 2\exp\left[-(\varphi - S)/av\pi\right],$$

was obtained, where a is a constant determined experimentally [132]. The dependence of the yield of sputtered negative ions on the work function, with surface cesium deposition, for bombardment with neon ions has been investigated [133]. The dependence of the work function and the emission of H⁻ ions on the cesium deposition thickness is shown in Fig. 4.68. There is a very sharp dependence of H⁻ ion emission on the work function of the surface.

Detailed studies of the probability of reflection in the form of positive and negative ions from surfaces with different work functions have been carried out [134]. Figure 4.69 shows the dependence of the work function and conversion efficiency of reflected H⁻ hydrogen ions on the degree of cesium surface coating.

Fig. 4.68 Dependence of
the surface work function
and the yield of Ne+
sputtered H⁻ ions on
cesium deposition
thickness (deposition
time). (Reproduced from
Yu [128], BNL- 50727,
edited. By Th. Sluyters and
C. Prelec)

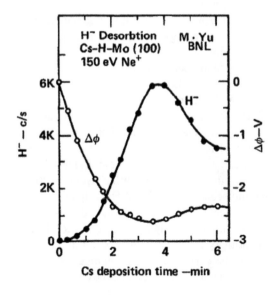

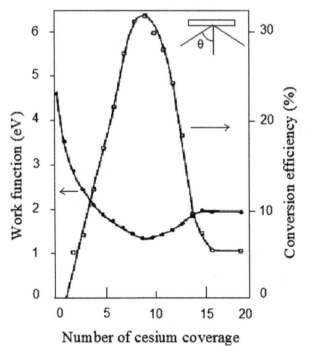

Fig. 4.69 Dependence of the surface work function and conversion efficiency on cesium deposition thickness (deposition time). (Reproduced from Los and Geerlings [134])

The dependence of the output of negative ions reflected from cesium coated surfaces on the energy of the bombarding ions H_2^+, H_3^+, D_2^+, D_3^+ has been investigated [136].

Correlation between surface work function and backscattered D^- ion yield. The thickness of Cs layer on nickel increases as more Cs is evaporated to the nickel target is shown in Figs. 4.70 and 4.71.

The formation of negative ions by hydrogen and deuteron ions backscattering from alkali metal and cesiated surfaces was studied by a group at LBL [135]. The energy of incident particles varied in the 0.15–4-keV range. It was found that the backscattering negative ion yield, R, has a larger value for alkali metals with larger atomic number and lower surface work function. During collisions with the surface, the molecular ions H_2^+ and H_3^+ (D_2^+, D_3^+) dissociate and give two or three times greater yield of H^- (D^-) ions than that of the H^+ (D^+) ions with the corresponding equivalent energy. Figure 4.72 shows R^- dependences on the equivalent energy of the primary D ions incident normal to a nickel-cesium surface. The figures to the left of the curves denote the number of cesium portions injected into the nickel. Figure 4.66 to the right denote the corresponding change of the Ni-Cs surface work

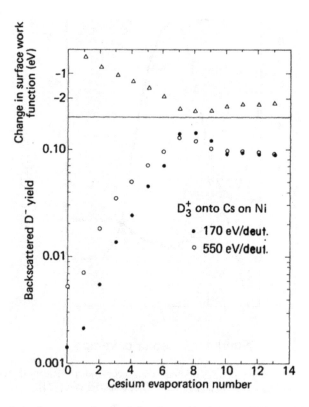

Fig. 4.70 Correlation between surface work function and backscattered Dion yield. The thickness of Cs layer on nickel increases as more Cs is evaporated to the nickel target. (Reprinted with permission from Schneider et al. [135]. Copyright 1981 American Physical Society)

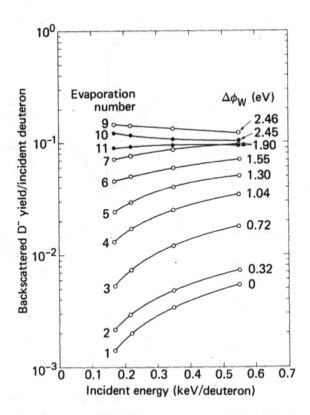

Fig. 4.71 Backscattered D$^-$ yield. Cs-Ni bombarded by D$_3^+$ ions. (Reprinted with permission from Schneider et al. [135]. Copyright 1981 American Physical Society)

function. For optimal cesium coverage on nickel ($\phi \sim 1.6$ eV) reaches the value of ~0.14 and is slightly diminished with an increase in the primary ion energy within 170–550 eV/deuteron range. The extrapolation of these data to the region of lower primary ion energies 100–20 eV/deuteron through the use of the MARLOWE code gives greater values of $R- \sim 0.24$–26. For low cesium coverage on nickel, the $R-$ value increases with the energy of incident particles (Fig. 4.72). For pure nickel, NISEC has a value $R- = 0.1\%$ at the energy of the primary ions D$^+$ 170 eV/deuteron and increases 1.5 times due to small cesium coverage of ~0.1 monolayer. In the case of grazing incidence of hydrogen ions, the particle reflection coefficient is greater. The values $R- = 0.35$ were detected for hydrogen ions with an energy 400 eV/ nucleon and 82″ incident angle with respect to the normal to the W-Cs surface ($\phi \sim 1.45$ eV) [31]. The cited high values of $R- \sim 0.243$ may be realized in the reflection of fast hydrogen ions and atoms from the surfaces of cathodes, anodes, or special emitters covered with cesium and biased negatively relative to the SPS plasma with a potential between 100 and 200 V.

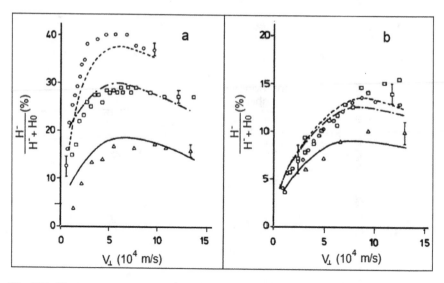

Fig. 4.72 The H⁻ fraction of the scattered particles as a function of the normal velocity. The left-hand side gives the results for a $W(110)$ surface covered with half a monolayer of cesium ($\varphi = 1.45$ eV) and the right-hand side for a full monolayer ($\varphi = 2.15$ eV). Measurements are indicated by (o) for 400 eV ($v_{\mathrm{II}} = 2.8 \times 10^5$ m/s), ($\square$) for 1000 eV ($v_{\mathrm{II}} = 4.5 \times 10^5$ m/s), and ($\triangle$) for 2000 eV ($v_{\mathrm{II}} = 6.3 \times 10^5$ m/s). The calculations are indicated by the dashed lines, dotted dashed lines, and solid lines, respectively. (From Ref. [2]. Reproduced from Berkner et al. [136])

The dependence of the conversion efficiency for H⁻ ions on the normal component of escape velocity of the particle, for an optimal cesium coating and for a thick layer of cesium, is shown in Fig. 4.72.

Figure 4.73 shows the experimental and extrapolated dependence of the yield of negative ions on the work function of the surface.

The formation of H⁻ ions in the reflection of hydrogen ions from the surface of tungsten 100 coated with cesium has been investigated [137]. Figure 4.74 shows the fraction of reflected negative ions $I^-/(I^- + I)$ as a function of angle of reflection β, for reflection from cesium-coated $W(110)$. The angle of incidence is 45° (filled symbols) or 70° (open symbols).

Reflection in the form of H⁻ ions upon bombardment of $LaB_6(100)$ and Cs/Si(100) surfaces with H⁺ ions of energy 100 eV has been investigated [138]. Up to 30% of the bombarding ions on the LaB_6 (100) surface and up to 55% on the Cs/Si (100) surface were scattered in the form of H⁻ ions.

The effectiveness of H⁻ ion formation upon bombardment of various surfaces sprayed with cesium was explored [139]. Figure 4.75 shows the yield of negative ions from various cesiated materials as a function of energy of bombarding particles.

In all these papers, high (tens of percent) conversion efficiencies were obtained for negative sputtered and reflected ions. But the dependence on work function is very sharp. An acceptable efficiency is obtained when the work function decreases below 2.5 eV. The dependence of H⁻ ion current on surface work function upon

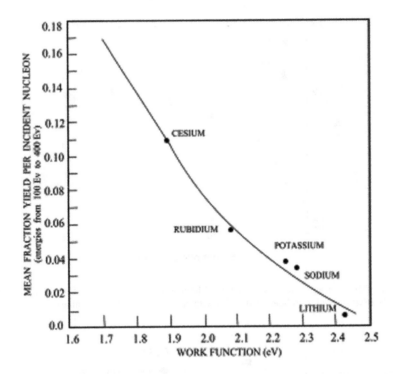

Fig. 4.73 Experimental and extrapolated dependence of the output of reflected negative ions on the work function of the surface

Fig. 4.74 Fraction of negative ions reflected from cesium-coated $W(110)$, $I^-/(I^- + I)$, as a function of angle of reflection β. The angle of incidence is 45° (filled symbols) or 70° (open symbols). (Reproduced from van Amersfoort et al. [137])

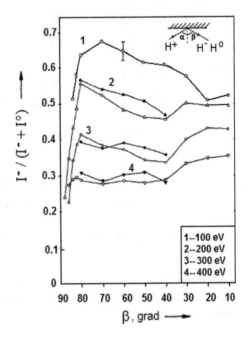

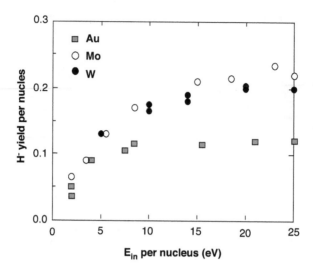

Fig. 4.75 Yield of negative ions as a function of bombarding particle energy for various cesiated materials. (Reproduced from Seidl et al. [137] with permission of AIP Publishing)

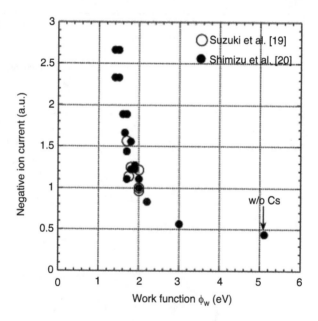

Fig. 4.76 Dependence of H⁻ ion current on work function upon bombardment of surfaces in a hydrogen plasma. (Reproduced from Shimizu et al. [141])

bombardment of surfaces in a hydrogen plasma is shown in Fig. 4.76 [140, 141]. The dependence of negative ion production efficiency on work function has been investigated [142].

4.13 Implementation of Surface Plasma Production of Negative Ion Beams

The characteristics that we have considered of the reflection of bombarding particles, adsorbate sputtering, and the features of electron exchange provide a basis for optimizing processes 3 and 4 (as described in Sect. 4.10) in the negative ion formation sequence. These processes are specific for surface-plasma sources and secondary emission (sputtering) sources. Process 2 is specific for SPSs. The remaining processes of the sequence in one way or another have to be performed in positive ion plasma sources, in plasma, and in charge-transfer and thermionic sources of negative ions.

To increase the H^- ion formation efficiency by capture of electrons at the electron affinity level of sputtered and reflected particles, it is necessary to minimize the work function of the emitting electrode whenever possible. Without special tricks, the work function of refractory metals decreases by cesium deposition to values of around $\varphi = 1.6$ eV.

For such work function values, the formation of H^- ions by reflection is most effective at a bombarding atom energy $W = 100$ eV. Taking into account the patterns of reflection and electron exchange under such conditions, the formation of H^- ions should be ensured with a conversion coefficient $K_r^- = 0.2$–0.3 when the bombardment is normal to the surface. When bombarded with molecular ions H_2^+ and H_3^+, K_r^- can be 2–3 times larger (for the same values $\beta^- = 0.5$). Bombardment at an angle to the normal should increase the probability of reflection K_r and the emissivity K_r^-, but at large angles the ionization coefficient may decrease because of the decrease in escape velocity v. Apparently an energy of 10^2 eV is close to the optimum for formation of H^- ions by sputtering of the adsorbate.

The results of the experiments considered in the previous sections show that the emission of H^- ions with an effective negative ion secondary emission coefficient of $K^- = J^-/J^+ = 0.2$–0.6 is realized in SPSs at considerable bombarding particle flux intensities. Taking into account the mass composition of the bombarding hydrogen ions, the values of K^- obtained should be considered consistent with theoretical predictions.

Note that these emission factors are defined as the ratios of current densities of negative and positive ions, but in fact the emission of H^- ions is due to the bombardment of the surface by fast neutral particles. For $\beta^- < 1$, a part of the bombardment flux is reflected in the form of fast atoms. Fast negative ions are converted into fast atoms upon charge exchange and electron transfer. The presence of a significant number of fast atoms has been confirmed experimentally by spectroscopic studies of the planotron discharge plasma [143]. Fast atoms can initiate the emission of H^- ions in subsequent collisions with surfaces. Due to such repeated use of bombarding particles, the effective values of K^- can increase substantially.

Apparently, the possibilities for minimizing the work function of the surface are far from exhausted. By creating and dynamically maintaining more complex surface structures using electronegative oxygen additives, fluorine or effective emitters (lanthanum hexaboride) with optimal surface cesium concentration, the emitter work function can be reduced to $\varphi < 1$ eV. When optimizing the bombardment conditions for reflection and sputtering, and taking into account the repeated use of bombarding particles, one can count on ensuring the emission of one or even more H^- ions by an accelerated bombarding ion with energy of tens of eV with a relatively small energy release at the emitting electrode.

The studies considered in Sect. 4.10, show that plasma is a necessary component of the SPS. It promotes the preservation of good emission properties of electrodes in the harsh intense bombardment conditions. The good retention of cesium and other heavy components at the emitting electrode makes it possible to hope that the emission properties of electrodes can be maintained close to optimum even under steady operating conditions.

Optimizing the cesium supply is an important requirement for implementation of the SPS. The available cesium supply options give satisfactory results. Supplying cesium from cesium-containing substances (mixtures of cesium chromate or bichromate with titanium, calcium, chlorides, cesium fluorides, or cesium alloys with bismuth) and heating these substances with the discharge is the simplest way; it is important to monitor the steadiness of cesium release. This method allows stable operation of the sources during research. A more flexible adjustment is achieved when cesium is fed from a container with independent heating. In this case, one can use ampoules of the listed cesium-containing substances or metallic cesium. This method allows very long operation of the SPS in a high-current mode. An important standby for reducing the consumption of cesium is to decrease the electrode temperature by cooling. It is necessary to fine-tune the temperature regimes of different electrodes in the SPS with the supply of cesium. The difference between the results of a number of SPS studies and from optimal results seems to be due to such mismatches. Dissociation of the working substance in the plasma and the interaction of the plasma with the electrodes facilitate the adsorption of particles of the working substance into states from which they are efficiently sputtered as negative ions.

Naturally, options for the implementation of processes 3 and 4 (of Sect. 4.10) do not exhaust all opportunities, and there is room for further optimization.

The efficiency of emission of negative ions $K^- = 0.2$–0.8, which is controlled in a number of ways, to obtain ion fluxes of H^- with current density $J^- = 1$–5 A/cm^2, close to the limiting values for electrostatic formation systems, requires fluxes of bombarding particles with current density $J^+ = 2$–15 A/cm^2 at an energy of tens to hundreds eV. To date, the creation of a particle flux with such parameters is possible only by their generation and acceleration in the plasma. The rational implementation of these processes (processes 1 and 2) of the sequence of SPM largely determine the resulting efficiency of obtaining H^- ion beams. The way of implementing processes 1 and 2 (again, see Sect. 4.10) determines the destruction of negative ions when they are transported to the beam formation system (in the implementation of process 5) and the efficiency of gas utilization in the SPS.

The implementation of these processes in the SPS with planotron electrode configuration was relatively successful. In this case, positive ions are generated in a narrow gap between the $H^-$and electron emitter (the central plate of the cathode) and the anode with emission hole by a discharge with electrons reflexing in the magnetic field. The resulting positive ions are transported at low gas density to the emitter by the electric field of the gas discharge gap and are accelerated in the near-electrode layer to the emitter. Effective transport of H⁻ takes place by their acceleration in the emitter near-electrode layer and by minimizing the plasma thickness and gas layer thickness between the emitter and the beam formation system. With realization of these processes in regimes with maximum beam intensity, a significant quantity of negative ions formed is destroyed during transport and beam formation. In connection with this, these extreme modes should realistically be used only in cases when it is important to obtain a negative ion beam with maximum emission current density and brightness. To ensure high H⁻ ion beam production efficiency in such sources, it is sensible to use regimes providing emission current densities up to $J = 1$–2 A/cm², which are enough for operational high-current injectors. The negative ion generation efficiency increases dramatically with geometric focusing.

In SPSs with large gaps between the cathodes and the beam formation system (tube-discharge duoplasmatrons, Penning gas discharge cells, tandem sources), negative ions formed at the cathode cannot reach the beam formation system without charge-exchange and destruction. In these cases, negative ions escaping from the source begin to be formed by emission from electrodes adjacent to the emission aperture and from the walls of the emission aperture itself, initiated by bombardment by fast ions and neutral particles formed by neutralization of accelerated ions. In this way, a second method for obtaining streams of bombarding particles is realized. The resulting negative ion formation efficiency in this case, as a rule, is lower than when the entire flow of negative ions is completely used, formed during all bombardment events. However, in these SPSs, with special optimization of the electrode emission properties, which provide the main contribution to negative ion formation, the production efficiency of negative ion beams can be maintained at a high level. Using elements of the beam formation system as H⁻ ion emitter, a weakening of ion flux decreases when the beam is formed, so that the intensity of the produced beams is proportional to the discharge current over a wide range of variation. The SPM allows for a wide variety of methods for generating and accelerating bombarding particles compatible with the effective implementation of the remaining processes for obtaining negative ions in the SPS.

The data in Sect. 4.5, show that for an SPS with combined electrode functions, when optimizing the electrode emission properties, the efficiency of generation of positive ions deteriorates due to the decrease in energy of fast electrons. The data on generation of primary positive ions in higher-voltage SPS discharges and experience on obtaining beams from plasma sources of positive ions with thermal cathodes show that in an PS with separated functions, it is possible to reduce the energy cost for the generation of primary ions to a level close to the theoretical limit $P^+_{min} = 0.3$ keV/ion. Very importantly, at the same time, it is possible to count on a significant reduction in initial hydrogen density and a corresponding increase in the SPS gas utilization efficiency. As already noted in Sect. 4.10 with complete optimization of the formation process of H⁻ ions in the SPS, one can count on the reduction of energy cost of an H⁻ ion to a level $P^- = 1$ keV/ion. The value $P^- = 10$ keV/

ion already attained does not worsen the injector efficiency, but further reduction is important to simplify the technical implementation of high-current sources with a large resource of work.

An important role is played by the last sixth process (number 6 of Sect. 4.10)—formation and primary acceleration of the negative ion beam. Naturally, the basic laws of ion optics remain valid here. To form the beam, it is possible to use a system close to those developed recently for the formation of intense positive ion beams. However, the extraction and beam formation for negative ions is complicated by a number of circumstances that have an overriding influence on the ways of realizing this process.

The extraction of negative ions from the gas discharge chamber and the formation of beams in the SPS, as well as in volume plasma sources, is complicated by the presence of accompanying electrons. In plasma source designs that have been developed, the flux of accompanying electrons is attenuated by displacement of the emission hole away from the discharge axis, the use of a tube discharge, and the extraction of negative ions across a magnetic field [3, 4]. However, in all these cases, the electron flux still remains many times greater than the negative ion flux.

More radically, the electron flux is limited by extraction of negative ions across the magnetic field through an emission slit with small length along the magnetic field [144]. In the SPS with the emission slit oriented across the magnetic field, the current in the high-voltage gap of the beam formation system proves to be comparable to the H⁻ ion beam current formed under optimal conditions [12, 145, 146] due to the removal of electrons, interception of the beam by the extraction electrode, secondary emission, ionization and charge exchange, cesium release, and other factors. Good filtration of electrons from the negative ion flux is because of a significant difference in the plasma flow velocity along and across the magnetic field. The attenuation of the electron-ion plasma flow in its motion across the magnetic field between absorbing surfaces perpendicular to the magnetic field has been considered by a number of authors [3, 147]. In the case of interest to us, the electrons can quickly escape along the magnetic field to absorbing surfaces spaced by a distance d and slowly diffuse perpendicular to the magnetic field. The diffusion flux of plasma across the magnetic field is proportional to the density gradient. In the stationary case, the rate of change of the flow along the diffusion direction is equal to the flux density of plasma along the magnetic field and proportional to the plasma density in each section. Under such conditions, the plasma density along the channel must decrease exponentially:

$$n_e(x) = \exp-(x/a),$$

with a characteristic density decrease scale of

$$a = \left(D_\perp d / \xi\right)^{1/2},$$

where $D_\perp$ is the diffusion coefficient of the plasma across the magnetic field, d is the distance from the absorbing surface, and ξ is the coefficient of proportionality between the plasma density along the magnetic field and the average plasma density in the given section. In the case of diffusion motion of the plasma along the magnetic field with diffusion coefficient $D_\parallel$,

$$a = d\left(D_\perp / D\right)^{1/2},$$

and with free plasma escape along the magnetic field, which is more appropriate for the situation in question,

$$a = \left(dD_\perp / v_s\right)^{1/2},$$

where v_s is the plasma flow velocity along the magnetic field. The filtration should be effective for sufficiently strong magnetic fields B for which the electron cyclotron frequency ω_{ce} is larger than the collision frequency τ_e^{-1}. For conditions of interest to us, $\omega_{ce}\tau_e \gg 1$, and thus the coefficient of transverse diffusion is

$$D_\perp = \left(1 + \omega_{ce}^2\tau_e^2\right)^{-1},$$

which should rapidly decrease with increasing magnetic field. If the wall thickness of the emission slit h is comparable to the width of the emission slit along the magnetic field, the scale of electron density decay can be reduced to a value $a < h$ with increase in the magnetic field, so that the electrons of the plasma flux passing through the emission slit will be collected on the slit walls and negative ions with large Larmor radius will pass through the emission slit practically freely. The limitation of plasma flow due to plasma escape along the magnetic field to the walls has been observed in positive ion sources [3, 148], and this effect is manifested more strongly with decrease in the length of the emission aperture along the magnetic field. In real SPS designs with emission slit oriented across the magnetic field, the filtration of electrons from the negative ion flux is complicated by a number of factors. From sources with an emission slit for which $h \sim d$, for small magnetic fields, it is possible to extract and form electron beams with a current density of tens of Amps/cm^2. As the magnetic field increases, the electron flux rapidly decreases, so that the current in the extractor circuit becomes comparable to the current of the negative ion beams generated, but with a further increase in the magnetic field, the parasitic current can increase. The magnitude of the parasitic current is affected by changes in discharge regime, plasma parameter fluctuations, cesium supply, secondary emission properties of electrode materials, gas density, etc. However, with good choice of the emission slit size, the magnetic field configuration and intensity, and appropriate selection of other parameters, it is possible to achieve a satisfactory purification of the negative ion flux from electrons.

Fig. 4.77 Electrode
configuration in a negative
ion beam formation system
for accelerators.
(Reproduced from
Dudnikov [1])

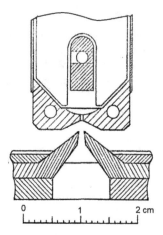

The investigation of the regularities of formation of negative ion beams is at a very early, semiempirical stage. However, in a number of SPS designs, it has been possible to obtain H$^-$ ion beams with good ion-optical characteristics [4, 5], comparable to the corresponding parameters for proton beams. An SPS system for forming H$^-$ ion beams for accelerator injection is shown in Fig. 4.77.

In addition to the features of the beam formation systems already described, note that the main column of a discharge with fast electrons must be separated from the emission slit by a magnetic filter so that fast electrons with large Larmor radius do not fall into the emission gap. To prevent discharges in the high-voltage accelerating gap of the beam-forming system, the magnetic field configuration must eliminate the conditions for electron confinement and oscillation.

These conditions must also be met when using multi-aperture beam formation systems, as necessary for obtaining intense beams of negative ions. From an SPS with a planotron gas discharge cell configuration and operating in pulsed mode (Fig. 4.31), it was possible to extract an H$^-$ ion beam of current 1 A per cm of emission slit length through a sectioned emission slit with a total width of 3 mm. At the same time, the extraction voltage $U_0 = 60$ kV could be supported at the gap of the beam-forming system with a minimum electrode spacing of 3 mm.

Experiments to optimize the processes considered were attempted in an SPS with separated electrode functions [12]. In these sources, positive ions were generated by an independent discharge with a separate electrode system, and a special electrode was used as the H$^-$ ion emitter. The energy of the bombarding particles extracted from the plasma into the emitter was regulated by changing the potential difference between this electrode and the gas discharge plasma. A schematic of one option for an SPS with separated functions is shown in Fig. 4.78.

Positive ions are formed in a flat layer of a gas discharge plasma with transverse dimensions 0.12×3 cm^2 formed by establishing a discharge with electron reflexing in a magnetic field between hollow cathodes with cavities in the form of long

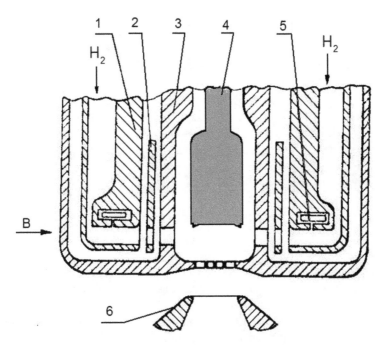

Fig. 4.78 Schematic of an SPS with separated functions. (Reproduced from Dudnikov [1]). (1) Cathodes of the plasma generation system, (2) bounding diaphragms, (3) anode, (4) independent emitter, (5) cavities for cesium, (6) extraction electrode

narrow slits and coaxial anode windows 3. The distance between cathodes in this SPS is 1.2 cm. Hydrogen is supplied by a pulse valve in the cathode cavity. To increase the efficiency of electron emission in the cathode cavity, cesium was released from cesium chromate tablets with titanium in container 5, when heated. The discharge voltage can be regulated over a wide range by changing the cesium release. Stable operation of pulsed discharges obtains up to a discharge current of 450 A. The emitter 4 is installed with an adjustable gap Δ between its working surface and the boundary of the gas discharge plasma layer (negative values of Δ correspond to its penetration into the plasma). Cesium for the emitter was fed from a container with independent heating. An emission hole with dimensions up to 0.3×3 cm^2 is located opposite the emitter working surface. To reduce the flow of accompanying electrons, the emission slit along the magnetic field is divided into three separate narrow slits by a jalousie (a shutter with a row of slats) with transverse dimensions 0.1×0.5 mm^2. The gas discharge chamber is biased negatively at the extraction voltage. The extraction electrode 6 is grounded. The above concepts were also confirmed when working with this SPS. Without the emitter and with a positive potential on the emitter, comparatively intense H$^-$ ion beams were extracted from the source with emission density up to 2 A/cm^2 by formation of H$^-$ ions at the electrodes near the emission slit. With increase in the potential of emitter 4 up to -100 V relative to the plasma, the H$^-$ intensity increased by a factor of two

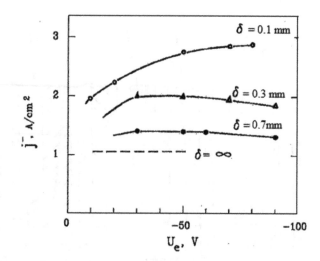

Fig. 4.79 Dependence on the voltage of an independent emitter of the current density of ions H⁻ for different positions of the emitter relative to the plasma. (Reproduced from Dudnikov [1])

($\Delta \sim 0.1$ mm). The dependence of the H⁻ ion current density on the voltage of the independent emitter for different emitter positions relative to the plasma is shown in Fig. 4.79. In high-current mode, H⁻ ion beams with emission density up to 5.4 A/cm² were obtained by optimizing the emission properties of the emitter and the energy of bombarding particles.

In 1989, Lawrence Berkeley National Laboratory (LBL), Berkeley, CA [149], following Novosibirsk results [3, 4], reported a substantial increase in negative ion current by seeding small amount (~1 g) of cesium vapor into a volume production type negative ion source. This increase of negative ion current was later reproduced by many laboratories [5, 6]. Moreover, it was observed that the negative ion current density remained high even when the source pressure was reduced to 1 Pa. By seeding cesium in the JAERI multiampere negative ion source, negative ions of 10 A was successfully extracted at the current density of 380 mA/cm² [150]. It is thought that cesium in the volume production type sources acts similarly to that in surface-plasma-type negative ion sources, which was developed in Novosibirsk and are widely used in high-energy accelerators [5, 6]. In those sources, a "converter or cathode" is biased negatively to attract positive ions from the discharge plasma. Then negative ions are produced on the surface of the converter, of which the work function is lowered by a thin layer of cesium. These negative ions are accelerated off the surface to the energy (hundreds of Volts) corresponding to the bias potential. The analysis of the processes involved using quantum mechanics and plasma physics shows that the negative ion yield reaches a maximum when the work function is minimized, and the work function on tungsten surfaces has a minimum value with a Cs coverage of 0.6 monolayers (on a tungsten crystal 110 surface). Experiments carried out at the FOM-Institute for Plasma Physics, the Netherlands, demonstrated that a ion could be formed by bombarding a 110 crystal tungsten surface coated with cesium with ions and showed that the probability of formation from bombardment could be very high, 75% [2]. However, in the volume production type negative ion sources, without any specific target like converter and high bias voltage, Cs covers the whole interior surface. KEK and JAERI measured the work function of the plasma grid and the negative ion current while introducing Cs into the source gradually. The result of the negative ion current and work function measurement is shown in Fig. 4.72. The result indicates that the negative ion current increases as the work function of the plasma grid surface is lowered. This suggests that negative ions produced on the plasma grid surface are the additional ions extracted when the ion source is cesiated. After seeding a certain amount of Cs, it was observed that the negative ion current increased as the plasma grid temperature, up to 250–300 °C. Obtained negative ion current as function of the work function of the plasma grid surface that the cesium coverage, and hence, the work function is optimized by varying the plasma grid temperature. Furthermore, a study of the transport of negative ions in the plasma found inside these sources showed that only or produced very close to the plasma grid (~15 mm) can reach an extraction aperture. The abovementioned experiments and calculations led to the accepted model of negative ion production under Cs-seeded conditions in magnetic multipole and RF ion sources. It is considered that the negative ions are produced on the surface of the plasma grid as follows: atomic hydrogen or positive ions approach the plasma grid, but the ions capture an electron from the metal before reaching the surface, so that only neutral particles arrive at the surface. As the atom approaches the surface, the affinity level of the atom broadens and shifts to a lower level and broadens due to the interaction between the atom and its mirror image induced in the met al. If the level overlaps, as can be the case with a low work function surface, with the Fermi level of the metal surface, a resonant charge transfer can take place, i.e., an electron can transfer to the atom. As atomic hydrogen or protons can form negative ions on a cesiated surface, it is likely that dissociation of molecules is one of the key processes for negative ion production in the cesiated sources.

The KAMABOKO source has produced ion beams of 2 A (300 A/cm^2) at an operating pressure of 0.1 Pa, which meets the requirements of the ITER negative ion source.

As tungsten evaporation from the high-temperature discharge cathode accumulates on the Cs film deposited on the ion source plasma grid surface, contemporary H$^-$/D$^-$ ion sources excite plasma with MHz frequency range radio frequency (RF) power. The ELISE test facility achieved 251 A/m^2 current density for H$^-$ ion extracted current and challenging for higher extracted current density for D$^-$ with better beam homogeneity. The 1-MV full-ITER-power-level NBI injector test stand MITICA (megavolt ITER injector and concept advancement) started activity testing the performance. The RNIS device is a high-temperature filament cathode-driven arc discharge source and has aspects different from the RF ion source as suggested by Fantz et al. [151]. The operation of an arc discharge source with deuterium may increase the Cs consumption rate, as the Cs consumption rate of the arc discharge source is believed to be proportional to the tungsten cathode evaporation rate, and the tungsten loss from the cathode filament will be enhanced by the larger sputtering yield by deuterium ions. Thus, one has to note that an isotope effect like the change in Cs recycling may appear differently between an arc discharge source and an RF discharge source. Some maintenance requirements more complicated than the ones discussed in the past by Hemsworth et al. [152] may be required for the neutral beam-heating system. The RF source was chosen since 2007 as the reference source for ITER, which solves the Cs problem of enhanced sputtering yield of filaments by deuterium. The RF source is used in the test facility ELISE where Nocentini et al. [153] challenge operation of long plasma pulses of up to 1 h. As they extracted beams for 10 s in every 180-s interval, the extracted ion current tended to decrease slightly from one extraction pulse to the next, while the co-extracted electron currents increased substantially especially in deuterium operation. In certain extraction pulses, the co-extracted electron current changed very dynamically and decreased during the beam-on phase. They attribute this in part to the influence of back-streaming positive ions which impact on the source back-plate and sputter Cs from its surface, therefore affecting the source condition. Obviously the sputtering of Cs from the back-plate by the back-streaming deuterium ions is stronger than that due to hydrogen positive ions, which explains the observed isotope effect.

4.14 Observation of an Ion-Ion Plasma

The SPS with cesiation at the National Institute for Fusion Science (NIFS) was observed to be positive/negative ion-ion plasma [154]. After starting the Cs seeding, the electron and the H⁻ densities are drastically changed near the PG. It is found that the electron densities both measured with millimeter-wave interferometer and a Langmuir probe are decreased, while the H⁻ density measured with CRD is increased, as the supplied Cs is increased.

With the Cs seeding, the production of the negative ions is much enhanced through the surface plasma production on the PG, and the H⁻ density exceeds the electron density near the PG. As the Cs seeding proceeds, it is observed in the I-V curve of a Langmuir probe located near the PG that the negative saturation current is decreased, while the positive saturation current is not changed. That means that the dominant negative-charged particles are replaced to negative ions from electrons. Finally, the negative and the positive saturation currents become nearly equal, and the I-V curve shows quite symmetric, as shown in Fig. 4.80. From this observation, it is recognized that an ion-ion plasma, which consists of mainly positive and negative ions, is produced in the extraction region.

Reviews of development of the surface plasma method of negative ions and development of SPS are presented in [6, 25–27, 29, 31, 33, 155–166].

Fig. 4.80 I-V curve in
Langmuir probe
measurement for an
ion-ion plasma in the
extraction region

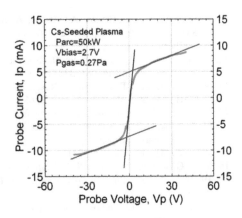

References

1. V.G. Dudnikov, Surface Plasma Method of Negative Ion Production, Dissertation for doctor of Fis-Mat. nauk, INP SBAS, Novosibirsk (1976). В.Г. Дудников, "Поверхностно-плазменный метод получения пучков отрицательных ионов", Диссертация на соискание учёной степени доктора физ.-мат. Наук, ИЯФ СОАН СССР, Новосибирск (1976)

2. V.L. Komarov, A.P. Strokach, *Preprint NIIFA, D-0282* (Leningrad, 1976). В. Л, Комаров, А. П. Строкач, Препринт НИИЭФА, Д-0282, Ленинград (1976)

3. V. Dudnikov, Method of negative ion obtaining, USSR Patent 411,542, 10/III (1972), http://www.findpatent.ru/patent/41/411542.html; Dudnikov, V.G., Technique for producing negative ions, AN SSSR, Novosibirsk. Inst. Yadernoj Fiziki. В.Г. Дудников, Способ получения отрицательных ионов, Авторское свидетельство, М. Кл.Н 01 J 3/0,4, 411542, заявлено 10/III (1972)

4. Y. Belchenko, G. Dimov, V. Dudnikov, Physical principles of surface plasma source operation, in *Symposium on the Production and Neutralization of Negative Hydrogen Ions and Beams, Brookhaven, 1977,* (Brookhaven National Laboratory (BNL), Upton, NY, 1977), pp. 79–96; Y. Belchenko, G. Dimov, V. Dudnikov, Physical principles of surface plasma source method of negative ion production, Preprint IYaF 77-56, Novosibirsk (1977); Ю. Бельченко, Г. Димов, В. Дудников, Физические основы поверхностно плазменного метода получения пучков отрицательных ионов, препринт ИЯФ 77-56, Новосибирск (1977), http://irbiscorp.spsl.nsc.ru/fulltext/prepr/1977/p1977_56.pdf

5. V. Dudnikov, *Negative Ion Sources* (NSU, Novosibirsk, 2018). В. Дудников, *Источники отрицательных ионов*. НГУ, Новосибирск (2018)

6. V. Dudnikov, *Development and Applications of Negative Ion Sources* (Springer, 2019)

7. B.M. Smirnov, *Atomic Collisions and Elemental Process in Plasma* (Atomizdat, Moscow, 1968). Б.М. Смирнов, *Атомные столкновения и элементарные процессы в плазме,* Москва, Атомиздат (1968)

8. B.M. Smirnov, Doklady AN SSR, 92 (1965). Б. М. Смирнов, Доклады АН СССР, №, 92 (1965)

9. G.E. Derevyankin, V.G. Dudnikov, P.A. Zhuravlev, Electromagnetic shutter for a pulsed gas inlet into vacuum units. Pribory i Tekhnika Eksperimenta **5**, 168 (1975)

10. V. Kuchinsky, V. Mishakov, A. Tibilov, A. Shukhtin, V Allunion conference on physics of atomic collision, Abstracts, Uzhgorod, 189 (1972). В. Кучинский, В. Мишаков, А. Тибилов, А. Шухтин, V Всесоюзная конференция по физике атомных и электронных столкновений, тезисы докладов, Ужгород, стр. 189 (1972)

11. V. Kuchinsky, V. Mishakov, A. Tibilov, A. Shukhtin, Opt. Spectrosc. **39**, 1043 (1965). В. Кучинский, В. Мишаков, А. Тибилов, А. Шухтин, Оптика и спектроскопия, 39, 1043 (1965)

12. Y.I. Belchenko, V. Dudnikov, et al., Ion sources at the Novosibirsk Institute of Nuclear Physics. Rev. Sci. Instrum. **61**, 378 (1990)

13. N.D. Morgulis, A.P. Przhonsky, Rus. Plasma Phys. **2**, 683 (1976). Н. Д. Моргулис, А. М. Пржонский, Физика плазмы, 2, 683 (1976)

14. Y. Belchenko, Emission of negative ions from high current discharge, Dissertation on Phys-Math. Nauk, INP Novosibirsk (1974). Ю. И. Бельченко, "Эмиссия отрицательных ионов из сильноточных разрядов", диссертация на соискание ученой степени кандидата физ. мат. наук, Новосибирск (1974)

15. M.D. Gabovich, *Physics and Technology of Plasma Ion Sources* (Atomizdat, Moscow, 1972). М.Д. Габович, "*Физика и техника Плазменных источников ионов*", Москва, Атомиздат (1972)

16. Y.I. Belchenko, G.I. Dimov, V.G. Dudnikov, Emission of an intense flow of negative ions from surfaces bombarded by fast particles of a discharge. Isvestiya AN SSR **37**, 2573 (1973). Ю.И. Бельченко, Г. И. Димов, В. Г. Дудников, Известия АН СССР, серия физическая, 37,2573 (1973)

17. Y.I. Belchenko, G.I. Dimov, V.G. Dudnikov, Formation of negaive ins in a gas discharge. Doklady AN SSSP **231**, 1283 (1973). Ю.И. Бельченко, Г. И. Димов, В. Г. Дудников, А. А. Иванов, Доклады АН СССР, 213, 1283 (1973); Препринт ИЯФб 81-72 (Новосибирскб 1972)

18. G.I. Dimov, G.V. Roslyakov, V. Ya, Savkin, PTE **4**, 29 (1973). Г. И. Димов, Г. В. Росляков,В. Я. Савкин, ПТЭ, 4, 29 (1977)

19. P.J.M. van Sommel,·K.N. Leung, K.W. Ehlers, H – production on poly- and monocrystalline converters in a surface-plasma ion source. J. Appl. Phys. **56**, 751 (1984). https://doi.org/10.1063/1.334005

20. V. Dudnikov, Relevance of volume and surface plasma generation of negative ions in gas discharges. AIP Conf. Proc. **763**(1), 122–137 (2005)

21. Y.I. Bel'chenko, G.I. Dimov, V.G. Dudnikov, H^{--} ions from a crossed-field discharge. Sov. Phys. Tech. Phys. **18**, 1083 (1974)

22. Y.I. Belchenko, G.I. Dimov, V.G. Dudnikov, Plasma- surface source of negative ions. Sov. Phys. Tech. Phys. **20**(1), 40–43 (1975)

23. Y.I. Belchenko, G.I. Dimov, V.G. Dudnikov, *Surface: Plasma Source of Negative Ions* (ICIS 1975, California Univ, Berkeley, 1975)

24. Y. Belchenko, G. Dimov, V. Dudnikov, Neutral beam injector with intense surface plasma negative ion source. Nucl. Fusion **14**(1), 113–114 (1974)

25. V.G. Dudnikov, 20 years of cesium catalysis for negative ion production in gas discharges. Rev. Sci. Instrum. **63**(4), 2660–2668 (1992)

26. V. Dudnikov, Thirty years of surface plasma sources for efficient negative ion production. Rev. Sci. Instrum. **73**(2), 992–994 (2002)

27. V. Dudnikov, Forty years of surface plasma source development. Rev. Sci. Instrum. **83**(2), 02A708 (2012)

28. V. Dudnikov, R.P. Johnson, Cesiation in highly efficient surface plasma sources. Phys. Rev. Spec. Top. Accel. Beams **14**(5), 054801 (2011)

29. V.G. Dudnikov, Surface-plasma method for the production of negative ion beams. Physics-Uspekhi **62**(12), 1233 (2019)

30. V.G. Dudnikov, Some effects of surface-plasma mechanism for production of negative ions, in *Second Symposium on the Production and Neutralization of Negative Hydrogen Ions and Beams, Brookhaven, 1980*, (Brookhaven National Laboratory (BNL), Upton, NY, 1980)

31. V. Dudnikov, Forty-five years with cesiated surface plasma sources. AIP Conf. Proc. **1869**(1), 030044 (2017)

32. V. Dudnikov, 30 years of high-intensity negative ion sources for accelerators, PACS2001. Proceedings of the 2001 Particle Accelerator Conference (2001)

33. V. Dudnikov, Progress in the Negative Ion Sources Development, report FRBOR01 Proceedings of RUPAC2012, Saintl-l Petersburg, Russia (2012)

34. L.Y. Abramovich, B.N. Klyarfeld, Y.N. Nastich, ZTF **39**, 125 (1969). Л. Ю. Абрамович, Б. Н. Клярфельд, Ю. Н. Настич,ЖТФ, 39,125 (1969)

35. J. Backus, J. Appl. Phys. **42**, 1477 (1977)

36. A. Ortykov, R.R. Rakhimov, A. Kashetov, *Second Allunion Symposium Interaction of Atomic Particles with Solid* (Moscow, 1972), p. 54. А. Ортыков, Р. Р. Рахимов, А. Кашетов, в сборнике Второй Всесоюзный симпозиум Взаимодействие атомных частиц с твердым телом, Москва (1972), стр. 54

37. J.A. Ray, C.F. Barnett, J. Appl. Phys. **42**, 3260 (1971)

38. J.R. Hiskes, A. Karo, M. Gardner, J. Appl. Phys. **47**, 3888 (1976)

39. V.L. Komarov, V.S. Kuznetsov, L. Saksagansky, *Proc. Reports Allion Workshop for Engineering Problems of Controlled Fusion*, vol 1 (NIIFA, Leningrad, 1975). В. Л. Комаров, В. С. Кузнецов, Г. Л. Саксоганский и др. В сборнике Доклады Всесоюзного совещания по инженерным прблемам УТС, НИИЭФА, Ленинград, т. 1 (1975)

40. Y.I. Belchenko, V.G. Dudnikov, *Preprint INP 77–56* (Novosibirsk, 1977). Ю. И. Бельченко, В. Г. Дудников, Препринт ИЯФ 77-56, Новосибирск (1977)

41. Y.I. Belchenko, V.G. Dudnikov, *4 Allunion Conference for Interaction of Atomic Particles with Solid*, vol 3 (Kharkov, 1976), p. 180. Ю.И. Бельченко, В. Г. Дудников, 4 Всесоюзная Конференция по взаимодействию атомных частиц с твердым телом, Харьков, , т. 3, стр. 180 (1967)

42. J. Backus, J. Appl. Phys. **30**, 1866 (1959)

43. V.G. Dudnikov, E.G. Obrazovsky, G.I. Fiksel, *Preprint INP 77–50* (Novosibirsk, 1977). В. Г. Дудников, Е. Г. Образовский, Г. И. Фиксель, Препринт ИЯФ 77-50, Новосибирск (1977)

44. C.-S. Weng, J. Apll. Phys. **486**, 1477 (1977)

45. V.S. Borodin, V.P. Gofmeister, Y.M. Kogan, K. Kovallin, ZTF **38**, 1814 (1968) В. С. Бородин, В. П. Гофмейстер, Ю. М. Коган, Г. Коваллин, ЖТФ, 38, 1814 (1968)

46. J. Laulainen, S. Aleiferis, T. Kalvas, et al., Hydrogen plasma induced photoelectron emission from low work function cesium covered metal surfaces. Phys Plasmas **24**, 103502 (2017)

47. O.B. Firsov, ZTF **26**, 445 (1956). О. Б. Фирсов, ЖТФ, 26, 445 (1956)

48. N.P. Stakhanov, M. Atomisdat. Physical Basis of Thermoemission Converters (1973). *Физические основы термоэмиссионного преобразования энергии*, ред. Н. П. Стаханов, М. Атомиздат (1973)

49. B.Y. Moizhes, G.E. Picus, *Thermoemission Converters and Low Temperature Plasma* (M. Nauka, 1973). *Термоэмиссионные преобразователи и низкотемпературная плазма*, ред. Б. Я. Мойжес, Г. Е. Пикус, М. Наука (1973)

50. A.I. Morozov et al., ZTF **42**, 54 (1972). А. И. Морозов и др. ЖТФ, 42, 54 (1972)

51. A.N. Apolonskii, Y.I. Belchenko, G.I. Dimov, V. Dudnikov, Gas efficiency of surface-plasma sources of negative hydrogen ions. Pisma v Zhurnal Tekhnischeskoi Fiziki **6**, 86–89 (1980)

52. W. Maus-Friedrichs, M. Wehrhahn, S. Dieckhoff, V. Kempter, Coadsorption of Cs and hydrogen on W(110) studied by metastable impact electron spectroscopy. Surf. Sci. **237**, 257 (1990)

53. В.С. Бородин, В.П. Гофмейстер, Ю.М. Коган, Г. Коваллин, ЖТФ **38**, 1814 (1968). V. S. Borodin, V. P. Gofmeister, Yu. M. Kogan, K. Kovallin, ZTF, 38, 1814 (1968)

54. G.Y. Pivovarov, L.A. Saminsky, *Thechnology Processes of Electrovacuum Industry* (M, 1975), p. 192. Г. Я. Пивоварова, Л. А. Саминский, *Технологические процессы электровакуумного производства*, М., стр. 192 (1975)

55. Y. Belchenko, V.I. Davydenko, G.E. Derevyankin, A.F. Dorogov, V.G. Dudnikov, Sov. Tech. Phys. Lett. **3**, 282 (1977)

56. G.I. Dimov, G.E. Derevyankin, V. G., A 100 mA negative hydrogen-ion source for accelerators. Nucl. Sci. **NS24**(3), 1545 (1977)

57. V. Dudnikov, P. Chapovsky, A. Dudnikov, Cesium control and diagnostics in surface plasma negative ion sources. Rev. Sci. Instrum. **81**, 02A714 (2010)

58. U. Fantz, C. Wimmer, Optimizing the laser absorption technique for quantification of cesium densities in negative hydrogen ion sources. J. Phys. D. Appl. Phys. **44**, 335202 (2011)

59. R. Gutser, D. Wunderlich, U. Fantz, Dynamics of the transport of ionic and atomic cesium in radio frequency-driven ion sources for ITER neutral beam injection. Plasma Phys. Control. Fusion **53**, 105014 (2011)

60. A.A. Aliev, Z.A. Isakhanov, Z.E. Mukhtarov, M.K. Ruzibaeva, Stimulation of secondary negative ion emission by implantation of alkaline ions into the surface layer of a solid followed by heating. Tech. Phys. **55**(1), 111 (2010)

61. M. Wada, Measurements of work function and surface conditions in cesiated negative ion sources. Rev. Sci. Instrum. **92**, 121502 (2021). https://doi.org/10.1063/5.0071522

62. R. Gutser, C. Wimmer, U. Fantz, Rev. Sci. Instrum. **82**, 023506 (2011)

63. U. Fantz, R. Gutser, C. Wimmer, Rev. Sci. Instrum. **81**, 02B102 (2010)

64. R. Friedl, U. Fantz, AIP Conf. Proc. **1655**, 020004 (2015)

65. R. Friedl, U. Fantz, J. Appl. Phys. **122**, 083304 (2017)

66. S. Cristofaro, R. Friedl, U. Fantz, Plasma Res. Express **2**, 035009 (2020)

67. R.G. Wilson, J. Appl. Phys. **37**, 3161 (1966)

68. U. Fantz, C. Wimmer, Optimizing the laser absorption technique for quantification of caesium densities in negative hydrogen ion sources. J. Phys. D. Appl. Phys., IOP Publishing **44**(33), 335202 (2011)

69. R. Friedl, U. Fantz, Fundamental studies on the Cs dynamics under ion source conditions. Rev. Sci. Instrum. **85**, 02B109 (2014). https://doi.org/10.1063/1.4830215

70. S. Cristofaro, R. Friedl, U. Fantz, Correlation of Cs flux and work function of a converter surface during long plasma exposure for negative ion sources in view of ITER. Plasma Res. Express **2** (2021). https://doi.org/10.1088/2516-1067/abae81

71. A. Heiler, R. Friedl, U. Fant, *Ultra-Low Work Function of Cesiated Surfaces and Impact of Specific Hydrogen Plasma Species* (NIBS 2022, Padova, Italia, 2022)

72. J. Berner, C. Wimmer, U. Fantz, *Work Function Measurements in BATMAN Upgrade Using LEDs Revealing Remarkably Low Values* (NIBS 2022, Padova, 2022)

73. B.X. Hana, M.P. Stockli, R.F. Welton, et al., Recent performance of the SNS H- ion source with a record long run. AIP Conf. Proc. **2373**, 040004 (2021). https://doi.org/10.1063/5.0057408

74. A. Ueno et al., *Beam Intensity Bottleneck Specification and 100 mA Operation of J-PARC Cesiated RF-Driven H⁻ Ion Source* (NIBS 2018, Novosibirsk, Russia, 2018)

75. SASAO, Mamiko et al., *Plasma Electrode Materials for Cs-Free Negative Hydrogen Ion Sources* (NIBS 2022, Padova, 2022)

76. U.A. Arifov, A.H. Ayukhanov, Isvestiya AN Us. CCR, Seriya Fis-Mat. Nauk **6**, 34 (1961). У.А. Арифов, А.Х. Аюханов, Изв. АН Уз. ССР, серия физ-мат. наук, 6, 34 (1961)

77. V.E. Kron, J. Appl. Phys. **34**, 3523 (1962)

78. E. Bender, G. Dimov, M. Kishinevskii, *Proc. 1 Allunion Seminar for Secondary Ion-Ion Emission* (Kharkov, 1975) (deponir VINITI N 2783-75 dep) p. 119-131, preprint INP 75-9, Novosibirsk (1975) Е. Д. Бендер, Г. И. Димов, М. Е. Кишиневский, Труды 1 всесоюзного семинара по вторичной ионно-ионной эмиссии, Харьков, 1975 (Рук. Деп. ВИНИТИ No 2783-75 деп.) Стр. 119-131. Препринт ИЯФ-75-9, Новосибирск (1975). Bender E D, Kishinevskii M E, Morozov I I, Preprint No. 77-47 (Novosibirsk: Budker Institute of Nuclear Physics, SB USSR Acad. Sci., 1977)

79. J.A. Pillips, Phys. Rev. **91**(2), 455 (1953)

80. G.I. Zhabrev, V.A. Kurnaev, V.G. Telkovskii, *Proc. Allunion Workshop on Engineering Problems of Controlled Fusion*, vol 4 (NIIFA, Leningrad, 1975), p. 138. Г. И. Жабрев, В. А, Курнаев, В. Г. Тельковский, Доклады Всесоюзного совещания по инженерным проблемам УТС, НИИЭФА, Ленинград, ,т. 4, стр. 138 (1975)

81. G.I. Zhabrev, V.A. Kurnaev, V.G. Telkovsky, Depon. VINITI, 7515–73. (1973). Г. И. Жабрев, В. А, Курнаев, В. Г. Тельковский,Деп. Рук. ВИНИТИ, 7515–73

82. G.I. Zhabrev, V.A. Kurnaev, V.G. Telkovsky, *Second Allunion Symposium on Interaction of Atomic Particles with Solid* (Proceedings, Moscow, 1972), p. 112. Г. И. Жабрев, В. А, Курнаев, В. Г. Тельковский, Второй Всесоюзный Симпозиум по ваимодействие атомных частиц с твердым телом, Москва, Сборник докладов, стр. 112 (1972)

83. J. Bohdansty, J. Roth, M.K. Siuha, W. Ottenberger, Int. Conf. Surface Effects in Controlled Fusion Devices, San Francisco, pp. 21–36 (1976)

84. V.A. Kurnaev, V.G. Telkovskii, *Interaction of Atomic Particles with Solid*, vol 1 (Kiev, Naukova Dumka, 1974), p. 67. В. А, Курнаев, В. Г. Тельковский, *Взаимодействие атомных частиц с твердым телом*, Киев, Наукова думка, , часть 1, стр. 67–70 (1974)

85. G.I. Zhabrev, V.A. Kurnaev, V.G. Telkovskii, *Interaction of Atomic Particles with Solid*, vol 1 (Kharkov, 1976), p. 54. Г. И. Жабрев, В. А, Курнаев, В. Г. Тельковский, *Взаимодействие атомных частиц с твердым телом*, Харьков, часть 1, стр. 54–57 (1976)

86. M.T. Robinson, *The Third Nat. Conf. Atomic Collision with Solids* (Kiev, 1974)

87. V.M. Sotnikov, *Interaction of Atomic Particles with Solid* (Kharkov, 1976), p. 58, 60. В. М. Сотников, *Взаимодействие атомных частиц с твердым телом, Харьков*, 1976, стр. 58–59, 60–63

88. Y.V. Gott, Y.N. Yavlinsky, *Interaction of Slow Particles with Solid and Plasma Diagnostics* (M. Atomizdat, 1973). Ю. В. Готт, Ю. Н. Явлинский, *Взаимодействие медленных частиц с веществом и диагностика плазмы*, М. Атомиздат (1973)

89. O.B. Firsov, ZTF **40**, 83 (1970). О. Б. Фирсов, ЖТФ, 40, 83 (1970)

90. S.K. Erents, G.M. Mjcracken, J. Appl. Phys. **44**, 3139 (1973)

91. F.L. Arnot, Proc. Roy. Soc. **137** (1938)

92. F.L. Arnot, Nature **141m**, 1011 (1938)

93. L.N. Dobretsov, V.N. Lepeshinskaya, I.E. Bronshtein, ZTF **22**, 961 (1952). Л. Н. Добрецов, В.Н. Лепешинская, И. Э. Бронштейн, ЖТФ, 22, 961 (1952)

94. M.E. Kishinevskii, Zh. Tekh. Fiz. **45**, 1281 (1975).; translated in Sov. Phys. – Tech. Phys. 20, 799 (1975). М. Е. Кишиневский, ЖТФ, т. 45, в., 6, 1281, 1975, препринт ИЯФ 116–73, Новосибирск (1973)

95. M.E. Kishinevskii, translated in Sov. Phys. – Tech. Phys. 20, М. Е. Кишиневский, ЖТФ, 48, 73 (1978). Препринт ИЯФ, 76-18, Новосибирск (1976). 799 (1975). M. E. Kishinevskii, Interaction of atomic particles with solid, p. 22, Kharkov, 1976.М. Е. Кишиневкий, "К вопросу о вторичной отрицательно-ионной эмиссии", Взаимодействие атомных частиц с твердым телом, с. 22, ХАРЬКОВ 7-9 ИЮНЯ 1976 г. Zh. Tekh. Fiz. **48**, 1281 (1978)

96. L.M. Kishinevsky, Bulleten AN SSSR, cer. Physics **38**, 392 (1974). Л. М. Кишиневский, Изв. АН СССР, сер. Физич., 38, №-,392 (1974)

97. L.M. Kishinevsky, Y.A. Vinokurov, *Proc. Ion Bombardment-Method of Investigation of Property of Solid* (FAN, 1979), p. 89. Л. М. Кишиневский, Я. А. Винокуров, в сборнике Ионная бомбардировка-метод исследования свойств поверхности, ФАН, 1975, стр. 89

98. R.K. Janev, Surf. Sci. **45**, 609 (1974)

99. R.K. Janev, Surf. Sci. **47**, 583 (1975)

100. M. Kaminsky, *Atomic and Ionic Collision on Material Surface* (M. Mir, 1967). М. Каминский, *Атомные и ионные столкновения на поверхности металлов*, М. Мир (1967)

101. B.A. Trubnikov, Y.N. Yavlinsky, JETF **52**, 1638 (1967). Б. А. Трубников, Ю. Н. Явлинский, ЖЭТФ, 52, 1638 (1967)

102. S.S. Shechter, JETF **7**, 750 (1937). С.С. Шехтер, ЖЭТФ, 7, 750 (1937)

103. Y.N. Demkov, V.N. Ostrovsky, *Method of Zero Radius Potential* (L. Leningrad university, 1975). Ю. Н. Демков, В. Н. Островский, *Метод потенциала нулевого радиуса в атомной физике*, Л. Ленинградский университет (1975)

104. J.R. Hiskes, A. Karo, Preprint UCRL-79512, (1977)

105. H.L. Cui, Resonant charge transfer in the scattering of hydrogen atoms from a metal surface. J. Vac. Sci. Technol. A **9**(3), 1823 (1991)

106. I.K. Gainullin, V.G. Dudnikov, Theoretical investigation of the negative ionization of hydrogen particles on metal surfaces with low work function. Plasma Res. Express **2**, 045007 (2020)

107. H.H. Brongersma, M. Draxler, M. de Ridder, P. Bauer, Surf. Sci. **62**, 63 (2007)
108. I.K. Gainullin, Phys. Rev. A **95**, 052705 (2017)
109. P. Liu, L. Yin, Z. Zhang, et al., Phys. Rev. A **101**, 032706 (2020)
110. T. Hecht, H. Winter, A.G. Borisov, J.P. Gauyacq, A.K. Kazansky, Faraday Discuss. **117**, 27 (2000)
111. I.K. Gainullin, Moscow University Phys. Bull. **74**, 585 (2019)
112. I.K. Gainullin, M.A. Sonkin, Phys. Rev. A **92**, 022710 (2015)
113. I.K. Gainullin, M.A. Sonkin, Comp. Phys. Commun. **188**, 68 (2015)
114. I.K. Gainullin, Comp. Phys. Commun. **210**, 72 (2017)
115. S. Majorosi, A. Czirják, Comp. Phys. Commun. **208**, 9 (2016)
116. Y. Fu, J. Zeng, J. Yuan, Comp. Phys. Commun. **210**, 181 (2017)
117. H.J. Lüdde, M. Horbatsch, T. Kirchner, Eur. Phys. J. B **91**, 99 (2018)
118. C. Wang, D.J. Ding, Chinese Phys. B **28**, 083101 (2019)
119. I.K. Gainullin, Phys. Rev. A **100**, 032712 (2019)
120. J.S. Cohen, G. Fiorentini, Phys. Rev. A **33**, 1590 (1986)
121. E.V. Chulkov, V.M. Silkin, P.M. Echenique, Surf. Sci. **437**, 330 (1999)
122. P.J. Jennings, R.O. Jones, M. Weinert, Phys. Rev. B **37**, 6113 (1988)
123. M. Seidl, H.L. Cui, J.D. Isenberg, H.J. Kwon, B.S. Lee, S.T. Melnychuk, J. Appl. Phys. **79**, 2896 (1996)
124. I.K. Gainullin, V.G. Dudnikov, *Theoretical Calculation of the Hydrogen, Deuterium, Tritium and Muonium Negative Ionization Probability on the Low Work Function Surfaces* (ICIS 2021, 2022)
125. V. Dudnikov, A. Dudnikov, Ultracold muonium negative ion production. AIP Conf. Proc. **2052**, 060001 (2018). https://doi.org/10.1063/1.5083774
126. S. Cristofaro, R. Friedl, U. Fantz, Negative hydrogen and deuterium ion density in a low pressure plasma in front of a converter surface at different work functions. Plasma **4**, 94–107 (2021). https://doi.org/10.3390/plasma4010007
127. M. Wada, K.H. Berkner, R.V. Pyle, J.W. Stearns, Photoelectric work function measurement of a cesiated metal surface and its correlation with the surface produced H⁻ ion flux. J. Vac. Sci. Technol. A **1**, 981 (1983)
128. M.L. Yu, Work function dependence and isotope effect in the production of negative hydrogen ions during sputtering of adsorbed hydrogen on Cs covered Mo (100) surfaces, in *Proc. Symp. Production and Neutralization of Negative ions and beams*, (BNL, Upton, NY, 1977), p. 48
129. R.K. Janev, S.B. Vojvodic, Survival probabilities of hydrogen negative ions emerging from alkali-coated W and Ni surfaces. Phys. Lett. **75a**, 348 (1980)
130. J.K. Norskov, B.I. Lundqvist, Secondary-ion emission probability in sputtering. Phys. Rev. B **19**, 5661 (1979)
131. B. Rasser, J.N.M. Van Wunnik, J. Los, Theoretical models of the negative ionization of hydrogen on clean tungsten, cesiated tungsten and cesium surfaces at low energies. Surf. Sci. **118**, 697 (1982)
132. M. Bacal, M. Wada, Negative ion production by plasma-surface interaction in caesiated negative ion sources. AIP Conf. Proc. **1515**, 41 (2013)
133. M.L. Yu, Phys. Rev. Lett. **40**, 574 (1978)
134. J. Los, J.J.C. Geerlings, Charge exchange in atomic-surface interaction. Phys. Rep. Rev. Sect. Phys. Lett. **190**(3), 133–190 (1990)
135. J. Schneider, K.H. Berkner, W.G. Graham, et al., Phys. Rev. B **23**, 941 (1981)
136. K.H. Berkner, K.N. Leung, R.V. Pyle, A. Schlachtear, S. Tearms, J. W., Phys. Lett. **64A**, 217 (1977)
137. P.W. van Amersfoort, J.J.C. Geerlings, L.F. Tz, A. Kwakman, E.H.A. Hershcovitch, Granneman, et al., Formation of negative hydrogen ions on a cesiated W (110) surface; the influence of hydrogen implantation. J. Appl. Phys. **58**, 3566 (1985)

138. R. Souda, E. Asari, H. Kawanowa, T. Suzuki, S. Otani, Capture and loss of valence electrons during low energy H^+ and H^- scattering from $LaB_6(100)$, Cs/Si(100), graphite and LiCl. Surf. Sci. **421**, 89 (1999)

139. M. Seidl, H.L. Cui, J.D. Isenberg, et al., Negative surface ionization of hydrogen atoms and molecules. J. Appl. Phys. **79**(6), 15, 2896 (1996)

140. Y. Suzuki, M. Hanada, Y. Okumura, M. Tanaka, Measurement of Work Function of a Plasma Grid in a Cesium Seeded Negative Ion Source, Japan Atomic Energy Research Report No. 92–168 (1992) (in Japanese)

141. T. Shimizu, T. Morishita, M. Kawagi, M. Hanada, T. Iga, T. Inoue, K. Watanabe, M. Wada, T. Imai, Japan Atomic Energy Research Institute, Report JAERI-Tech 2003–006, March 2003. T. Morishita, M. Kashiwagi, M. Hanada, et al., Mechanism of Negative Ion Production in a Cesium Seeded Ion Source, Jpn. J. appl. Phys., 40, 4709 (2001)

142. V. Seidl, H.L. Chui, D.I. Isenberg, et al., Surface production of negative hydrogen ions. AIP Conf. Proc. **287**, 25 (1992)

143. R. Prelec, Z.W. Stenberg, Fisika **9**(1) (1977)

144. Y.I.Belchenko,G.I.Dimov,V.G.Dudnikov,ZTF**43**,1720(1973).Ю.И.Бельченко,Г.И.Димов, В. Г. Дудников, ЖТФ, 43, 1720 (1973)

145. Y.I. Belchenko, G.I. Dimov, V.G. Dudnikov, ZTF **45**, –68 (1975). Ю. И. Бельченко,Г. И. Димов, В. Г. Дудников,ЖТФ, 45, 68 (1975)

146. Y.I. Belchenko, G.I. Dimov, V.G. Dudnikov, Surface plasma negative ion source, Proc. Sec. Symp. On Ion Source and Ion Beams, Berkeley, (LBL Rep. No 3399) p. VIII-1 (1974)

147. A.V. Zharinov, Sov. At. Energy **7**, 220 (1957). А. В, Жаринов, Атомная энергия, 7, 220 (1957)

148. N.N. Semashko et al., *VI-2 in Proc. Symp. Ion Sources and Beams* (LBL-3399, Berkeley, 1974)

149. K.N. Leung, O.A. Anderson, C.F. Chan, et al., Development of an advanced "volume" H sources for neutral beam application. Rev. Sci. Instrum. **61**(9), 2378–2382 (1990)

150. T. Inoue, M. Hanada, M. Mizuno, Y. Ohara, Y. Okumura, Y. Suzuki, M. Tanaka, K. Watanabe, Development of a multi-ampere H- ion source at JAERI, in *Proc. 6th Int. Symp. Production and Neutralization Negative Ions and Beams, Upton, NY*, AIP Conf. Proc, vol. 287, (1992), pp. 316–325

151. U. Fantz, H.D. Falter, P. Franzen, E. Speth, R. Hemsworth, D. Boilson, A. Krylov, Rev. Sci. Instrum. **77**, 03A516 (2006)

152. R. Nocentini, U. Fantz, M. Froeschle, B. Heinemann, W. Kraus, R. Riedl, D. Wuenderlich, Fusion Eng. Des. **123**, 263 (2017)

153. D. Wunderlich, R. Riedl, U. Fantz, B. Heinemann, W. Kraus, NNBI team 2018, Plasma Phys. Control. Fusion **60**, 085007 (2018)

154. Y. Takeiri, K. Tsumori, M. Osakabe, et al., Development of intense hydrogen-negative-ion source for neutral beam injectors at NIFS. AIP Conf. Proc. **1515**, 139 (2013). https://doi.org/10.1063/1.4792780

155. V. Dudnikov, Development of a surface plasma method for negative ion beams production. J. Phys. Conf. Ser. **2244**, 012034 (2022). https://doi.org/10.1088/1742-6596/2244/1/012034

156. Y.I. Belchenko, V.I. Davydenko, G.E. Derevyankin, G.I. Dimov, V.G. Dudnikov, et al., Ion sources at the Novosibirsk Institute of Nuclear Physics. Rev. Sci. Instrum. **61**(1), 378–384 (1990)

157. Y.I. Bel'chenko, G.I. Dimov, V.G. Dudnikov, *Physical Principles of the Surface Plasma Method for Producing Beams of Negative Ions* (BNL, 1977)

158. Y.I. Belchenko, G.I. Dimov, V.G. Dudnikov, A.S. Kupriyanov, Negative ion surface-plasma source development for fusion in Novosibirsk. Revue de physique appliquée **23**(11), 1847–1857 (1988)

159. V.G. Dudnikov, *Some Effects of Surface-Plasma Mechanism for Production of Negative Ions* (Second BNL symposium, 1980)

160. V. Dudnikov, Modern high intensity H- accelerator sources, ArXiv preprint ArXiv:1806.03391 (2018)

161. Y.I. Belchenko, G.E. Derevyankin, G.I. Dimov, V.G. Dudnikov, Studies of surface-plasma negative ion sources at Novosibirsk. J. Appl. Mech. Tech. Phys. **28**(4), 568–576 (1987)
162. Y.I. Belchenko, G.I. Budker, G.E. Derevyankin, G.I. Dimov, V.G. Dunikov, et al., *High Current Proton Beams at Novosibirsk* (Proceedings of the X International Conference on High Energy Accelerators ..., 1977)
163. V. Dudnikov, *30 Years of High-Intensity Negative Ion Sources for Accelerators* (PACS2001. Proceedings of the 2001 Particle Accelerator Conference (Cat. No ...), 2001)
164. V. Dudnikov, Development of charge-exchange injection at the Novosibirsk Institute of Nuclear Physics and around the World, arXiv preprint arXiv:1808.06002 (2018)
165. V. Dudnikov, Relative Contribution of Volume and Surface-Plasma Generation of Negative Ions in Gas Discharges, Proceedings of the 2005 Particle Accelerator Conference, pp. 2482–2484 (2005)
166. V.G. Dudnikov, Practical aspects of surface plasma sources operation. AIP Conf. Proc. **380**(1), 237–240 (1996)

Chapter 5
Surface Plasma Negative Ion Sources

Abstract Topics covered in this chapter include: Surface plasma H^- ion sources for accelerators, design of surface plasma H^- ion sources for accelerators, formation of H^- ion beams in surface plasma sources for accelerators, surface plasma sources with Penning discharge for microlithography, semiplanotrons, geometric focusing, semiplanotrons for accelerators, semiplanotrons with spherical focusing for continuous operation, compact surface plasma sources for heavy negative ion production, development of surface plasma sources worldwide, large volume surface plasma sources with self-extraction, large volume surface plasma sources for accelerators, large volume surface plasma sources for heavy ion production, surface plasma sources for intense neutral beam production for controlled fusion, RF surface plasma sources for ITER (International Thermonuclear Experimental Reactor), neutral beam injector with RF (radio frequency) SPS (surface plasma source) development at Novosibirsk, EF (electron fence) surface plasma source for neutral beam injectors in China, RF surface plasma sources for spallation neutron sources, carbon films in RF surface plasma sources with cesiation, poisoning and recovery of converter surfaces, RF surface plasma sources with external antenna, RF surface plasma sources with solenoidal magnetic field, testing RF surface plasma sources with saddle antenna (SA) and magnetic field, estimation of H^- ion beam generation efficiency, RF surface plasma source operation in continuous mode, RF surface plasma sources at CERN (European Council for Nuclear Research), surface plasma sources at J-PARC (Japan Proton Accelerator Research Complex), RF surface plasma sources at Chinese SNS (Spallation Neutron Source), and surface plasma sources for low energy neutral production.

© The Author(s), under exclusive license to Springer Nature   223
Switzerland AG 2023
V. Dudnikov, *Development and Applications of Negative Ion Sources*, Springer
Series on Atomic, Optical, and Plasma Physics 125,
https://doi.org/10.1007/978-3-031-28408-3_5

5.1 Surface Plasma H^- Ion Sources with Cesiation for Accelerators

H^- ion beams with parameters sufficient to meet the needs of a number of applications discussed in Sect. 2.7, "Charge-Exchange Injection into Accelerators and Accumulator Rings" [1–3], were obtained in the early stage of SPS investigation [4, 5]. We thus began the development of SPS designs adapted for specific applications.

We first turned to the original concern—creation of a source for accelerators, capable of providing efficient application of charge-exchange technology in accelerator technology—and first of all, charge-exchange (stripping) injection of protons into accelerators and storage rings.

The choice of source parameters was guided by the needs of the linear accelerator of the meson factory at the Institute of Nuclear Research of the USSR Academy of Sciences (Troitsk) [6, 7]: pulse intensity ~100 mA, pulse duration 0.1 ms at a repetition rate up to 100 Hz, and normalized emittance < 0.5 π mm.mrad. Beam production from a planotron SPS ensured the necessary brightness and an emission current density of several amperes/cm^2. However, experience acquired working with ion sources at the Novosibirsk Institute of Nuclear Physics showed that fluctuations in the parameters of the emitting plasma cause a radical deterioration of the ion-optical characteristics of intense beams.

Discharges without fluctuations (plasma noise) were obtained in SPSs with planotron electrode configuration only at high hydrogen density, when H^- ion generation efficiency decreases substantially and becomes comparable to the efficiency of beam production from the SPS with Penning electrode configuration, in which discharges without fluctuations are more easily established.

For ion sources used in accelerator technology, beam brightness is an important quality characteristic. We had hoped that an increase in brightness by elimination of plasma fluctuations would more than compensate for a reduction in emission efficiency, so it was decided to use Penning cell for H^- ion production [8]. In SPS versions already developed, the beams were extracted through a long narrow emission slit oriented across the magnetic field so as to improve the filtration of accompanying electrons from the ion flux and to increase the perveance. The source ion-optical system forms a beam with two planes of symmetry, with quasi-parallel cross-sections along the slit and with wedge-shaped cross-section boundaries in the perpendicular direction. In this connection, the problem arises of transforming such a wedge-shaped beam into a quasi-parallel and symmetric beam [9]. One solution to this problem is beam formation using the additional focusing field of a segment of a bending magnet with inhomogeneous field. By placing the source so that the emission gap is in the median plane of the magnet, the wedge beam can be converted into a bundle with any configuration of four-dimensional phase volume by appropriate choice of the decay index n and the azimuthal extent of the magnet φ. Initially, we chose the most simple variant of beam formation: conversion of the wedge beam into a beam that is quasi-parallel in both transverse directions due to bending through $\varphi = 90°$ in the magnet with a decay of field along the radius $n = 1$.

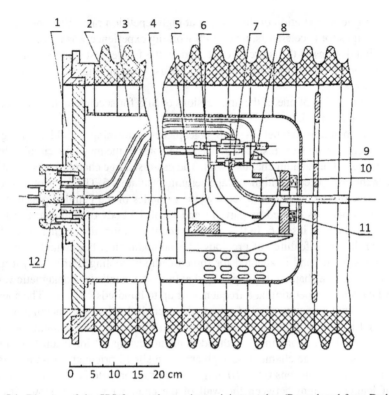

Fig. 5.1 Diagram of the SPS for accelerator in pre-injector tube. (Reproduced from Dudnikov [5]). 1, high-voltage accelerator tube flange; 2, tube insulators; 3, external source shield; 4, support rod; 5, yokes of the source electromagnet; 6, sealed electromagnet coils; 7, gas source chamber; 8, high-voltage source insulators; 9, extractor plates; 10, rotary focusing magnet pole; 11, Rogovski belt; and 12, feedthrough insulators

The proposed scheme for matching the source with the accelerator system of a standard linear accelerator pre-injector is shown in Fig. 5.1 [5, 8].

The beam must pass through the bending magnet in a space charge compensated state, and after exiting the magnet yoke and the screen, be picked up by the strong electric field of the first accelerating gap and the contracted field of the main gap of the accelerator tube. A slight change of high voltage ratio at these gaps can be adjusted to some extent by the focal length of the system. For a more radical change in optical characteristics, other configurations of bending magnet can be used.

5.2   Design of Surface Plasma H^- Ion Sources for Accelerators

By the time the work began on the source for accelerators in 1972, the production of H^- ion beams with an intensity of 100 mA was no longer a problem. It remained a problem to form symmetric quasi-parallel beams with high brightness and to ensure long-term operation of the sources at high repetition rate. When developing source designs for accelerators, much SPS working experience was used.

The first designs of the source were tested on the former test stand with an electromagnet and a chamber pumped by an H-5 diffusion pump (Fig. 4.3) and using the old power system.

The construction of one of the first versions of SPS for accelerators, described in [8], is shown in Fig. 5.2.

Poles of bending magnet 14 with pole pieces 3 forming a field in the gas-discharge cell region were installed between the poles of the magnetic circuit which passes through the vacuum chamber wall. The gas-discharge chamber was installed between the pole pieces on high-voltage ceramic insulators 5. The walls of a groove in a massive molybdenum rod 9 serve as the cathodes of the Penning plasma cell. An anode window is formed by parts 11 and 12 of the body of chamber 4 and the stainless steel anode insert 6. In this electrode configuration, the discharge cell has preserved the basic features of previous SPS embodiments, but cathode mount 8 was made stronger and is adapted for forced cathode cooling. Puffs of hydrogen were fed through a channel developed by that time as a small electromagnetic valve 10 [10] capable of operating at a frequency of up to hundreds of Hertz. The source could work with a cesium supply from anode insert 11 by discharge heating of tablets placed in it containing a mixture of cesium chromate with titanium or with cesium supply from a container 7 with independent heating. H^- ion beam 1 emerged from the gas-discharge chamber through emission slit 13 oriented across the magnetic field with dimensions 0.5×10 mm^2. The extraction field is concentrated in a gap of length 1.5 mm between the wall of the chamber with emission slit and grounded extraction electrode 2. The configuration of the walls of the emission slit and the construction of the beam formation system were repeatedly varied, but we

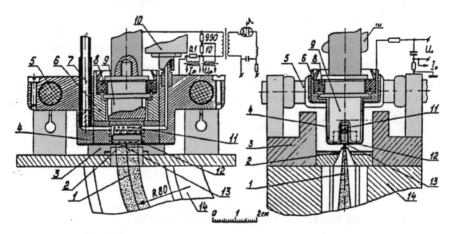

Fig. 5.2 Schematic of an SPS for H^- ions with Penning electrode configuration. (Reproduced from Dudnikov [5]). 1, H^- ion beam; 2, extractor electrode; 3, magnet pole tips; 4, gas-discharge chamber; 5, high-voltage insulators; 6, anode insert; 7, heated cesium container; 8, cathode insulator; 9, cathode; 10, pulse valve; 11, anode insert with cesium container; 12, anode protrusion wall; 13, emission slit; and 14, bending magnet

eventually settled on a simple two-electrode beam formation system. The distance between the cathodes was varied from 4 to 12 mm, and the transverse vertical dimension of the anode window was 2–8 mm. As the size of the gas-discharge cell increases, the ratio of beam current to discharge current decreases, and with a decrease in the anode window, a large initial hydrogen density and high magnetic field strength are needed to ignite the discharge. For design reasons, we stopped at a distance between the cathodes of 5 mm and anode window dimensions of 3×15 mm^2. Discharges in the source had characteristics almost identical with previous discharges. It was soon possible to obtain symmetric beams with a current of 0.1 A at the output of the bending magnet. Oscillograms displaying the operation of this source with various discharge power systems are shown in Fig. 5.3. Estimates

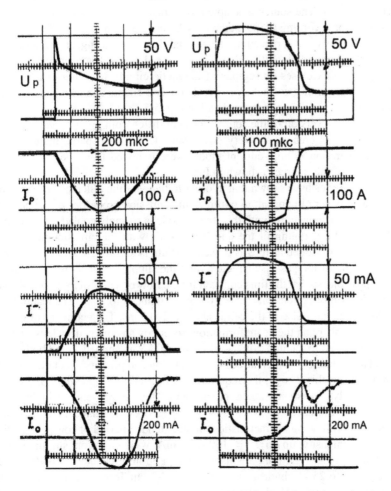

Fig. 5.3 Oscilloscope characterizing the operation of the SPS when the discharge is powered by half-sinusoidal current pulses (left) and by rectangular voltage pulses (right). (Reproduced from Dudnikov [5])

of the ion-optical divergence characteristics of thin beamlets formed from the beam by small holes in the collector have shown that the emittance requirements are fulfilled with a margin if there are no fluctuations in the discharge. When fluctuations in the discharge appear, the ion-optical characteristics deteriorate significantly.

This embodiment of SPS without forced cooling worked at a repetition rate of up to 17 Hz with pulse duration 0.6 ms. At frequencies of 50 and 100 Hz, it could work continuously for only a few minutes due to overheating. The pumping speed provided by a H-5 diffusion pump with liquid nitrogen trap was sufficient for the operation of the source at a frequency up to 100 Hz.

Following extensive studies, the SPS design was brought to a state that provides long-term operation with the desired beam parameters.

The general layout of one of the latest versions of the SPS for accelerators is shown in Fig. 5.4. The source is adapted to be installed in the acceleration tube of the pre-injector according to the matching scheme shown in Fig. 5.1. The magnetic field is excited by coils 7, placed in a sealed stainless steel casing, and water cooled. A bending magnetic field with a decay index of radius $n \sim 1$ is formed by poles 6. The magnetic flux closes on regions of magnetic circuit 8 serving as a magnetic shield. The design of gas-discharge chamber 1 and elements of the beam-forming systems 2 and 3 are shown in more detail in Fig. 5.5.

Basically, this version of the source does not differ from the first one (Fig. 5.2), but the design differences are very significant. Thin-walled parts, melted during long-term operation in high-current mode, were removed from the gas-discharge cell. A photo of this source is shown in Fig. 5.6.

Previous studies have shown that in order to increase the efficiency of formation of H^- ions, it is very important to optimize the emission properties of the emission slit electrode. In connection with this, the cathode is strongly cooled by water flow, and for controlled cooling of the anode electrode with the emission slit through channels 16, a controlled airflow is passed. Cesium is fed from heated container 17.

The volume of the gas-discharge cell and the supply channels is minimized as much as possible, so that the time constant for the flow of a puff of hydrogen through the emission slit of dimensions 0.5×10 mm^2 is reduced to 1 ms. The leading edge of the hydrogen gas pulse in the cell is about 0.2–0.3 ms. The geometry of the gas-discharge chamber and the beam formation system on an enlarged scale is shown in Fig. 5.4. A pulsed extraction voltage of $U_0 = -30$ kV is applied to the body of the gas-discharge chamber, which affects the formation and primary acceleration of the H^- ion beam. In this design, the requirements learned in all previous SPS studies have been fairly consistently taken into account. These improvements in the source and the creation of adequate systems to ensure its operation have made it possible to achieve the following characteristics for an extended period of source operation: beam pulse intensity up to $I^- = 100$–150 mA with emission slit dimensions 0.5×10 mm^2, pulse duration 0.25 ms with a repetition frequency up to 100 Hz, H^- ion energy up to 25 kV, hydrogen consumption ~ 1 cm^{3}Torr per pulse, and cesium consumption ~ 0.1 g for 100 h of operation.

Oscillograms illustrating the operation of this source with corresponding power supply oscillograms are shown in Fig. 5.7. Operation of this source at a pulse repetition rate of 400 Hz has been tested.

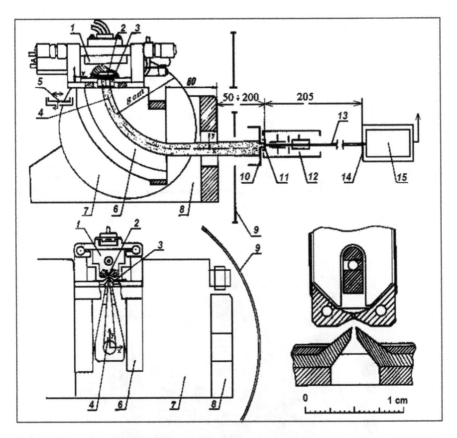

Fig. 5.4 General layout of SPS for accelerator injection and system for recording the distribution of H^- ions along the transverse direction of the focused beam. (Reproduced from Dudnikov [5]). 1, gas-discharge chamber; 2, emission slit; 3, extractor electrodes; 4, $-H^-$ ion beam; 5, movable collector; 6, pole of bending, focusing magnet; 7, sealed electromagnet coils; 8, magnetic yoke; 9, source screen; 10, collector; 11, a small collector; 12, system for scanning ion beamlets; 13, ion beamlet in the drift space; 14, screen of secondary electron multiplier with analyzing aperture; and 15, secondary electron multiplier

Cathode sputtering by the discharge does not limit the lifetime of the cesiated Penning discharge SPS. The cesiated cathode surface is shielded from sputtering by the discharge. The source lifetime is limited by sputtering of the upper part of the Penning window shown as item 3 in Fig. 5.5 by back-streaming positive ions which bombard the anode window surface. Sputtered metal is deposited to the cathode surface. The lifetime of the Penning discharge SPS can be increased by fabricating the upper part of the Penning window from tungsten. To increase the lifetime of Penning discharge SPS, it is possible to braze to the hollow anode tungsten plate of ~1 mm thick; a sputtering yield of W by hydrogen ions with energy 17 keV, $Y \sim 10^{-4}$ atom/ion; the sputtering yield of Mo by hydrogen ions with energy 17 keV $Y \sim 1.2$ 10^{-3} atom /ion; and sputtering yield of SS by hydrogen ions with energy 17 keV $Y \sim 3 \ 10^{-3}$ atom /ion. It should decrease a sputtering rate by back accelerated ions at 12 times relative molybdenum anode and 30 times relative stainless steel anode.

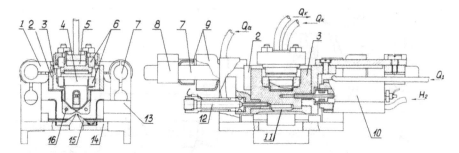

Fig. 5.5 Design of negative ion PD SPS for accelerators (Dudnikov-type source). (Reproduced from Dudnikov [5]). 1 and 8, mechanical supports; 2, gas-discharge chamber; 3, anode insert; 4, cathode; 5, cathode cooler; 6, cathode insulators; 7, high-voltage insulators; 9, screens; 10, hydrogen valve; 11, emission slit; 12, heated cesium container; 13, magnet pole tips; 14, base; 15, extractor electrode; and 16, cooling channel for anode wall with emission slit

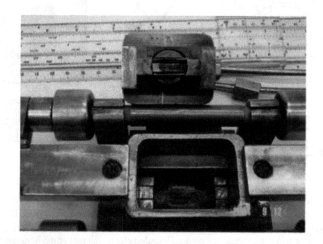

Fig. 5.6 Photo of SPS with Penning discharge. The cathode and anode inserts are visible. (Reproduced from Dudnikov [5])

5.3   Formation of H^- Ion Beams in Surface Plasma Sources for Accelerators

The beam ion-optical characteristics are especially important when using these beams in accelerator technology. In this connection, let us dwell on the determination of the ion-optical characteristics of the formed beams. A schematic of the setup is shown in Fig. 5.4.

In the notation of Fig. 5.4, the magnetic field is directed along the X axis, and emission slit 2 lying on the median plane of the magnet is oriented along the Y axis perpendicular to the magnetic field. In this version of the magnetic system, it was possible to guide the extracted beam at different radii ($R = 50$–90 mm). Because of the large radial aperture, the weight of this magnet turned out to be quite large

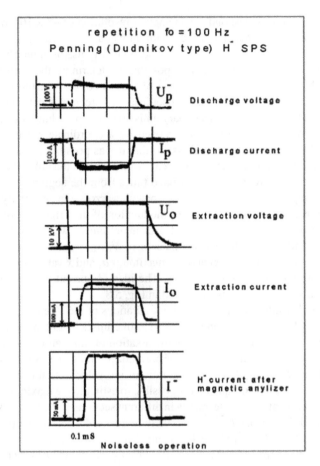

Fig. 5.7 Oscillograms characterizing the beam behavior in the SPS for accelerators. (Reproduced from Dudnikov [5])

(about 40 kg). When studying the ion-optical characteristics, the beam intensity I_1^- was recorded by collector 5 located immediately after the extraction gap. The intensity of the formed beam I_2^- after the magnet yoke 8 and screen 9 was recorded by collector 10. The current density distribution $J^-(x,y)$ was determined from the intensity of the beamlet selected by an opening of area 1 mm² in collector 10 and recorded by a small collector 11. A hole with dimensions of 0.05 × 0.05 mm² in collector 11 selected an ion beamlet whose divergence determines the local spread of ion transverse velocities. The system used made it possible to determine the ion distribution function with respect to the slope angles of the trajectories $f(\alpha_x,\alpha_y)$ at various points of the beam cross section. To do this, we determined the current density distributions in the selected beamlets after their expansion due to the ion transverse velocities in the drift space. Secondary emission multiplier 15 located in the screen 14 recorded the beamlet intensity of ions passing through the hole in the screen with dimensions 0.1 × 0.1 mm². The distance between the selecting and analyzing diaphragms of 20 cm corresponds to an angular resolution of ~5 × 10⁻⁴ rad. An ion beamlet was scanned across the screen with an electrostatic scanning system. By recording the oscillograph signal from the secondary emission multiplier and signals from the scanning system, it was possible to determine the ion distribution function with respect to the slope angles of the trajectories on the oscilloscope screen. The scale of the picture on the oscilloscope screen was determined by the displacement of the image on a secondary emission multiplier luminescent screen.

The entire registration system was moved by a coordinate mechanism along three mutually perpendicular directions, so that it was possible to view the entire cross section of the beam formed at different distances from the yoke of the magnet. The measurements were carried out at fixed times from the beginning of the beam pulse. Either rapid scanning (10 μs) of a beamlet across the secondary emission multiplier screen with analyzing hole or modulation of the brightness of the oscilloscope beam with a slow development over many pulses was used. By distributions, $J^-(x,y)$ and $f(\alpha_x,\alpha_y)$ defined boundaries of $V_x(\alpha_x,x)$ and $V_y(\alpha_y,y)$, where the maximum α_y and y values $f(\alpha_x,x)$ and the maximum α_x and x values $f(\alpha_y,y)$ differ═ from zero within oscillography accuracy. The idea of current density distribution over the beam's cross section is illustrated in Fig. 5.8.

Figure 5.8 also shows characteristic configurations of the V_x and V_y regions. For the magnitude of emittances E_x and E_y, the areas of ellipses that fit the corresponding regions V_x and V_y were divided by π. In the notation of this figure, $E_x = \Delta\alpha_x\Delta_x$ and $E_y = \Delta\alpha_y\Delta_y$. The corresponding values of normalized emittances are $E_{nx} = \beta E_x$ and $E_{ny} = \beta E_y$, where $\beta = (2eU_o/Mc^2)^{1/2}$ is the ratio of ion velocity to the speed of light. For known values of the emittance, it is possible to estimate the maximum values of the local half-width of transverse energy in different sections of the beam, $\Delta W_x = Mc^2 E_{nx}^2/2(\Delta x)^2$ and $\Delta W_y = Mc^2 E_{ny}^2/2(\Delta y)^2$, including recalculated to the plane of the emission slit $\Delta W_{ox} = 2Mc^2 E_{nx}^2/\delta^2 \sim 40$ eV and $\Delta W_{oy} = 2Mc^2 E_{ny}^2/l^2 \sim 3$ eV, where l is the emission slit length and δ the slit width. From these parameters, one can judge the effect of the optics of the beam formation system and processes in the transport space on the ion-optical characteristics of the produced beams. Varying the electrode shape in the beam formation system, and even changing to a three-electrode system, had little effect on the ion-optical characteristics of the formed beams, but even a slight symmetry violation in the manufacture and installation of the elements of the beam forming system caused beam deterioration—displacement of the emission gap by just 0.1 mm relative to the plane of symmetry of the extraction electrodes causes a noticeable deflection of the beam. In this regard, measures were taken for accurate positioning of the extraction system components and maintaining the geometric dimensions of the beam formation system during source operation.

There is a substantial difference in the ion-optical characteristics of beams generated by discharges with different levels of plasma fluctuations. As mentioned above, significant fluctuations over a wide frequency spectrum (10^5–10^7 Hz) are typical of discharges in Penning and planotron plasmas with low hydrogen feed rate and high magnetic field strength. With an increase in the size of the anode window, the lower boundary of the region of ignition and maintenance of the discharge, as shown in Fig. 4.33, shifts to a region of lower magnetic field strength B, so that maintenance

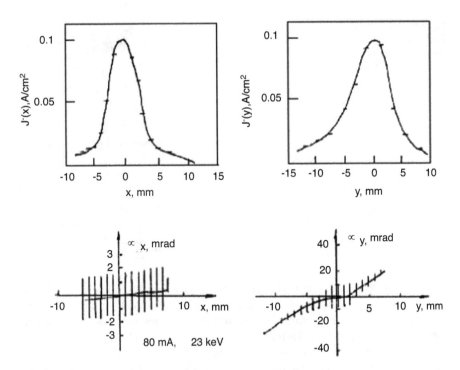

Fig. 5.8 Current density distribution (upper) and emittance configuration (lower) in the formed ion beam. (Reproduced from Dudnikov [5])

of a discharge without fluctuations becomes possible over the operating range of hydrogen density. When the operating point moves through this diagram (Fig. 4.33), a discharge with chaotic fluctuations transfers into a discharge without detectable fluctuations through an intermediate mode with coherent oscillations and without fluctuations. Typical voltage oscillograms in different discharge operating modes are shown in Fig. 5.9.

Fluctuations in the plasma parameters increase the collision frequency of electrons by increasing the transverse conductivity and lead to redistribution of discharge current. The frequency of coherent oscillations, ~18 MHz, depends very little on the operational parameters, hydrogen density, cesium feed, magnetic field intensity, and discharge current, although changes in these parameters strongly affect the transition between regimes.

The causes of fluctuations have been considered repeatedly in connection with the study of positive ion sources with discharges in a magnetic field [1, 11, 12]. There is no complete understanding of this problem up to now. Because of the particularities of beam formation systems used in positive ion sources, the magnetic field intensity can as a rule be varied only over a narrow range. As far as we know, fluctuations of plasma parameters have so far not been eliminated in the operating regimes of plasma positive ion sources with a magnetic field. Discharges without fluctuations are easily obtained without cesium, since in this case the coefficient of

Fig. 5.9 Oscillograms of
discharge voltage (**a**) with
noise, (**b**) with coherent
oscillations, and (**c**)
without noise.
(Reproduced from
Dudnikov [5])

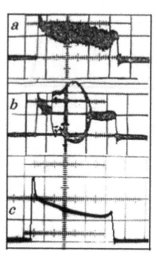

secondary electron emission is <0.1 and the current across the magnetic field is carried mainly by ions and plasma electrons. When cesium is added, or thermoemitters are used, the coefficient of secondary electron emission increases to 3, and the current across the magnetic field is transferred by high energy cathode electrons, which have much smaller collision cross sections. Since fluctuations in the plasma parameters greatly impair the beam ion-optical characteristics, sources with multi-aperture beam formation systems for generating intense positive ion beams have switched to using emitting plasma in "quieter" discharges without a magnetic field or in utilizing a "magnetic multicusp wall."

Studies of the vibrational properties of discharges in the SPS were undertaken, but unambiguously interpreted results have not yet been obtained. It is possible to give only general considerations explaining the transition between regimes. Apparently, they are associated with a change in transverse conductivity of the plasma column. For fixed magnetic field strength B, the transverse conductivity $\sigma_\perp$ as a function of the electron scattering frequency τ_e^{-1} is

$$\sigma_\perp \sim \tau_e / \left(1 + \omega_{ce}^2 \tau_e^2\right)$$

and thus increases proportionally to τ_e for $\omega_{ce}\tau_e << 1$ and decreases proportionally to τ_e^{-1} for $\omega_{ce}\tau_e >> 1$ and has a maximum at $\omega_{ce}\tau_e = 1$. If the gas density is low, the particle collisional scattering frequency for plasma electrons, τ_e^{-1}, is less than the electron cyclotron frequency ω_{ce}, and the occurrence of fluctuation is favorable for transport of the discharge current. For high gas density, the additional scattering of electrons by fluctuations must reduce the transverse conductivity of the plasma, so that the appearance of fluctuations turns out to be "unprofitable." From these considerations, it follows that the condition for transition between modes with and without fluctuations, $\omega_{ce}\tau_e \sim 1$, is a reasonable choice given the rather uncertain electron scattering cross sections for hydrogen ions and satisfactorily agrees with the experimental limit of the transition diagram shown in Fig. 4.33. This hypothesis is confirmed by the suppression of discharge noise by a small addition of heavy gas. Experience with plasma sources shows that the production of discharges without noise when using thermal cathodes is facilitated if only part of the cathode voltage is applied to the anticathode. In discharges with cold cathodes, this method of eliminating fluctuations has not been tested.

At the Institute of Nuclear Research (Troitsk, Russia) [13], the Penning source is used to produce a 100 Hz, 250 μs beam pulse. The beam is extracted from a 15×0.6 mm^2 slit at 20 kV. For a 50 mA beam with fluctuations, the normalized emittances in the x and y planes are 0.4 and 0.7 π mm mrad.

5.4 Surface Plasma Sources with Penning Discharge for Microlithography

On the basis of previous developments, an SPS with a Penning discharge for microlithography and having record low emittance was developed [14–16]. An illustration of this source is shown in Fig. 5.10. It consists of cathode 1, anode with insert with an emission aperture 1 mm in diameter, extraction electrode 3, grounded electrode 4, cathode insulators 6, and cathode cooler 7. A magnetic field for the Penning discharge is established by permanent magnet 5 and a yoke. The discharge burns between the cathode and the anode. H^- ions are extracted from the plasma by the extraction electrode. The discharge is fed with cesium from a heated container.

The discharge voltage is 100 V and the discharge current is up to 50 A. An ion beam with current up to 1 mA is extracted through the emission hole of diameter 1 mm. This source was delivered to the University of Maryland and was further investigated there [15, 16].

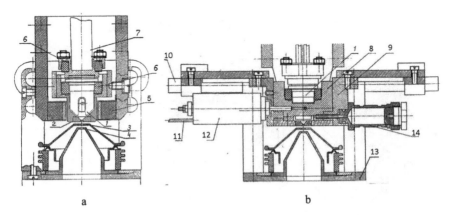

Fig. 5.10 SPS with Penning discharge for microlithography (**a**) cross section along magnetic field and (**b**) cross section perpendicular to magnetic field. (Reproduced from Dudnikov [5]). 1, cathode; 2, insert with emission aperture; 3, extraction electrode; 4, ground electrode; 5, permanent magnet; 6, cathode insulators; 7, cathode cooler; 8, anode insert; 9, source body; 10, high-voltage insulator; 11, gas tube; 12, gas valve; 13, base plate; and 14, cesium oven

5.5   Semiplanotron: Geometric Focusing

The generation efficiency of negative ion beams could be significantly increased by geometric focusing of the negative ions produced. For the first time, effective geometric focusing was accomplished in the SPS that we called the semiplanotron [17, 18]. A schematic of the first semiplanotron is shown in Fig. 5.11.

The semiplanotron SPS consists of a cathode 1 with a semicylindrical groove 2 on one side, an anode 3, cathode holders 4 with a cooling channel 5, and cathode insulators 6; there is a recess at the beginning of the cathode for ignition of discharge 7. The semiplanotron SPS is attached to the magnet between pole 8 with magnetic inserts 9 and extractor 10. A discharge in E x B crossed fields burns in semicylindrical groove 2, adjacent to the emission slit. The end of the cathode is skewed to interrupt the closed electron drift. H^- beam 11 is extracted from the discharge through the emission slit.

The geometric focusing scheme is illustrated in Fig. 5.12 [19]. Negative ions emitted by a cathode are accelerated in the near-cathode potential drop along a normal to the surface. If the surface is cylindrical or spherical, negative ions are focused to the center of curvature. In experiments on the degree of focusing, one could judge by the sputtering trace on the anode. With transverse dimension of the groove 3 mm, the etched track had a width of 0.8 mm. The discharge voltage was typically 100 V.

The emission characteristics of the discharge are shown in Fig. 5.13. The H^- ion current was over 200 mA with a discharge current less than 40 A, using an emission slit of dimensions 1×10 mm^2.

The D^- ion current is a half the H^- ion current because heavy deuterium ions form a denser plasma that destroys more negative ions during transport through the plasma. The dependence of H^- ion current and D^- ion current on discharge current

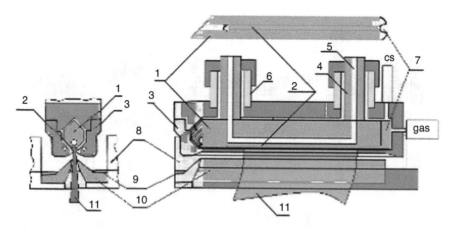

Fig. 5.11 Schematic of the first semiplanotron SPS. (Reproduced from Belchenko [18]). 1, cathode, 2, semicylindrical groove; 3, anode; 4, cathode holders; 5, cooling channel; 6, cathode insulators; 7, recess for discharge ignition; 8, magnet pole; 9, magnetic insert; 10, extractor; and 11, H^- beam

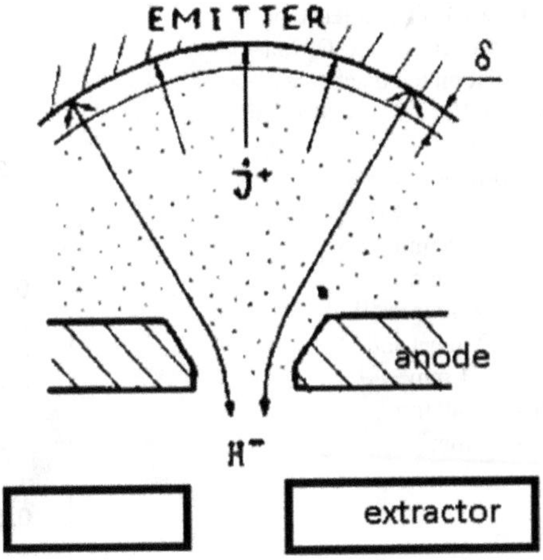

Fig. 5.12 Geometric focusing system. (Reproduced from Belchenko [18])

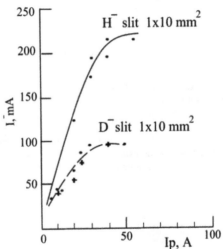

Fig. 5.13 Dependence of H^- ion current and D^- ion current on discharge current for a 1×10 mm^2 emission slit. (Reproduced from Belchenko [18])

for a number of emission slit dimensions is shown in Fig. 5.14. Beams of H^- ions are obtained with current up to 0.9 A for a discharge current of 100 A.

A significant increase in H^- ion beam current was produced by a multislit semi-planotron [20]. A schematic of a five-slit semiplanotron is shown in Fig. 5.15.

The dependence of the H^- ion beam current on discharge current for this source is shown in Fig. 5.16. The H^- ion beam current reaches 4 A for a discharge current of 400 A.

The surface area of the five emission slits is $5 \times 0.8 \times 35$ mm^2. At a discharge current of 10 A, the intensity of the H^- ion beam is up to 0.5 A. The change in the dependence at a discharge current of about 100 A is associated with the disappearance of molecular ions. The energy cost of an H^- ion in the early stages is 1.5 keV/ion.

Fig. 5.14 Dependence of H^- ion current and D^- ion current on the discharge current for several different emission slits. (Reproduced from Belchenko [18]). H^- (1 – 0.72×45 mm², 2 – 0.5×41 mm², 3 – 1×40 mm², 4 – 1×20 mm²) and D^- (5 – 0.5×41 mm², 6 – 1×40 mm², 7 – 1×20 mm²)

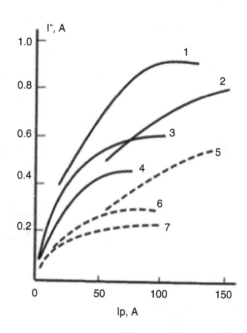

Fig. 5.15 Schematic of a multislit semiplanotron. (Reproduced from Belchenko [18]). 1, cathode; 2, anode; 3, pole magnet; 4, emission slits; and 5, electrode extraction system

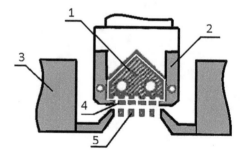

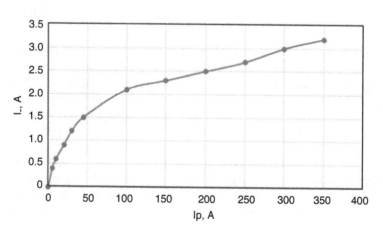

Fig. 5.16 Dependence of H^- ion beam current on discharge current for a multislit semiplanotron. (Reproduced from Belchenko [18])

A further increase in energy and gas efficiency was associated with introducing two-dimensional (spherical) geometric focusing [21]. A schematic of a source with spherical focusing is shown in Fig. 5.17.

The source consists of a cathode with spherical dimples on the surface, an anode 2 with emission holes, and a multi-aperture extraction electrode 3. A photograph of this SPS is shown in Fig. 5.18. Figure 5.19 shows the dependence of the H^- beam current on discharge current. Up to 11 A of H^- ions were obtained from this source. The maximum H^- ion current density in the emission hole was 3 A/cm^2, and mean negative ion density in the beam was 180 mA/cm^2.

Honeycomb SPS with cathode emission area 3×20 cm^2 and 600 cathode indentations, arranged in the hexagonal order, is shown in Fig. 5.20. Multislit extraction electrode consists of 11 rods, equipped with the immersed ohmic heaters. To

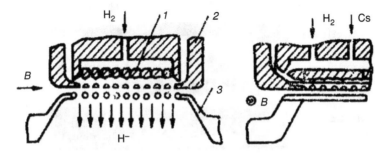

Fig. 5.17 Schematic of a semiplanotron with spherical focusing. (Reproduced from Belchenko [18]). 1, cathode with spherical dimples; 2, anode with emission holes; and 3, multi-aperture extraction electrode

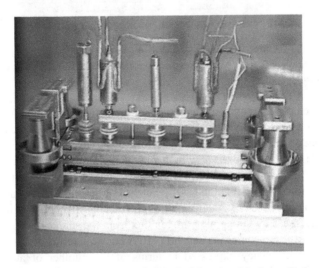

Fig. 5.18 Photograph of a semiplanotron with spherical focusing. Maximum H^- ion current is 12 A. There are two pulse valves for gas supply and two containers with cesium tablets. (Reproduced from Belchenko [18])

Fig. 5.19 Dependence of
H^- ion current on
discharge current in a
semiplanotron with
spherical focusing.
(Reproduced from
Belchenko [18])

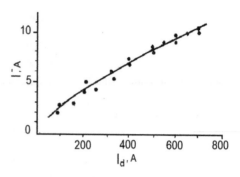

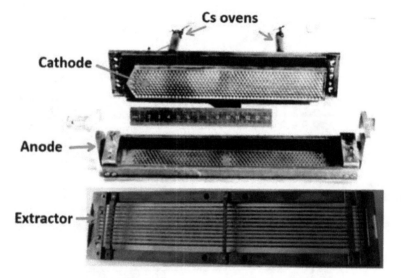

Fig. 5.20 Honeycomb SPS with cathode emission area 3×20 cm^2 and 600 cathode indentations, arranged in the hexagonal order. Multislit extraction electrode consists of 11 rods, equipped with the immersed ohmic heaters

enhance the electric strength of the extended extraction gap, the extracting rods were made of profiled molybdenum and heated by immersed ohmic heaters.

In the honeycomb source, it proved possible, due to the fourfold decrease in the emission hole area (in comparison with that in the multislit semiplanotron), to reach a more uniform distribution of current emission density over the cathode surface and obtain H$^\sim$ ion beams with high gas efficiency. For example, in the case of extraction of a beam with a 2.5 ± 3-A current through 100 emission holes with a total area of 0.5 cm^2, the measured integral hydrogen flux from the source during a discharge pulse was 3×10^{19} molecules per second (~10 A equiv. in terms of atoms); in other words, the gas efficiency expressed in terms of the number of atoms reached 25 ± 30%. The high emission density of the negative ion beams thus obtained reduced power released at the electrodes, low consumption of working substances (hydrogen and cesium), and shielding of the main working surface of the cathode from external fast particle fluxes, as well as the possibility of a manyfold increase in beam current due to a greater number of cells, taken together stimulated further investigations and modernization of honeycomb SPSs. Modeling and experimental studies of the efficiency of ballistic focusing in honeycomb SPSs were undertaken. A three-dimensional MODA code using the main principles of the MARLOWE code was developed at BINP (Budker Institute of Nuclear Physics) [22, 23] for the quantitative assessment of factors influencing ballisticfocusing. The formation of negative ions by proton sputtering and reflection from hydrogen-saturated targets was simulated; the sputtering and reflection coefficients were calculated together with energy spectra of sputtered and reflected particles. The effect of the near-cathode discharge voltage drop, the shape and thickness of this layer, the composition of ion current onto the cathode, and the geometry of gas-discharge electrodes on the effectiveness of negative ion accumulation in SPS emission holes have been established.

More detailed investigations of geometric focusing have been described in [24]. A schematic of a spherically focused SPS is shown in Fig. 5.21. The high-current glow discharge with unclosed electron drift was localized in a gap between the cathode plane and the anode cover. The side cathode projections provide for electron oscillation along external magnetic field lines. Hydrogen supply and a small addition of cesium to the discharge were fed from external containers via cathode channels. Most negative ions were produced on the cathode surface, bombarded by an intense positive ion flux. The cathode work function was decreased by cesiation. A number of spherically concaved indentations were fabricated on the cathode emitting surface for geometric focusing of cathode-produced negative ions to the emission holes.

The indentation concavity radius was 3 mm and the average depth 0.4 mm. The indentations were arranged on the cathode surface in an orthogonal matrix with 3 mm "mesh size." The height of the honeycomb cathode side projection was decreased down to 0.8–0.6 mm for compensation of discharge enhancement in the E x B drift direction. For comparison, the SPS was also tested with a flat cathode emitting surface. Conical emission holes with inner diameter 0.8 mm were drilled at the negative ion focusing points on the anode cover (Fig. 5.21). The emitting cathode surface was of area approximately 10 cm^2. The total area of the 120 emission holes was 0.6 cm^2. A steady-state extraction voltage of up to 15 kV was applied to the anode body. The multislit extractor was made of thick molybdenum wires, tensioned with springs at their ends. A set of short high-voltage pulses (up to 1 ms) with repetition rate up to 100 Hz was used for fast observation of the model emission properties.

The negative ion beam current was measured by Faraday cup collectors, and the extracted beam was mass analyzed. The H^- ion energy spectrum was investigated using a 90°·electrostatic analyzer. A hybrid cathode with concave and flat indentations (Fig. 5.22) was tested for negative ion generation efficiency. In this case, the anode cover with emission hole was shifted by an external driver with respect to the cathode indentations over a distance of 9 mm. A positive ion current was also

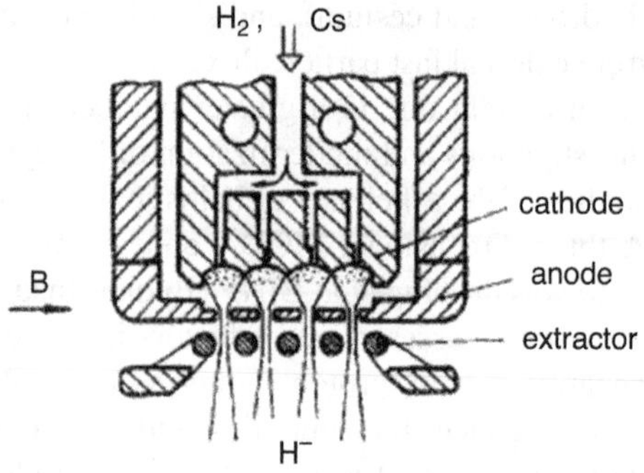

Fig. 5.21 Schematic of a
spherically
focused SPS. (Reproduced
from Belchenko [18])

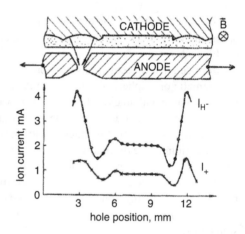

Fig. 5.22 H^- beam current
I^- and positive ion current
I^+ as a function of emission
hole position. (Reproduced
from Belchenko [18])

extracted from the SPS using the opposite extraction voltage polarity. The typical
"steady-state" discharge voltage was 90–80 V.

With reduced pulse duration or lower discharge current, the steady-state voltage
was 100–150 V and had little influence on the H^- yield in the H/Cs discharge mode.
The main part of the extracted negative ion beam was H^- ions, but in the high discharge voltage start-up period of the pulse, the heavy ion component was up to 10%
of total ion beam. Thus, the initial excess in negative ion current was due to the
heavy ion component. This component decreased down to 1–0.3% after the beginning of the pulse. The increase in the negative ion yield with increasing discharge
current is shown in Fig. 5.23 for honeycomb (GF) and flat cathodes in the H/Cs
mode and for the hydrogen pulse mode. The vertical line in Fig. 5.23 corresponds to
the negative ion current drop during over pulse duration. The honeycomb H^- output
was 3–3.5 times greater than that for the flat cathode and 6–7 times greater than that
for the hydrogen mode with activated electrodes. The H^- yield of the honeycomb
SPS was 1.1–0.9 A for a discharge pulse length 0.35 s and 90 A discharge current
(short pulse extraction). Similar dependencies were observed with the steady-state
extraction system. Thus, an H^- beam with intensity 0.5–0.45 A and at pulse duration
0.25 s was obtained. In this case, the total current in the extraction circuit was <1.5 A.

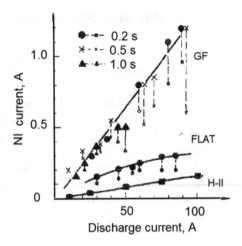

Fig. 5.23 Negative ion beam current as a function of discharge current. (Reproduced from Belchenko [18])

Negative ion beams were filtered of accompanying electrons by a source transverse magnetic field. These electrons were effectively dumped to the magnet poles. The absence of electron accumulation and multiplying significantly increased the electric field hold-off strength and supported the fast high-voltage recovery of after nondestructive breakdown. The H^- yield was maximum at minimal hydrogen density in all discharge modes. Direct measurement of the hydrogen flow from the source, operating in the H/Cs mode, gave the value 0.5–0.7 Torr.L /s, and the gas efficiency of H^- beam production was 12–17%. Figure 5.22 shows the H^- yield I^- and the extracted positive ion current I^+ for various emission hole positions relative to cathode indentations (H^-II). With the emission hole situated at the GF (geometric focus) point, the H^- output was 3–3.5 times greater than that for the "minimum" point and twice that for the flat indentation. The value of I^+ was three times lower than for H^-, and its magnitude also decreased by a factor of 2.5–3.5 with an emission hole shift from the GF point to the "minimum" point (discharge current of 90 A).

The energy spectrum of ions extracted from the GF point (H/Cs mode) consisted of H^- ions produced at the cathode surface (80–85%), and the remaining 15–20% was H^- produced in the near-anode plasma or on the emission hole surface. At the "minimum" point, the "anode" part of the H^- spectra increased by 1.5–2.0 times, while the "cathode" part decreased by approximately 10 times. The average energy of the "cathode" H^- group was 175 eV at the GF point (discharge voltage 150 V). This energy increased up to 240 eV during extraction through the "minimum" point due to disappearance of the cathode "sputtered" H^- ion fraction. The "anode" fraction of the H^- beam at the "minimum" point was 2–3 times greater than the "cathode" fraction. There were three groups of ions observed in the H^- energy spectra for the hydrogen mode with activated electrodes. These groups have energies of 0.15–0.25 eU_d, 0.5–0.9 eU_d, and 1.3–1.8 eU_d. The only H^- group with "anode" potential was observed for the hydrogen mode.

5.6   Semiplanotron SPS for Accelerators

For use in accelerators, a compact semiplanotron SPS of simple design was developed [25–27]. A schematic of this semiplanotron is shown in Fig. 5.24 and a photograph in Fig. 5.25.

The source consists of anode 2 with emission slit 1, cathode 3 with a semicylindrical groove and a cathode notch 5 for ignition of the discharge, cathode insulator 4, and extractor 6. The source is attached between the magnetic poles 7 with magnetic inserts. The dependence of H^- ion current on discharge current for various configurations of the cathode and the emission slit is shown in Fig. 5.26. The curves are N-shaped. For low discharge current, H^- ions formed at the cathode are extracted. As the discharge current rises, fast H^- ions are destroyed (by electron loss in

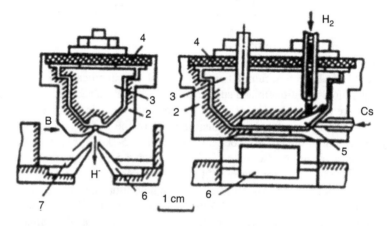

Fig. 5.24 Schematic of semiplanotron SPS for accelerators. (Reproduced from Derevyankin [27], with permission of G. Derevyankin). 1, emission slit; 2, anode; 3, cathode with semicylindrical groove; 4, cathode insulator; 5, cathode notch for ignition of discharge; 6, extractor; and 7, magnetic poles with magnetic inserts

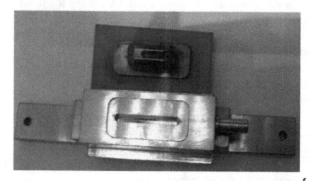

Fig. 5.25 Photograph of the dismantled semiplanotron. The cathode with semicylindrical groove is visible: (Reproduced from Derevyankin [27], with permission of G. Derevyankin)

Fig. 5.26 Dependence of the H^- ion current on the discharge current for various configurations of the cathode and the emission slit. Curve 7 is for SPS with Penning discharge. (Reproduced from Derevyankin [27], with permission of G. Derevyankin)

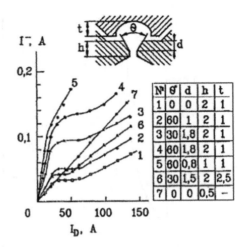

N	ϑ	d	h	t
1	0	0	2	1
2	60	1	2	1
3	30	1,8	2	1
4	60	1,8	2	1
5	60	0,8	1	1
6	30	1,5	2	2,5
7	0	0	0,5	–

collisions with plasma particles), and fast neutrals initiate the emission of H^- ions from the walls of the emission slit but with less efficiency (curve 7 is for a Penning discharge). For an open emission slit of dimensions 0.5×10 mm², a beam of intensity 100 mA is obtained for a discharge current of 10 A and discharge voltage 100 V. The ion cost is 10 keV/ion (100 mA/kW).

Beam characteristics are shown in Fig. 5.27. The beam broadens significantly if the intensity exceeds a critical limit determined by the extraction voltage, and in this case, the emittance increases substantially also [28]. This discharge can be noiseless. The H^- beam had a current of 80 mA, energy of 20 keV. Slit dimensions were 0.5×10 mm². Normalized emittance was $\varepsilon_{xnrms} = 0.05\ \pi$ mm.mrad and $\varepsilon_{ynrms} = 0.2\ \pi$ mm.mrad, $kT_x \sim 27$ eV and $kT_y \sim 2$ eV.

5.7 Semiplanotrons with Spherical Focusing for Continuous Operation

A semiplanotron with spherical focusing was developed for continuous operation [29]. A schematic of this SPS is shown in Fig. 5.28. It consists of cathode 1, with hollow cathode 3 with channels for supply of working gas and cesium, anode 2, cathode insulator 4, and extractor 5. The SPS is installed between the poles of magnet 6. The discharge is supported as a hollow cathode in a mixture of working gas with cesium.

Plasma drifts in the E x B crossed fields along a semicylindrical groove to a spherical dimple at the cathode end, where emission of negative ions is initiated, focused to the emission aperture. Figure 5.29 shows the dependence of the H^- beam current on discharge current. In a long-term stationary regime, up to 2.5 mA of H^- ions were obtained using an emission aperture 1 mm in diameter.

The emittance of the H^- beam at an extraction voltage of 25 keV is shown in Fig. 5.30.

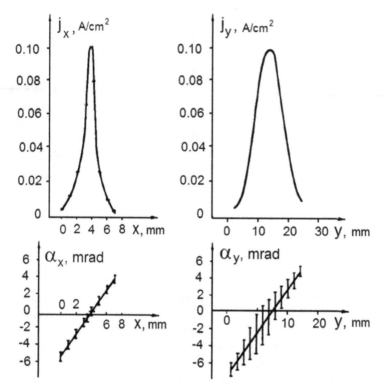

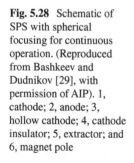

Fig. 5.27 Distribution of the H^- beam current density distribution and emittance. (Reproduced from Derevyankin [27], with permission of G. Derevyankin)

Fig. 5.28 Schematic of SPS with spherical focusing for continuous operation. (Reproduced from Bashkeev and Dudnikov [29], with permission of AIP). 1, cathode; 2, anode; 3, hollow cathode; 4, cathode insulator; 5, extractor; and 6, magnet pole

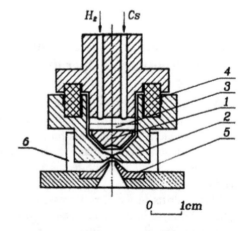

The mass spectra of beams when operating with hydrogen and with hydrogen plus added ammonia are shown in Fig. 5.31. Up to 2 mA of heavy ions were obtained by adding the oxygen-containing gases, water vapor, or ammonia. A small hole (0.2 mm in diameter) was drilled in the cathode by back-streaming energetic

Fig. 5.29 Dependence of
the beam current H^- on the
discharge current.
(Reproduced from
Bashkeev and Dudnikov
[29], with permission from
the AIP Publishing)

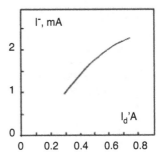

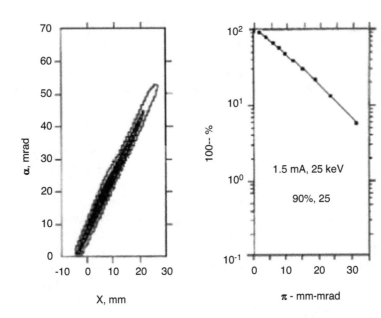

Fig. 5.30 Emittance of the H^- beam at an energy of 25 keV and dependence of emittance on percent of beam intensity

positive ions. Without cesium, the discharge voltage was 400–600 V. When fed with cesium, it decreased to 100 V.

Further studies of this electrode configuration were carried out at the Fermi National Accelerator Laboratory [30, 31]. The investigated configuration is shown in Fig. 5.32.

This SPS consists of hollow cathode 1 with feed channel 2 for cesium and feed channel 3 for hydrogen supply, a semicylindrical groove 7 for plasma E x B drift in crossed fields, and a spherical dimple 9 at the end. The cathode insulators 4 isolate cathode 1 from anode 5 with emission aperture 8. The SPS is located between the poles of magnet 10 with extractor 11. In embodiment (a), the hollow cathode discharge could be ignited to the top of the anode. Then the plasma drifts to the insulator without getting onto a spherical dimple and without emitting negative ions. In

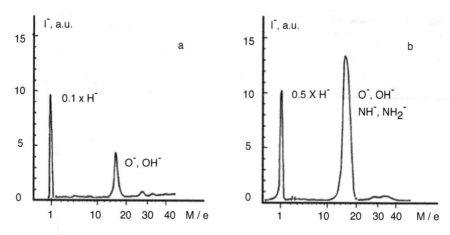

Fig. 5.31 Mass spectra of beams when operating with (**a**) hydrogen and (**b**) hydrogen with ammonia. (Reproduced from Bashkeev and Dudnikov [29], with permission from the AIP Publishing)

case b, this case is excluded. In variant c, a more effective plasma drift to the spherical dimple is used.

The general design of an SPS with spherical focusing is shown in Fig. 5.33. It consists of an isolated flange 3 with gas supply tube 1, vacuum feedthrough 2, cylindrical insulator 4, cooled flange 6 with high-voltage feedthrough 5, copper holders 7 of platform 17, cesium container 8, cathode 9 with cathode insulators 10, and anode 11. The SPS is located between the poles of magnet 12 together with extraction electrode 13 and grounded extractor 14. The magnetic field is generated by permanent magnets 15. The SPS is attached by high-voltage insulators 16 to the platform-base 17. The ion beam 18 is formed from the discharge by the extractor voltage. The radius of the spherical dimple on the cathode is 3.5 mm. A photo of this SPS is shown in Fig. 5.34a. A photograph of a semiplanotron SPS with spherical focusing is shown in Fig. 5.34b.

The dependence of the H^- ion current density on the extraction voltage for the three SPS variants using an 0.4 mm aperture is shown in Fig. 5.35.

The burning voltage for a discharge in hydrogen is 400–500 V. When cesium is added, it decreases to 80 V. At a current up to about 0.9 A, the H^- ion emission current density, J^-, is up to 1 A/cm². The current density production efficiency $F^- = J^-/P = 12$ A/cm².kW is much greater than for the best proton sources where $F^+ = J^+/P = 0.25$–0.05 A/cm².kW.

The sputter-etched trace of the focused H^- ion flux in a spherical focusing SPS is shown in Fig. 5.36, indicating good geometric focusing of the H^- ion flux. The hollow cathode discharge can be noise-free if the hollow cathode is filled with molybdenum wires of 0.3 mm diameter.

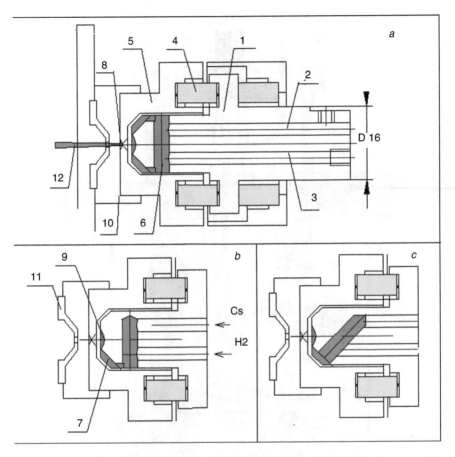

Fig. 5.32 (**a**) Hollow cathode SPS open on both sides, (**b**) hollow cathode SPS open on one side, and (**c**) hollow cathode SPS open on one side at 45°. (Reproduced from Vadim Dudnikov et al. [31], with permission from the AIP Publishing). 1, cathode; 2, cesium supply channel; 3, hydrogen supply channel; 4, cathode insulators; 5, anode; 6, hollow cathode; 7, semicylindrical groove for plasma drift; 8, emission aperture; 9, spherical dimple; 10, magnet pole; 11, extractor; and 12, ion beam

5.8   Compact Surface Plasma Sources for Heavy Negative Ion Production

SPSs for heavy negative ion production have been described [32]. A generalized schematic of this source and the beam diagnostics are shown in Fig. 5.37. It consists of high-voltage flange 1 with gas inlet and electrical feedthroughs and cylindrical insulator 2 connected to the vacuum chamber 3. The SPS itself consists of gas-discharge cathode chamber 4, anode 5, emitter 6, high-voltage extraction insulators 7, magnetic system 8, and base plate 9 with extractor.

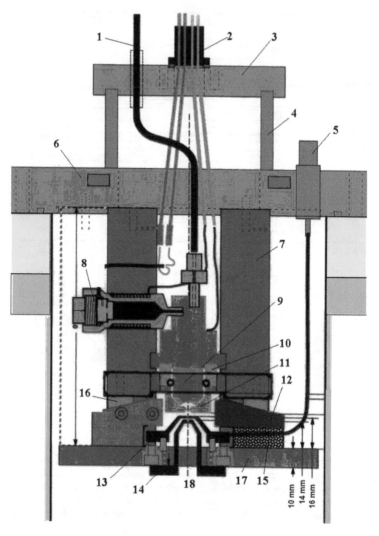

Fig. 5.33 Construction of a semiplanotron SPS with spherical focusing. (Reproduced from Vadim Dudnikov et al. [31] with permission from the AIP Publishing). 1, gas supply tube; 2, vacuum feedthrough; 3, flanges; 4, insulators; 5, high-voltage feedthrough; 6, cooled flange; 7, copper platform-cooled holders; 8, cesium container; 9, cathode; 10, cathodic insulator; 11, anode; 12, pole magnet; 13, pulling electrode; 14, grounded extractor; 15, permanent magnet; 16, high-voltage insulators; 17, platform-base; and 18, ion beam

The ion beam is formed from the plasma by the extraction voltage and recorded on the silicon or aluminum implantation plate 12 located in collector 13. Secondary electron emission is suppressed by grid 11 and permanent magnet 14. An emittance meter 15 was used to measure the emittance. Magnetic analyzer 16 was used to record the ion mass spectrum.

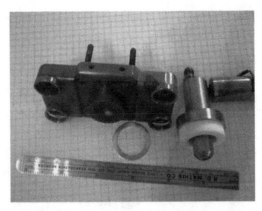

Fig. 5.34 (**a**) Photograph of SPS with spherical focusing. The cathode with cesium container and cathode insulator is on the right. (Reproduced from Vadim Dudnikov et al. [31], with permission from AIP Publishing). (**b**) Photograph of assembled of a semiplanotron SPS with spherical focusing

A more detailed schematic of a compact SPS for heavy ion beam production is shown in Fig. 5.38. This ion source consists of anode 1 with anode insulator 3, hollow cathode 2, hemispherical emitter 4, front plate 5 with emission aperture 6, gas-discharge chamber cathode 8, holder-coolers of the discharge chamber 9, insulator for the emitter holder 10, remitter holder-cooler 11, gas supply tube 12, cesium source 13, emitter insulator 14, and emitter screens 15. A focused flux of negative ions 7 emitted from the surface of the emitter 4 is extracted through emission hole 6. Discharges in

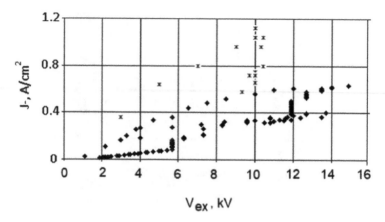

Fig. 5.35 Dependence of H^- ion current density on extraction voltage for three variants of SPS, with an aperture of 0.4 mm. (Reproduced from Vadim Dudnikov et al. [31], with permission from AIP Publishing)

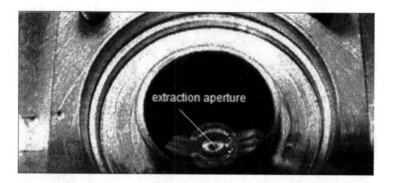

Fig. 5.36 Sputter-etched trace of the focused flow of H^- ions in an SPS with spherical focusing

hydrogen, nitrogen, krypton, and xenon have been tested. The best results for the generation of heavy negative ions were obtained with a xenon discharge. A discharge in xenon is initiated at a discharge voltage of 0.4–0.5 kV. When cesium is added, the discharge voltage decreases to 180 V. A stable discharge can be maintained up to a current of 2.5 A. The plasma drifts in crossed fields to the emitter surface biased to 300 V. The plasma potential is close to the anode potential of 180 V. Positive ions of energy 180 + 300 eV bombard the emitter surface. Sputtered secondary negative ions are accelerated from the surface by the same voltage and are focused into the emission hole. Negative ions are extracted by extraction voltage of up to 15 kV, accelerated and formed into a beam which is recorded on an Si or Al implantation plate positioned in the collector. The diameter of the emitter is 12 mm, with radius of curvature 15 mm.

The implanted target was subsequently analyzed by secondary ion mass spectroscopy (SIMS) to determine the implantation dose. Primary experiments were carried out with a copper emitter. The main focus was on a lanthanum hexaboride emitter to produce negative boron ions.

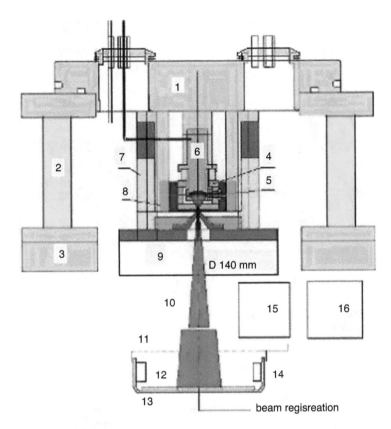

Fig. 5.37 General scheme of an SPS for production of heavy ions. 1, high-voltage flange with gas inlet and electrical feedthroughs; 2, cylindrical insulator; 3, vacuum chamber; 4, gas-discharge cathode chamber; 5, anode; 6, emitter; 7, high-voltage extractor insulators; 8, magnet; 9, base plate with extractor; 10, ion beam; 11, electron suppression grid; 12, implantation plate; 13, collector; 14, permanent magnet; 15, emittance meter; and 16, magnetic analyzer with mass spectrum recording

Activation of the lanthanum hexaboride emitter with increasing discharge current is shown in Fig. 5.39. After each increase in discharge current, a prolonged increase in the emission current is observed. The maximum B_2^- ion current was 0.9 mA at a discharge current of 2.4 A. Stable long-term operation was attained with a negative ion current of 0.6 mA.

An autograph of the B_2^- beam on the emittance meter plate is shown in Fig. 5.40. The plate has a matrix of holes of diameter 0.5 mm spaced by 3 mm. The beam has an elliptical cross section with axes of length 10 mm and 7.5 mm. After much work, a sizeable amount of sputtered cathode material is formed, but this does not interfere with normal operation of the SPS, since the material is collected at the bottom of the cathode box.

Depending of SIMS count of implanted B into Al foil on sputtering time is shown in Fig. 5.40b.

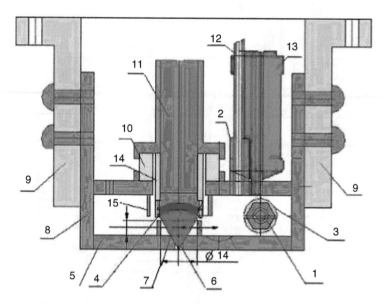

Fig. 5.38 Schematic of compact SPS for producing heavy negative ion beams. 1, anode; 2, hollow cathode; 3, anode insulator; 4, hemispherical emitter; 5, front plate with emission hole; 6, emission hole; 7, focused negative ion flux; 8, gas-discharge cathode chamber; 9, holder-coolers of discharge chamber; 10, insulator of emitter holder; 11, emitter holder; 12, gas supply tube; 13, cesium source; 14, emitter isolator; and 15, emitter screens

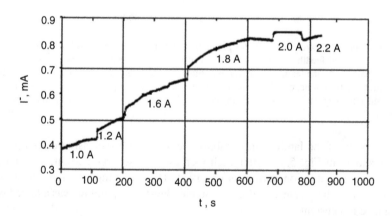

Fig. 5.39 Activation of the SPS with lanthanum hexaboride emitter with increasing discharge current. (Reproduced from Vadim Dudnikov and J. Paul Farrell [33], with the permission from AIP Publishing)

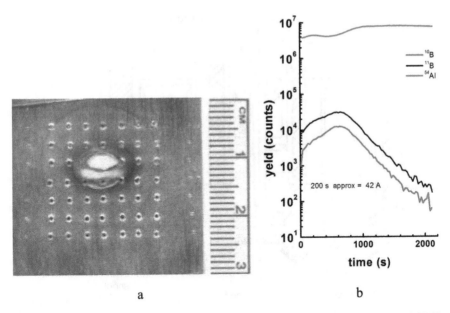

a   b

Fig. 5.40 (**a**) Autograph of the B_2^- beam on the emittance meter plate. (**b**) Depending of SIMS count of implanted B into Al foil on sputtering time. (Reproduced from Vadim Dudnikov and Paul Farrell [33], with permission from AIP Publishing)

5.9   Development of Surface Plasma Sources Worldwide

After 1974, it became possible for USSR researchers to communicate with foreign colleagues (although not everyone was allowed to publish). Contact between the Novosibirsk group (the Institute of Nuclear Physics (INP), Novosibirsk, Russia) and the Brookhaven National Laboratory (BNL) group (Th. Sluyters and K. Prelec in particular) contributed to the development of the SPS at BNL. They were given a copy of a planotron SPS, on the basis of which magnetron and Penning SPSs were developed at BNL [34–36]. A diagram of a BNL duoplasmatron with tubular discharge is shown in Fig. 5.41. It was possible to extract up to 60 mA of H^- ions from this source after adding cesium, but with a total current of 1.2 A of extractor current, at a discharge current of 150 A and discharge voltage 80 V in 1 ms pulses. From the ion energy spectrum, it was determined that a group of H^- ions are produced on the central rod by the surface plasma mechanism. Up to 0.9 A of H^- ions were extracted from a magnetron SPS planotron with a current of co-extracted electrons up to 2 A, at a discharge current 260 A and discharge voltage 150 V, in 10 ms pulses.

From an SPS with Penning discharge, up to 0.32 A of H^- ions were extracted with total extractor current 0.55 A, at a discharge current of 80 A and discharge voltage 220 V, with 4 ms pulse duration. Up to 0.14 A of D^- ions were extracted from a deuterium discharge with total extractor current 0.4 A, at a discharge current of 40 A, and discharge voltage of 420 V, with 4 ms pulse duration.

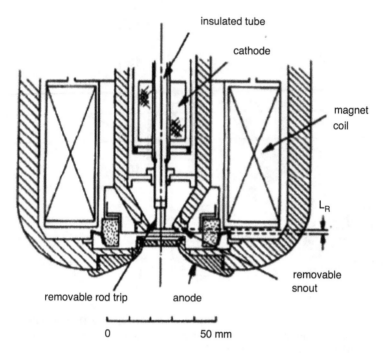

Fig. 5.41 BNL duoplasmatron SPS with tubular discharge. (Reproduced from Kobayashi [37])

After the invention of geometric focusing at the Institute of Nuclear Physics (Novosibirsk), an SPS with geometric focusing was developed at BNL [38, 39]. An asymmetric magnetron-planotron SPS is shown in Fig. 5.42. The radius of curvature of the groove is 3.7 mm. The dependence of the H^- ion current on discharge current for an asymmetric magnetron, for a regular magnetron with cylindrical groove, and for a magnetron with planar cathode is shown in Fig. 5.43. The energy expense for generation of H^- ions in the asymmetric magnetron was decreased from 33 keV/ion to 8 keV/ion for a flat cathode magnetron. (In the semiplanotron SPS, the energy price was reduced to 10 keV/ion and the gas efficiency was increased up to 30%.)

A multislit magnetron SPS with geometric focusing was developed in BNL for continuous operation and is shown in Fig. 5.44.

A magnetron was proposed with plasma generation by a hollow cathode discharge with an independent negative ion emitter with geometric focusing.

In 1976, BNL sold a prototype of a planotron-type SPS, shown in Fig. 5.45, to the Fermi National Laboratory for $40 k.

Based on this prototype SPS, C. Schmidt at Fermilab developed a design for the planotron (magnetron) shown in Fig. 5.46a [41–43]. The magnetron cathode is made of molybdenum, the body is the anode made of stainless steel, and the ceramics are processable MACOR. Cesium is supplied from a heated metal container with cesium ampoule. A 90° bending magnet with field decay index $n \sim 1$ is used for one-dimensional beam focusing, as is used in the Novosibirsk SPS with Penning

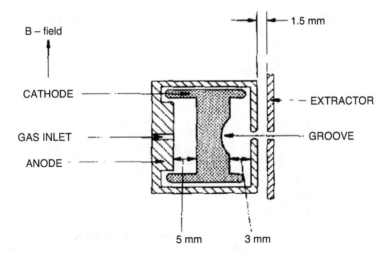

Fig. 5.42 Schematic of an SPS with asymmetric grooved magnetron (planotron). The radius of curvature of the groove is 3.7 mm. (Reproduced from Alessi et al. [39])

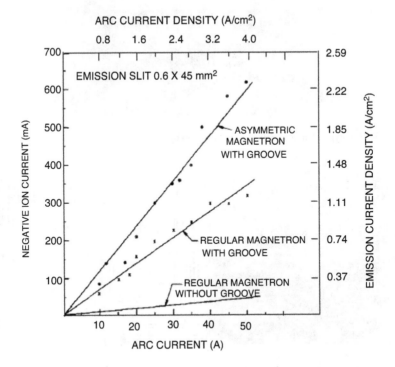

Fig. 5.43 Dependence of the H^- ion current on discharge current in an asymmetric magnetron SPS, a regular magnetron SPS with a cylindrical groove, and a magnetron SPS with a flat cathode. (Reproduced from Alessi et al. [39])

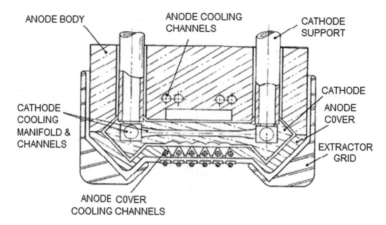

ANODE BODY

ANODE COOLING
CHANNELS

CATHODE
SUPPORT

CATHODE
COOLING
MANIFOLD &
CHANNELS

CATHODE

ANODE
COVER

EXTRACTOR
GRID

ANODE COVER
COOLING CHANNELS

Fig. 5.44 Multislit magnetron SPS with geometric focusing for continuous operation. (Reproduced from Prelec [40], with permission from AIP Publishing)

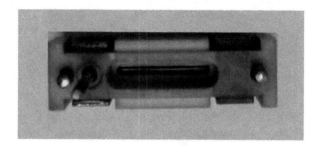

Fig. 5.45 Prototype planotron SPS sold to Fermilab by BNL

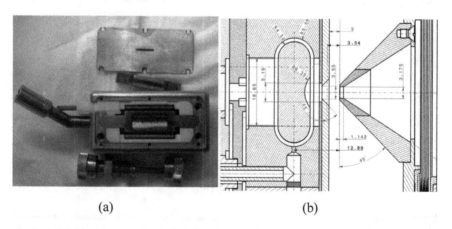

(a)   (b)

Fig. 5.46 (**a**) Magnetron-planotron SPS developed at Fermilab. (Reproduced from Schmidt and Curtis [41]) (**b**). Drawing of magnetron-planotron from BNL, producing 100 mA with discharge current 10 A, voltage 130 V

discharge. A drawing of the BNL magnetron-planotron with spherical focusing is shown in Fig. 5.46b. Magnetrons producing beams are operating with 0.5% duty factors or less. This limit is attributed to gas loading of the vacuum system and maintenance of the proper cesium coverage on the cathode surface, which is slowly sputtered away when the discharge is on. Operating in pulsed mode minimizes the gas load and maximizes the time cesium has to condense on the cathode surface. Gas injection is accomplished utilizing a piezoelectric valve, mounted close to the discharge area. With pulsed operation, gas efficiencies of around 6% for the grooved cathode design have been achieved. An external boiler charged with liquid cesium is held between 70 C and 150 C allowing cesium to continuously creep into the system. The cesium consumption at DESY (Deutsches Elektronen-Synchrotron) is approximately 0.8 mg per day at a plasma duty factor of 0.063%. The consumption rates at the other accelerator labs are slightly higher, when scaled with df. Magnetron lifetimes are on the order of 6–9 months, and typically the sources are turned off for maintenance rather than failing. The principle failure modes are listed in [43] and include build-up of cesium hydride near the gas inlet restricting gas flow and build-up flakes of Mo on the anode around the extraction aperture, which may dislodge limiting the aperture, shorting the cathode, or impeding the plasma. Operation of the BNL circular aperture SPS was recently summarized [41]. Typical operation parameters are an arc voltage of $U_d = 140$ to 160 V, arc currents of $I_d = 8$–18 A, and a 0.5% df (7.5 Hz, 700). The extracted current is 90–100 mA at 35 keV. The source power efficiency was significantly improved when cathode geometrical focusing was introduced. The BNL magnetron SPS is the highest efficiency 75 mA/kW of any operating H^- accelerator source. The beam emittance, after transport through a two-solenoid space-charge neutralized low energy beam transport (LEBT) system and acceleration to 750 keV in a 200-MHz RFQ, is $\varepsilon_{nrms} = 0.4 \pi$ mm mrad. Simple scaling of this magnetron suggests the df could be extended to 4.5%, although lifetime reduction should be expected because of increased component wear.

For the Run-2020 of RIC, the second magnetron source was introduced into operation [44]. While it was used in identical (spare) magnetron body, several upgrades were implemented. The pulsed valve operation is essential for both source gas efficiency and the stable source operation at the best performance. It was used the original BINP, Novosibirsk valve [10] as a prototype and made improvements: primarily the replacement of rubber compression parts with miniature springs. This change improved valve temperature stability and simplified valve assembly and tuning. The valve outlet hole was reduced to 0.13 mm. As a result, reliable valve operation was obtained and somewhat reduced total gas flow. Also a new power supply for the valve (with current stabilization and pulse duration control) was developed. This helped to increase beam pulse duration to 1000 μs. The heaters for the Cs supply to the discharge chamber to ensure steady flow with minimal Cs consumption were modified. A new arc-discharge power supply improved beam current stability and reduced the beam current noise. The average arc-discharge power is quite low at ~10–15 W; therefore, an additional heater for the magnetron body to maintain the magnetron body temperature at 160–180 °C was used, which is optimal for the magnetron operation. Finally, a complete new set of power supplies for the source was built. The power supplies have a compact design, placed in the bottom part of the source bench in a standard 60 cm wide rack just 90 cm tall. At 7 Hz, repetition rate was used just one turbo-molecular pump of 1000 L/s pumping speed to operate the source. The source ion-optical system with longer ceramic insulators and larger protection cups to reduce Cs and sputtered tungsten deposition to the insulators was modified. The custom-made Macor insulators were replaced with conventional ceramic insulators. This design eliminated leakages along the ceramic insulators, which developed after extended operation. The extraction voltage operational range to at least 40 kV (maximum voltage of the power supply) was increased. Electron current is less than 50% of the ion current. It was measured as the difference between the total HV extractor power supply current and the H– ion beam current. The second modified magnetron source produced 120 mA H– ion current at 12 A arc-discharge current. At 18 A arc-discharge current and 36 kV extraction voltage, the source produced 130 mA current. About 110 mA current was transported for injection to the RFQ in the LEBT with the two focusing solenoids, and 85 mA was accelerated to 750 keV in the RFQ (see Fig. 5.47).

The gas injection into the LEBT line reduced the space charge neutralization time to less than 50 μs and increased total integrated current by about 15%. A pulsed electro-magnetic valve for the xenon gas injection was used. The best results were obtained at the optimal pressure in the LEBT of about 3×10^{-6} torr. The feasibility of the source pulse length increase to 1000 μs for the ongoing linac intensity upgrade with the longer pulse duration was tested. In the original design, the magnetron

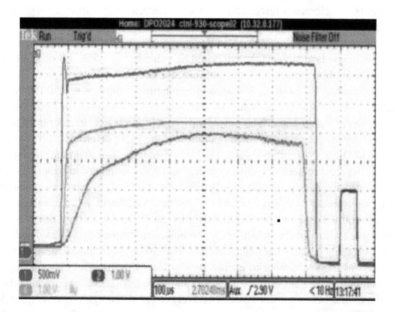

Fig. 5.47 The BNL magnetron SPS operation at 18 A arc-discharge current. The blue trace is the 130 mA H^- ion current, the magenta trace is the 110 mA current injected into the RFQ, and the red trace is the 85 mA current measured after the RFQ

body was thermally insulated by the Macor plate and overheating and current degradation at 1000 us pulse duration were observed. Macor ceramic was replaced with stainless steel plate which transferred heat to the source flange and added additional forced air cooling to the flange. As a result, a reliable long-term operation was demonstrated at arc-discharge current of 12 A and H– ion beam intensity of 100 mA. The original magnetron arc chamber was designed at FNAL (Fermi National Accelerator Laboratory) for the slit ion extraction system and has a rectangular shape with many intricately shaped Macor and molybdenum parts. This design makes it difficult to develop the source for higher duty factor, which will require the introduction of a cooling line. BNL are working on a complete redesign of the magnetron arc chamber to a circular geometry, with provision for an additional cooling line.

Figure 5.48 shows the erosion of material on a BNL magnetron SPS that successfully operated for 2 years: the cathode has a hole of 1.8 mm^2 close to the center of its spherical focusing dimple and the anode cover plate shows marks in the vicinity of the extraction hole spread in an area of 6.2 mm^2, which is not influent for magnetron operation. This damage is produced a back accelerated positive ions of Cs^+ and H_2^+. Estimation of sputtering of cathode and anode magnetron SPS was presented in [45].

But estimation of cesium density was incorrect, because during discharge, cesium is strongly ionized and cannot escape the discharge chamber as shown in Fig. 4.47 from [46]. Figure 4.47 shows a typical oscillogram of the cesium ion current from the collector of the mass spectrometer, illustrating changes in the cesium atom flux from the source in time at a high (~1000 K) planotron cathode temperature, in conjunction with oscillograms of discharge current I_p and discharge voltage U_d. One can see that cesium atoms leave the source mainly after the end of the discharge pulse. Cesium release during the pulse is small, since cesium is highly ionized and the extraction voltage blocks the escape of cesium ions.

Alignment of the H^- beam with the accelerating tube at the Fermilab pre-injector is shown in Fig. 5.49. This SPS provided all the needs of the Fermilab accelerator

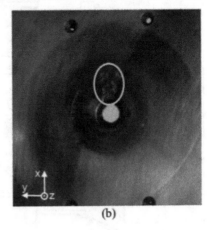

(a)   (b)

Fig. 5.48 Wear traces on the (**a**) cathode and (**b**) the anode cover plate of BNL's magnetron. The location of the traces on the cathode and anode cover plate is indicated by a circle and an ellipse (red and green), respectively

complex for 35 years and led to building SPSs for other laboratories. In 1986, after invention of geometric focusing at Novosibirsk, a semicylindrical groove was located on the cathode [47], which allowed reduction of the discharge current from 150 A to 35 A with a corresponding increase in the attainable continuous operation time.

Fermilab supplied magnetrons to the Argonne National Laboratory (ANL) [48], to BNL [49], to DESY [50, 51], and to Chinese synchrotron. When moving to an RFQ injector, after invention in Novosibirsk BINP geometrical focusing, BNL switched to a magnetron cathode modification with a spherical dimple for three-dimensional focusing to a cylindrical emission aperture of diameter 2.8 mm [52]. An H^- current of 120 mA was obtained at a discharge current of 14 A and discharge voltage 130 V. The lifetime of the source at a duty cycle of 0.5% was 8 months (3 A.hours). When Fermilab transitioned to RFQ injection, they also switched to spherical focusing similar to BNL [53].

5.9.1 Surface Plasma Sources at the Los Alamos National Laboratory

In 1972, the Los Alamos National Laboratory (LANL) began to develop a negative ion source for their linac [54]. Only 2.5 mA of H^- ions could be obtained from the charge-exchange source. After publication of the results obtained with a SPS with

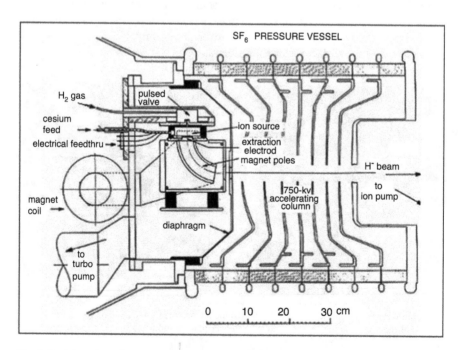

Fig. 5.49 Adaptation of the SPS magnetron H^- beam with an accelerating tube in the pre-injector at Fermilab. (Reproduced from Schmidt and Curtis [41])

Penning discharge at INP Novosibirsk [8, 13], LANL decided to reproduce these results. At this time, Ronald Reagan announced the Strategic Defense Initiative (Star Wars) program. One of the elements of Star Wars was to be beams of neutrals in space, which could be obtained from high-quality beams of negative ions [55, 56].

In 1976, P. Allison, of Los Alamos, learned Russian and traveled with his wife across Japan and along the trans-Siberian railway to Novosibirsk. They spent about a month in Novosibirsk, and we were allowed to familiarize him with our SPS work. After returning to Los Alamos, he received a grant of $50 M for reproduction of the "Dudnikov-type source" and the acceleration of H^- ions [57, 58]. The Los Alamos version of the Dudnikov-type SPS source with Penning discharge is shown in Fig. 5.50 (Reproduced from Allison [58]).

At LANL, the Penning source technology has been applied to numerous linac commissioning exercises. The first Penning experiments at LANL used a 10×0.5 mm^2 emission slit at 0.5% (7 Hz, 0.7 ms) to 2% (40 Hz, 0.5 ms) df. These two pulsed modes gave 108 mA and 30–40 mA currents, respectively. For 100 mA

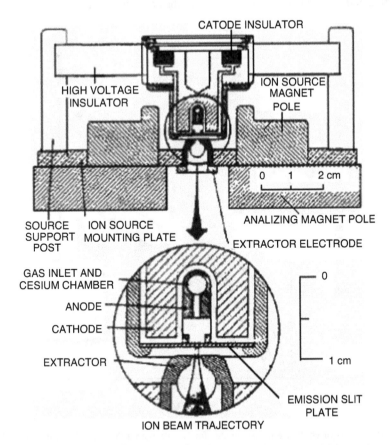

Fig. 5.50 Los Alamos version of SPS with Penning discharge (Dudnikov-type source). (Reproduced from Allison [58])

beam, estimated normalized rms emittances were 0.2 π mm mrad and 0.03–0.04 π mm mrad in the long and narrow slit dimensions, respectively. Experiments with circular apertures showed that a factor of three emittance reductions could be achieved with a factor of three reductions in the discharge voltage noise. For the low noise cases and $R = 1$ mm, normalized emittances of 0.10 and 0.08 π mm mrad were measured in the transverse planes for 35 mA current at 15 keV beam energy. The beam fraction for these emittances was 63%.

The results of measurements of emittances in the SPS with Penning discharge (PD SPS) and the asymmetric magnetron (AM) are shown in Fig. 5.51 [59]. The PD SPS with current 79 mA has a transverse ion temperature of 5 eV along the slit and 840 eV across the slit. The asymmetric magnetron SPS [56] with current 40 mA has transverse ion temperatures 22 eV (along the slit) and 5650 eV (across the slit). The brightness of the PD SPS is an order of magnitude higher than the brightness of the AM SPS. The high transverse temperature of ions across the emission gap is associated with heating of the "overcooled" beam by expansion of the ion beam in this direction and insufficient resolving power of the emittance meter.

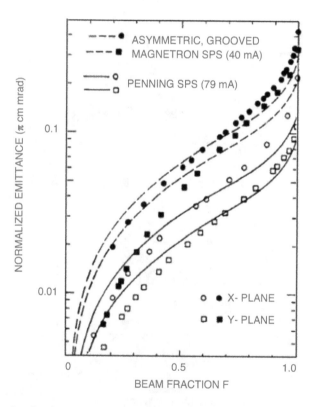

Fig. 5.51 Results of emittance measurements for an SPS with a Penning discharge and an SPS with asymmetric discharge magnetron. (Reproduced from Vernon Smith Jr. and Paul Allison [59])

The Penning source measurements discussed above use a 90 bending magnet immediately following extraction. In 1983, a significant improvement in the Penning discharge SPS occurred at LANL. In order to reduce the 20-keV beam transport length, the 90 bending magnet was replaced with a permanent magnet assembly. This led to an 8.1° bend for 22 keV source that was named the small angle source (SAS). This step resulted in a slit emission source with 150 mA current at 0.5% df. The rms normalized emittances in the transverse planes were found to be 0.27 π mm mrad and 0.04 π mm mrad in the long and short slit dimensions, respectively. Subsequent circular aperture SAS development with $R = 1.25$ mm gave rms normalized emittances of 0.053×0.056 π mm mrad in the transverse planes with 82 mA current. The duty factor was 0.5%.

An injector was developed at LANL that preserves high beam brightness. A schematic is shown in Fig. 5.52 [60]. A strong electric field picks up the beam after a minimal drift. The angular spread of the beam from an SPS with Penning discharge is shown in Fig. 5.53. The effective transverse ion temperature along the slit is 3 eV and 24 eV across the slit. Near the SPS, the emittances are 0.27 π mm.mrad along the slit and 0.04 π mm.mrad across the slit. After the accelerator column, emittances are 0.23 π mm.mrad along the slit and 0.19 π mm.mrad across the slit.

The SPS embodiments that were larger by four and eight times compared to the original size of the Dudnikov source were developed and tested in LANL. The LANL group carried out measurements of the ion temperature of the beam extracted from an SPS with a Penning discharge that was eight times larger than the original size [61]. The measurements were determined from the angular broadening of the beam extracted through a narrow slit (along the slit).

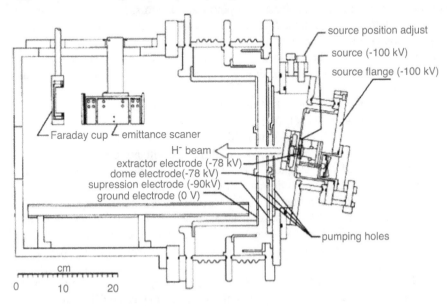

Fig. 5.52 LANL injector, which preserves high beam brightness. (Reproduced from Alisson and Sherman [60], with permission from AIP Publishing)

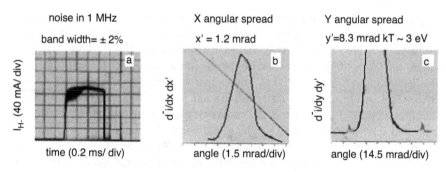

Fig. 5.53 Angular spread of the beam from the SPS with Penning discharge. (Reproduced from Alisson and Sherman [60], with permission from AIP Publishing)

These embodiments are shown in Fig. 5.54. With low discharge current up to 2 A, the transverse ion temperature was 0.1–0.3 eV, and at high discharge current up to 400 A, the transverse ion temperature was 0.7–1.3 eV. This creates the prerequisites for obtaining beams with very high brightness. A similarity theory for Penning discharges was developed by the Los Alamos team [63].

At LANL, sources were designed and constructed applying plasma scaling laws and increasing two of the source dimensions by a factor 4 (the 4X source), which reduced the cathode power load from 16.7 to 2.24 kW/cm^2, while increasing the H^- current from 160 to 250 mA [58]. The emittance product ε_x, ε_y increased by a factor 4, but the increase was not distributed to the x and y emittances with the same ratio as the increase in outlet slit apertures, with the y emittance being slightly larger than the value demanded for most high-power accelerators. It is further suggested that this 4X source could run at 5% duty factor and 105 mA H^- current, while keeping the source lifetime to about a 2-month operation. In tests, the discharge was successfully maintained at 6% duty factor without extraction. Ion current measurements with a circular aperture result in an appreciable decrease in the average current density.

A study of the emittances of H^- ion beams from four SPSs with Penning discharge has been reported [64]. With an emission aperture diameter of 5.4 mm for a 100 mA beam from a discharge with noise >20%, an emittance of 0.22 π mm.mrad was obtained, and for a 67 mA beam from a discharge with noise <1%, an emittance of 0.11 π mm.mrad was obtained. The potential of this quiescent beam was measured at different Xe gas densities and different ion trap potentials and reported in [65]. The beam potential becomes positive for Xe density greater than 5×10^{11} cm^{-3} and sharply increases to 10 V for Xe density greater than 7×10^{11} cm^{-3}. The emittance of the beam after transport over 36 cm increased from 0.12 π mm.mrad to 0.23 π mm.mrad with a decrease in the Xe density from 2.2×10^{12} cm^{-3} to 0 (with up to 1.5×10^{12} cm^{-3} of hydrogen remaining).

LNAL developed an 8X source with hot water cooling of the cathode and anode close to the discharge surfaces, based on the same scaling as the 4X source. Initial

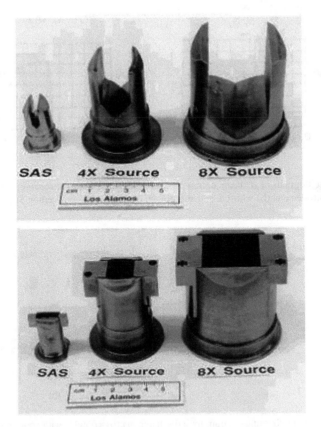

Fig. 5.54 Scaled-up SPSs with Penning discharge (PD SPSs) developed at LANL. Cathodes and anodex. (Reproduced from Sherman et al. [62])

pulsed operation is reported; however, full CW operation was never achieved due to funding limitations.

A noiseless discharge was obtained by adding a small admixture of nitrogen [62, 66]. The oscillograms of the current of the H^- ion beam from Penning discharges with and without added nitrogen are shown in Fig. 5.55. Addition of a small amount of nitrogen (1%) makes the discharge noise free and reduces emittance. From the four SPSs with emission slit 2.8×11 mm^2, an H^- ion beam was extracted with current 250 mA and emittances 0.15×0.29 π mm.mrad. (across the slit and along the slit).

Suppression of noise by the addition of a heavy gas is explained by an ignition/burning diagram for a discharge in crossed fields with coordinates of gas density n and magnetic field B [67] as shown in Fig. 5.56a. The discharge burns at a gas density higher than some critical n and at a magnetic field B higher than a critical B.

For gas density above a critical value n^*, the discharge becomes noise-free. The transverse mobility of electrons in a magnetic field is given by the formula:

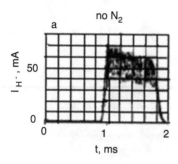
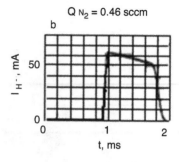

Fig. 5.55 Oscillograms of H^- beam current from an SPS with Penning discharge: (**a**) without the addition of nitrogen and (**b**) with 1% added nitrogen. (Reproduced from Vernon Smith Jr. [66])

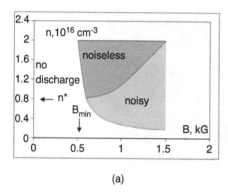
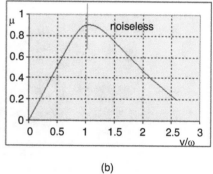

(a)   (b)

Fig. 5.56 (**a**) Ignition/burning diagram for a discharge in crossed fields with coordinates n and B, showing noisy and noise-free regions of operation. (**b**) Dependence of the electron mobility, μ, on the relative scattering frequency ν/ω_{ce}

$$\mu = e\nu / m\left(\nu^2 + \omega_{ce}^2\right)$$

where ν is the electron scattering frequency and ω_{ce} is the Larmor frequency of the electrons represented in Fig. 5.56b. For $\omega_{ce} < \nu$, excitation of noise increases the collision frequency and reduces the mobility, which is unprofitable for current transmission. When $\omega_{ce} > \nu$, the occurrence of noise increases the mobility of electrons, which is beneficial for current flow, and instability develops. Thus a small addition of a heavy gas increases the electron collision frequency and suppresses noise in the discharge, increasing the brightness of the beam.

An attempt to obtain intense H^- ion beam from a Dudnikov-type source with lanthanum hexaboride cathodes without cesium has been reported [68]. A beam of H^- ions with an emission current density up to 350 mA/cm^2 (from an emission slit 0.5×10 mm^2) at a discharge current of 55 A without cesium was obtained, which is much less than produced from a planotron without cesium (see Fig. 4.22).

5.9.2 Surface Plasma Sources at the Rutherford Appleton Laboratory

Development of an SPS with Penning discharge (SPS PD) was commenced in 1979 at the Rutherford Appleton Laboratory (RAL), UK, to provide charge-exchange injection into the rapid cycling synchrotron of ISIS, a powerful neutron generator [69, 70]. The SPS device developed at RAL is shown schematically in Fig. 5.57, and a photograph is shown in Fig. 5.58a. The dimensions of the gas-discharge chamber and the sector magnet are identical with the Dudnikov-type source shown in Fig. 5.2. Figure 5.57 shows a sectional view of the RAL Penning source with cathode, anode insert, discharge region, and ion beam extractor. Cesium metal flows from a 3 gram reservoir to the discharge from an external oven. The ISIS source operates at 50 Hz, 200 μs beam (1% df) while the discharge pulse width is 400–650 μs long (2–3.2% df). The operating current is 45 mA. While operating at this df, the source lifetime is up to 40–50 days. The beam is extracted from the plasma at 17 kV through a 10×0.6 mm^2 slit. The beam is then transported by a $n = 1$, 90° dipole magnet with an 8 cm bending radius. The dipole magnet provides the magnetic field for the Penning discharge (2.3 kG), co-extracts electron suppression, separates other negative ion impurities, and transforms the slit beam into a spatially round beam. Upon leaving the magnet box, the beam energy is increased to 35 keV. Recent RAL source development work has led to rms normalized emittances of 0.62 and 0.73 mm in the horizontal and vertical planes.

A photograph of melted cathode deposit from bombarded by back accelerated positive ion anode insert is shown in Fig. 5.58b [71]. Penning SPS lifetime can be improved by fabrication anode insert from tungsten. It is possible to braze a Tungsten plate of ~1 mm thick to the upper part of anode window. Deposit from anode insert shielded cesium on cathode surface and increase necessary cesium deposition.

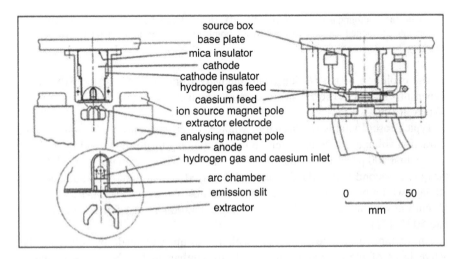

Fig. 5.57 Schematic of the RAL SPS Penning discharge. (Reproduced from Gear and Sidlow [69])

a

b

Fig. 5.58 (a) Photograph of the RAL SPS PD. (Reproduced from Sidlow et al. [70]). (b) Photograph of melted cathode deposit from bombarded by back accelerated positive ion anode insert

The evolution of the H^- ion beam intensity from the RAL of the SPS PD is shown in Fig. 5.59. The H^- beam current was increased from 3 mA in 1981 to 40 mA in 1991.

Typical oscillograms of discharge voltage, discharge current, H^- ion current, extraction voltage, and total current in the extraction gap are shown in Fig. 5.60.

The first 200 μs of the discharge is noisy, and so the H^- beam is extracted during the quieter second-half of the discharge. The period of continuous operation is up to 50 days with a pulsed of H^- ion current of 45 mA and duty cycle 2.5%. Later, the H^- current was increased to 70 mA at a pulse duration up to 1.5 ms and repetition rate 50 Hz [72].

An attempt to operate the source without cesium was made by replacing the active faces of the molybdenum cathode by lanthanum hexaboride. Although a

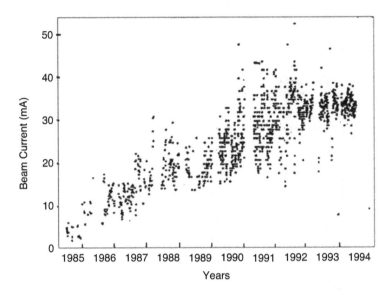

Fig. 5.59 Evolution of the H^- ion beam current from the RAL SPS PD. (Reproduced from Sidlow et al. [70])

high-current, low-impedance arc was easily obtained, only a very low current H^- beam could be extracted. There was no improvement with the addition of cesium. The use of tantalum in place of molybdenum for the source components was tried in the hope of increasing the source operating lifetime. A stable high-current arc could not be established using a tantalum cathode, and results with a tantalum anode were inconclusive. In the only case in which a satisfactory beam was obtained, the tantalum anode was found to have become severely distorted after only a few days of operation.

However, the existing 1X Dudnikov source design could not deliver the full 2 ms, 50 Hz, 60 mA beam requirements without a significant droop in beam current over the 2 ms pulse length. During the long pulse length, the temperature of the electrode surfaces facing the plasma increases to above the range for optimum H^- production. By increasing the surface area of the electrodes, this transient rise in electrode surface temperature can be reduced. Previous researchers [59] demonstrated beam currents up to 250 mA [60] by scaling a Penning SPS. A new 2X scaled source was developed [73] to deliver the full duty cycle requirements for FETS, to produce higher beam currents up to 150 mA, and to yield longer lifetimes for ISIS operations when run at lower discharge currents. The plasma chamber of the 2X source has doubled the linear dimensions of the previous 1X design. A cross section through the second prototype 2X source in RAL is shown in Fig. 5.61.

The full 2 ms 50 Hz 10% duty factor has been achieved with a "noisy" 65 mA beam. Penning SPS discharges operate in two distinct modes: "clean" and "noisy." Initial tests showed that the source would run "clean" with high cesium fluxes. Quartz crystal microbalance deposition rate measurements at different points in the

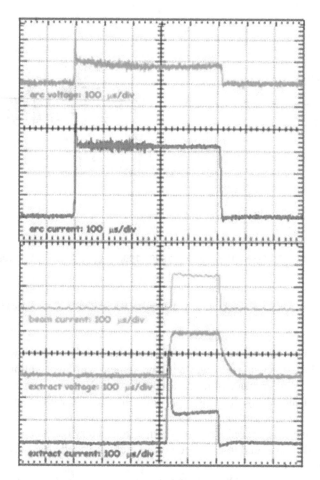

Fig. 5.60 Oscillograms of discharge voltage, discharge current, H^- ion current, and extraction voltage and total current in the extraction gap, for the RAL SPS PD. (Reproduced from Sidlow et al. [70])

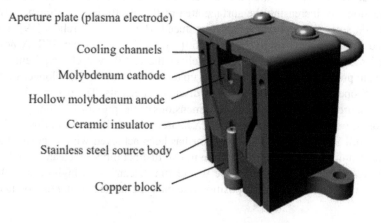

Fig. 5.61 A cross section through the second prototype 2X source in RAL

test vessel showed increased cesium densities, but the largest increase was measured at the end of the vessel downstream from the source aperture. The fourfold increase in emission aperture area (from 0.6 × 10 mm to 1.2 × 20 mm) means that more cesium vapor escapes the source and thus a higher cesium oven temperature is required to maintain the required cesium density in the source itself. A standard 0.6 × 10 mm aperture to fit on the 2X source is being manufactured to test if operation in "clean" mode is possible at lower cesium fluxes. Other geometry modifications are also being considered [74].

Based on the RAL SPS, an SPS with Penning discharge was developed for the Chinese spallation neutron source [75, 76]. Copies of oscillograms characterizing the operation of the source are shown in Fig. 5.62. The discharge in this SPS is very noisy, so the emittance of the H^- beam is large.

Possibilities for improving Penning discharge SPS were discussed in [66]. Development of test stand for development of Penning discharge SPS with improved characteristics was presented in [77].

5.9.3 Surface Plasma Sources at the Oak Ridge National Laboratory (ORNL)

A modified calutron SPS was developed at Oak Ridge as described in the following [78]. A schematic of Oak Ridge Penning SPS is shown in Fig. 5.63. It consists of direct heated cathode, anode, and biased converter.

The gas feed is in the filament region. Energetic electrons stream along the magnetic field (1 kG) and are reflected by an electrically isolated plate—the reflector.

Thus plasma is produced everywhere along this electron ribbon which is 0.3 cm wide by 1.3 cm deep by 15 cm long. A molybdenum converter surface on which

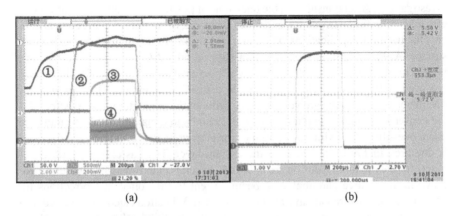

(a)   (b)

Fig. 5.62 (**a**) Oscillograms characterizing the operation of the Penning SPS for the CSNS. (Reproduced from Liu Sheng-Jin et al. [75]). 1, pulse valve; 2, discharge current; 3, pulse extraction voltage 17 kV; and 4, current in the extraction gap 240 mA. (**b**) H^- ion current 50 mA

Fig. 5.63 Schematic of
Oak Ridge Penning SPS

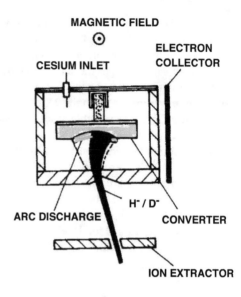

cesium is deposited extends along one side of the plasma ribbon and is separated
from it by about 0.2 mm. The converter power supply is used to accelerate positive
ions through the sheath onto the cesiated converter; ions pass quickly through the
thin ribbon to the converter surface. In this manner, secondary negative ions are
produced on the converter surface and are accelerated by the converter bias voltage
U_g. An independently controlled cesium supply is used to inject cesium onto the
converter. Negative ion losses are minimized by the geometry of the converter,
plasma ribbon, and extraction surface. This geometry can be easily varied to optimize the negative ion output. In all measurements of the extracted current, the values given here are magnetically analyzed, whole beam values along with the
extracted current density J_H^-. This type of test arrangement is well suited to beam
investigations. After considering the questions of positive to negative ion conversion
efficiency, extraction geometry, and beam transport, the extracted beam current density for beamline applications was estimated to be 100 mA/cm² at the extraction
surface. A 90 mA/cm² at the extraction slit and 60 mA/cm² at the collector have
been achieved for 100 ms pulses at 25 kV extraction. Current pulses of 5 s have been
achieved at 45 mA/cm² with a 0.4 cm² slit area. The current signal is low initially
due to application of the arc voltage alone without the converter voltage. After 10 s,
the converter voltage is applied. The output climbs to a maximum and then falls to
about one-fifth the maximum before the converter voltage is removed. The decrease
is due presumably to depletion of the cesium layer. The gas efficiency measurements for the modified calutron are about 4% which is determined by measuring gas
flow into the source and analyzing beam current to the collector. Corrections for
beam losses during transit have not been included. Whealton calculates one order of
magnitude improvement in gas efficiency for the modified calutron with a spherically focused converter at BINP, Novosibirsk [21].

5.9.4 SW Surface Plasma Sources at the Budker Institute, Novosibirsk

The development of an SPS with Penning discharge for continuous operation was initiated at the Budker Institute of Nuclear Physics (BINP) (Novosibirsk, Russia) in 2003 for the injector for the VITA (Vacuum Insulation Tandem Accelerator) [79, 80] for neutron capture cancer therapy. An illustration of an SPS with Penning discharge for continuous operation at BINP is shown in Fig. 5.64.

The source consists of a cathode with channels for the supply of gas and cesium, an anode with anode collar and an emission aperture up to 5 mm in diameter, an extraction electrode, an accelerating electrode, and magnetic field for the Penning discharge and electron emission suppression. A photograph of the source is shown in Fig. 5.65. The dependence of the H^- ion current and the total current in the extraction gap on the discharge current, for emission apertures of 5 mm and 3 mm, is shown in Fig. 5.66a. With a 5 mm diameter aperture, the H^- ion current reaches 25 mA at a total current of 120 mA for a discharge current of 11 amperes. The discharge power is ~1 kW. The dependence of H^- ion current on extraction voltage is shown in Fig. 5.66b [82, 83]. Over time, the current in the extraction gap begins to increase. The growth of this current is suppressed by controlled atmospheric airflow through channel A, as was suppressed in [28, 29]. A photograph of an H^- ion beam with current of 25 mA is shown in Fig. 5.67.

A data acquisition system for unattended control of the source has been developed. Various scenarios with fine-tuning of the source parameters are supported by autopilot. The source start by autopilot is illustrated in the picture of the control software application window shown in Fig. 5.68.

All power supply subsystems for the source are equipped with computer interface modules, which then provide source data adjustment and recording. Fiber optic links are used for data exchange between the source, the rack, and the main computer. Low capacitance multicore HV cable was designed and used for connection between the source and rack. The computer control system provides automatic source start and maintains the necessary source parameters during long-term operation according to the prescribed scenario.

A third, flanged version of CW Penning SPS for accelerator use was developed in the 2010–2013 period. To reduce the co-extracted electron flux, an anode collar was placed around the emission aperture (Fig. 5.64). The collar separates the dense discharge plasma from the emission plane and supplies the magnetic filtering of electrons, like the collar in magnetron sources. The adoption of a collar improves the electron filtering with only a small decrease of the extracted H^- beam. The reduction in the accompanying electron flux diminishes the erosion of extraction electrode and sputtering of the anode cover by backscattered ions. The sputtering by the back-streaming ions is lower in the presence of collar because of decreased electron and cesium fluxes to the extraction gap. A detachable tip of the extractor electrode was introduced so as to more easily change the deteriorated tip. As a result, the source lifetime was improved and the extractor electrode repair procedure

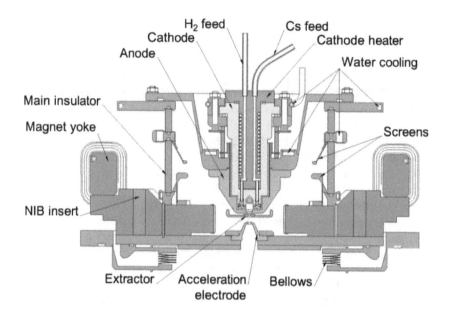

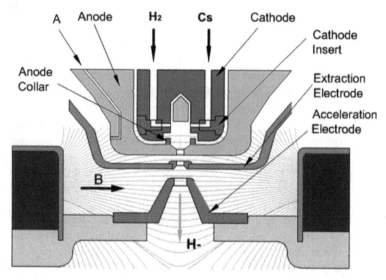

Fig. 5.64 Schematic of SPS with Penning discharge for continuous operation at the Budker INP, Novosibirsk. Cross sections along magnetic field. (Reproduced from Yuri Belchenko et al. [81], with permission from AIP Publishing)

was simplified. The magnet system was rearranged. For the previous magnet system configuration with the magnet yoke installed outside the high-voltage isolator, a portion of co-extracted electrons produced erosion and cracking of the high-voltage insulator. The insulator lifetime is greatly improved by installing the magnet system

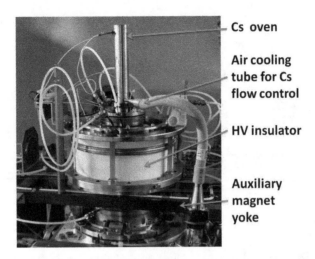

Fig. 5.65 Photograph of SPS with Penning discharge for continuous operation. (Reproduced from Yuri Belchenko et al. [81], with permission from AIP Publishing)

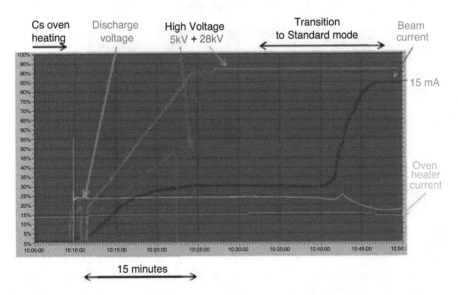

Fig. 5.66 Window of the ion source control program. Source start with discharge ignition, IOS electrode conditioning, and negative ion production activation are shown

yoke inside the high-voltage insulator and elimination of electron drift to the insulator. The new magnet configuration provides a smaller magnetic field <0.06 T at first and reduces the H^- production efficiency and increases the co-extracted electron current. The advanced permanent magnets improve the H^- production rate as well.

A comparative analysis of the H^- ion generation efficiency in SPSs with different discharge geometries was carried out [84, 85]. Schematics of the SPSs

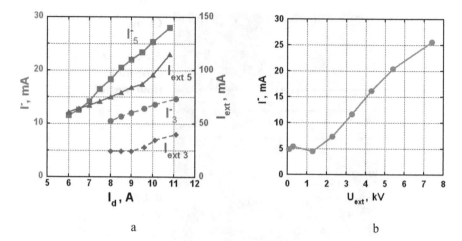

a   b

Fig. 5.67 (**a**). Dependence of H^- ion current, I^-, and total current in the extraction gap, I_{ext}, on discharge current I_d, for SPS with Penning discharge with apertures of 5 mm and 3 mm. (**b**) Dependence of H^- ion current I^- on extraction voltage U_{ext} for SPS with Penning discharge. (Reproduced from Yuri Belchenko et al. [81], with permission from AIP Publishing)

Fig. 5.68 Photograph of H^- ion beam with current 25 mA from SPS with Penning discharge. (Reproduced from Yu. Belchenko et al. [83], with permission from AIP Publishing)

compared—with Penning, planotron, and semiplanotron electrode systems—are shown in Fig. 5.69. The dependence of the H^- ion current on the discharge current for these three SPSs is shown in Fig. 5.70.

A discharge in Penning geometry with hollow cathodes has high level of fluctuations.

A dependence of H^- ion current and of beam fluctuation intensity on extraction voltage is shown in Fig. 5.71 [81]. The intensity of fluctuations at first decreases from 20 to 5% but then increases again to 13%. Irreducible noise is associated with relaxation of the cesium coating caused by pulsed superheating of the cathode surface in the discharge.

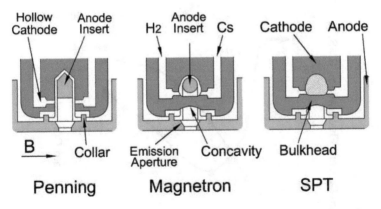

Fig. 5.69 Schematics of SPS with Penning, planotron, and semiplanotron electrode systems, BINP. (Reproduced from Yu Belchenko et al. [84]), with permission from AIP Publishing)

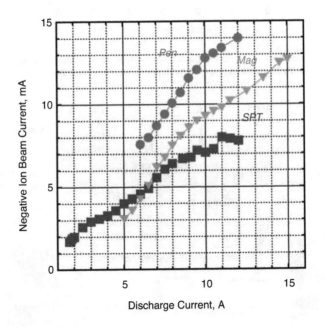

Fig. 5.70 Dependence of H^- ion current on discharge current for SPSs with Penning, planotron, and semiplanotron electrode configurations. (Reproduced from Yu Belchenko et al. [84], with permission from AIP Publishing)

The Penning discharge SPS can be easily converted into a spherical focusing semiplanotron with improved negative ion generation efficiency and improved cathode cooling, as shown in Figs. 5.64 and 5.72.

A semiplanotron with spherical focusing is shown in Fig. 5.72. The shapes of the anode and cathode are shown in Fig. 5.46b. The working gas is fed through the H_2 channel and cesium through channel Cs. The discharge burns in a hollow cathode.

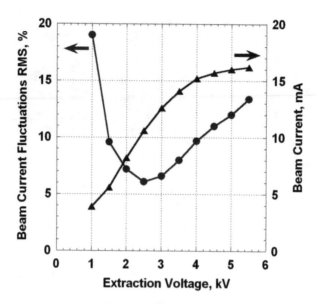

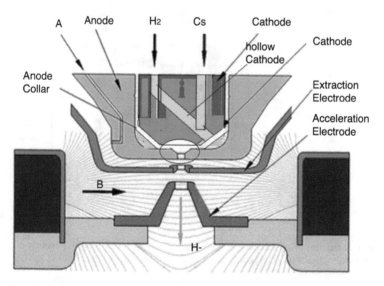

Fig. 5.71 Dependence of H^- ion current and of beam fluctuation intensity on extraction voltage. (Reproduced from Yu Belchenko et al. [84], with permission from AIP Publishing)

Fig. 5.72 Design of semiplanotron with spherical focusing (cathode is rotated on 90°)

Plasma drifts in crossed ExB fields along a semicylindrical groove to a spherical focusing surface. Negative ions emitted by the spherical surface are focused into the emission hole and pulled out by the extraction electrode. The dimensions of the spherical recess are shown in Fig. 5.46. The rest of the dimensions of the cathode and anode remain the same as those of the SPS with a Penning discharge. The ion current H^- 10 mA should be obtained at a discharge current of 1 A and a voltage of 130 V (power ~ 130 W). Ion current H^- 20 mA should be obtained at a discharge current of 2 A and a voltage of 130 V (power ~ 260 W). Ion current H^- of 40 mA should be obtained at a discharge current of 4 A and a voltage of 130 V (power ~ 520 W).

An injector with negative ion beam 90° bend and pre-acceleration to the energy of up to 130 keV was developed in BINP, Novosibirsk, for H^- beam injection into VITA tandem accelerator with vacuum insulation. The scheme and photo of the 130 keV injector and its diagnostics are shown in Figs. 5.73 and 5.74. This injector uses the upgraded H^- SPS with 15 mA current yield. The vacuum box with 90° bending magnet is used to separate the 33 keV negative ion beam from accompanying particles before entering the acceleration tube. The box was equipped with a high-speed turbomolecular pump protected from particle streams from the source by a water-cooled jalousie. Accelerating the beam to a higher energy of 130 keV before injection into the tandem reduces the effect of space charge and improves the beam transport through the tandem at a wide range of beam parameters and tandem voltages. Beam focusing by the 90° bending magnet and by the acceleration tube permits operation without the additional magnetic lens in the LEBT. A larger 100 mm diameter of the transport channel and an additional 200 l/s pump in the LEBT support enhanced vacuum and decrease H^- stripping losses during transport.

Studies of H^- beam production, acceleration, and transport at different vacuum conditions in the transport channel were performed [86]. About 95% beam transmission was achieved. The presence of secondary electrons with

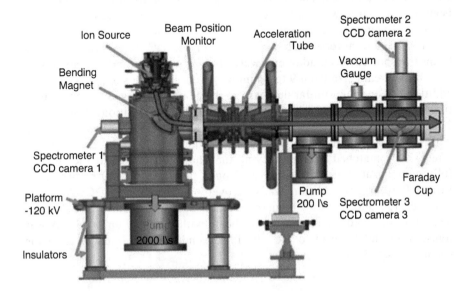

Fig. 5.73 Test stand with 130 keV injector and its diagnostics

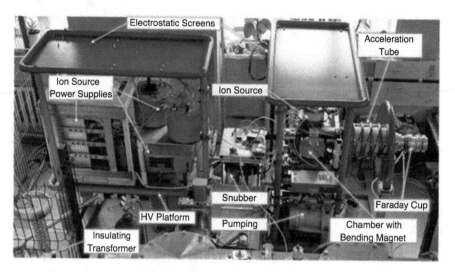

Fig. 5.74 Photo of the 130 keV injector with preliminary 90° beam turn

a current of ~0.4 mA was recorded, which were produced in the acceleration tube and co-accelerated with the negative ion beam. The presumable origin of secondary electrons is H^- stripping and secondary emission from the acceleration tube electrodes. The secondary electrons were deflected from the beam by a magnetic filter, installed at the transport tube sides. Reliable operation of the ion injector with ion energy of 133 keV and current of 14.5 mA was achieved. The influence of gas addition to compensate the ion beam space charge was studied extensively. In the present case, hydrogen, argon, and xenon injection into the 0.8-m-long transport tube were tested. The resultant transverse size and current of the DC 33 keV negative ion beam transported to Faraday cup were studied using optical diagnostics. A drop of the beam size from 9 to 6 mm was recorded with small $\sim 3 \cdot 10^{-6}$ Torr addition of xenon. A similar decrease of the beam size was obtained with 10 times larger addition of argon and $\sim$ 40 times larger addition of hydrogen. The measured beam full width at half maximum (FWHM) was compared with the value calculated by COMSOL for the beam with no space charge effect. They matched well, signifying that the space charge of the beam is fully compensated at the residual hydrogen pressure of $2 \cdot 10^{-5}$ Torr in the LEBT. Further gas addition leads to overcompensation of DC H− beam space charge and to beam focusing. Xenon addition is more effective for beam focusing and transport. The H^- beam with intensity up to 13 mA was transported and focused to the LEBT exit under a decreased level of the beam losses (<5%).

5.10   Large Volume Surface Plasma Sources with Self-Extraction

A large volume SPS (LV SPS) with self-extracted H^- beam [87] was developed by Ehlers and Leung at the Lawrence Berkeley National Laboratory (LBNL) in 1981. A schematic of this large volume SPS is shown in Fig. 5.75. This LV SPS consists of a large (37.5 liter) plasma chamber with permanent magnet multipole magnetic wall, heated cathodes, cooled molybdenum converter for producing negative ions, cesium supply system, and an output aperture with magnets to suppressing electrons. The converter dimensions are 8×25 cm². The rear side of the converter and the converter holders are screened with ceramics. The output aperture has a size of 3×25 cm². A photograph of this LV SPS is shown in Fig. 5.76. Hydrogen is injected into the chamber to a pressure of 10^{-3} Torr, and a discharge is initiated between the heated cathodes and the wall of the gas-discharge chamber, with a voltage of 80 V. A

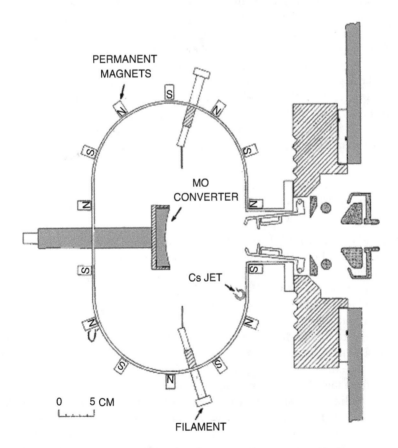

Fig. 5.75 Schematic of a large volume SPS with beam self-extraction. (Reproduced from Leung and Ehlers [87], with permission from AIP Publishing)

Fig. 5.76 Photograph of
the LBNL large volume
SPS with beam self-
extraction. (Reproduced
from Leung and Ehlers
[87], with permission from
AIP Publishing)

potential of −200 V is applied to the converter. Positive ions are accelerated by this
potential and bombard the converter, causing the emission of negative ions.

The emitted negative ions are accelerated by this potential difference and are
directed to the emission aperture, where they are accelerated by the extraction voltage of 40 kV. At a discharge current of 145 A, the converter current is 20 A. Without
cesium, the negative ion current was 20 mA, with 98% being heavy impurities with
high electron affinity. When cesium was added, the negative ion current rapidly
reached 1 A, with an impurity ion level of 1%. The extracted beam current with and
without Cs is shown as a function of time in Fig. 5.77. The gas efficiency of this
source is estimated as 13%.

5.11 Large Volume Surface Plasma Sources for Accelerators

On the basis of the LBNL large volume source, a LV SPS with a converter was
developed for the Los Alamos linear accelerator [88]. A schematic of this source is
shown in Fig. 5.78. A large gas-discharge chamber with multipole magnetic wall
has a diameter of 17.8 cm and a height of 12.8 cm. Two heated cathodes of diameter
1.5 mm and length of 20 cm support a discharge with a voltage of 90 V, generating
a plasma with density up to 3×10^{12} cm^{-3}. A cooled converter of diameter 5 cm and
at a potential of up to −300 V emits negative ions, accelerates them, and focuses
them to the emission aperture of diameter 6.4 mm. The distance from the converter
emitter surface to the emission aperture is 8.25 cm. Up to 15 mA of H^- ions were
extracted from this SPS at a duty cycle of up to 10%.

The normalized emittance of the beam was 0.13 π cm.mrad. This source is characterized by a high consumption of cesium, up to 1 g per day, whereas in other
SPSs, the cesium consumption is less than 1 mg per hour. Along with negative ions,
up to 20% electron current is also extracted. In 1985, a similar source was developed for the linear accelerator injector at the KEK (Japan) accelerator complex [89].

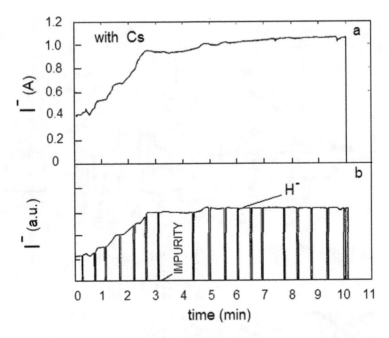

Fig. 5.77 H^- ion current as a function of on-time for the LBNL large volume SPS: (**a**) with cesium and (**b**) without cesium. (Reproduced from Leung and Ehlers [87], with permission from AIP Publishing)

This SPS with converter is shown in Fig. 5.79. The source consists of a large gas-discharge chamber with multipole magnetic wall, two lanthanum hexaboride direct-heated cathodes, cooled molybdenum converter of diameter 5 cm and radius of curvature 14 cm, cesium source, and an emission aperture. The cathode current is 130 A, the converter voltage −500 V, the discharge current 29 A, and the discharge voltage 137 V. The H^- ion beam current is 21 mA, with an electron current of 90 mA. The service life of the cathode is more than 1000 h. At a discharge current of 40 A, up to 40 mA of H^- ions were extracted. To ensure stable operation, a long discharge conditioning process is necessary.

An attempt to increase the beam current of the source for the Los Alamos linear accelerator was undertaken in [90–92]. Figure 5.80 shows a photograph of this SPS with a converter. It consists of a plasma chamber with multipole magnetic wall 20 cm in diameter and 23 cm in height; a cooled molybdenum converter of diameter 3.8 cm and radius of curvature 12.7 cm, with radial extraction; two tungsten cathodes of diameter 1.5 mm; and an output aperture with magnets for suppressing electrons. The source managed to extract up to 20 mA of H^- ions.

A schematic of the LBNL SPS for the Los Alamos linear accelerator is shown in Fig. 5.81a. The source produced an H^- ion current of 40 mA. It consists of a gas-discharge chamber with multipole magnetic wall, cooled molybdenum converter of diameter 5.3 cm and radius of curvature 13.7 cm, axial ion extraction, cesium feed system, six tungsten cathodes of diameter 1.5 mm, and an output aperture with

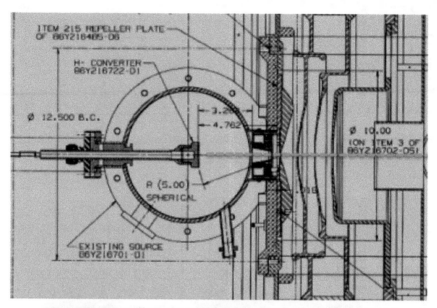

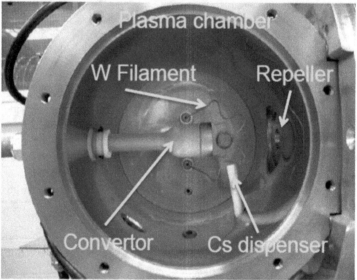

Fig. 5.78 Schematic of SPS with converter for the Los Alamos Linear Accelerator and photograph. (Reproduced from York et al. [88])

magnets for electron suppression. This source could produce 40 mA of H^- ions with a duty cycle of 12%. Unfortunately, the emittance of the beam turned out to be more than expected, and it did not find application on the linac.

The dependence of extracted H^- beam current on discharge current typically shows a strong saturation because of H^- destruction in a thick plasma layer in the

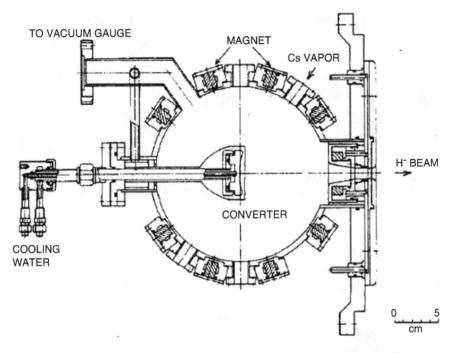

Fig. 5.79 SPS with converter for the linear accelerator at KEK. (Reproduced from Takagi et al. [89])

Fig. 5.80 Photograph of LBNL SPS with converter for Los Alamos linear accelerator. (Reproduced from Wengrow et al. [90])

discharge. The H^- generation efficiency and H^- beam current can be increased by decreasing the thickness of the plasma layer between converter surface and emission aperture. As proposed by Dudnikov, it is possible to improve the beam characteristics by small modification of this converted SPS as shown in Fig. 5.81 [93]. A

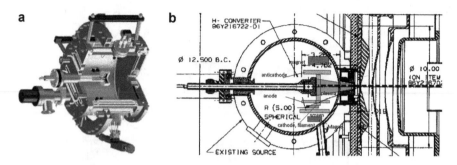

Fig. 5.81 (a) Schematic diagram of LBNL large volume SPS with converter and axial extraction for the Los Alamos linear accelerator; H^- ion current was 40 mA. (Reproduced from Thomae et al. [91]). (b) Modified LANSCE converter SPS with a heated cathode Penning discharge

Penning discharge is used in front of the converter, shifted to emission aperture. The same discharge chamber with hot filament located in front of the anode with slit for electron passage and an anticathode for reflection of electrons. The magnetic field for the Penning discharge is created by permanent magnets. The decrease of plasma and gas between converter and emission aperture can decrease the H^- beam loss and increase the extracted beam intensity by up to a factor of two. The Penning magnetic field can be created by permanent magnets on the discharge chamber wall, as used for forming the multicusp magnetic field.

5.12   Large Volume Surface Plasma Sources for Heavy Ion Production

In 1988, Alton and coworkers at Oak Ridge used the SPS with converter as developed at KEK, shown in Fig. 5.79, to produce heavy negative ions [94–96]. The molybdenum converter was replaced by a similar shaped sputter electrode made from the material whose ions are wanted. Xenon was used as the working gas. Cesiation was used to optimize negative ion emission. A voltage of up to −600 V was applied to the sputter electrode. Ion beams of 5.5 mA Au⁻, 5.1 mA Ni⁻, 8.2 mA Cu⁻, and 4.5 mA O⁻ were produced from sputter electrodes of gold, nickel, copper, and copper oxide, respectively. The intensities of negative ion beams of various elements obtained from this sputter SPS are given in Table 5.1. The beam currents obtained are orders of magnitude greater than those obtained in sputter sources with bombardment of cesium ions [94].

An SPS with converter and RF plasma generation was developed at the Oak Ridge National Laboratory [97, 98]. A schematic of this source is shown in Fig. 5.82. The plasma is generated by an RF discharge excited by an internal antenna, and ions are generated on the cesiated spherical surface of a sputter electrode with diameter up to 25 mm biased at a negative potential; ions are then accelerated to the emission aperture and formed into a beam by the extractor. Working gas is xenon. The

Table 5.1 Heavy negative ion beam current from sputter negative ion source. List of total heavy negative ion beam intensities (peak) from high brightness surface plasma negative ion sources

Sputter material	Sputter target voltage (kV)	Target geometry	Total peak intensity (mA)	Species (%)
Ag	0.94	s	6.2	Ag-(91)
Au	0.44	s	10.3	Au-(73)
Bi	0.94	s	2.7	Bi-(6) and O-(42)
C	0.94	s	6	C-(36) and C_2^-(58)
Co	0.94	s	6	Co-(85)
Cu	0.44	s	8.2	Cu-(77)
CuO	0.44	f	4.5	Cu-(40) and O-(60)
GaAs	0.94	f	3.7	As-(20) and As_2^- (52)
GaP	0.94	f	1.8	P-(44)
Mo	0.44	s	30	O-(67)
Ni	0.44	s	6	Ni-(87)
Pd	0.94	s	7.6	Pd-(69)
Pt	0.94	s	8.1	Pt-(71)
Si	0.94	s	6	Si-(75)
Sn	0.94	s	3.6	Sn-(67)

s spherical, and *f* flat

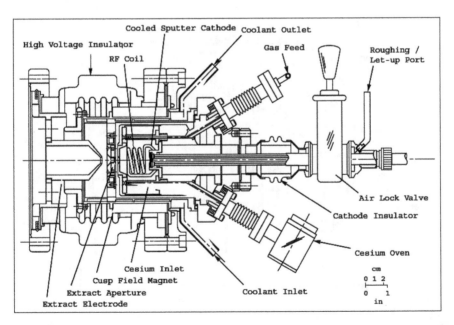

Fig. 5.82 SPS with converter and RF plasma generation, ORNL. (Reproduced from Alton et al. [97])

following ion currents were obtained: C$^-$ (0.61 mA), F$^-$ (0.1 mA), Si$^-$ (0.5 mA), S$^-$ (0.5 mA), P$^-$ (0.125 mA), Cl$^-$ (0.2 mA), Ni$^-$ (0.15 mA), Cu$^-$ (0.23 mA), Ge$^-$ (0.125 mA), Se$^-$ (0.05 mA), Ag$^-$ (0.07 mA), Au$^-$ (0.25 mA), and Pt$^-$ (0.125 mA). Normalized emittance at the 80% level was 7.5 mm.mrad $(MeV)^{1/2}$.

An SPS with RF discharge and sputter electrode for obtaining heavy ions was developed by Ishikawa et al. [99] at Kyoto. An illustration of this source is shown in Fig. 5.83. The plasma is generated by an RF discharge using an interior stainless steel antenna of diameter 50 mm. Negative ions are generated on the spherical surface of a sputter electrode of diameter 42 mm and radius of curvature 70 mm, biased to a negative potential of up to −1 kV. Negative ions are accelerated toward the emission aperture of diameter 10 mm and formed into a beam by an extractor with diameter 14 mm. A transverse magnetic field of 370 Gauss is established in the emission aperture. Working gas is xenon. The RF frequency is 13.56 MHz. With a discharge power of 200 W, the sputter electrode current is up to 200 mA at a Xe pressure of 10^{-3} Torr. The dependence of the sputter electrode current on Xe pressure is shown in Fig. 5.84.

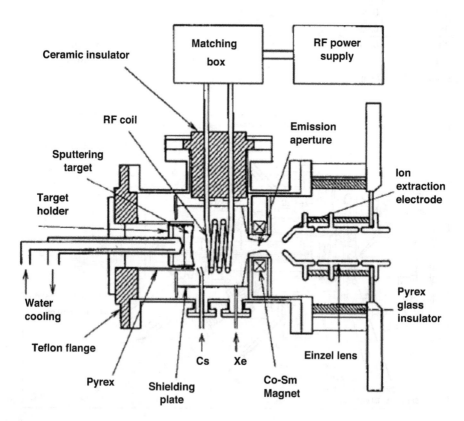

Fig. 5.83 SPS for heavy negative ions with RF discharge and sputter electrode, Kyoto. (Reproduced from Tsuji et al. [99])

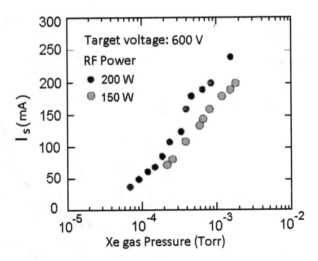

Fig. 5.84 Dependence of sputter electrode current on Xe pressure, Kyoto. (Reproduced from Tsuji et al. [99])

Fig. 5.85 Dependence of total copper ion current on cesium oven temperature. (Reproduced from Tsuji et al. [99])

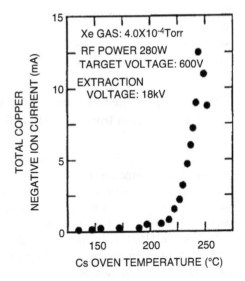

The dependence of the total copper ion current on cesium temperature with a copper sputter electrode is shown in Fig. 5.85; the maximum target current is 280 mA. The dependence of the total of carbon ion current on cesium temperature with a graphite sputter electrode is shown in Fig. 5.86.

The maximum generation efficiency for various ions by sputtering in a xenon plasma with optimal cesium coating is given in Table 5.2.

Fig. 5.86 Dependence of total of carbon ion current on cesium oven temperature. (Reproduced from Tsuji et al. [99])

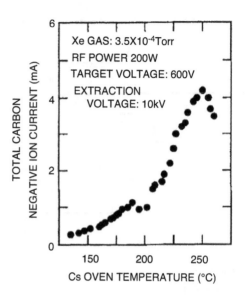

Table 5.2 Maximum negative ion production efficiencies with xenon sputtering and optimal cesiation

Negative ion	C^-	Si^-	Cu^-	Ge^-	Mo^-	Ta^-	W^-
Efficiency (%)	18	15	12	13	0.5	1.6	8

5.13 Surface Plasma Sources for Intense Neutral Beam Production for Controlled Fusion

The large helical device (LHD) in Japan, a large superconducting stellarator for controlled fusion experimental work, requires the injection of intense neutral beams to heat the confined plasma. Neutral beam sources of this kind are based on negative ions; intense negative ion beams are formed which are then charge-exchanged into intense beams of neutral atoms [100].

In 1990, in [101], following Novosibirsk's example, cesium was added to the tandem negative ion source, and a large increase in the emission of negative ions was obtained with a simultaneous significant decrease in the current of accompanying electrons. The dependence of H^- ion current on discharge current for discharges with and without cesium is shown in Fig. 5.87. In subsequent experiments, it was shown that the increase in negative ion emission is due to activation of the surface plasma mechanism of negative ion formation on the plasma electrode, the work function of which is lowered due to the adsorption of cesium.

The neutral beam injectors with cesiation arc discharge SPS are reliably working in LHD from 1998. A schematic of the high-current SPS for negative ion production is shown in Fig. 5.88a (cross section perpendicular to magnetic field). Cross section along magnetic field is presented in Fig. 5.88b. The cesium supplied by three cesium

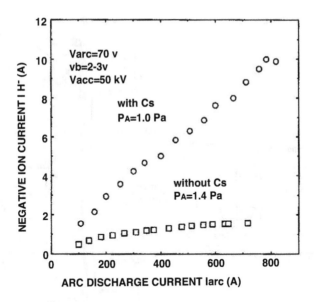

Fig. 5.87 Dependence of the H^- ion current from a tandem SPS on discharge current, for discharges with and without cesium. (Reproduced from Okumura et al. [101])

ovens. The H^- ion beam energy is 150 keV and the H^- beam current is 40 A. The discharge chamber is 35 cm × 145 cm with depth 21 cm and is equipped with a multipole magnetic wall and an external magnetic filter. The chamber is surrounded by a multipole magnetic wall. Magnetic filter is 50 Gauss.

The 25 tungsten cathode filaments maintain the discharge by a power supply of more than 200 kW. Three cesium ovens regulate the flow of cesium by their temperature. The accelerating system is composed of 3 electrodes about 25 cm × 125 cm with 770 emission holes separated into 5 sections, at small angles to a common focus at 13 meters. The plasma electrode is made of molybdenum, and the extraction electrode and ground electrode are made of oxygen-free copper. A drawing of the extraction and ground electrodes is shown in Fig. 5.88 left. Permanent magnets are mounted in the extraction electrode to suppress the flow of accompanying electrons. The different deflection of beams from neighboring emission holes is compensated by displacement of the holes in the extraction electrode. A general view of the LHD injection system with neutral beam injectors based on negative ion SPS is shown in Fig. 5.89.

In order to develop a surface plasma negative ion source for the LHD-NBI system, a 1/3-scale ion source was constructed [102]. A rod-type magnetic filter was adopted in the source. The properties of cesium-seeded discharges and the characteristics of H^- beam extract were investigated.

The egg-box cell was designed initially for the following three reasons. The first is enhancement of the H^- current by increasing the low work function area near the extraction region. If the flux of H° onto the cesiated surface is sufficient, the population of H^- ions is proportional to the cesiated surface area. The second is to bias the

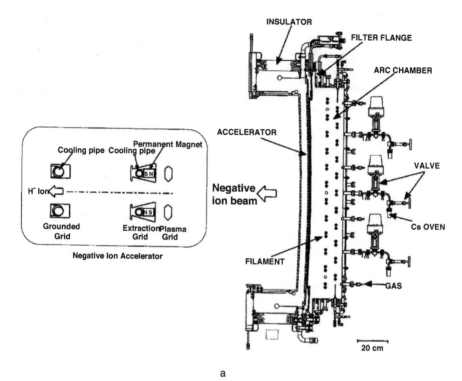

a

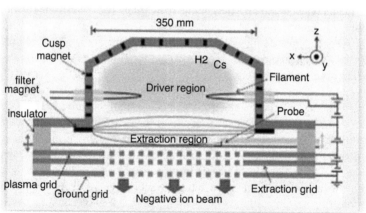

b

Fig. 5.88 (a) Schematic of large volume negative ion SPS with arc discharge. Left is schematic of extraction system with co-extracted electron magnet deflection. Cross section perpendicular to magnetic field. (Reproduced from Koneko et al. [100]). (b) Cross section along the magnetic field

cell potential to the plasma. If the cell can be biased negatively to the arc plasma, the local proton flux toward the cell increases, and H^- will be produced from protons through the process of double charge exchange. The H^- produced is repulsed from

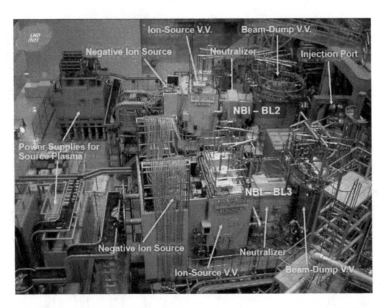

Fig. 5.89 General view of the LHD neutral beam injection system based on SPS. (Reproduced from Koneko et al. [100])

the cell wall and can flow into the extraction region. The third is to reduce the electron current by removing electrons trapped by the magnetic filter field. On installing a wall structure near the grid, the electrons trapped by the filter field collide with the wall, and the electron density is expected to be reduced. A schematic diagram of the egg-box cell is shown in Fig. 5.90. The cell has a lattice form. Each unit cell of the lattice is a 13 mm square box 7 mm in height and surrounds each aperture of the plasma grid. The total area increases 6.4 times with the cell. The cell is installed between the magnetic filter rods and the plasma grid, 4 mm from the rods and 1.5 mm from the grid. The cell is electrically insulated from the chamber, the filter rods, and the plasma grid and can be biased to them. The filter rods are usually connected with the plasma chamber. In pure hydrogen discharge, there is no difference in the total H^- current and extraction current with or without the egg-box cell. After cesium seeding, the total H^- current does not change and only the extracting current is reduced, as shown in Fig. 5.91.

The extraction current is reduced by a factor of 1/2 with the cell, under the same arc discharge conditions. The H^- current does not change in spite of increasing the area of the reaction surface. The extraction current decreases as Vbias increases from negative to positive values. The total H^- current has maximum at Vbias = 0 V. These dependences are similar to those without the egg-box cell, where the bias voltage is applied between the plasma chamber and the plasma grid; however, the bias dependences are different when the cell is connected to the plasma chamber and the plasma grid is biased with respect to them.

Cesium seeding into a SPS ion source results in enhancement of the H^- current, reduction of the electron current, and decrease in the operating pressure. A

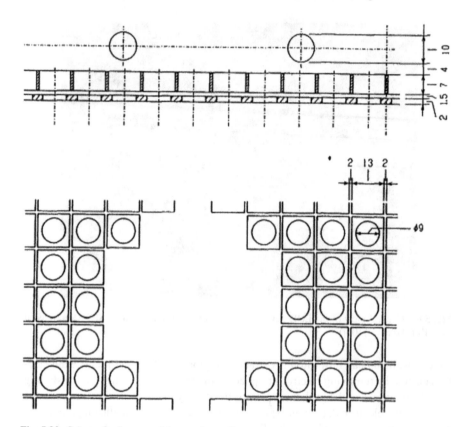

Fig. 5.90 Schematic diagram of the egg-box cell

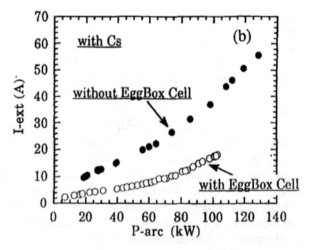

Fig. 5.91 Dependences on arc power extraction current I-ext with the egg-box cell and without the egg-box cell (Q); $V_{acc} = 21.0$ kV, $V_{ext} = 4.0$ kV, and $p = 1.2$ Pa

H^- current of 10.2 A has been extracted from a 1/3-scale SPS negative ion with cesium seeding at a current density of 37 mA cm^{-2}.

Improvement of deuterium injection power in the negative ion-based NBIs (n-NBIs) for the large helical device (LHD) was reported. Co-extracted electron current at acceleration of deuterium negative ions (D^- ions) limits the injection power. The electron current is reduced by decreasing the extraction gap, and the injected D^- current evaluated from the injection power increased from 46 to 55 A. Greater electron reduction was achieved by installing a structure named an "electron fence" (EF), with which D^- beam power was successfully improved from 2.0 MW to 3.0 MW. A schematic of electron fence (EF) installed in extraction system is shown in Fig. 5.92.

The injection power in three configurations—without EF and with EF of 5 mm and 7 mm distance from the plasma grid (PG) surface—has been compared in both cases of hydrogen and deuterium operations, and it was found that the configuration with the EF of 5 mm distance was the best to satisfy the performance for both hydrogen and deuterium injections. Although the co-extracted electron current is reduced in the negative ion sources applied for JT-60SA and ITER by utilizing the PG filter, it is possible to achieve more effective electron reduction by combining the PG filter and the EF [103].

Neutral beam injectors with energy up to 500 keV based on the SPS are designed to heat the plasma and maintain the current in the JT-60 Tokamak [104] in Japan. An illustration and photograph of these injectors are shown in Fig. 5.93, and an illustration of the overall injector system for the JT-60 Tokamak is shown in Fig. 5.94. The nominal energy of the beam is 500 keV and the nominal D^- current is 22 A. The

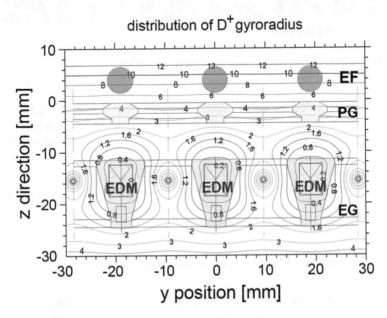

Fig. 5.92 Schematic of electron fence (EF) installed in extraction system

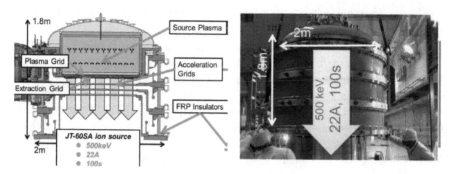

Fig. 5.93 Diagram and photograph of SPS-based injectors for JT-60 Tokamak. (Reproduced from Hanada et al. [104])

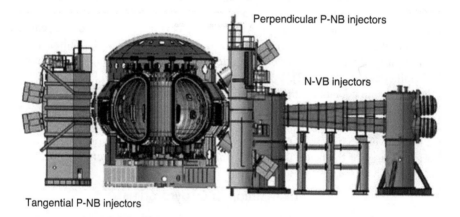

Fig. 5.94 Overall H-based neutral beam injection system for the JT-60 Tokamak. (Reproduced from Hanada et al. [104])

magnetic filter for electron suppression is formed by a current of 3–5 kA through the plasma grid. The dimensions of the SPS are 2 m diameter by 1.7 m height. The size of the molybdenum plasma electrode is 45 cm × 110 cm, divided into 5 segments with 216 apertures. A three-step acceleration is performed by a six-electrode beam formation system. A multipole magnetic wall and minimized surface-to-volume ratio enable the discharge to be maintained with heated cathodes at a low gas pressure of <0.3 Pascal.

5.14   RF Surface Plasma Sources for ITER

The international fusion experiment ITER [105] will be equipped with two powerful neutral beam injection (NBI) beamlines for heating and current drive (power per beamline: up to 16.5 MW) with the option to add a third one in a later stage [106].

A diagnostic injector with low power (2.2 MW) will be required to diagnose the He ash content using charge-exchange resonance spectroscopy [107]. These NBI systems are based on the surface plasma generation H^- (D^-) in a low-temperature, low-pressure plasma, on electrostatic extraction and acceleration of negative hydrogen ions, followed by neutralization in a hydrogen gas target to inject fast neutral particles into the fusion device. The particles shall have an energy of 870 keV and 1 MeV for hydrogen and deuterium, respectively, for the heating neutral beams (HNBs), whereas only hydrogen at 100 keV is foreseen for the diagnostic neutral beam (DNB). The ion sources have to operate at a pressure of 0.3 Pa or below in order to keep stripping losses in the seven-stage accelerator system below 30%. The requirements for the HNBs in terms of current extracted from one large ion source are very challenging, namely, 57 A for 3600 s in deuterium operation and 66 A for 1000 s in hydrogen (corresponding to current densities of 286 and 329 A/m^2, respectively). The DNB should deliver even 77 A in hydrogen (corresponding to 391 A/m^2) at a 3 s on 20 s off time with a repetition rate of 5 Hz. The inevitably co-extracted electron current has to be kept below the ion current in order to avoid damages to the second grid of the extraction system, to which the electrons are deflected by embedded magnets. An extraction voltage of 9–10 kV is envisaged to achieve these parameters. For a proper beam transport, the homogeneity of the accelerated beam, composed of 1280 beamlets, has to be better than 90% down to the scale of single beamlets. Additionally, the core of the accelerated beamlets (carrying ≈85% of the power) needs to have a divergence of less than 7 mrad, while significantly higher divergences of 15–30 mrad are allowed for a halo carrying the remaining ≈15% of the beamlet power. The ion source itself has a rectangular shape with a width of 0.9 m and a height of 1.9 m, accommodating the 1280 apertures with 14 mm diameter resulting in an extraction area of 0.2 m^2. The RF-driven concept (inductively coupled plasma source, ICP) has been chosen for plasma generation using a frequency of 1 MHz and a total power of up to 800 kW delivered by four RF generators. In order to generate sufficient negative ions at the required low pressure, the surface conversion process is utilized, for which cesium is evaporated into the source [108]. By covering the molybdenum-coated copper plasma grid (the first grid of the extraction system) with cesium film, its work function is reduced, which enhances the negative ion production [109], accompanied by a simultaneous decrease of the amount of co-extracted electrons. The requirements for the operational parameters of the ion source and accelerator are very challenging and by far exceed those of the NBI systems based on negative ions (NNBI [negative ion-based neutral beam injector]) at existing fusion devices JT-60 U, JT-60SA [110], and LHD [111]. The ion sources at those systems are based on filament discharges (arc discharges) and are typically operating only for a few seconds at their nominal heating power. For reduced power, pulses of up to several tens of seconds have been demonstrated. In preparation for the NBI system of JT-60SA, pulses of up to 60 s have been obtained with an accelerated current density of 190 A/m^2, but only for a small extraction area (nine apertures). For RF sources, ITER-relevant ion current densities have been demonstrated at smaller test facilities for a beam duration of a few seconds. The responsibility for the full HNB beamline is equally shared between Europe and

Japan. Focus on the accelerator technology for 1 MeV is given to Japan, whereas Europe focuses on the ion source. As the ITER NBI requirements have not yet been met simultaneously, the ITER Organization (IO), together with the European Domestic Agency Fusion for Energy (F4E), has defined an R&D roadmap toward the NBI systems for ITER. An important step in the size scaling of RF sources from the prototype source (1/8 size) to the full ITER size is the half-size ITER source at the test facility ELISE (Extraction from a Large Ion Source Experiment), which is in operation at IPP, Garching. The full-size source will be commissioned and operated at the European Neutral Beam Test Facility (NBTF) in Padua, Italy. The NBTF hosts a test facility for the ITER NBI ion source (source for the production of ions of deuterium extracted from a radio frequency plasma, SPIDER), which went into operation in 2018. The second test facility that represents the full prototype beamline of the HNB of ITER (Megavolt ITER Injector and Concept Advancement, MITICA) is currently planned to go into full operation in 2024.

ITER-India is responsible for the DNB and has also developed a dedicated R&D roadmap: starting with a 1/8 ITER prototype source at the test facility ROBIN (replica of BATMAN-like source in India), which went into operation in 2011; an intermediate step using a 1/4 ITER source—the twin source—has been started recently. Finally, a full prototype DNB is under preparation including the long transmission line to the fusion device. Accompanying programs for the ion source and the accelerator technology are undertaken at NIFS and at QST in Japan, together with supporting activities at universities all over the world.

As an intermediate step between ITER and a commercial power plant, activities started in several countries to design a DEMO (DEMOnstration reactor), for which NBI is also foreseen as an option. Such a system would be based on a similar source concept to benefit from the experience gained at ITER. The prototype source has been tested first at the BATMAN (Bavarian Test Machine for Negative Ions) test facility, in operation at the IPP Garching, with focus on the demonstration of the ion source parameters for short pulses (up to 4 s extraction) in hydrogen and deuterium, accompanied by extensive plasma diagnostics to get insights into the underlying ion source physics. In 2018, the test facility was upgraded to focus on the beam properties (Batman UpGrade, BUG). The typical size of the total extraction area is about 0.01 m². Long pulses, that is, steady-state operation, were demonstrated at the test facility MANITU, which was equipped with the prototype source. The modular design of the prototype source has been proven at RADI, a source size scaling experiment without extraction system. The half-size ITER source with extraction went into operation in 2012 at the test facility ELISE with a total extraction area of about 0.1 m². A detailed overview of the work performed at IPP can be found in [106]. At ELISE, the ITER parameters could be demonstrated in hydrogen, that is, more than 90% of the required extracted ion current for 1000 s was achieved, limited only by technical constraints regarding the available RF power and the high-voltage power supply. The achievement of the target parameters in deuterium for long pulses is still pending as the co-extracted electrons are limiting the ion source performance: the amount of the co-extracted electrons, their dynamics during long pulses, and their inhomogeneity in vertical direction are significantly stronger in deuterium than in hydrogen. Compared to the 1 MeV of the ITER beamlines, the negative ions are accelerated at these test facilities to significantly smaller energies (e.g., at ELISE up to 60 keV), with the consequence that demonstrating the low beamlet divergences planned for ITER is intrinsically not possible. The capture addresses key issues of the development of the negative ion sources covering the aspects of effective plasma.

The development of negative ion sources and extractors for long-pulse and high-power ion beams is pursued in fusion research institutes worldwide, for example, IPP Garching (Germany), Consorzio RFX (Padova, Italy), JAEA, and NIFS (Japan). At the Max Planck Institute of Plasma Physics, Garching, Germany, a project was proposed to develop a high-frequency SPS as a basis for the large neutral beam injectors needed for the generation of 1 MeV D^- ion beams with current 40 A or H^- beam of 73 A for ITER [112, 113]. The development stages for the RF SPS for ITER are shown in Fig. 5.95. The ELISE SPS is being developed; it is half the size of the ITER SPS, generating 20 A of H^- or D^- at a beam extraction voltage of 60 kV for 10 s.

The SPIDER SPS is being developed at Padova, Italy—a full-size ITER SPS with 40 A of H^- or D^- with an extraction voltage of 100 kV for 3600 s [114]. The next stage is the MITICA SPS, also being developed at Padova as a full-size ITER RF SPS generating 40 A of H^- or D^- ions at 1 MeV for 3600 s and finally the full-size ITER source (1.0×2.0 m^2 with an extraction area of 0.2 m^2), being tested at the Neutral Beam Test Facility at RFX. A diagnostic SPS is being developed at Gandinabad, India, to produce H^- ions with energy 100 keV and current 40 A for 3600 s.

A schematic of a high-frequency SPS with an expander and suppression of accompanying electrons is shown in Fig. 5.96. It consists of a high-frequency plasma generator (1 MHz at 100 kW) with an Al$_2$O$_3$ ceramic gas-discharge chamber and a cooled Faraday screen, expansion chamber with multipole magnetic wall,

· ELISE (IPP Garchin): Half-size ITER-type source in cw operation with 60kV/10s beam extraction.
  → to asses spatial uniformity of negative ion flux, validate or alter source concept
· SPIDER (RFX,Padua): Full size ITER source with full extraction voltage 100keV 3600s → to validate or alter source and extractor
· MITICA(RFX, Padua): Full size ITER source, 1 MeV, 3600s
  → to validate or alter a accelerator and beamline components
· DNB source test facility (Ghandingar, India), Full size ITER source, 100Kev,3600s

Cryo Pump Cabrimeter

Gate Valve

Ion Source

NBTF in
Padua, Italy

ELISE, 2012+ · · · · · SPIDER, 2017+ · · · MITICA 2021+ · · · · · ITER NBI

Validate or alter **source concept** · Gain experience with operation of large sources · Validate or alter accelerator and beam line componets

Fig. 5.95 Stages of development of RF SPSs for ITER. (Reproduced from Kraus et al. [112], with permission from AIP Publishing)

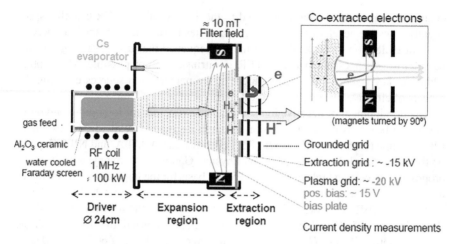

Fig. 5.96 Schematic of high-frequency SPS with an expander and suppression of accompanying electrons. (Reproduced from Kraus et al. [112], with permission from AIP Publishing)

external magnetic filter, cesium vaporizer, plasma electrode, extraction electrode, and a ground electrode.

The structural scheme of the RF BATMAN (Bavarian Test Machine for Negative Ions) at the IPP Garching, Germany, with an expander and suppression of the accompanying electrons is shown in Fig. 5.97. The diameter of the plasma generator is 24 cm, the height is 15 cm, and the expansion chamber is $32 \times 59 \times 23$ cm^3. The plasma electrode is 306 cm^2 with 406 apertures.

The dependence of the ion beam current density on the extraction voltage for different emission areas is shown in Fig. 5.98 for the BATMAN source. Permanent magnets are mounted in the extraction electrode to suppress the accompanying electron flow. Extraction voltage is up to 10 kV, and accelerating voltage is up to 60 kV. The emission current density is up to 32 mA/cm^2; the total H^- current was 9.7 A.

Coextracted electrons are inevitable for negative ion-based NBI systems. A filter field on the order of a few mT in the extraction region magnetizes the electrons but not the ions and reduces both the electron temperature and density in front of the plasma grid. This improves the coextracted electron-to-ion ratio, which should be smaller than one, by suppressing negative ion destruction and reducing the amount of electrons that are available to be extracted. To avoid wastefully accelerating the co-extracted electrons to full energy, they are deflected onto the extraction grid in the extraction gap. The horizontal deflection of the electrons is caused by a vertical magnetic field which alternates direction row-wise, created by permanent magnets embedded in the extraction grid [115]. These permanent deflection magnets are called CESM (co-extracted electron suppression magnets). The alternating magnetic field also creates an unwanted row-wise zigzag deflection of the ions, which should be corrected in order to satisfy the beamlet misalignment requirements.

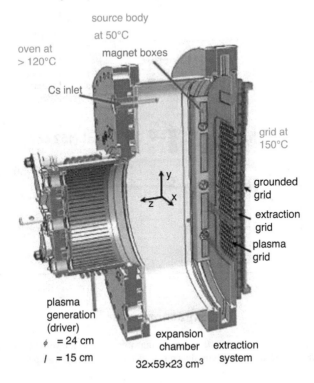

Inductively coupled ion source

Fig. 5.97 Structural outline of the RF BATMAN SPS with expander and suppression of accompanying electrons. (Reproduced from Kraus et al. [112], with permission from AIP Publishing)

In order to perform experiments with a magnetic deflection correction system in BUG, the test facility will be equipped with the ITER-HNB plasma grid and extraction grid geometry and the standard BUG grounded grid. The extraction gap of 6 mm is kept identical to the ITER-HNB design. The acceleration gap is set to 12 mm, since this leads to an almost identical modeled optimum Uacc/Uext ratio as the standard BUG configuration, which has an acceleration gap of 15 mm. In addition, the 12 mm acceleration gap reproduces the proof-of-principle demonstration of the deflection correction system in National Institute of Thermonuclear Study (NITS). The configuration with the ITER-HNB plasma grid and extraction grid is called MITICA-like extraction (MLE). MITICA is the full prototype injector of the ITER-HNB. Figure 5.99 shows the standard BUG grid configuration together with the new grid system. Calculations are reported for the divergence optimum at an extraction potential of 5 kV and an acceleration potential of 26 kV. The simulations are below the experimental 45 kV limit, to provide buffer to scan around the optimum in the experiment. Standard BUG has a horizontal beamlet deflection of 11 mrad in these conditions. In the BUG-MLE configuration, the deflection should be corrected in the divergence optimum. The magnetic deflection correction does not

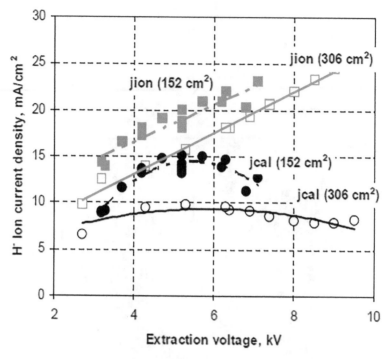

Fig. 5.98 Dependence of the H^- ion beam current density on extraction voltage for different emission areas in the BATMAN SPS. (Reproduced from Krauset al. [112], with permission from AIP Publishing)

Fig. 5.99 The standard BUG plasma grid (PG), extraction grid (EG), and grounded grid (GG) shown in blue, together with the BUG MITICA-like extraction grid system (red)

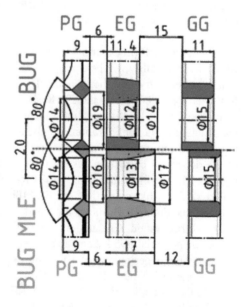

depend on the total beam energy. Because the standard BUG grounded grid is used, the distance between apertures is 20×20 mm (horizontal x vertical). This is different from the 19×21 mm in NITS and the 20×22 mm in the ITER-DNB and HNB. The different aperture spacing in BUG with respect to the NITS configuration is one of the reasons the magnet design from NITS is not directly transferable to BUG. Magnetic deflection correction depends on the ratio of the magnetic field strength upstream and downstream of the extraction grid; the absolute field strength upstream of the extraction grid is in principle a free parameter in the magnet design. There is a minimum field strength required to dump the electrons on the upstream surface of the extraction grid, which is between 40 and 50 mT for the BUG-MLE geometry at 10 kV extraction potential: the most demanding condition in terms of co-extracted electrons. The magnet design is based on VACOMAX 225 HR, the Sm_2Co_{17} material used for the deflection magnets in BUG and other IPP test facilities. Because the ADCM strengthen the field upstream of the extraction grid, keeping the standard BUG CESM magnet size and adding the ADCM will result in a higher peak vertical field than the standard on-axis value of 62 mT. The RFX-QST NITS 2017 campaign that demonstrated complete deflection correction had low peak field strengths of 46 mT. For the BUG-MLE design, it is chosen to imitate the standard BUG magnetic field. This results in a reduced CESM size of $20 \times 3.4 \times 6.8$ mm (width x height x depth), so that BUG-MLE with ADCM magnets has a comparable peak field strength upstream of the extraction grid compared to the standard BUG, which has $20 \times 5 \times 6$ mm CESM size. The width is the size in horizontal direction, the height is the size in the vertical direction, and the depth is the size in the axial direction.

In the RFX-QST NITS design, the CESM and ADCM have the same depth, resulting in very thin ADCM. This is a design choice, several ADCM shapes can be used to correct the deflection, although the magnet volume varies slightly since the field topology changes with the magnet shape. For BUG-MLE, it was chosen to have $2 \times 16.2 \times 4$ mm ADCM magnets, since these are more robust mechanically. Additionally, the grid manufacturing process is easier since the magnet grooves have a more symmetrical aspect ratio. Figure 5.100a shows the extraction grid seen from downstream, that is, the field weakening side. Small vertical correction magnets in between the horizontal deflection magnets form an asymmetric magnetic flux structure. Because the polarity of the correction magnets aligns with the deflection magnets, the correction magnets are repelling the deflection magnets, which is a safeguard to ensure correct assembly. A distribution beam current density on collector with ADCM and without ADCM is shown in Fig. 5.100b. With ADCMH (compensated zigzag), beam divergency is ~12 mrad, and without ADCMWT (adaptive continuous Morlet wavelet transform) (uncompensated zigzag), the beam divergency is ~28 mrad (ITER requirement at 870 kV is 7 mrad). Figure 5.101 shows the vertical magnetic field component for the NITS 2017 configuration, the standard BUG configuration, and BUG-MLE with and without ADCM. In the BUG-MLE configuration with ADCM, the integral of the vertical field upstream of the magnets is 1.8 times the downstream vertical field integral.

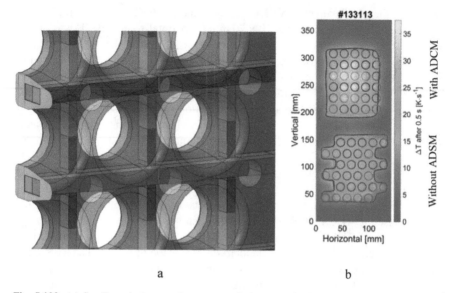

a   b

Fig. 5.100 (**a**) Small vertical correction magnets in between the horizontal deflection magnets form a Halbach array: an asymmetric magnetic flux structure. The extraction grid is seen from the downstream, that is, field weakening side. Cooling channels not shown; grid colors show manufacturing steps by electrodeposition. (**b**) Compensation of zigzag deflection. Successful in a wide operational regime

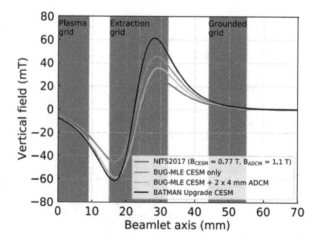

Fig. 5.101 The on-axis vertical magnetic field for various experimental configurations. The BUG-MLE grids are indicated by gray blocks

Lower filling pressure strongly increases beamlet divergence (biggest step: 0.3 Pa → 0.4 Pa). Physics understanding is an ongoing task. Ion source operation at 0.4 Pa would be beneficial [116].

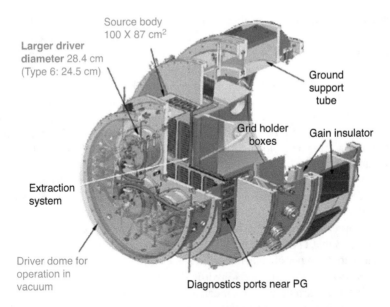

Source body
100 X 87 cm²

Larger driver
diameter 28.4 cm
(Type 6: 24.5 cm)

Ground
support
tube

Grid holder
boxes

Gain insulator

Extraction
system

Driver dome for
operation in
vacuum

Diagnostics ports near PG

Fig. 5.102   Construction scheme of SPS ELISE. (Reproduced from Wünderlich et al. [117])

The construction scheme of the ELISE SPS is shown in Fig. 5.102. It consists of four RF plasma generators of diameter 28.4 cm, a holder for the extraction system, an expansion chamber with diagnostic ports, a plasma electrode, an extraction electrode, and a ground electrode. The chamber is pumped out by two cryogenic pumps at a pumping speed of $2 \times 350,000$ l/s. A magnetic filter is created by current through the plasma electrode. There are four RF plasma generators each of 180 kW. H^- ion current is 20 A. The Faraday screen, plasma electrode, and potential electrode are made from copper and coated with 3 μm of molybdenum.

The RF ion source test facility ELISE (Extraction from a Large Ion Source Experiment) is part of the R&D roadmap of the European ITER domestic agency F4E for the ITER neutral beam injection systems. ELISE is dedicated to demonstrate the ITER requirements in terms of accelerated negative hydrogen ion current densities (230 A/m² H^-, 200 A/m² D^-) at an electron-to-ion ratio of less than one, for a source of the same width but only half the height of the ITER source $(0.9 \times 1$ m²). The extraction area is 0.1 m², consisting of 640 beamlets with diameter 1.4 cm each, and including the calculated stripping loss for negative ions in the extraction system of 30%, a current of 33 A for H^- and 28.5 A for D^- has to be extracted at a source pressure of 0.3 Pa. Another challenging requirement concerns the beam duration and beam homogeneity: beams of duration up to 3600 s have to be achieved, and deviations in the uniformity of the large beam of less than 10% are required. Negative ions are created via the surface conversion process, that is, the conversion of mainly atoms and positive ions at surfaces with low work function, for which cesium is evaporated into the source. The ELISE test facility went into operation in November 2012 with its first maintenance phase after 2 years of operation. Using hydrogen, a stable 1 h plasma discharge with repetitive 10 s beam pulse extraction was demonstrated, with 9.3 A extracted current and an electron-to-ion ratio of 0.4 using 20 kW RF power for each of the four drivers only, which is less than a quarter of the available RF power. At 45 kW RF power per driver and thus half of the available RF power, a stable 400 s plasma discharge with extracted beam pulses of 18.3 A (same duty cycle) and an electron-to-ion ratio of 0.7 could be achieved. Challenges for long pulse operation are the cesium dynamics in the source and the stability of the co-extracted electron current, the latter being the limiting parameter for the power load on the extraction grids and thus for the source performance.

Photograph of the plasma grid of SPS ELISE is shown in Fig. 5.103a.

Dependences of the emission current density and electron-to-ion ratio on RF power with cesiation and without cesiation are shown in Fig. 5.103b [119].

Good progress has been achieved at the ELISE test facility, where it was demonstrated in 2018 that ITER-relevant 1000 s pulses are possible in hydrogen. Similar results have not been possible up to now in deuterium [120]. The reason is a strongly pronounced isotope effect between hydrogen and deuterium with regard to the co-extracted electron current, which is much higher (typical values for this increase are ≈10 for otherwise unchanged operational parameters of the ion source) in deuterium, and their temporal increase is much more pronounced. Hence, the amount of co-extracted electrons limits the applied RF power or extraction voltage and, thus, the source performance. To date, in deuterium, only ≈66% of the ITER target for the extracted negative ion current density has been achieved over the required of ~1 h. Thus, it is essential to perform general physical investigations that can help

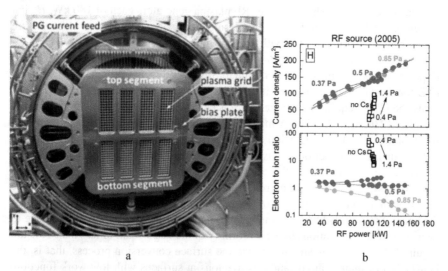

a   b

Fig. 5.103 (**a**) Photograph of the plasma grid of SPS ELISE. (Reproduced with permission from Heinemann et al. [118]. Copyright 2018 Elsevier Science Publishers B.V). (**b**) Dependences of the emission current density and electron-to-ion ratio

develop measures for reducing and stabilizing the co-extracted electrons. One candidate for such measures is modifications of the strength and the topology of the total magnetic field used to tune the plasma parameters in front of the extraction system: the magnetic field topology is given, on the one hand, by the magnetic filter field aligned in the horizontal direction of the ion source to reduce the electron temperature and density close to the grid system. The second component is the electron deflection field, which is aligned vertically alternating the polarity row by row of the multiple aperture grid system. To generate this field, magnets are embedded in the second grid, the extraction grid, such that the electrons impinge on this grid and prevent their further acceleration. In the plasma region close to the extraction apertures, both fields have a comparable strength with the consequence that both influence the extracted negative ions via the extraction probability and the co-extracted electrons. Due to the fact that the magnetic filter field is created by a current IPG flowing through the first grid of the extraction system, the plasma grid (PG), the modification of its strength is an easy task to perform in the experiment. It has been shown already that, for deuterium, higher filter fields are required than for hydrogen to better control the amount of co-extracted electrons. In contrast, the magnets creating the deflection field are fixed in the experimental setup and cannot be modified easily. Numerical models simulating the transport of negative ions and electrons inside the plasma toward the meniscus (the boundary surface between the plasma and the beam) are, thus, a valuable tool. Fluid models can be applied to get a global picture of the plasma dynamics. However, these models usually ignore the extracted beam and plasma sheath formed in front of the PG. To check the influence of the deflection field on the particle trajectories, particle tracking codes can be used, but these codes do not contain a self-consistent description of the plasma. Additionally, both fluid models and particle tracking codes fail to describe the physics of the meniscus, which is defined as the equipotential surface dividing plasma and beam. For a full and self-consistent description of plasma, sheath, meniscus, and beam, particle-in-cell Monte Carlo collision (PIC-MCC) models need to be applied. The results of two-dimensional simulations are extensively presented in the literature, using either some sort of parameter scaling or scaling of physical constants in order to reduce the computation time. Since the filter field and deflection field are perpendicular to each other, the structure is fully three dimensional, and application of three-dimensional models is mandatory.

The three-dimensional electrostatic particle-in-cell model ONIX18–21 is applied to a small computational domain in direct vicinity of one extraction aperture. This paper focusses the magnetic field configuration at the test facility ELISE using the plasma parameters of hydrogen discharges. The independent contribution of the filter field and the deflection field to the transport and extraction of negative ions and electrons is investigated. Possible measures for reducing the co-extracted electron current without reducing too much the extracted negative ions are identified.

The self-consistent three-dimensional PIC-MCC code ONIX is applied for investigating the effect of the magnetic field topology on the plasma dynamics in the vicinity of one extraction aperture with the geometry and magnetic field topology of the ion source at the ELISE test facility. The effect of the magnetic filter field and the electron deflection field on the spatial distribution of the electron and positive ion density as well as on the particle fluxes onto the PG with focus on the extraction of negative ions and co-extracted electrons is studied. The complex three-dimensional structure of the total magnetic field close to the extraction aperture, resulting from the superposition of the filter field and the electron deflection field, results in a drift of the electrons trapped in the filter field and additionally in a non-uniform depth of the virtual cathode over the PG surface, and consequently, a non-homogeneous flux of surface produced negative ions into the plasma, having an impact on the structure of the meniscus. In a parametric study, the intensity of the filter field and of the deflection field was increased by a factor of two independently. When increasing the strength of the filter field, no relevant difference in the co-extracted electron current or extracted negative ion current is observed. In the experiment, the interplay of the filter field with the bias applied to the PG is used as a knob to reduce the amount of co-extracted electrons. The latter, however, is a challenge to implement in the code having such a small domain, that is, small reference surfaces, and is a task treated in the near future. For the strengthened deflection field, a significant reduction of the co-extracted electron current is observed. It is demonstrated that this reduction is related to the transport of magnetized electrons along the field lines of the deflection field: the field lines connect the PG at both sides of the extraction aperture. Magnetized electrons following these field lines can be reflected several times by the potential difference at the PG surface until their kinetic energy is sufficient to overcome the potential, and consequently, they are lost at the PG. Besides studies on variations of input plasma parameters with changes of the magnetic field, the next steps pertain to the implementation of the electron losses along the field lines of the magnetic filter. At present, the periodic boundary conditions ignore the loss of electrons at the sidewalls of the ion source. These studies will be done both in hydrogen and in deuterium since the latter typically shows much more co-extracted electrons and a stronger increase of these electrons over time. The identification of a magnetic field topology reducing the amount of co-extracted electrons without reducing the negative ions and its realization in the experiment would be a significant step toward demonstrating the ITER NBI target values for deuterium and the success of the ITER deuterium and deuterium-tritium campaigns.

To study the effect of a floating BP in a high RF power scenario, the maximum RF power (PRF = 75 kW/driver) was used with the following parameters: Ibias = 1 A, decreased magnetic filter field to a value of I_{PG} = 2.2 kA, higher U_{ex} = 10.5 kV, and U_{acc} = 40 kV with an achieved averaged values of j_{ex} = 256 A/m^2 and $j_{e,t}$ = 76 A/m^2 and $j_{e,b}$ = 97 A/m^2. The time traces of je are almost symmetric and increase during the pulse, while j_{ex} stays constant. Compared to the standard case, the strong increase in j_e with time is clearly reduced but still present since the increase during the pulse is related to the cesium redistribution caused by the plasma. Consequently, the floating BP gives a simple method to decrease the electron currents and to influence their symmetry. Investigations toward 1-h pulses are ongoing [122].

Size scaling process of the RF-driven ion sources, from the prototype source used at BUG via the ELISE ion source to the source for the ITER NBI beamline, is

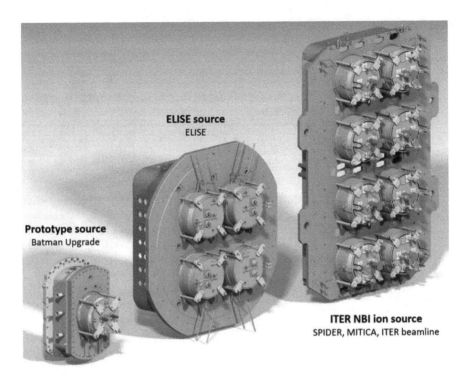

Fig. 5.104 Size scaling process of the RF-driven ion sources: from the prototype source used at BUG via the ELISE ion source to the source for the ITER NBI beamline. (Reproduced from Riccardo Nocentini et al. [121])

shown in Fig. 5.104. The construction scheme of the overall neutral injector with RF SPS with cesiation for ITER is shown in Fig. 5.105. It will consist of RF SPSs for the generation of 70 A of H^- ions and 39 A of D^- ions, with an ion acceleration system up to 1 MeV, gas charge-exchange neutralizer, residual ion receivers, calorimeter for beam intensity measurement, and gate valve and bellows. The estimated cost of three injectors is ~one billion US dollars.

The RF ion source is composed of a rectangular expansion chamber, three cesium (Cs) ovens, and eight cylindrical drivers attached to the rear wall of the expansion chamber. A "bias plate" and the first grid of the extractor (the plasma grid) form the front wall of the source. Each driver consists of a ceramic cylinder mounted with its axis horizontal, a Faraday screen inside the cylinder, a metal back plate that closes both the rear end of the Faraday cage and the ceramic cylinder, a coil wound around the ceramic cylinder, and a directly heated tungsten "starter" filament. The fields generated inside the Faraday cage by the RF applied to the coils together with electrons from the starter filament ionize the gas and create plasma inside the Faraday cage. Once initiated, the starter filaments can be turned off as the RF power alone suffices to sustain the plasma. The field from permanent magnets embedded in the back plate of the driver minimizes plasma loss to the backplate. Plasma flows from

Two beam lines
16.7 MW per beam line, one source
Pulse Length: 3600 s

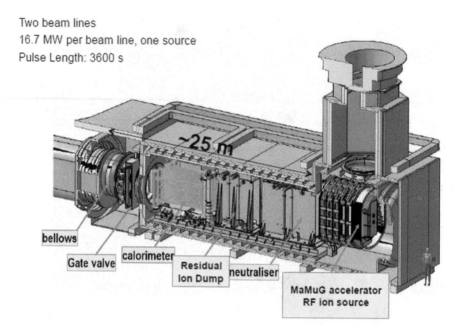

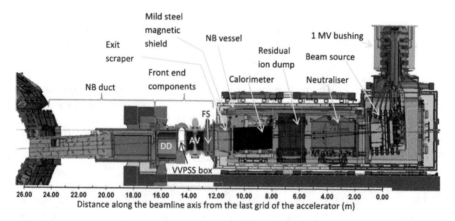

Fig. 5.105 Construction diagram of the neutral beam injector for ITER, using an RF SPS with cesiation. (Reproduced from Hemsworth et al. [106])

open end of the driver into the expansion chamber. Cesium (Cs) from the ovens is injected into the expansion chamber and deposited on the walls of the expansion chamber, the plasma grid (PG), and the bias plate. It is necessary to have Cs on the PG as that reduces its work function and allows negative ions to be more easily created when atoms and positive ions from the plasma in the expansion chamber impinge on the PG. An essential feature of the ion source is the electron current. All the metal surfaces facing the plasma inside the ion source are made of copper (Cu) that is coated with molybdenum. The sputtering yield for D^+ and D_2^+ incident on Mo is very much lower than that for Cu, so the Mo coating minimizes any sputtering which could otherwise contaminate the Cs layer on the PG. In general, the coating is achieved by plasma vapor deposition and the layer thickness is $\approx 5 \times 10\text{--}6$ m. The exceptions are the back plates of the expansion chamber and the drivers. On those, an explosion bonding technique will be used to create a coating 1 mm thick, which is calculated to be the thickness needed to ensure that the coating is not compromised by sputtering due to back-streaming positive ions within the foreseen lifetime of ITER.

The extractor and accelerator consist of sets of multi-aperture grids at various potentials between −1 MV and 0 V. Each grid is supported from the upstream and downstream grid by sets of cylindrical ceramic "post insulators" situated around the periphery of the grid support structure. Each end of each post insulator is enclosed within an electrostatic shield so that electric field at the triple point (the junction between the ceramic, the metal, and the vacuum) is minimized as it is known that high fields at the triple point can induce high-voltage breakdowns. The grounded grid (GG) is supported from a transverse metal beam above the beam source in the beam source vessel (BSV) and below by the beam source tilting mechanism. The lower support can be moved to tilt the whole beam source about the upper support beam in order to tilt the accelerated beam upward or downward by the required ±9 mrad. The extractor consists of two grids, the PG and the EG, followed by five acceleration grids. When producing a 1 MeV beam, the PG is at −1 MV, the EG at $\approx - 0.99$ MV, and the following acceleration grids at −0.8, −0.6, −0.4, −0.2, and 0 MV. The last grid is usually referred to as the "GG." A negative ion exiting the ion source through one of the apertures in the PG is accelerated in the resultant electric field toward the EG, and it will then pass through the corresponding aperture in the EG. The ion is then further accelerated toward the next acceleration grid and so on until it passes through the final grid, which is at ground potential. Magnetic fields in the extractor and accelerator deflect electrons so that few manage to pass through the downstream grids, thus minimizing the useless acceleration of electrons. All the grids are ≈ 1.6 m high and ≈ 0.8 m wide. They consist of four segments stacked vertically, each ≈ 0.4 m × ≈ 0.8 m, for alignment and manufacturing reasons. Each grid has 1280 apertures, 320 per segment. The apertures are arranged in 16 rectangular arrays, each with 5 vertical columns of 16 apertures. In the case of the EG and the acceleration grids, except the 0 kV grid, it is required to embed permanent magnets in the grids. The channels to accommodate the magnets are formed in a similar way to the cooling channels but on the downstream side of the base plate. The mechanical design fulfils several different requirements for the most heated components, particularly, maximum temperature (200 °C), to keep acceptable mechanical properties in the copper, and an acceptable stress distribution in the copper.

"Electron suppression magnets" (ESMs), which are embedded in the EG, are arranged in horizontal rows between the rows of apertures. They are polarized parallel to the beam direction, and the polarity alternates between adjacent rows. This layout produces a vertical component of the induction field (By) characterized by two symmetric peaks of opposing polarity along the axis of each beamlet. The field upstream of the EG is calculated to be sufficient to deflect the co-extracted electrons onto the EG. Unfortunately, the magnetic field of the ESMs would also induce an undesired "zigzag" deflection of the extracted and accelerated ions, the direction of which alternates from aperture row to aperture row. To a first approximation, the angular deflection is proportional to the net value of the integral of By along the ion trajectory from the starting point, that is, the meniscus at the PG aperture, to the point at which the field becomes negligible. That deflection, which would be $\approx \pm 3$ mrad, must be corrected in order to avoid beam loses downstream of the accelerator. That is achieved by embedding a second set of magnets in the EG (referred to as ADCM below) which are placed vertically beside each aperture of the EG and mag-netized along the vertical direction. Their effect is to increase the magnetic field produced by the ESMs on the upstream side of the EG, and to decrease it on the downstream side, which is calculated to eliminate the net deflection of the ions [123].

Referring to Fig. 5.105, MITICA (Megavolt ITER Injector and Concept Advancement) is a full ITER-size prototype neutral beam injector for ITER, to be constructed at Padua, Italy. A three-dimensional view of the MITICA power supply system is shown in Fig. 5.106 from [124]. A detailed description of this power sup-ply is presented in [125].

The SPIDER ion source has been designed to generate a current of H^- ions up to 60 A and accelerate them up to 100 keV. SPIDER RF SPS with cesiation. It is shown in Fig. 5.107. Additionally, the ratio between the co-extracted electrons and nega-tive ions has to be below 1, and a beam inhomogeneity within 10% has to be attained.

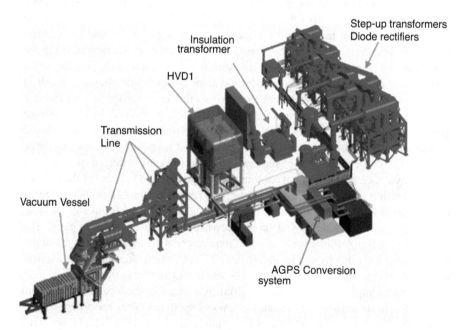

Fig. 5.106 Three-dimensional view of the MITICA power supply system. (Reproduced from Toigo et al. [124])

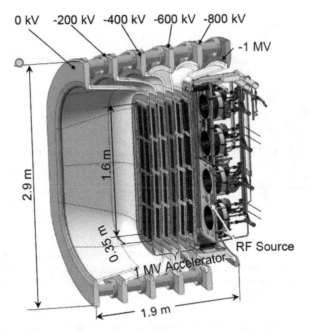

Fig. 5.107 SPIDER RF SPS with cesiation

A magnetic filter field is produced by a current flowing through the plasma electrode in the vertical direction with the aim of reducing electron temperature and
density in the extraction region, so as to enhance the survival probability of H^- ions
while reducing the co-extracted electrons. Another strategy to reduce the co-
extracted electrons is to apply a small bias voltage to the plasma grid with respect to
the ion source body. In addition to this, a so-called bias plate is installed in the vicinity of the plasma grid, which can also be biased with respect to source body.
However, the application of a bias voltage affects the beam inhomogeneity by modifying the ExB plasma drift within the ion source. Recently, the filter field configuration of SPIDER was modified so as to improve the plasma diffusion toward the
extraction region. In addition, the voltage ratings of the bias power supplies were
increased. Both these modifications permit to efficiently lower the electron co-
extracted current. The present contribution describes such improvements while
focusing on the effects they have on the beam inhomogeneity [126].

The high ionization degree of cesium and the consequent possible negative
impact of electric fields close to the PG on the flux of cesium impinging the PG is
the motivation behind attempts to increase the fraction of neutral cesium impinging
on the PG. Tests have been carried out of cesium evaporation by means of a so-
called cesium shower close to the PG, that is, in a plasma region where the ionization probability of cesium is reduced due to the low electron temperature. The
design of the cesium shower is similar to the single-oven multi-nozzle cesium evaporation system presented in Novosibirsk BINP [127].

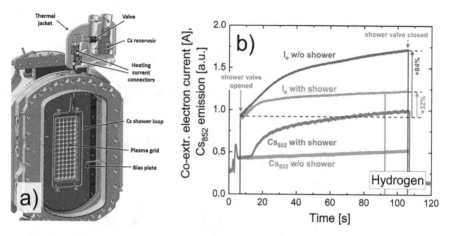

Fig. 5.108 (**a**) Cesium shower loop close to the BUG PG together with the cesium reservoir and the solenoid valve for quickly starting or stopping additional cesium evaporation. (**b**) Time traces of the co-extracted electron current and the Cs852 emission taken during two 100 s extraction pulses in hydrogen: with and without additional cesium evaporation from the cesium shower started at the beginning of the pulse

Figure 5.108a shows the cesium shower installed at BUG, consisting of a stainless steel pipe (inner diameter 3 mm) arranged to a loop around the PG. Both ends of the pipe are connected through a solenoid valve to a high-pressure cesium reservoir. The temperature of the cesium reservoir determines the evaporation rate, which can be stopped by closing the valve. The orifices for the evaporation are located on the two vertical segments of the shower loop and are directed toward the PG. Besides reducing the ionization degree of cesium impinging the PG, the shower provides a controllable cesiation of the PG because it relies on the direct evaporation and not on the plasma-assisted redistribution. Results of initial tests of the cesium shower, performed in hydrogen, are shown in Fig. 5.108b for two 100 s extraction pulses: time traces of the co-extracted electron current and the emission of the neutral cesium emission line at 852 nm measured close to the PG (labeled Cs_{852} in the figure). Evaporation from the standard cesium oven at the source back plate was active in both cases, but the valve of the cesium shower was opened at the beginning of the second pulse. A significant increase of the cesium emission is observed for the second pulse after around 10–20 s, which is the typical reaction time of the cesium shower. This increase indicates a nonnegligible additional cesium flow toward the PG provided by the shower. In parallel, the increase of the co-extracted electron current is significantly reduced after 10–20 s, and it reaches a considerably lower value compared to the reference case. The overall increase is reduced from 84% to 32% after 100 s while the extracted negative ion current is not affected by the additional cesium evaporation. These initial results demonstrate that the shower concept is a promising solution for stabilizing the dynamics of the co-extracted electron current. Although it can be expected that the much more direct operation of the shower compared to the standard oven results in a reduced cesium evaporation rate needed for achieving a comparable neutral cesium flux onto the PG, due to the preliminary character of the performed experiments, up to now, no statements on the cesium evaporation rates from the shower can be made. Further investigations, including also long-term tests, are planned for the future [128].

The RF-driven large negative ion source ELISE can be operated at a filling pressure of ≈0.2 Pa under discussion for DEMO. Further decreasing the pressure can cause issues regarding the sustainment and stability of the plasma. The onset of observed oscillations in the kHz range and the low-pressure operational RF limit strongly depends on the magnetic field topology. In particular, the magnetic field in or close to the drivers plays a crucial role, most probably by its direct impact on the transport and the drift of magnetized charged particles. As a theoretical description of these oscillations is not yet available, the results presented within this paper are intended to motivate theoretical investigations and to define some of the needed input parameters. It has to be kept in mind that the possibility to generate a plasma at 0.2 Pa does not necessarily mean that a high extracted negative ion current density can be achieved for such low pressures. For both RF- and arc-driven ion sources, the amount of co-extracted electrons strongly increases when reducing the filling pressure, in particular in deuterium. Additionally, the typically observed increase in time of these co-extracted electrons gets much steeper when reducing the filling pressure, which is of particular importance for DEMO where pulse lengths of up to 7200 s are envisaged. It may be possible to solve these issues by introducing alternative cesium evaporation concepts, for example, evaporating cesium very close to the plasma grid. Alternative geometries of the ion source or the RF driver, like the racetrack-shaped RF drivers used in the current European DEMO design, may have an effect on the low pressure operational limit. Thus, currently planned experiments using a racetrack driver at ELISE, following successful initial tests of this type of RF driver performed at the smaller prototype source, are of high relevance for DEMO [129].

While due to the frequency of the RF generators (1 MHz), the plasma intrinsically oscillates in the MHz range, additional strong oscillations in the kHz range are observed for both hydrogen and deuterium in the signals of several diagnostics (mainly electrical measurements). A distinct disturbance is seen, for example, in the electron saturation branch of Langmuir probes, but also optical measurements like cavity ring-down spectroscopy can be disturbed for a filling pressure around or below 0.2 Pa [130].

The cesium conditioning of the full-size ITER NBI prototype source SPIDER was performed for the first time [131]. The experimentation covered technical aspects, related to the commissioning and first operation of the cesium ovens, cesium diagnostics, and the gradual increase of high-voltage acceleration. It also covered the investigations of cesiation procedure in SPIDER, and of the source performances in terms of extracted negative ion current density and co-extracted electrons, in the present configuration with limited number of extraction apertures. In this regard, the source performances in the five phases of the campaign are summarized. In the present operating conditions for SPIDER, the first campaign with cesium has highlighted the following aspects: in pulsed operation, higher Cs injection has the same role as longer pause between plasma pulses. The plasma duration has a detrimental effect on the Cs effectiveness. Emission current density J^- ~ 120 A/cm^2 with electron-to-ion ratio of ~2 was produced in first cesiation run. The minimal angular divergence at 0.3 Pa measured by the emittance scanner was ~12 mrad at low RF power and beam energy of 37 keV; it decreases lower at higher pressure, with minor effects from bias currents and filter field. The measurement was performed at optimal perveance, requiring the reduction of RF power to 23 kW/driver in order to accommodate the present limit of acceleration voltage (about 45–50 kV).

In diagnostic neutral beam injector with cesiated RF surface plasma source, ROBIN achieved necessary beam parameters: RF powers up to 80 kW coupled, emission current density >300 A/m^2, H^- beam energy >40 keV, Cs consumption ~10–12 mg/h, and extracted current ratio $e^-/H^- \leq 1$ [132].

5.15   Neutral Beam Injector with Cesiated RF SPS Development at Novosibirsk

The development of a large neutral beam injector using an RF SPS with cesiation was initiated in 2011 at the Budker Institute of Nuclear Physics (BINP), Novosibirsk, Russia [127, 133–137]. The high-voltage negative ion-based neutral beam injector is under development at the Budker Institute of Nuclear Physics (BINP). According to the BINP injector scheme, the primary 120 keV beam of negative ions is transported from the source through the low energy beam transport (LEBT) section to the single aperture multi-electrode acceleration tube. LEBT section tank contains the beam bending magnets and cryopumps. The LEBT section purifies the negative ion beam from the parasitic fluxes of gas, cesium, and secondary particles outgoing the ion source and protects the source from the secondary particles back-streamed from the accelerator. The ion source and LEBT section are placed on the platform, biased to potential up to −880 kV with respect to the ground (Fig. 5.109).

The accelerated negative ions are neutralized in a plasma target. The non-neutralized fractions of the beam are separated from the neutrals and directed to the ion beam energy recuperators. A schematic of the overall device is shown in Fig. 5.108a, b. It consists an RF SPS 1, gate valve 2, separating magnets 3, supporting insulator 4, accelerating tube 5, gate valve 6, neutralizer target 7, magnetic analyzer and residual ion receiver 8, gate valve 9, cryopump 10, internal tank 11, cryopump12, and quadrupole lenses 13. In this scheme, the high-voltage accelerating system is separated from the RF SPS by a beam transport path with good pumping and a beam shift of 40 cm.

The design parameters of the injector are energy up to 1 MeV at a power up to 5 MW. The RF SPS should produce 9 A of H^- ions. It is proposed to use a plasma or photon neutralizer target and recovery of residual ion beams. A schematic of the RF SPS negative ion source for the BINP neutral injector is shown in Fig. 5.110a. It consists of RF plasma generators 1, expansion chamber with multipole magnetic wall and cesium system 2, magnetic filter 3, electrodes of the extraction and

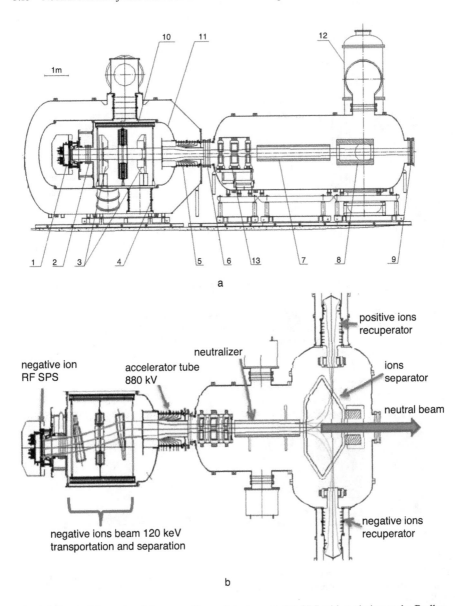

Fig. 5.109 (**a**) Schematic of the neutral beam injector with RF SPS with cesiation at the Budker Institute of Nuclear Physics, Novosibirsk (cross section parallel to magnetic field); (**b**) cross section perpendicular to magnetic field. (Reproduced from Sotnikov [136], with permission of O. Sotnikov). 1, RF SPS; 2, gate valve; 3, separating magnets; 4, supporting insulator; 5, accelerating tube; 6, gate valve; 7, neutralizer target; 8, magnetic analyzer and residual ion receivers; 9, gate valve; 10, cryopump; 11, internal tank; 12, cryopump; and 13, quadrupole lenses

acceleration system 4, ceramic insulators 5, and deflection magnet 6. The electrodes of the extraction and acceleration system have 142 emission apertures each 18 mm in diameter, for a total of 361 cm² of emission area. To obtain 9 A of H^- ions, an

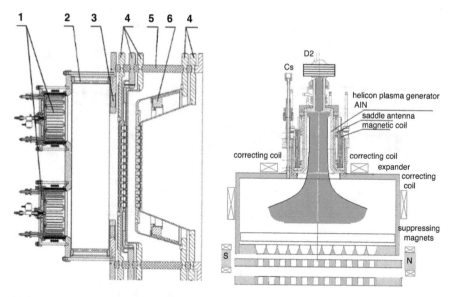

Fig. 5.110 (**a**) Schematic of the RF SPS negative ion source for the neutral beam injector at BINP, Novosibirsk. (Reproduced from Sotnikov [136], with permission of O. Sotnikov). 1, RF plasma generators; 2, expansion chamber with multipole magnetic wall; 3, magnetic filter; 4, electrodes of the beam extraction and acceleration system; 5, ceramic insulators; and 6, deflector magnet. (**b**) Schematic of the RF SPS negative ion source with solenoidal magnetic field for the neutral beam injector

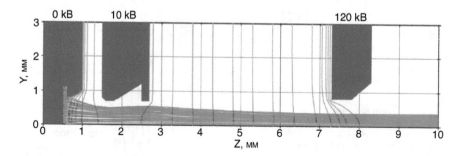

Fig. 5.111 Calculated trajectories of H^- ions through a single aperture of the extractor and accelerating system; beamlet current is 60 mA. (Reproduced from Sotnikov [136], with permission of O. Sotnikov)

emission current density of 25 mA/cm^2 is surface plasma source. A schematic of RF SPS negative ion source with solenoidal magnetic field for the neutral beam injector is presented in Fig. 5.110b [137].

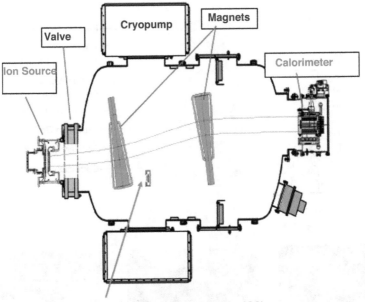

- • Faraday Cup current (1.6 m from the IOS)

- **By calorimeter (4 m from the source)**

Fig. 5.112 Layout of the test stand for development at BINP, Novosibirsk, of an RF SPS with transport channel to the accelerator tube. (Reproduced from Sotnikov [136], with permission of O. Sotnikov)

The calculated trajectories of H^- ions through a single aperture (i.e., one single beamlet) of the extractor and accelerating system are shown in Fig. 5.111. The beamlet ion current is 60 mA.

The program calls for the development, initially, of a prototype RF SPS with energy 120 keV and current 1.5 A. The layout of the test stand is shown in Fig. 5.112. It consists of a vacuum vessel, RF SPS, gate valve, analyzing magnets, cryopump, calorimeter, and Faraday cup.

A schematic of the prototype RF SPS is shown in Fig. 5.113a. It consists of a plasma generator inside a ceramic chamber with Faraday screen, RF antenna and igniter, expansion chamber with multipole magnetic wall, magnetic filter, plasma electrode with cesium distributor (Fig. 5.113b), extraction electrode with an electron-suppressing magnetic field (Fig. 5.113c), and beam accelerating electrode with a correcting magnetic field (Fig. 5.113d). The plasma electrode has 21 emission apertures each 1.6 cm in diameter for a total extraction area of 42 cm^2; the emissive zone is 14 cm in diameter with an area of 220 cm^2. The plasma electrode and the extractor are temperature-controlled to more than 100° C. The extraction voltage is up to 12 kV with a gap of 5 mm. The beam accelerating voltage is up to 108 kV with a gap width of 49 mm. The magnetic field is 120–180 Gauss. The design parameters of the beam are 1.5 A at 120 keV.

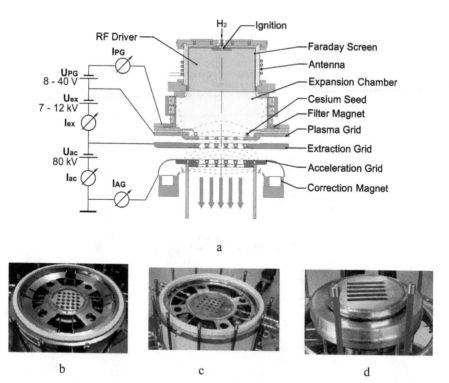

a

b   c   d

Fig. 5.113 (**a**) Schematic of the prototype RF SPS at the Budker Institute of Nuclear Physics; (**b**) photograph of plasma electrode; (**c**) photograph of extraction electrode; (**d**) photograph of accelerating electrode. (Reproduced from Sotnikov [136], with permission of O. Sotnikov)

Cesium is injected into the distribution tube from an external oven with pellets and ejected through small apertures facing the plasma grid emission region. The tube is heated by an internal thermocable. Steady cesium ejection from the tube is provided by a single cesium release, and ~ 0.5 G of cesium seed was enough to enhance the H^- yield for a 2-month-long experimental cycle. Cesium recovery from cesium compounds coating the cold source walls is efficiently done by the RF discharge as well. This was confirmed experimentally by the recovery of good H^- yield after a number of RF discharge pulses following Cs pollution by air filling or by the occasional water leak.

The escape of cesium from the source plasma chamber occurs mainly during pauses between pulses. This leads to gradual accumulation of cesium on the high-voltage IOS (ion-optical system) electrodes. Cesium deposition on the electrode surfaces enforces electron emission and desorption processes, and it also degrades the high-voltage hold-off capability of the source. As a result, conditioning of large cesiated negative ion sources to the designed voltage takes many days. Decreased cesium consumption in the BINP RF source facilitates the IOS electrodes' high-voltage conditioning and improves the high-voltage hold-off by heating/cooling the separate electrodes with hot fluid circulation. The plasma grid and extraction grid were heated/cooled by circulation of heat transfer fluid (Marlotherm) through channels in the electrodes. A Lauda thermal stabilization system is located on the 120 kV platform and supplies hot fluid circulation through internal channels in the electrodes and is connected to the source by thermally shielded hoses.

Electrode heating enhances surface diffusion and thermal desorption. As a result of increased cesium redistribution over the hot electrodes, the accumulation of cesium at the edges of the IOS electrodes and its uncontrolled desorption under bombardment by accelerated negative ions, accompanying electrons and back-streaming positive ions, is suppressed, and improved high-voltage hold-off of the IOS was achieved. The optimal concavity of the magnetic filter field in the extraction and acceleration gaps of the IOS (shown by dotted lines in Fig. 5.113a) is important too. This favors high-voltage improvement by preventing electron oscillations and avalanche in the IOS magnetic field.

The results of the development program have been reported [138]. Oscillograms of the H^- ion beam current I_b, the accompanying electron current I_e, and the current of the extraction electrode I_{ex} are shown in Fig. 5.114a. Figure 5.114b shows the dependence of beam current I_b, co-extracted electron current I_e, and plasma grid current I_{PG} on the plasma grid bias voltage U_{PG} (empty data points for $p = 0.3$ Pascal and $P_{RF} = 32$ kW; filled data points for $p = 0.4$ Pascal and $P_{RF} = 34$ kW). The co-extracted electron current I_e is markedly decreased at bias voltage $U_{PG} > 15$ V.

At an RF discharge power of 36 kW, the H^- ion current is 1.2 A for a beam energy of 90 keV, and a hydrogen pressure of 0.4 Pascal is produced. The plasma

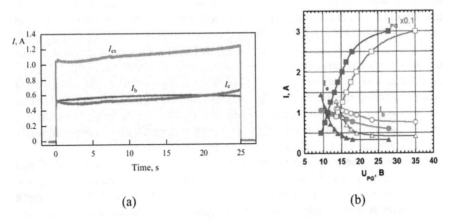

(a)   (b)

Fig. 5.114 (a) Oscillograms of emission currents of the source in long pulses: I_{ex}, current in extracting voltage rectifier circuit; I_b, IOS output beam current; and I_e, current of accompanying electrons. RF driver with Faraday shield: RF discharge power PRF 17 kW, extracting voltage U_{ex} 7 kV, accelerating voltage U_{ac} 75 kV, and hydrogen pressure in discharge chamber 0.35 Pa. (Reproduced from Sotnikov [136], with permission of O. Sotnikov). (b) Ion beam current I_b, co-extracted electron current I_e, and plasma grid current I_{PG}, as a function of plasma grid bias voltage U_{pg},V. Empty data points: $p = 0.3$ Pascal, $P_{RF} = 32$ kW; filled data points: $p = 0.4$ Pascal, $P_{RF} = 34$ kW. (U_{pg} is the plasma grid voltage relative to the plasma at high current, ~30 A) electrode potential is 9.7 V. An H^- ion beam current of up to 0.9 A is transported to a calorimeter at a distance of 3.5 m.

The source operation in pulses with duration up to 25 s was successfully tested with the Faraday screen installed. Higher RF field amplitude was necessary to deposit RF power into discharge in this case, and the RF power deposited in the plasma was limited at the lower end to 26 kW (due to RF power system voltage restrictions). It was found that the presence of the Faraday screen increases the positive plasma potential by augmenting electron diffusion to the metallic Faraday screen. The increased plasma potential of the RF discharge with Faraday screen enhances the value of the optimal plasma grid bias needed to suppress the accompanying electron flux. The H^- ion beam current I_b at the exit of the beam formation system is stable during the pulse. The total current I_{ex} in the extraction circuit, characterizing the sum of the ion beam current and the accompanying electron current, was two times greater than the ion beam current. The differential current $I_e = I_{ex} - I_b$, corresponding to the current of co-extracted electrons, increased by about 20% over the 25 s pulse duration. The stability of the ion beam current shows only a small change in cesium coverage of the plasma grid emission area over the 25 s pulse. The increase in co-extracted electron current toward the end of the pulse could be produced by a gradual depletion of cesium coverage on the expansion chamber and driver walls, which in turn produces a change in plasma potential and increases the co-extracted electron current. An essential feature of the BINP negative ion-based neutral beam injector is the installation of a low energy beam transport (LEBT) section between the negative ion source and the accelerator. It is necessary to prepare the beam before entering the accelerator region: to clean it of accompanying gas, cesium, fast neutrals, electrons, etc. and to move the ion source away from the positive ion beam that back-streams from the accelerator. The H^- ion beam production and beam transport through the LEBT section was studied using the test stand schematically shown in Fig. 5.95. The layout of the LEBT section was similar to that of the BINP injector and consisted of a large vacuum tank equipped with two cryopumps and two dipole magnets used for a sideway shift of the negative ion beam away from the source axis. Intense cryopumping of the vacuum tank reduces negative ion stripping and secondary particle production in the beam transport line and in the accelerator. Beam characterization was performed using a movable Faraday cup and a beam calorimeter mounted at the LEBT exit.

Preliminary heating of the plasma and extracting electrodes to 150–250 °C and their cooling with a hot coolant during pulse passing are secured by pumping the high-temperature coolant through channels drilled in the IOS electrodes using a commercial thermal stabilization system (Lauda). IOS electrode heating and cooling by pumping a hot coolant, the use of an IOS transverse magnetic field with bent lines of force, and cesium feeding directly onto the emitter surface through the distribution/accumulation tube are conceptually new elements introduced in the construction of an RF source designed at BINP. Studies performed on the BINP experimental bench demonstrated the working efficiency of these innovations. They stabilized the work of the source in the hydrogen-cesium regime with low cesium

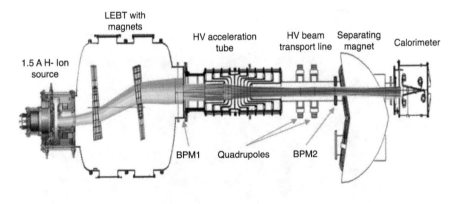

a

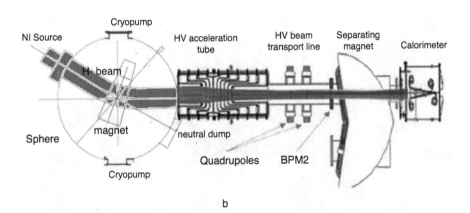

b

Fig. 5.115 (a) Schematic of HV test stand in BINP. H^- ion trajectories shown are calculated with the COMSOL software. (Reproduced from Sotnikov [136], with permission of O. Sotnikov). (b) Schematic of preferable cleaning accelerating beam from parasitic particles

consumption and required a high-voltage strength of IOS gaps. Stable 1.2 A H^- ion beams at the exit from the IOS had an energy of 90 keV in 2 s pulses with the RF discharge power of 34 kW and initial hydrogen pressure of 0.4 Pa. Emission current density of the beam during extraction was 28 mA/cm². The current of accompanying.

A schematic of HV test stand in BINP is shown in Fig. 5.115a. H^- ion trajectories shown are calculated with the COMSOL software. A schematic of preferable cleaning accelerating beam from parasitic particles is shown in Fig. 5.115b.

Electron current did not exceed 1 A. The size of the H^- ion beam transported to a distance of 1.6 m was 30 × 25 cm².

a

b

Fig. 5.116 (a) HV platform of NI NBI BINP (top view). (b) HV platform of NI NBI BINP (side view). (Reproduced from Sotnikov [136], with permission of O. Sotnikov)

HV platform of NI NBI BINP (top view) is shown in Fig. 5.116a. HV platform of NI NBI BINP (side view) is shown in Fig. 5.116b. A high-voltage accelerating tube, connecting HV platform with grounded platform, is visible [139].

The arrangement of the BINP HV test stand is shown in Fig. 5.117. The H− source with subsystems and LEBT are installed at the HV platform, negatively biased with respect to the ground. The source subsystems include the RF power supply, ion beam extraction and acceleration power supplies, thermo-stabilization system, and hydrogen and cesium feeding systems. The primary power at the

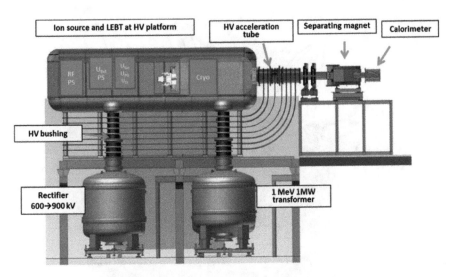

Fig. 5.117 The arrangement of the BINP HV test stand for investigation of high energy neutral beam production. (Reproduced from Sotnikov [136], with permission of O. Sotnikov)

platform is provided via the 1 MW isolation transformer. As is shown in Fig. 5.103, each of the acceleration tube electrodes is powered by separate sections of HV rectifiers. The last electrode of the acceleration tube is connected to the HEBT tube and grounded. The HV rectifiers are installed in the tanks filled with gaseous SF_6 and are equipped with the HV bushings. The rectifiers are connected in series to the acceleration tube electrodes. The primary power is supplied by 0.5÷3 kV/1 kA/2 kHz invertors.

In case #1, the extraction voltage of 6.5 kV provides optimal beam transport efficiency for the beam with current 0.65 A. About 31% of the beam was measured by FC, located at 2.5 m from the source. It shows about 32% transmission of the beam to the entrance of acceleration tube through the ⌀ 200 mm diaphragm at 4.45 m. In case #2, the source extraction voltage was increased to 10 kV at the fixed total beam energy 85 kV. In this case, the FC registered current fraction decreases from 31% to 25%, and the beam transport efficiency to the distance of 4.45 m decreases from 32% to 21%, which corresponds to the increase in the beam divergence. Case #3 demonstrates the transport efficiency measurements at 3.8 m distance for the beam current, extraction voltage, and total energy similar to case #1 with optimized beam angular divergence for total energy 84 keV. The measurements for the experimental case #4 were done at the higher total beam energy of 93 keV and higher extraction voltage 10 kV. It shows higher beam transport efficiency 60% compared to 50% for the 84 keV case.

The full-scale RF H− SPS with the projected total negative ion beam current 9 A, energy up to 120 keV, and pulse duration up to 100 s is constructed at BINP. The view of the source is shown in Fig. 5.118. It includes design features of the 1.5 A source: (a) the grids of IOS are kept hot by thermal fluid circulating through the

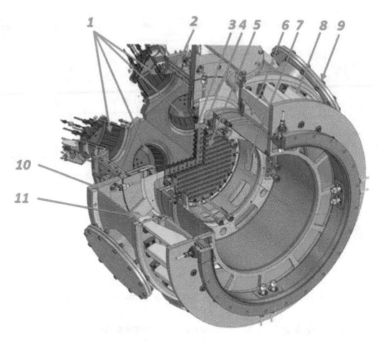

Fig. 5.118 The cross-sectional view of the 9 A RF SPS negative ion source. 1, RF drivers; 2, expansion chamber; 3, plasma grid; 4, extraction grid; 5, acceleration grid and positive ion suppression grid; 6, HV vacuum ceramics; 7, support of acceleration grid; 8, external HV supports; 9, flange for additional pumping; 10, HV vacuum chamber; and 11, internal HV ceramic supports. (Reproduced from Sotnikov [136], with permission of O. Sotnikov)

internal channels machined in the grids; (b) the magnetic field lines in the IOS extraction and acceleration gaps are concaved; and (c) the primary deposition of cesium is done via two distribution tubes adjacent to the PG periphery. New features introduced in the 9 A source design are as follows: (a) use of four RF drivers to irradiate the increased emission area (b) enforced pumping system with 1000 m^3 s^{-1} cryopump and four additional turbomolecular pumps installed at IOS periphery (9 in Fig. 5.118) with capacities of 2.7 m^3 s^{-1} each; and (c) the assembly of extraction and acceleration grids with the use of internally installed HV insulator supports.

Inductively driven RF discharges (4 MHz, power up to 40 kW each) are excited by external antennas in four RF drivers. Drivers are mounted on the back flange of the expansion chamber at an angle of 26° to the axis of the source. The oblique drivers provide uniform distribution of the plasma over the emission area of PG. To enhance the surface generation of NI, the cesium is deposited to the PG surface with the help of two distribution tubes. The negative ions are extracted through 142 emission apertures in PG with ⌀ 14 mm and accelerated to the energy up to 120 keV by four-electrode IOS. The projected emission current density of single beamlet is >30 mA cm^{-2}. To reduce the co-extracted electron flux and to lower NI stripping by electrons, a filter with dipole magnetic field is created near the plasma electrode. All co-extracted electrons are deflected by transverse magnetic field and collected to the conical chamfers at the downstream side of the EG apertures or to downstream side of the AG and its support.

5.16 Research Activities of RF-Based Surface Plasma Source in the ASIPP (in China)

The Comprehensive Research Facility for Fusion Technology (CRAFT) is a large scientific program that is preferentially deployed for the construction of major national science and technology infrastructures. A negative ion-based neutral beam injector (NNBI) with beam energy of 400 keV, beam power of 2 MW, and beam duration of 100 s is one of the projects. A radio frequency (RF)-based negative beam source was designed for the CRAFT NNBI system. In order to understand the physics and pre-study the engineering problems for RF negative beam sources, a prototype source with a single driver was developed.

Recently, this source was tested on a test facility with RF plasma generation, negative ion production, and extraction. The long pulsed plasma discharge and negative ion beam extraction with a three-electrode accelerator was achieved successfully. The extracted ion current density is 153 A/m² with Cs injection, and the ratio of electrons to negative ions is around 0.3. It lays good foundations for the R&D of negative ion sources for the CRAFT NNBI system. The details of design and experimental results of the beam source are given in this paper [140].

Schematic diagram of the prototype of negative ion RF SPS in the ASIPP is shown in Fig. 5.119 (left). A schematic diagram of extraction system is shown in Fig. 5.119 (right). This design is similar to the RF BATMAN SPS shown in Fig. 5.96.

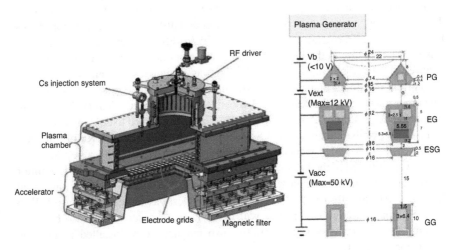

Fig. 5.119 Schematic diagram of the prototype of negative ion RF SPS in the ASIPP (left) and schematic of extraction system (right)

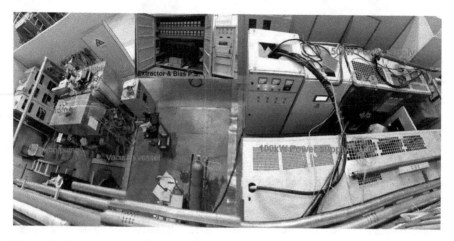

Fig. 5.120 Photograph of the RF SPS test facility in ASIPP

In order to test the performance of the ion source, a test facility was developed and is shown in Fig. 5.120. It contains a vacuum vessel, a gas pumping system, a water cooling system, a 100 kW RF power supply operating at 1 MHz, a matching network, a −16 kV, a 20 A beam extraction high-voltage power supply, and a diagnostic system. Recently, the RF ion source achieved a long pulsed discharge of 1000 s with RF power of 47 kW and a short pulse of 60 s with 80 kW on the RF ion source test bed. Subsequently, the RF ion source was tested with negative ion production and extraction with cesium (Cs) injection. As a result, a 105 s long pulse of negative ion beam was extracted with a negative ion current density of 153 A/m^2. The details about the ion source design and experimental results are presented.

Photograph of the RF SPS test facility in ASIPP is shown in Fig. 5.120.

The plasma generator contains an RF driver and a plasma expansion chamber. The RF-driven plasma is generated in the driver and diffuses to the expansion chamber to form a uniform plasma. The Faraday shield (FS) is a key part of the plasma driver. The structure of the ITER-like FS is very complicated and not easy to manufacture, so a three-channel water-cooled FS was designed and manufactured. The FS has 200 mm inner diameter and 210 mm external diameter and is 150 mm tall. The inner diameter of the rear helical water channel is 6 mm, whereas the sidewall water channel is 4 mm × 2 mm. Three water channels are used to cool the back plate and two side walls of the FS. The FS was installed in the source for RF plasma performance tests. In order to measure the plasma parameters, a moveable Langmuir probe was installed in the bottom of the plasma expansion chamber. The Langmuir probe can move into the chamber by about 16 cm. The electron density increased from 2×10^{16} m^{-3} to 9×10^{16} m^{-3} when RF power was increased from 6 kW to 52 kW: almost increasing linearly with RF power. In order to increase both the plasma density and plasma uniformity, cusp magnets were installed around the expansion chamber.

Without the cusp magnets, significant plasma is lost on the wall, so the electron density increased by a factor of three from 6×10^{16} m^{-3} to 1.6×10^{17} m^{-3} when the probe was moved from 18 cm to 29 cm. With the cusp magnets installed, the plasma is only lost along the cusp lines and the electron density increased by a factor of two, from 1.5×10^{17} m^{-3} to 2.6×10^{17} m^{-3}. Compared with the results without cusp magnets, the electron density increased about three times, and the density difference also decreased. Because the magnetic filter was installed, the electron temperature had almost no change, being close to 1 eV throughout. The results show that the cusp magnets can form good confinement. Unfortunately, the uniformity was bad due to only a single driver used: two RF drivers will be used for better uniformity in the future. Long pulsed discharge was also tested, with pulses up to 1000 s long at an RF power of 47 kW. The plasma was stable and the temperature rise of the RF driver under control, with a maximum temperature rise about 40 °C on the sidewall of the FS. A higher power of 80 kW was also tested with 60 s pulse length. The pressure of cooling water on the FS was 1 MPa and on other part was 0.4 MPa. A maximum temperature rise of about 48 °C was on the sidewall of the FS. These results verified the successful development of the FS. Considering the temperature rise on other parts, there may be two ways to solve this problem: either increase the cooling water pressure or add more cooling pipes and optimize their structure.

The negative ions were generated and extracted first with volume production and then with the addition of surface production by feeding the Cs into the source. In

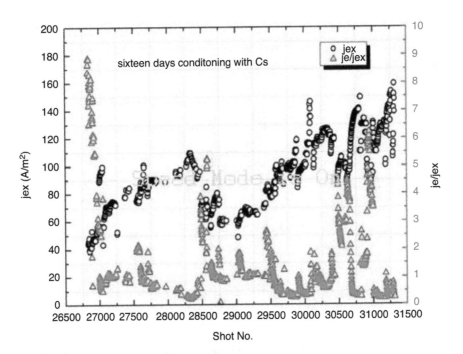

Fig. 5.121 The change of extracted negative ion current density and j_e/j_{ex} during Cs conditioning

order to enhance the production of negative ions, the temperature of the plasma expansion chamber was actively controlled to around 45 °C. The temperature of the PG was heated by plasma. Normally, several shots with long pulses of 100 s were needed to heat the PG from room temperature to more than 100 °C. Thereafter, the discharge duration was controlled actively to hold the temperature of the PG to around 150 °C. The experimental results when conditioning the ion source with Cs are shown in Fig. 5.121. The extracted ion current density j_{ex} started at 40 A/m² and increased day by day with Cs conditioning. It reached about 160 A/m² with 2 weeks' conditioning. The ratio of electron density to extracted negative ion density j_e/j_{ex} was about 10 to begin with and then decreased to 0.3 quickly. In general, 15 shots were needed to decrease the j_e/j_{ex} from several to 0.3.

A prototype RF ion source with a single driver and three electrodes for CRAFT NNBI system was developed and tested on the RF beam source test facility in the ASIPP. RF plasma generation, negative ion production, and beam extraction were studied. The characteristics of the RF plasma generation were investigated with a moveable Langmuir probe. Long pulses of 1050 s plasma with RF power of 47 kW were achieved. Negative ion production and extraction were investigated with both volume production and surface production. The characteristics of negative ion extraction were studied, resulting in stable long pulses of 105 s beam extraction achieved. The extracted ion density is 153 A/m² and the ratio of j_e/j_{ex} is around 0.3. This study was the first time that Cs was injected into the ion source and lays a good foundation of R&D for the CRAFT NNBI system.

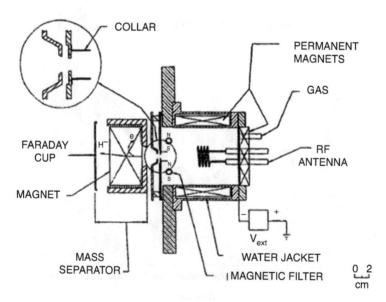

Fig. 5.122 Schematic of the RF negative ion source proposed for the SSC by LBNL. (Reproduced from Leung et al. [143], with the permission from AIP Publishing)

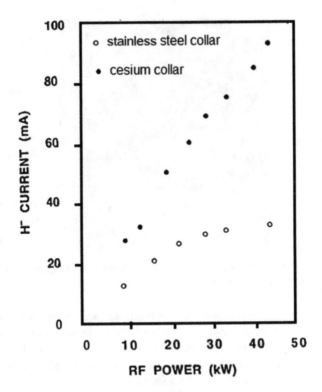

Fig. 5.123 Dependence of H^- current on RF power with and without cesium. (Reproduced from Leung et al. [143], with permission from AIP Publishing)

5.17   RF Surface Plasma Sources for Spallation Neutron Sources

In 1990, Leung and colleagues at LBNL, Berkeley, California, proposed an RF negative ion source with an internal RF antenna for the Superconductor Super Collider [141–143]. A schematic of this source is shown in Fig. 5.122. It consists of a copper plasma chamber of diameter 10 cm and depth 10 cm with a multipole magnetic wall, an internal solenoidal antenna covered with enamel insulation, a collar with magnetic filter and emission aperture of diameter 5.6 mm, an extraction electrode, and a mass separator. The RF discharge is supported by an RF amplifier at a frequency 2 MHz and power up to 50 kW.

   The dependence of the H^- ion current on RF power with and without cesium is shown in Fig. 5.123. Without cesium, the dependence saturates at an RF power of 30 kW. With cesium, there is no saturation even at maximum RF power of 50 kW, with the ion current reaching 90 mA. The accompanying electron current is 8–12 times greater than the negative ion current. The H^- ion current without cesium is relatively large. This is because the enamel that insulates the antenna has a high potassium content. Potassium is sputtered by the discharge and collected on the

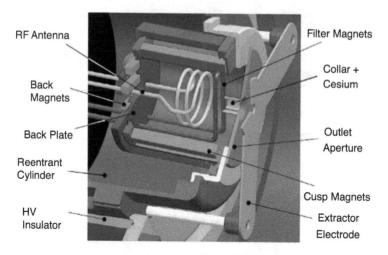

RF Antenna

Filter Magnets

Back
Magnets

Collar +
Cesium

Back Plate

Outlet
Aperture

Reentrant
Cylinder

HV
Insulator

Cusp Magnets

Extractor
Electrode

Fig. 5.124 The LBNL design for an RF SPS for the SNS. (Reproduced from Leunget al. [142], with permission from AIP Publishing)

collar, reducing the work function and providing surface plasma formation of H^- ions. The addition of cesium leads to a further decrease in work function and a further increase in surface plasma generation of negative ions. Cesium deposition was produced by heating of cartridges containing a mixture of cesium chromate with aluminum and zirconium from the SAES Getters.

After termination of the SSC project, this source was proposed as an alternative for an SPS with Penning discharge for the Spallation Neutron Source (SNS) [144]. Later this project became the main SPS for the SNS at the Oak Ridge National Laboratory [145]. The mechanical design of the RF SPS for SNS has been reported [146] and further described in [147, 148]. An illustration of this RF SPS for the SNS is shown in Fig. 5.124. The source consists of a gas-discharge chamber with water-cooled multipole magnetic wall, a solenoidal antenna of 2.5 turns of copper tubing coated with enamel insulation, a collar with emission aperture and cesium cartridges, a magnetic filter, a dipole magnet to suppress the flow of accompanying electrons, and an extractor electrode.

The design of the collar for this source, with cesium cartridges and an emission aperture 7 mm in diameter, is shown in Fig. 5.125. A heated cathode was used to ignite the pulsed RF discharge at a frequency of 2 MHz.

The dependence of the H^- ion current of RF power with and without cesium is shown in Fig. 5.126. Without cesium, the H^- current can be up to 36 mA with the accompanying electron current greater than this by a factor of 27. With the addition of cesium, the H^- ion current reaches 60 mA, with the accompanying electron current greater by a factor of two. The enamel that insulates the internal antenna has a high potassium content. Potassium is sputtered by the plasma and collected at the collar, enhancing the surface plasma formation of negative ions. The addition of cesium enhances this process.

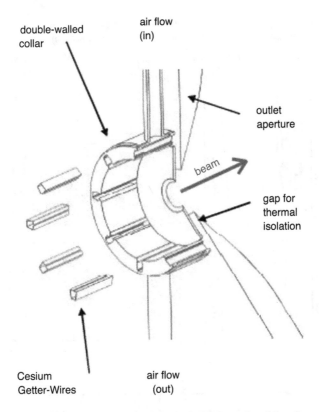

Fig. 5.125 Collar design for the LBNL RF SPS for the SNS. (Reproduced from Leung et al. [142], with permission from AIP Publishing)

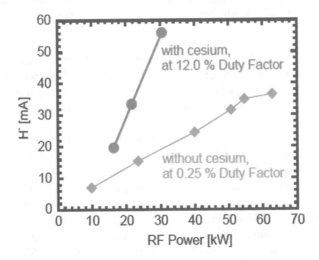

Fig. 5.126 Dependence of the H^- ion current in the LBNL RF SPS for the SNS on RF power with cesium and without cesium. (Reproduced from Leung et al. [142], with permission from AIP Publishing)

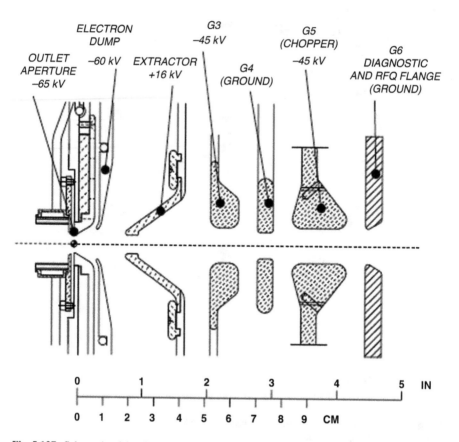

Fig. 5.127 Schematic of the electrostatic transport channel from the ion source emission aperture to the RFQ input, for the RF SPS for SNS. (Reproduced from Cheng et al. [149], with permission from AIP Publishing)

Without cesium, the duty cycle was limited to 0.25% because of the high accompanying electron current. With cesium, the duty cycle could be increased to 12%. The electrostatic low energy beam transport (LEBT) system for transporting the H^- beam from the ion source emission aperture to the RF accelerator input is described in [149].

A schematic of the electrostatic transport channel from the emission aperture to the RFQ is shown in Fig. 5.127. It consists of an output aperture, extraction electrode, dump for accompanying electrons, accelerating electrode, first electrostatic lens, grounded shield, second electrostatic lens, and the input diaphragm to the RFQ. The second lens is divided into four sectors, which can be supplied with independent voltages to deflect the beams and cut gaps to build up the voltage of the inflectors when the beam is extracted from the accumulator ring. The output aperture of the SPS is at a potential of −65 kV, the electron collector is at −57 kV, the accelerating electrode can be biased to +20 kV, and the lenses can be fed up to −47 kV.

Fig. 5.128 Construction of the LEBT for transporting the H^- beam in RF SPA for SNS to the RFQ. (Reproduced from Cheng et al. [149], with permission from AIP Publishing)

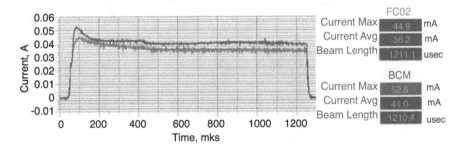

Fig. 5.129 Oscillograms of H^- ion current from a current transformer and from a Faraday cup. (Reproduced from Cheng et al. [149], with permission from AIP Publishing)

The design of the LEBT for transporting the H^- beam to the RFQ is shown in Fig. 5.128. The path is transparent for pumping gas from the source. The insulators are as far from the beam as possible.

Testing of this source at the SNS revealed that the SPS operation was unstable at high duty cycle [150, 151]. Figure 5.129 shows oscillograms of the H^- ion beam current as measured by a current transformer and by a Faraday cup. To ignite the pulsed discharge (2 MHz at 80 kW), a continuous discharge at frequency 13.56 MHz and power 200 W was used. The typical time dependence of the H^- ion beam current in an unstable mode is shown in Fig. 5.130.

After cesiation, the intensity of the beam increases to 30 mA but falls to 20 mA with a decrease rate of 4 mA/day. Studies at the SNS analyzed the behavior of cesium in the SPS and suggested a new collar design. Overcoming these difficulties

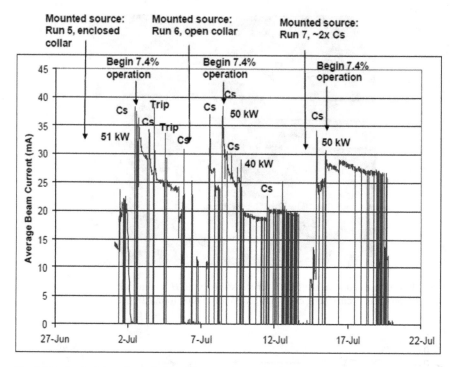

Fig. 5.130 Typical time dependence of the H^- ion beam current in an unstable mode. (Reproduced from Welton et al. [150], with permission from AIP Publishing)

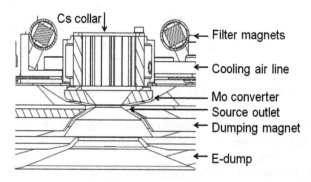

Fig. 5.131 Collar design with molybdenum cone converter for RF SPS for the SNS [503]. (Reproduced from Stockli et al. [152], with the permission from AIP Publishing)

is described in [152, 153]. The new collar design shown in Fig. 5.131 was developed. A molybdenum cone converter coated with cesium serves for the surface plasma formation of negative ions. A magnetic filter serves to separate fast electrons. A magnetic dipole is used to suppress accompanying electrons. The procedure for activating the collar and cesiation was changed. The time dependence of the H^- ion beam current and the RF power, in the stable operating regime, is shown in Fig. 5.132.

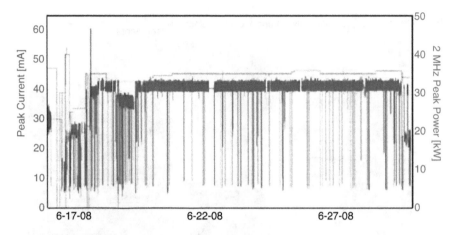

Fig. 5.132 Dependence of the H^- beam current and RF power on time. (Reproduced from Stockli et al. [152], with permission from AIP Publishing)

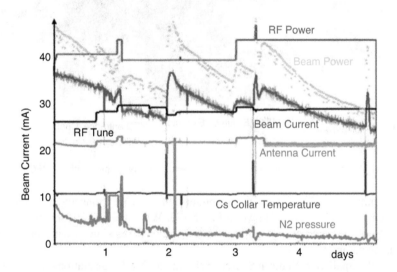

Fig. 5.133 Time dependence of the intensity H^- ions with minimal leakage in the RF SPS for SNS

The H^- ion current was stable for ten days. Later, the time of continuous operation of the RF SPS was increased to 116 days (9 A.hrs), with a cesium consumption of ~30 mg [154]. The service life of the internal antenna was increased by a thorough inspection before installation; this increased the price of the antenna to $5 k. The reasons for such low cesium consumption are discussed in [155]. There is an improvement in the stability of the cesium result after conditioning the collar with plasma discharge for 3 h instead of 0.5 h (with possible deposition of carbon film).

More thorough cleaning of the converter promotes an increase in the binding energy of cesium atoms to the surface. In a pure RF plasma, the energy of hydrogen ions is not sufficient to sputter heavy cesium atoms from the surface. At the same

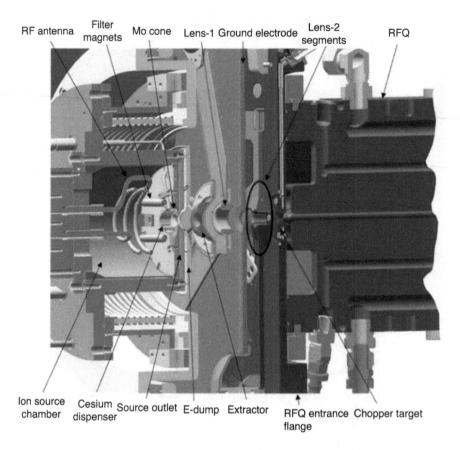

RF antenna | Filter magnets | Mo cone | Lens-1 Ground electrode | Lens-2 segments | RFQ

Ion source chamber | Cesium dispenser | Source outlet | E-dump | Extractor | RFQ entrance flange | Chopper target

Fig. 5.134 A cross-sectional view of the SNS H^- injector closely coupled with the RFQ entrance

time, minimal leakage causes cesium sputtering by nitrogen ions and a decrease in the intensity of H^- ions, as shown in Fig. 5.133. When the beam is intensified, the intensity of the beam at first increases but then decreases slowly, with unchanged discharge parameters.

Now the Spallation Neutron Source (SNS) at the Oak Ridge National Laboratory is an accelerator-based neutron facility that provides intense, short-pulse (<1 μs) neutron beams for scientific research and industrial applications [156]. The accelerator system of SNS includes a 65 keV H^- injector, a 2.5 MeV RFQ, a 1 GeV linac chain (DTL-CCL-SCL), and a proton accumulator ring. The H^- injector consists of an RF-driven SPS, with cesiation H^- ion source and a compact, two-lens electrostatic Low Energy Beam Transport (LEBT). The H^- injector provides the RFQ accelerator with high current, 1-ms-long H^- beam at 60 Hz with ~200–300 ns chopped for every ~1 μs. Figure 5.134 shows a cross-sectional model view of the H^- injector along with the entrance part of the RFQ accelerator.

The ion source is driven by two RF supplies separately matched to a 2.5-turn porcelain-coated, water-cooled, copper-tube antenna. High-power (tens of kW)

2 MHz RF, pulsed at 60 Hz with 1 ms pulse length, generates high-density plasma for beam production. A continuous, low-power (typically ~300 W) 13.56 MHz RF maintains a low-density plasma in the chamber to facilitate fast and reliable ignition of the high-density pulsed plasma. A (Cs_2CrO_4 + Zr, Al) solid reaction cesium dispenser system is utilized to release elemental Cs for ion source cesiation. A Mo cone attached to the cesium dispenser and fixed near the ion source outlet aperture enhances H^- yield through negative ion conversion on its cesiated surface. The co-extracted electrons are deflected toward the e-dump electrode by transverse magnetic field generated in the ion source extraction region with an array of permanent magnets imbedded in the outlet electrode. The effect of the electron dumping transverse magnetic field on the H^- beam can be compensated by mechanically tilting and offsetting the ion source against the LEBT axis. The second lens of the LEBT is split into electrically isolated four quadrants to allow beam steering and chopping. The beam output from the RFQ is continuously monitored with a beam current monitor toroid (BCM), and it is a commonly referred parameter for beam performance of the H^- injector in day-to-day operation. The LEBT output/RFQ input beam currents can be measured by deflecting the full beam onto the chopper target. Since this is a destructive measurement technique, it is primarily used to check the RFQ transmission performance when it is in question.

LANSCE and SNS are exploring the possibility of using the SNS RF H- SP source at LANSCE to increase the H^- beam current and the ion source lifetime while decreasing the start-up time. For this purpose, the SNS H^- source has been tested at a 10% duty factor by operating it at 120 Hz with 840 µs plasma pulses generated with ~30 kW of 2 MHz RF power and extracting ~25 mA around the clock for 28 days [157].

5.18   Carbon Films in RF Surface Plasma Sources with Cesiation

Another explanation for the lower cesium consumption is possible [158–160]. Good cesium retention can be associated with the formation of carbon films on the electrodes. The SNS and J-Parc internal antennas in the RF discharge are covered with a black film, as shown in Fig. 5.135a, b.

The results of elemental analysis of this antenna at ORNL are shown in Fig. 5.136, which shows the relative concentration of elements with depth (C, Si, O, and Na and Ti, Zr, and K). The concentration of C on the surface exceeds 50% and decreases with depth (1 min of sputtering corresponds to 20 nm of the depth). The sodium concentration of 13% is sufficient to provide for surface plasma formation of negative ions, and the potassium concentration of 8% is sufficient for the effective surface plasma formation of negative ions. These impurities can be sputtered by the discharge and collected in the form of ions on the converter surface, providing a means for the surface plasma formation of negative ions. A possible source of carbon is the alcohol used during ion source assembly.

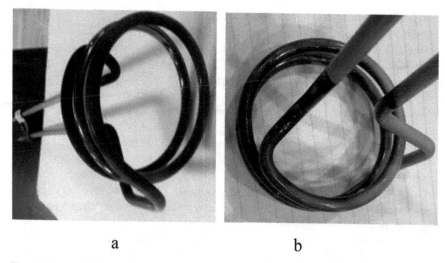

Fig. 5.135 (a) SNS internal antenna covered with black graphite and diamond-like carbon film; (b) J-Parc Internal antenna covered with black graphite and diamond-like carbon film. (Reproduced from V. Dudnikov [160])

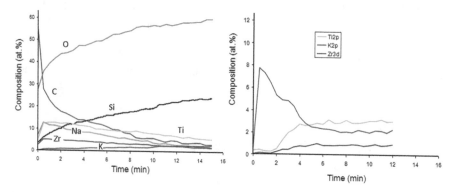

Fig. 5.136 Dependence of the concentration of elements on the surface of the antenna on depth. (Reproduced from Dudnikov [160])

The surface of the converter facing the plasma is the surface on which the black film is deposited, as shown in Fig. 5.137. It is deposited in the early stages of the converter's exposure to the discharge.

Black and transparent films on the surface of the heat shield exposed in the plasma are shown in Fig. 5.138. A black, high conductivity film is deposited at points of contact with weak plasma flow, and a transparent, blue diamond-like carbon (DLC) film, with a resistance of 10 Ohms between contacts that are 1 cm apart, is formed at regions of intense plasma flow.

The dependence of the graphite work function on the subsequent deposition of alkali metals does not have a minimum, as is the case for metals and

Fig. 5.137 Black film on the surface of the collar of the converter. (Reproduced from Dudnikov [160])

Fig. 5.138 Black and transparent films on the surface of the heat shield exposed to the plasma. (Reproduced from Dudnikov [160])

semiconductors, and has a large final work function, as shown in Fig. 5.139. The dependence of the work function of graphite on cesium deposition time is shown in Fig. 5.140. It also does not have a minimum and have the final work decrease 2.45 eV. With work function of clean graphite surface, 4.55 eV work function of cesiated surface can have work function 2.1 eV. Cesium is intercalated into the graphite film. For this reason, the probability of secondary emission of H^- ions from metals and semiconductors covered with an optimal cesium film can be larger than that for a graphite film, but the latter can hold cesium more tightly and can work at lower cesium consumption.

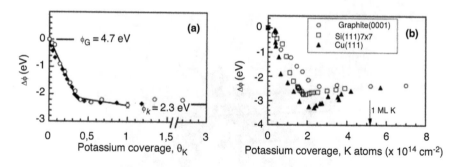

Fig. 5.139 Dependence of the work function of graphite, Si (111) and Cu (111), intercalated with potassium, on the potassium deposition time. (Reproduced from Osterlund et al. [161])

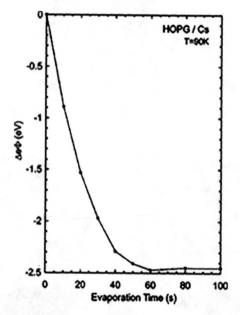

Fig. 5.140 Dependence of the work function of highly oriented pyrolytic graphite (HOPG) on cesium deposition time. 1 ML of Cs corresponds to an evaporation time of 350 s. (Reproduced from Davydov [162])

It is known that two-dimensional graphite films and pyrolytic graphite films can adsorb and retain alkali metals with high probability (adsorption coefficient ~ 1), up to a very high temperature. Intercalation of various elements in two-dimensional graphite films on metals was observed in 1981 [163] and described in detail in a review articles [164, 165]. It was shown that two-dimensional graphite films are formed on the surface of many metals that do not form bulk carbides (Pt, Ni, Re, Ir, Ru, Rh, and Pd) and metals with carbides (Mo, Ta, Nb, Ti, Zr, and Hf) without commensurability with being incorporated into the substrate, which occurs only for Ni (111). These results are explained by the high adsorption and extreme inertness of the graphite layers, which are connected to the substrate only by weak Van der Waals forces with the behavior of two-dimensional crystals [166, 167]. Carbon atoms in the layers are very strongly bound to each other, and the film has the same structure as in a graphite single crystal.

Thus atoms with low ionization potential (Cs, K, Na, etc.) are intercalated in two-dimensional graphite films. At room temperature, the flux of adsorbed atoms is divided into two almost equal parts. One half penetrates into the lattice and accumulates in intercalated states (thermal desorption α phase), and the second half remains adsorbed on the outer surface of the film (γ phase). The limiting concentration of electropositive atoms in both phases is of order a monolayer of 4×10^{14} cm^2 for cesium and potassium. The heating of such a system leads to consecutive thermal desorption of atoms: The γ phase atoms leave the surface completely at a temperature of $T \sim 700\text{--}800$ K. The intercalated atoms (α phase) are embedded in graphite layers, and this protects them from thermal desorption even at record high temperatures of $T \sim 2000$ K and higher. These atoms are desorbed only by thermal destruction of graphite islands.

The dependence of work function on cesium deposition time is similar to that shown in Fig. 5.141 [169], which was obtained for the deposition of cesium on molybdenum treated with plasma [168]. The work function decreases during the slow deposition of cesium to 2.3 eV, as in Fig. 5.142, and increases after the deposition is terminated. Probably, during the treatment of molybdenum in a plasma discharge, a carbon film is deposited and a cesium film deposited on the carbon film. It is interesting that without discharge treatment and cesium deposition, the work function is reduced only to 2.6 eV, much higher than the work function of a thick

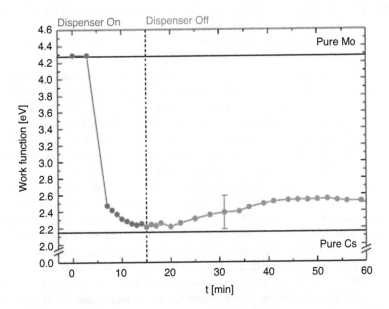

Fig. 5.141 Dependence of the work function of Mo on cesium deposition time. (Reproduced from Gutser et al. [168], with permission from AIP Publishing)

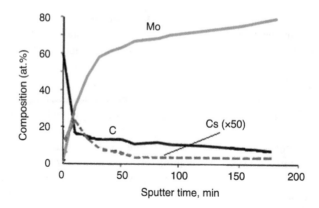

Fig. 5.142 Depth profiles of elements on the converter surface. Sputtering speed 15 nm/min. (Reproduced from Dudnikov [160])

cesium film of 2.14 eV. These measurements show that the work function of pure cesium is not achieved with intense cesium deposition for several minutes. This confirms the presence of electronegative impurities. An increase in the work function without cesium deposition confirms the deposition of impurities from residual gases. This behavior is very similar to the observation in [170], where cesium monolayer poisoning was detected with a short exposure in a higher vacuum of 10^{-6} Pa. There was no difference in the work function of Mo with deposited cesium and doped lanthanum, which should have a lower initial work function.

The black film on the collar converter, shown in Fig. 5.137, was analyzed at ORNL by low energy X-ray photo-spectroscopy. Depth profiles of the concentrations of elements of the converter are shown in Fig. 5.142. The surface is dominated by a carbon concentration of 60%, and Mo is dominant at 90% at depth. There is a high concentration of cesium.

For stable and reproducible production of surfaces with low work function, it is necessary to perform penetrating cleaning of the substrate by long-term heating to high temperature to remove bulk impurities. Fortunately, in the compact SPS, from the very beginning, it was possible to heat the cathodes up to a high temperature of 1500 K in the discharge. A discharge in air was used to purify the source of excess cesium and to purify it from carbon. Some procedures for obtaining low work function in systems with cesium similar to SPS are discussed in [171].

The increased secondary emission of H^- and D^- ions from graphite exposed to plasma has been studied and reported [172]. It is a question of emission from pure graphite, although the authors note that the graphite was surrounded by a screen of processed MACOR ceramic containing a large concentration of potassium. This potassium could be sputtered and transferred to the graphite, lowering its work function and initiating the surface plasma formation of negative ions.

The reflection of H^- ions and protons from a graphite surface coated with cesium or barium upon bombardment with H_2^+ ions has been investigated [173]. The effect of adsorption of cesium on HOPG (high-oriented pyrolytic graphite) and adsorption of barium on diamond films on the reflected particles upon bombardment with H_2^+

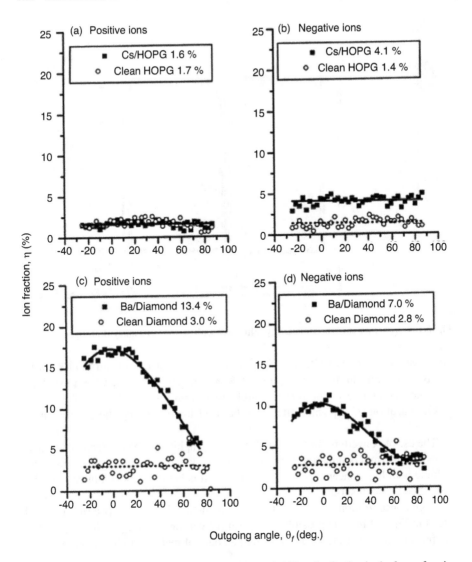

Fig. 5.143 Dependence on angle of incidence of the probability of reflection in the form of positive and negative hydrogen ions upon the bombardment by H_2^+ ions of carbon surfaces onto which cesium and barium have been deposited. (Reproduced from Gleeson and Kleyn [173])

ions of energy of 0.3 keV is shown in Fig. 5.143. Deposition of cesium on HOPG increases the probability of reflection in the form of H^- ions from 1.4% to 4.1% but does not affect the reflection of protons.

Deposition of barium on a diamond film increases the probability of reflection in the form of H^- ions from 2.8% to 7% and reflection in the form of protons from 3% to 13.4%. Reflection of H^+, H^0, and H^- ions from LaB$_6$ (100), Si (100)/Cs, graphite, and LiCl under bombardment by protons with energy 100 eV at an angle of 40° to the normal has been studied [174]. Up to 30% of protons upon reflection from LaB$_6$

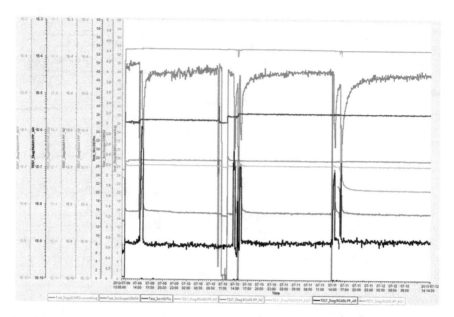

Fig. 5.144 Variation of H^- beam intensity (blue line) with periodic injection of nitrogen

(100) were converted to H^-, up to 56% protons upon reflection from Si (100) with an optimum concentration of Cs, up to 5% protons upon reflection from pyrolytic graphite, and up to 3.7% protons upon reflection from LiCl.

Optimization of deposited films can be used for improving the performances of SPSs.

The use of converter/emitter made from bulk graphite can be impractical because of its low thermal conductivity, and it can intercalate too much cesium in the bulk volume with low concentration on the surface. It is better to use a thin carbon film deposited to metal surface.

High concentration of CO and CO_2 is demonstrated by mass spectrum residual gases in Fig. 5.144. It is possible to deposit the carbon film on gas-discharge parts for preventing sputtering by adding small amount of methane to hydrogen gas.

5.19 Poisoning and Recovery of Converter Surfaces

The effect of using an admixture of heavier gases together with hydrogen on long-term persistent H^- beam generation was tested at the SNS using the RF SPS source with internal antenna. The dependence of ion source parameter during these experiments is shown in Fig. 5.144.

After long-term stable operation, when N_2 was injected, a rapid decrease of H^- beam current was observed and a slow decrease over the duration of the injection phase, followed by a fast recovery phase and then a slow recovery up to the previous level over ~2 h. During N_2 injection, a very strong increase in the mass 17 signal (ammonia NH_3) from the RGA was observed. NH_3 has a large dipole moment and can be strongly adsorbed to the cesiated surface, with an associated increase in the work function. The fast phase beam current decrease and recovery could be caused by adsorption and desorption of NH_3. The slow decrease of H^- beam current could be connected with desorption of cesium from the surface and thus increased work function. The slow recovery up to the nominal H^- beam current could be connected with recovery of optimal cesium surface concentration. The necessary cesium concentration can be delivered by Cs flow from the plasma volume and by diffusion from the volume of the carbon film on the converter cone. These experiments confirm the importance of plasma purity (absence of heavy impurities) and the potential

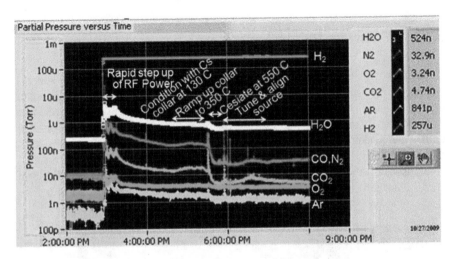

Fig. 5.145 SNS SPS out-gassing (CO_2 and CO densities are high)

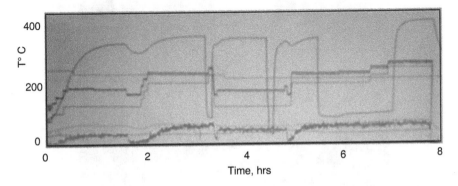

Fig. 5.146 Variation of converter temperature from 30 °C to 400 °C (brown) during SPS operation does not influence the beam intensity (blue, bottom)

for the recovery of optimal cesiation with very small admixture of cesium during the first cesiation.

The stability of H^- beam generation was significantly improved by using a longer activation period (~3.5 h instead of 2 h, at a temperature ~ 250 °C). High partial density of CO_2 and CO molecules was detected by the RGA during out-gassing of the SPS by a high-power discharge, as shown in Fig. 5.145. A carbon film can be deposited on the collar surface during this activation, with cesium intercalation into this film.

Variation of converter temperature from 30 °C to 400 °C during SPS operation is shown in Fig. 5.146. No influence of this temperature variation on the efficiency of H^- production is observed.

5.20   RF Surface Plasma Sources with External Antenna

RF sources with internal antenna, developed at LBNL, were presented as sources with an infinite service life. However, the service life of these sources turned out to be very limited. The enamel insulation of the plasma-contacting inner antenna was often damaged, and then the copper antenna itself was quickly sputtered (eroded) until water leaked into the vacuum chamber. An internal antenna with a well-developed leak is shown in Fig. 5.147.

Statistics on the service life of the internal antenna are shown in Fig. 5.148 of [175]. Nineteen percent of antennas are destroyed on the first day. Almost 50%

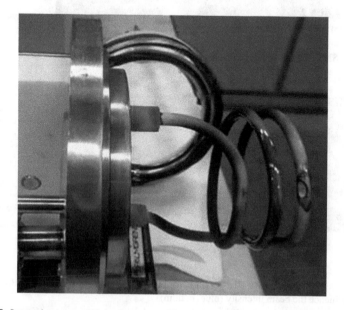

Fig. 5.147 Internal antenna with broken enamel insulation and with a water leak

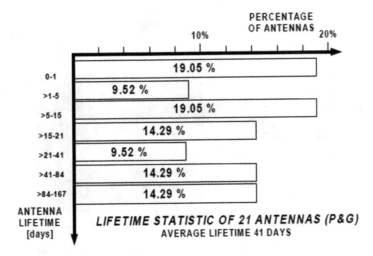

Fig. 5.148 Statistics on the service life of the internal antenna. (Reproduced from Peters [175])

Fig. 5.149 Photograph of destroyed internal antennas. (Reproduced from Peters [175])

work less than 2 weeks. The situation at the SNS was aggravated when moving to work with a high duty cycle of 6%. A photograph of destroyed antennas is shown in Fig. 5.149.

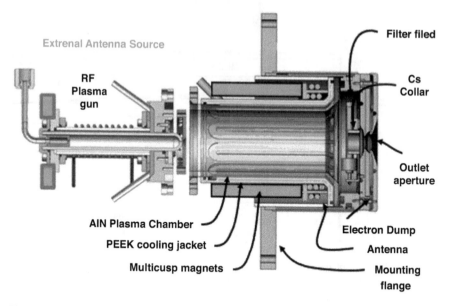

Fig. 5.150 Construction of high-frequency SPS with external antenna. (Reproduced from Welton et al. [176] with permission from AIP Publishing)

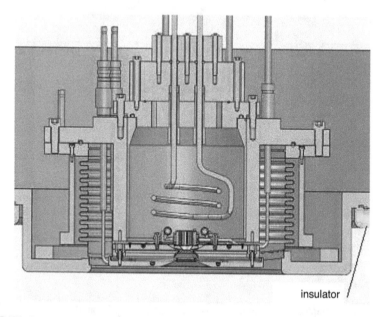

Fig. 5.151 Location of high-frequency RF SPS on a high-voltage insulator

Thus, it was decided at the SNS to start developing an RF SPS with an external antenna. First, a plasma chamber of alumina was tested, at DESY. But this could not work at high duty factor to insufficient thermal conductivity.

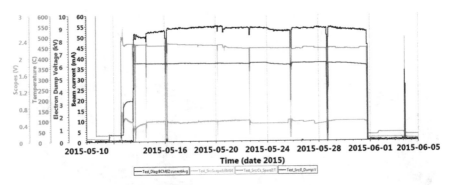

Fig. 5.152 Operation of high-frequency SPS with external antenna with H^- ion current of 56 mA for 40 days. (Reproduced from Welton et al. [176], with permission from AIP Publishing)

The design of an RF SPS with external antenna is shown in Fig. 5.150 [176, 177]. It consists of a plasma chamber made of aluminum nitride, surrounded by a cooling jacket made of PEEK (polyetheretherketone, an organic thermoplastic polymer), and a multipole magnetic wall. The antenna is wound around the cooling jacket with a copper tube insulated with heat-shrink tubing. On the left, a high-frequency plasma gun is connected to the chamber to generate the ignition plasma. To the right, the plasma electrode is connected to a cone-shaped converter, deposited with cesium, with an emission aperture 7 mm in diameter. A magnetic filter reflects energetic electrons. An electron collector gathers co-extracted electrons.

The location of the SNS RF SPS with cesiation on its high-voltage insulator is shown in Fig. 5.151. The diameter of the insulator, made of epoxy resin reinforced with glass fiber, is large, 0.7 m.

Various types of devices were tested for triggering of the high-power pulsed discharge, and eventually a high-frequency (13.56 MHz, 200 W) plasma gun was chosen with negative cathode voltage, as shown in Fig. 5.150.

For a long time, it was not possible to overcome a drop in H^- ion current with time in the RF SPS with external antenna. There was an opinion that this was due to the chamber being made of aluminum nitride, which is sputtered by the discharge. However, after degassing the chamber at high temperature, the current drop was overcome, and long-term stability of the H^- ion beam current was attained, as shown in Fig. 5.152 [178].

Production source #4 was tested on the ISTS with both the nominal $\phi \sim 7$ mm outlet aperture and then with the $\phi \sim 8$ mm aperture. An increase of ~25–30% (from 53 mA to 70 mA) in beam current was recorded with comparable power and higher H_2 flow.

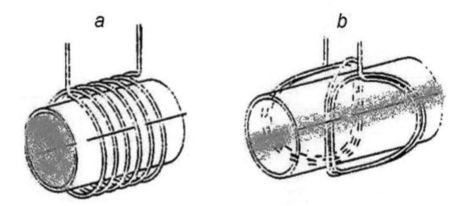

Fig. 5.153 Discharge (**a**) without solenoidal magnetic field and with solenoidal RF antenna and (**b**) with solenoidal magnetic field and with saddle-shaped antenna. (Reproduced from Vadim Dudnikov et al. [180], with permission from AIP Publishing)

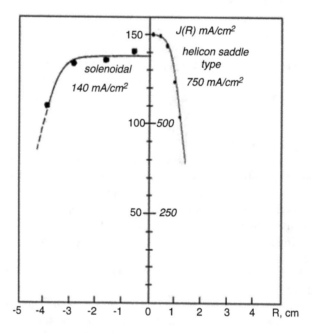

Fig. 5.154 Radial distribution of plasma density with a solenoidal antenna and without solenoidal magnetic field (on the left) and with a saddle-shaped antenna in a solenoidal magnetic field (on the right) at the same RF power (note the scale change). (Reproduced from Vadim Dudnikov et al. [180], with permission from AIP Publishing)

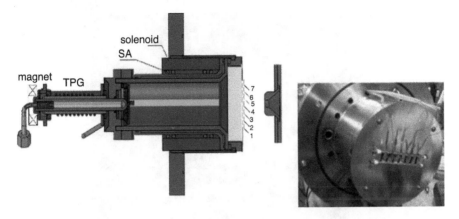

Fig. 5.155 RF plasma generator with a saddle antenna and a solenoidal magnetic field (left) and the device for determining the radial plasma density distribution (right). (Reproduced from Vadim Dudnikov et al. [180], with permission from AIP Publishing)

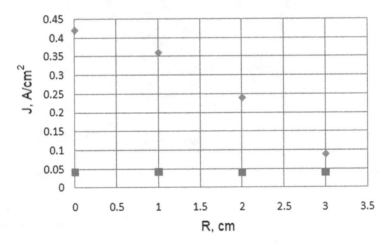

Fig. 5.156 Radial distribution of ion current density for a magnetic field of 50 G (squares) and for a field of 270 G (diamonds). (Reproduced from Vadim Dudnikov et al. [180], with permission from AIP Publishing)

5.21   RF Surface Plasma Sources with Solenoidal Magnetic Field

An RF SPS with a solenoidal magnetic field was proposed [179] to increase the efficiency of negative ion generation. To generate negative ions, the plasma needs to be exposed to a small area around the emission aperture, but in most existing RF SPSs, the plasma is uniformly exposed to the entire plasma electrode cross section, as shown in Fig. 5.153a. With a solenoidal magnetic field, a more peaked plasma density distribution is obtained, as shown in Fig. 5.153b.

Fig. 5.157 Design example of a saddle-shaped antenna with solenoidal magnetic field

The radial distributions of plasma density with solenoidal antenna (on the left), and with a saddle antenna in a solenoidal magnetic field (on the right) for the same RF power, are shown in Fig. 5.154 [180].

An RF plasma generator with saddle-shaped antenna and a solenoidal magnetic field and with an arrangement for determining the plasma density distribution along the radius is shown in Fig. 5.155. On the right side, the generator is closed by a plasma electrode in which holes have been drilled with a diameter of 1 mm in steps of 1 cm, in front of which there are collectors to which an extraction voltage of −3 kV is applied. Current to the collectors is measured by current transformers.

The radial distribution of ion current density for a magnetic field of 50 G and for a field of 270 G is shown in Fig. 5.156. At a field of 270 Gauss, the ion current density in the center is ten times greater than for the weak field, at the same RF power. Chen [181] explains the concentration of plasma density on the axis in a magnetic field by short-circuiting the flow of electrons and ions on the conducting plasma electrode. The on-axis concentration can be amplified by ion-electron secondary emission, enhanced by the addition of cesium.

One design option for a saddle-shaped antenna with a solenoidal magnetic field is shown in Fig. 5.157. In [182], a plasma generator with a saddle-shaped antenna

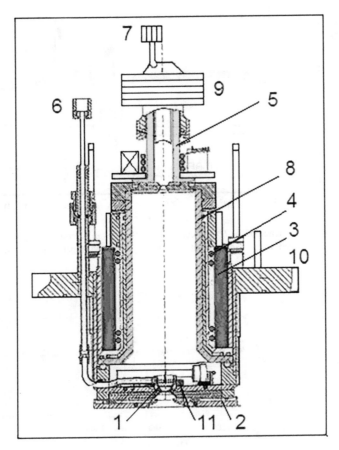

Fig. 5.158 RF SPS with saddle-shaped antenna and solenoidal magnetic field. (Reproduced from Vadim Dudnikov et al. [182], with permission from AIP Publishing). 1, collar converter with emission aperture and magnetic filter; 2, extraction electrode and electron dump; 3, solenoid; 4, saddle-shaped antenna; 5, igniting plasma gun; 6, external cesium source; 7, gas tube; 8, aluminum nitride plasma chamber; 9, permanent magnet; 10, vacuum flange; and 11, ferromagnetic insert

and a solenoidal magnetic field was attached to the plasma electrode of the high-frequency SPS as shown in Fig. 5.158.

A hollow cathode discharge was used to ignite the pulsed discharges. A cesium source used with an ampoule of metallic cesium, but the ampoule was not opened. The time dependence of the H^- ion current with the change in the discharge parameters is shown in Fig. 5.159.

For pulsed RF power 24 kW, the H^- ion current measured by a current transformer and by a Faraday cylinder was 8 mA. After 4 h of discharge heating the SPS and collecting cesium on the converter, the H^- current increased to 42 mA and remained stable at this level. A linear decrease over a period of ~25 h is associated with a decrease in the solenoidal magnetic field from 270 G to 50 G. The increase in the H^- ion current to 50 mA and to 60 mA is associated with an increase in the

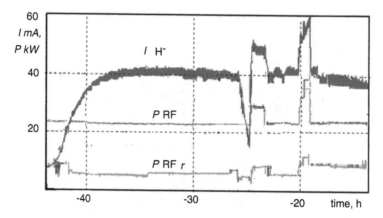

Fig. 5.159 Time dependence of the H^- ion current with changes in solenoidal magnetic field and RF power. (Reproduced from Vadim Dudnikov et al. [182], with permission from the AIP Publishing)

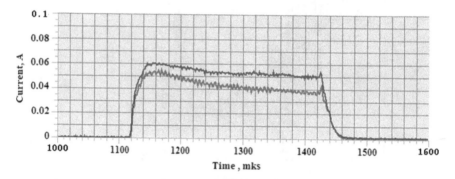

Fig. 5.160 Oscillograms of H^- ion current as monitored by a current transformer and by a Faraday cup. (Reproduced from Vadim Dudnikov et al. [182], with permission from AIP Publishing)

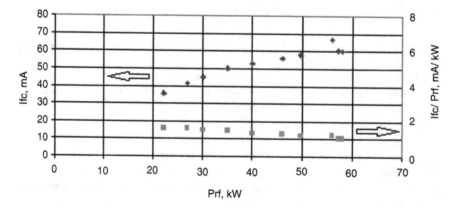

Fig. 5.161 Dependence of H^- ion current and energy efficiency on RF power. (Reproduced from Vadim Dudnikov et al. [182], with permission from AIP Publishing)

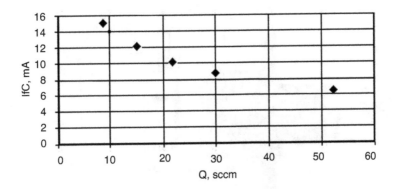

Fig. 5.162 Dependence of the intensity of the ion beam H^- on the gas flow rate

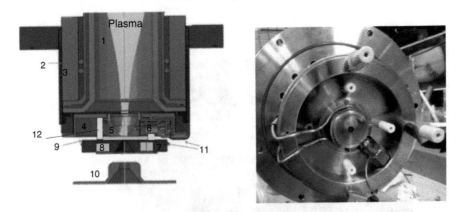

Fig. 5.163 Illustration of the next version of RF SPS with saddle antenna and solenoidal magnetic field (left) and photograph of the plasma electrode of the SPS (right). 1, aluminum nitride gas-discharge chamber; 2, RF saddle-shaped antenna; 3, solenoid; 4, plasma electrode; 5, conical collar converter; 6, cesium oven; 7, extraction electrode; 8, permanent magnet; 9, tungsten plate; 10, accelerating electrode; 11, cooling tubes; and 12, insulators

pulsed RF power to 30 and 40 kW. Oscillograms of the H^- current from a current transformer and from a Faraday cup are shown in Fig. 5.160.

The dependence of H^- ion current and energy efficiency on RF power is shown in Fig. 5.161. The H^- ion current reaches 67 mA and the energy efficiency decreases from 1.6 mA/kW to 1.2 mA/kW.

The dependence of the intensity of the ion beam on gas flow is shown in Fig. 5.162. At a gas flow rate of 20–30 sccm and above, a significant fraction of H^- ions is destroyed on the gas escaping from the emission aperture.

An illustration of a version of high-frequency SPS with a saddle-shaped antenna and solenoidal magnetic field and a photograph of the plasma electrode of the source are shown in Fig. 5.163. It consists of an aluminum nitride gas-discharge chamber 1, cooling jacket made of KEEP polymer, RF saddle antenna 2, solenoid 3, plasma electrode 4, conical collar converter 5, cesium oven 6, extraction electrode 7,

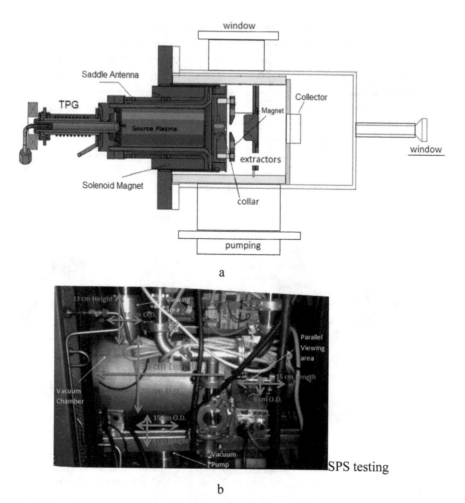

Fig. 5.164 (**a**) Placement of RF SPS in the vacuum chamber. (**b**) Photograph of vacuum chamber of stand for RF SPS testing. (Reproduced from V. Dudnikov et al. [183], with permission from AIP Publishing)

permanent magnets 8, tungsten plate 9, accelerating electrode 10, cooling tubes 11, and insulator 12.

Placement of the RF SPS in the vacuum chamber of a small test bench is shown in Fig. 5.164a. Photograph of vacuum chamber of stand for RF SPS testing is shown in Fig. 5.164b. One can observe the extraction and formation of the beam, through the upper window. Heating of the collector by the beam can be observed through the side window to the right.

The results of a simulation of H^- ion extraction from an RF SPS with a realistic magnetic field distribution are shown in Fig. 5.165. At the top of the figure is the distribution of transverse magnetic field with amplitude up to 0.1 Tesla.

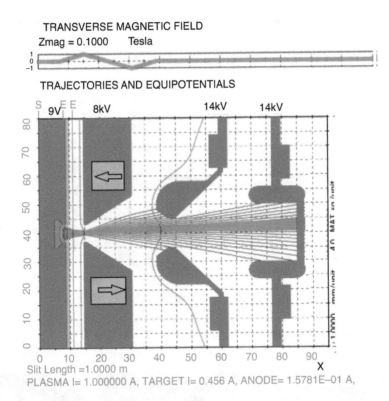

Fig. 5.165 Results of a simulation of the extraction of H^- ions from RF SPS. (Reproduced from V. Dudnikov et al. [183], with permission from AIP Publishing)

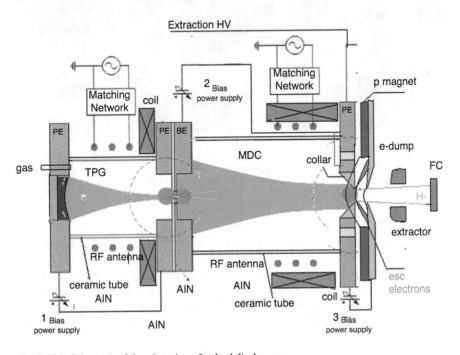

Fig. 5.166 Schematic of the triggering of pulsed discharges

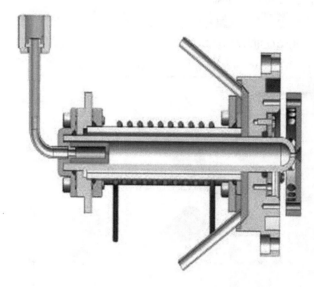

Fig. 5.167 Design of the triggering plasma gun

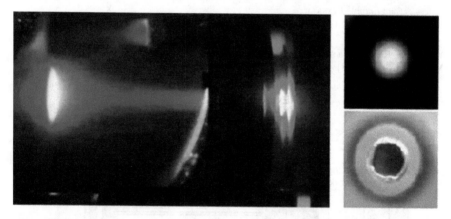

Fig. 5.168 Photograph of a beam of positive ions on a hot tantalum collector; the emission aperture is on the right side of the photograph. On the right side is a photograph of the glow of the back side of the collector and the collector after having been melted by the beam. (Reproduced from Dudnikov et al. [183], with permission from AIP Publishing)

Ignition of the pulsed discharges is done by a triggering plasma gun with high-frequency discharge (13.56 MHz, 200 W). A schematic of the triggering setup is shown in Fig. 5.166.

The design of the triggering plasma gun is shown in Fig. 5.167. The gas-discharge chamber is made of ceramics Al_2O_3 and is water cooled. On the left side is a tungsten cathode through which gas is supplied. The triggering plasma gun is connected to the main discharge chamber through an aperture of diameter 2 mm behind which

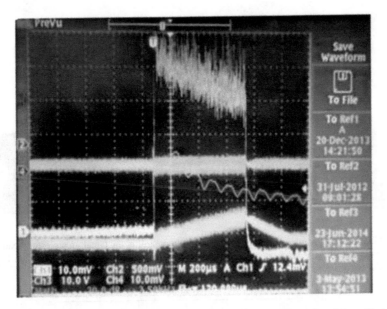

Fig. 5.169 Oscillogram of the positive ion current to the collector. (Reproduced from Dudnikov et al. [183], with permission from AIP Publishing)

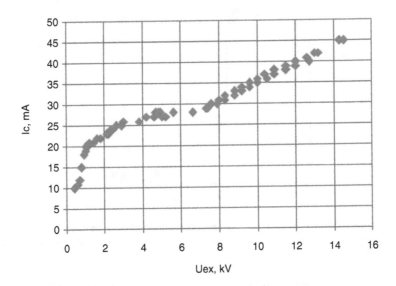

Fig. 5.170 Dependence of the positive ion current to the collector on the extraction voltage. (Reproduced from Dudnikov et al. [183], with permission from AIP Publishing)

is located a tungsten anode. The discharge is supported by high-frequency current in a spiral antenna, provided at 13.56 MHz at a power of 200 Watts. A potential of −300 V is applied to the cathode. The gun was tested in operation up to 1200 W.

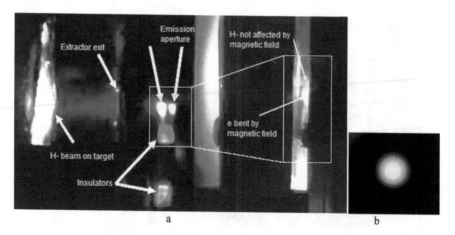

Fig. 5.171 (**a**) Photograph of the negative ion beam bombarding the tantalum collector, heated to high temperature by the H^- beam; (**b**) the back side of collector heated by negative ion beam. (Reproduced from Dudnikov et al. [183], with permission from AIP Publishing)

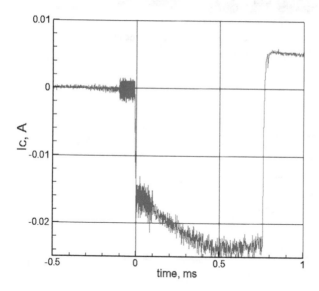

Fig. 5.172 Oscillogram of negative ion current to the collector. The beam current reaches 30 mA at an RF power to the plasma of 4 kW. (Reproduced from Dudnikov et al. [183], with permission from AIP Publishing)

The main discharge is supported by an Advanced Energy Apex 550 RF generator of power up to 5.5 kW. At the first stage, the extraction of positive ions was tested. A photograph of a beam of positive ions on a hot tantalum collector is shown in Fig. 5.168. The emission aperture is visible at the right.

An oscillogram of the positive ion current to the collector is shown in Fig. 5.169.

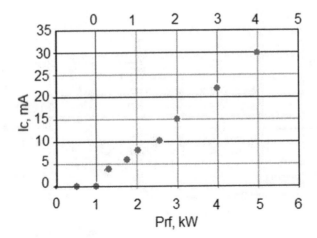

Fig. 5.173 Dependence of collector ion current on RF power. The lower scale is the RF power from the generator; the upper scale is the RF power into the plasma. (Reproduced from Dudnikov et al. [183], with permission from AIP Publishing)

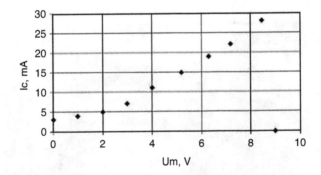

Fig. 5.174 Dependence of ion current to the collector on the solenoid voltage. (Reproduced from Dudnikov et al. [183], with permission from AIP Publishing)

The collector current is 50 mA at RF power to the discharge of 1.5 kW. The ion energy cost is 33 mA/kW or 30 keV/ion. The dependence of the positive ion current to the collector on the extraction voltage is shown in Fig. 5.170.

The extraction of negative ions was investigated by reversal of the polarity of the extraction and accelerating voltage. Primary cesium was produced by heating tablets of a mixture of cesium chromate and titanium by the discharge plasma. The tablets are located on a shelf in the plasma electrode. After cesiation, the negative ion current to the collector increased from 1 mA to 10 mA, and the current in the extraction gap decreased from 150 mA to 50 mA. A photograph of the negative ion beam and the tantalum collector heated to high temperature by the H^- beam is shown in Fig. 5.171a. The back side of collector heated by negative ion beam is

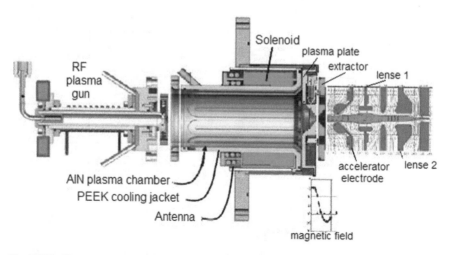

Fig. 5.175 Illustration of the RF SPS and beam transport path; the transverse magnetic field distribution is shown in the inset. (Reproduced from Dudnikov et al. [183], with permission from AIP Publishing)

Fig. 5.176 Test stand and control system for the RF SPS at the SNS. (Reproduced from Dudnikov et al. [183], with permission from AIP Publishing)

shown in Fig. 5.171b. The insulators luminesce in pink under electron bombardment.

The insert shows the trajectories of the accompanying electrons bent by the magnetic field. The tantalum collector is heated to a temperature to 1000 °C. An oscillogram of the negative ion current to the collector is shown in Fig. 5.172. The beam current reaches 30 mA at an RF power to the plasma of 4 kW. The dependence of collector ion current on RF power is shown in Fig. 5.173, and the dependence of collector ion current on the solenoid voltage is shown in Fig. 5.174. The collector current increases with increasing solenoidal magnetic field, but the discharge ceases at a solenoid voltage greater than 8.5 V because the electron drift velocity in this magnetic field is less than the critical velocity for molecular excitation.

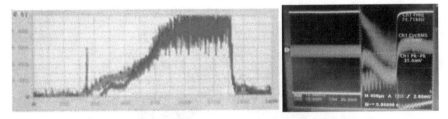

Fig. 5.177 Increase of beam current, accelerating electrode current, and lens current during cesiation. (Reproduced from Dudnikov et al. [183], with permission from AIP Publishing)

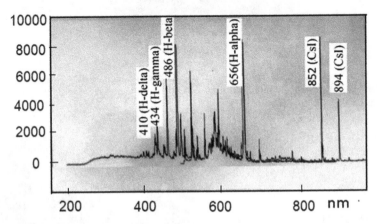

Fig. 5.178 Increase of cesium emission lines from the discharge upon cesiation. (Reproduced from Dudnikov et al. [183], with permission from AIP Publishing)

5.22   Testing RF Surface Plasma Sources with Saddle Antenna and Magnetic Field

Tests of the high-frequency SPS with saddle-shaped antenna were conducted on a large test bench at the SNS in 2016 [183]. The configuration of the RF SPS with the beam transport channel (LEBT) to the RFQ is shown in Fig. 5.175. The RF ion source consists of an AlN ceramic chamber with a PEEK (polyetheretherketone polymer) cooling jacket. An RF-assisted triggering plasma gun (TPG) is attached on the left side, and the plasma electrode with extraction system is attached to the right side. The discharge chamber is surrounded by a saddle (or solenoidal) antenna. A solenoid for the magnetic field is located around the antenna; this solenoid has no high conductivity shielding and introduces a large RF energy loss. The extraction system consists of the plasma electrode at a potential of up to −65 kV, extraction electrode at 8 kV or less, and accelerating electrode at up to +20 kV. The beam transport system consists of a first electrostatic lens at a potential of up to −47 kV, a ground shield, a second electrostatic lens at a potential up to −47 kV, and an input wall to the RFQ behind which is located in a current transformer and a Faraday cup

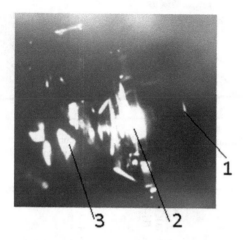

Fig. 5.179 Glow of the beam transport system electrodes during cesiation. (Reproduced from Dudnikov et al. [183], with permission from AIP Publishing). 1, emission aperture; 2, first electrostatic lens; and 3, second electrostatic lens

Fig. 5.180 White CsH powder that has not interacted with the plasma on the chamber surface. (Reproduced from Dudnikov et al. [183], with permission from AIP Publishing)

Fig. 5.181 Black film around the emission aperture of the conical converter. (Reproduced from Dudnikov et al. [183], with permission from AIP Publishing)

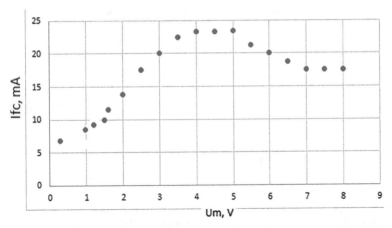

Fig. 5.182 Dependence of H^- ion current on solenoid voltage (i.e., magnetic field). (Reproduced from Dudnikov et al. [183], with permission from AIP Publishing)

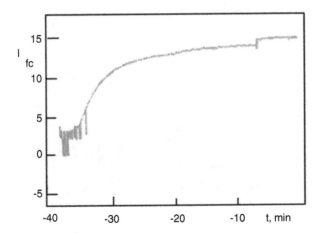

Fig. 5.183 Time dependence of H^- ion current upon cesiation. (Reproduced from Dudnikov et al. [183], with permission from AIP Publishing)

for measuring the beam current. The pulsed RF discharge is supported by RF current in the antenna generated by a 2 MHz RF generator with power up to 80 kW connected to the antenna through an isolation transformer and matching device.

A photograph of the SNS test stand with the control system is shown in Fig. 5.176.

The pulsed discharge is triggered by a plasma gun that operates at a frequency of 2 MHz and a power of 3.8 kW, with current in the antenna 120 A and voltage across the antenna 6.5 kV.

At a frequency of 13.56 MHz, the discharge can be started at power of 0.5 kW, at a current of 14 A, and at a voltage of 5 kV. Cesium is injected into the discharge by heating cesium chromate/titanium tablets by the plasma discharge and is accompanied by an increase in the negative ion current as shown in Fig. 5.177. The increase

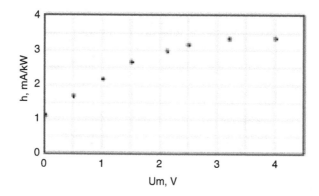

Fig. 5.184 Dependence of H^- generation efficiency on solenoid voltage (i.e., magnetic field). (Reproduced from Dudnikov et al. [184], with permission from AIP Publishing)

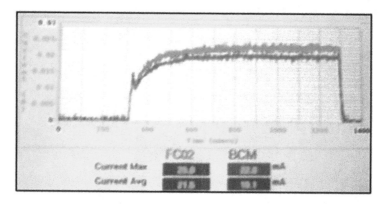

Fig. 5.185 Oscillogram showing a beam current of 25 mA. (Reproduced from Dudnikov et al. [184], with permission from AIP Publishing)

in the emission of cesium lines from the discharge with cesiation is shown in Fig. 5.178.

The glow of the electrodes of the beam transport system during cesiation is shown in Fig. 5.179. When opening the gas-discharge chamber and filling with argon, white CsH powder is visible on the surface of the chamber that has not interacted with the plasma; see Fig. 5.180.

Tablets of cesium chromate with titanium are visible in Fig. 5.180. A black film that has formed around the emission aperture of the conical converter is shown in Fig. 5.181.

The dependence of ion current on voltage across the solenoid (i.e., proportional to magnetic field strength) is shown in Fig. 5.182. The H^- ion current increases with solenoid voltage up to 4 V and then falls off.

After cesiation, the current in the extraction gap decreases from 120 mA to 8 mA. The time dependence of the H^- ion current during cesiation is shown in

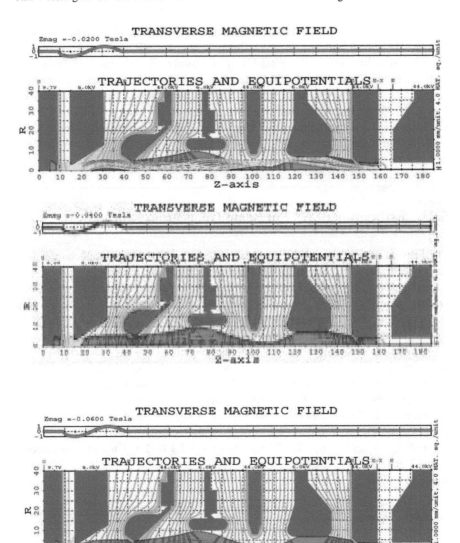

Fig. 5.186 Results of modeling of extraction and transport of the H^- ion beam for different transverse magnetic fields. The yellow lines are electrons; red lines are H^- ions. (Reproduced from Dudnikov et al. [183], with permission from AIP Publishing)

Fig. 5.183. The beam current increases from 3 mA to 13 mA at a constant RF power of 9 kW into the plasma.

The dependence of H^- generation efficiency on solenoid voltage is shown in Fig. 5.184.

An oscillogram indicating a beam current of 25 mA is shown in Fig. 5.185.

Extraction of the beam at a transverse magnetic field strength of 200 Gauss was tested. The beam current reached 70 mA for an RF generator power of 50 kW. Modeling using the PBGAN (PPG Biometric Generative Adversarial Network) code showed that at 200 Gauss, electrons can be focused and transported to the current transformer and Faraday cup. The simulation results are shown in Fig. 5.186. With a transverse field of 600 Gauss, the electrons are not captured in the ion beam. A test with a helium discharge showed that for a transverse field of 200 Gauss, electrons are trapped in the beam, and at 346 Gauss, electrons are not captured in the ion beam.

5.23   Estimation of H^- Ion Beam Generation Efficiency

The forward power from the RF generator is measured by a directional coupler and calculated by the following formula: $P_{rf} = 42.7 <I_{rms}>^2$ kW, where $<I_{rms}>$ is the rms current in V ($<I_{rms}>$ is read from the oscilloscope in Volts, and here we give these voltage values).

Before triggering the discharge, all the power is dissipated in the isolation transformer, antenna, solenoid, and matching network. For our case $<I> = 0.293$ V, 3.66 kW, antenna current $<I>_{ant} = 83.3$ A, and the antenna voltage $V = 6480$ V. The active resistance of network + antenna+ solenoid is $R = 2P/<I>_{ant}^2 = 2 \times 3660/(83.3)^2 = 1$ Ohm. This resistance is high because of RF loss in the solenoid. For a discharge with $<I> = 0.599$ V, the total power $P_{rf} = 15.3$ kW is dissipated in the discharge P_d, in the antenna + network P_{ant}, and in the surrounding antenna solenoid P_{sol}: $P_{rf} = P_d + P_{ant} + P_{sol}$. For antenna current $<I>_{ant} = 136$ A, $P_{ant} = R <I>_{ant}^2/2 = 10$ kW. $P_d = 15.3-10 = 5.3$ kW. For Faraday cup current $I_{fc} = 17$ mA, the efficiency of ion beam current generation is $h = 3$ mA/kW at a solenoid voltage $U_m = 2.11$ V.

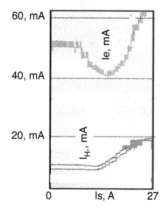

Fig. 5.187 Dependence of ion beam current and electron current on solenoid current. (Reproduced from Dudnikov et al. [184], with permission from the AIP Publishing)

For $\langle I \rangle$ = 0.872 V, P_{rf} = 34 kW, $\langle I \rangle_{ant}$ = 194.4 A, P_{ant} = 20 kW, and I_{fc} = 16 mA, then h = 16/14 = 1.14 mA/kW at U_m = 0.

For $\langle I \rangle$ = 0.963 V, P_{rf} = 41.7 kW. $\langle I \rangle_{ant}$ = 250 A, P_{ant} = 34.3 kW, P_d = 41.7–34.3 = 7.4 kW, and I_{fc} = 25 mA, then h = 25/7.4 = 3.37 mA/kW at U_m = 3.2 V.

The volume of the collar is 29 cm³, its mass is 290 g, and the specific heat of Mo is c = 0.255 J/g.K. The specific heat of the collar is 75 J/C. The collar cools after switching off the discharge at a rate of 0.7 °C/s. The power loss from the collar is 52 W (pulsed power 868 W from P_{rf} = 34.2 kW from the RF generator (14 kW in plasma) at U_m = 1.68 V). The Faraday cup current (mA) increases during cesiation from 3 mA to 14 mA at constant RF power of 40% of the maximum available power (9 kW into the plasma), as shown in Fig. 5.157. The efficiency of plasma generation and the stability of the discharge can be increased by control of the plasma potential [185]. The dependence of the *H⁻* production efficiency h (in mA/kW) on the solenoid voltage for pulsed mode operation at 2 MHz is shown in Fig. 5.183. It can reach h = 3.4 mA/kW at U_m = 4 V. The dependence of electron current I_e and collector current I_H^- on the solenoid current I_s are shown in Fig. 5.187 [184]. Fort low I_s, I_H^- is 8 mA and I_e is ~50 mA. With increased solenoid current, the electron current decreases to 40 mA, but the collector current increases. Increasing the solenoid current above 15 A causes the electron current to increase up to 60 mA and the

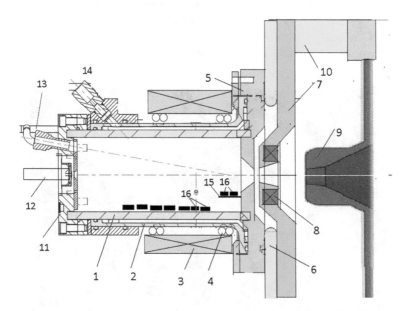

Fig. 5.188 Schematic of high-frequency SPS for continuous operation. (Reproduced from Dudnikov et al. [184], with permission from AIP Publishing). 1, gas-discharge chamber of aluminum nitride; 2, cooling jacket; 3, solenoid; 4, saddle antenna; 5, plasma electrode; 6, insulator; 7, extractor; 8, permanent magnets; 9, accelerating electrode; 10, main insulator; 11, rear flange; 12, gas supply tube; 13, window; 14, input-output cooling water; 15, shelf for cesium tablets; and 16, cesium tablets

collector current to increase to 20 mA at a solenoid current of 26 A (I_e is electron current +collector current).

Fort low I_s, I_H^- is 8 mA and electron current is ~40 mA. With increased solenoid current, the electron current decreases to 30 mA, but the collector current increases. Increasing the solenoid current above 15 A causes the electron current to increase up to 40 mA and the collector current to increase to 20 mA at a solenoid current of 26 A. Ratio of electron current to H-current becomes 2.

5.24   RF Surface Plasma Source Operation in Continuous Mode

For continuous operation, the RF SPS was simplified [184]. The triggering of the discharge can be carried out with increased gas pressure and then the gas supply can be reduced. Thus there is no need for a triggering plasma gun. A schematic of a high-frequency SPS for continuous operation is shown in Fig. 5.188. It consists of an aluminum nitride gas-discharge chamber 1, cooling jacket from PEEK 2, solenoid 3, saddle-shaped antenna 4, plasma electrode 5, extractor insulator 6, extractor 7 with permanent magnets 8, accelerating electrode 9, main insulator 10, rear flange 11 with gas supply tube 12 and window 13, cooling water input-output 14, and shelves for cesium tablets 15 with cesium tablets 16. The tablets pressed from a mixture of cesium chromate (bichromate) with titanium or zirconium and release pure cesium upon heating to a temperature of 300 °C (for bichromate) or 550 °C

Fig. 5.189 Photograph of the isolation transformer. (Reproduced from Dudnikov et al. [184], with permission from AIP Publishing)

Fig. 5.190 Photograph of the matching network. (Reproduced from Dudnikov et al. [184], with permission from AIP Publishing)

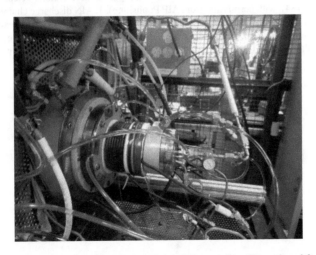

Fig. 5.191 Photograph of the simplified RF SPS for CW operation. (Reproduced from Dudnikov et al. [184], with permission from AIP Publishing)

(for chromate). The tablets are distributed in the gas-discharge chamber in regions with different discharge power, so that at first the cesium begins to evaporate from the most heated tablets and then from the less heated tablets at higher discharge power. Tablets emitting heat as a black body at a temperature of 925 K radiate a specific power of 4 W cm^{-2} (8 W/cm^2 from two sides). With an area of 150 cm^2 receiving of radiation, the power required to inject the cesium is ~1.2 kW.

The discharge is driven by RF current in the antenna from an RF generator (MKS Sure Power QL6513A-OF03) at frequency 13.56 MHz and power up to 6.5 kW, connected to the antenna through an isolation transformer and a matching network. A photograph of the isolation transformer is shown in Fig. 5.189, of the matching device

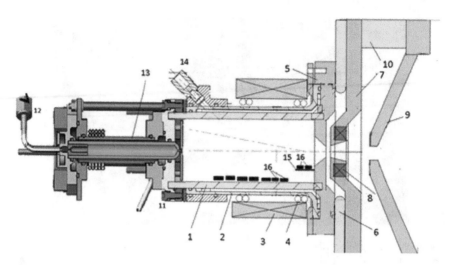

Fig. 5.192 Schematic of the simplified RF SPS for pulsed accelerators. (Reproduced from Dudnikov et al. [184], with permission from AIP Publishing). 1, AlN discharge chamber; 2, PEEK cooling jacket; 3, solenoid; 4, saddle antenna; 5, plasma electrode; 6, insulator; 7, extraction electrode; 8, permanent magnet; 9, ground electrode; 10, main insulator; 11, back flange; 12, gas tube; 13, triggering plasma gun; 14, cooling water inlet outlet; 15, shelf, and 16, cesium pellets

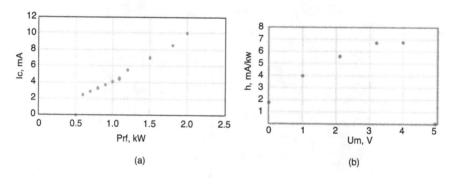

Fig. 5.193 (a) Dependence of H^- ion beam current on RF power, bottom scale RF power from generator, and upper scale RF power into plasma. (b) Dependence of H^- ion generation efficiency on solenoid voltage. (Reproduced from Dudnikov et al. [184], with permission from AIP Publishing)

in Fig. 5.190, and of the simplified RF SPS for continue operation in Fig. 5.191. A schematic of the simplified RF SPS for pulsed accelerators is shown in Fig. 5.192.

The dependence of the negative ion beam current on the power of the high-frequency discharge is shown in Fig. 5.193a. A beam current of 10 mA is obtained with an RF power into the plasma of 1.5 kW. The dependence of H^- generation efficiency on solenoid voltage is shown in Fig. 5.193b.

CW operation of the saddle antenna (SA) SPS was tested with RF power up to ~2 kW from the generator (~1.5 kW into the plasma) with ion current measured at

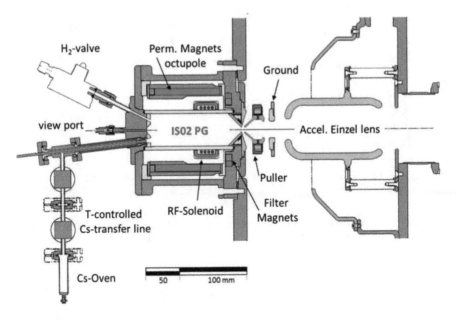

Fig. 5.194 Schematic of the RF SPS with cesiation at CERN. (Reproduced from Scrivens et al. [187])

the collector up to I_c = 10 mA. Long-term operation was tested with 1.8 kW from the RF generator (~1.3 kW into the plasma and 0.5 kW dissipated in the antenna and matching network) with collector current and extractor current of I_c = 9 mA, I_{ex} ~ 15 mA (U_{ex} = 8 kV, U_c = 15 kV). This mode of operation was tested for 50 days. After this test, the SA SPS was assessed as fully serviceable.

The collector current increases with increasing magnetic field up to U_m ~ 4 V and decreases with further increase of magnetic field because the plasma flux is then compressed to the emission aperture and interaction of plasma with the collar surface is decreased. The specific power efficiency of negative ion beam production in CW mode is up to S_{pe} = 20 mA/cm^2 kW. (In the existing RF SPS, the S_{pe} is about ~4–6 mA/cm^2.kW; in the TRIUMF filament arc discharge negative ion source, the best S_{pe} is about 2 mA/cm^2.kW. In the compact Penning discharge SPS, the S_{pe} is 100 mA/cm^2 kW.)

5.25 RF Surface Plasma Sources at CERN

To increase the brightness of the proton beam accelerated in the CERN accelerators, it was decided to use charge-exchange injection. For this purpose, Linac4 was developed for an energy of 160 MeV, which replaces the 50 MeV Linac2. For Linac4, it was decided to develop a source of negative ions. The required intensity

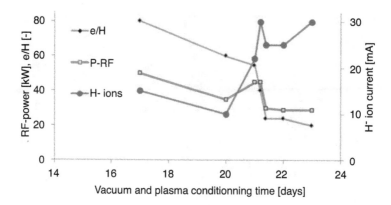

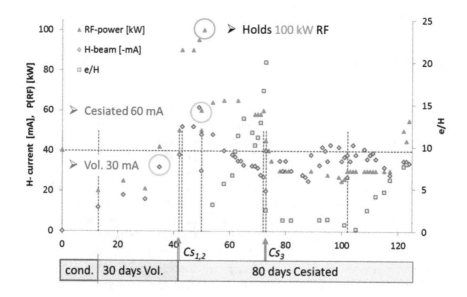

Fig. 5.195 Evolution of negative ion beam current, and ratio of accompanying electron current to negative ion current e/H^-, during an extended conditioning period. (Reproduced from Scrivens et al. [187])

Fig. 5.196 Results of the cesiation in RF SPS at CERN. (Reproduced from Scrivens et al. [187])

of the H^- ion beam was 80 mA, with emittance 0.25 π mm.mrad in 1 ms pulses with a repetition rate of 2 Hz. At first, it was planned to use an RF source without cesium, as developed at DESY, and to increase the RF power up to 100 kW [186]. But it turned out that only 10 mA of H^- ions were produced from it, with an accompanying electron current 300 times greater [187]. It was therefore decided to switch to the use of an RF SPS with cesiation. A schematic of the CERN RF SPS with cesiation is shown in Fig. 5.194 [188].

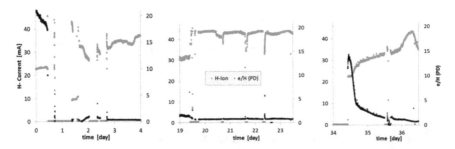

Fig. 5.197 Examples of stable operation of high-frequency SPS with cesiation. More detailed behavior of the current of negative ions and the e/H^- ratio during cesiation (left), during stable operation after the third cesiation (middle), and restoration of the cesium regime after a long interruption in operation (right). Red, H^- ion current (mA), and black, e/H^- ratio. The H^- ion current and the ratio of ion to electron current measured at the puller (extractor) electrode (dump electrode), e/H (PD), are shown. (Reproduced from Scrivens et al. [187])

The source consists of a plasma generator with Al_2O_3 discharge chamber, solenoidal antenna in epoxy insulation, multipole magnetic wall, gas valve, optical windows, cesium supply system, plasma electrode with conical converter, magnetic filter, extracting electrode grounded electrode, and accelerating electrostatic lens.

The evolution of the negative ion beam current and the ratio of accompanying electron current to negative ion current of e/H^- during conditioning are shown in Fig. 5.195. Over 20 days of discharge conditioning, the H^- ion current increased from 10 mA to 30 mA, the accompanying electron current dropped from 1 A to 0.6 A, and the e/H^- dropped from 100 to 20. This indicates that the discharge extracts easily ionizable impurities from the electrodes and ceramic, which are then transferred to the converter and catalyzed by surface plasma formation to form negative ions, as in the RF SPS at DESY.

Cesiation was performed by feeding cesium from an external source with an ampoule of metallic cesium. The results of cesiation in the CERN RF SPS are shown in Fig. 5.196.

As a result of cesiation, the negative ion current increased to 60 mA at an RF power of 95 kW, and the e/H^- ratio dropped to 1. For a long time, the H^- current was 40 mA at an RF power of 30 kW. The behavior of the negative ion current and the e/H^- ratio is shown in more detail in Fig. 5.197, during cesiation, during stable operation after the third cesiation, and during restoration of the cesiating regime after a long break in operation.

Improving the CERN RF SPS with cesiation has been described [189]. The cusp-free H^- source was conditioned in volume mode; after seven days, the H^- beam intensity reached 29 mA with an electron-to-ion ratio (e/H) of 41 ± 3 and an RF power yield of 0.8 mA/kW. The electron current contributes to space charge forces in the non-compensated beam extraction region. At low energy, an e/H value close to the H^- to electron velocity ratio $(m_H/m_e)^{1/2} = 42.87$ doubles the H^- beam-induced space charge. After cesiation, the electron-to-ion ratio dropped below 1, and a 62 mA H^- beam was extracted, at an RF power yield of 2 mA/kW. The peak current

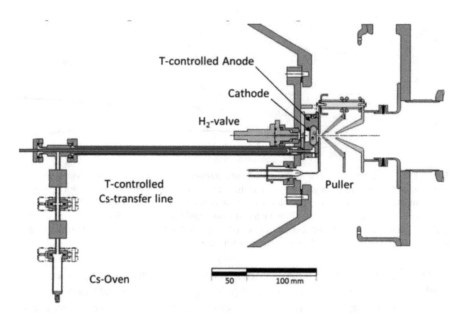

T-controlled Anode

Cathode

H_2-valve

T-controlled
Cs-transfer line

Puller

Cs-Oven

50   100 mm

Fig. 5.198 Schematic of adaptation of magnetron SPS to the CERN RFQ. (Reproduced from Scrivens et al. [187])

corresponds to plasma chambers equipped with Halbach offset octupole cusps and about 1 mA/kW RF power yields. Shortly after cesiation, the distance between the plasma electrode and the extractor electrode was reduced from 4.4 to 3.4 mm. Two measurement periods (7 and 9 h), dedicated to parameter scans of H_2 injection, RF power and RF frequency, during which the e/H was 3.5 and 4.5 provided 1800 sets of data. The coverage of operational parameter space provides experimental background for modeling, such as RF coupling studies. Automation of the source operation demands an operational phase space with identified boundaries within which the system reacts to parameter changes in a predictable and preferably smooth and continuous manner. From the data set, we can identify guidelines that will refine the control algorithms. For example, the RF-coupling is best matched for an RF phase equal to 0; the H^- yield is very responsive to hydrogen injection fluctuations (a 5% EM valve opening corresponds to a 30% variation of H^- beam intensity); and excessive hydrogen injection will increase plasma light emission but reduce the H^- beam intensity. The IS03 cesiated surface sources operate with monthly injection of typically 5 mg of metallic cesium, and during the cesiation process, the source is sealed off from the linac by closing a vacuum valve located in the low energy beam transport (LEBT) section. The IS03 source is re-cesiated once an e/H ratio of 8 is reached, corresponding to a 20% increase of the space charge forces in the extraction region. The SNS source plasma is maintained at lower density by a secondary low-power CW RF circuit to keep the plasma ignited between the 60 Hz high-power RF pulses, which keeps the Cs ionized and possibly reduces the loss rate through the plasma electrode aperture. At CERN's low repetition rate, the hydrogen is pulsed and the

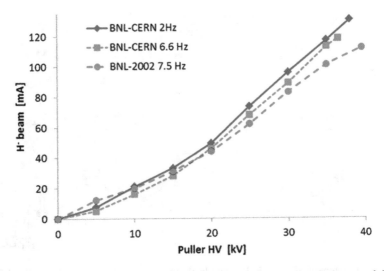

Fig. 5.199 Results of the CERN magnetron test at BNL. (Reproduced from Scrivens et al. [187])

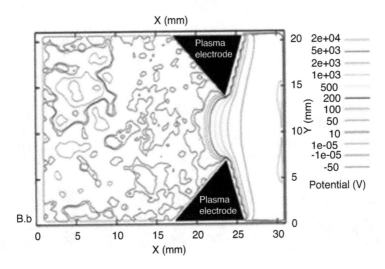

Fig. 5.200 Results of meniscus simulation. (Reproduced from Serhiy Mochalskyy et al. [190])

plasma extinguishes between pulses. Metallic cesium's high vapor pressure is likely to contribute to Cs losses from the plasma chamber through the plasma electrode aperture. Larger amounts of cesium could be injected monthly; disposing of a Cs inventory suitable to compensate losses keeps a low *e/H* as observed on an IS02 prototype.

As a backup ion source for CERN, the use of a magnetron-planotron SPS from BNL was considered. The adaptation of the BNL magnetron SPS at the CERN test stand is shown in Fig. 5.198.

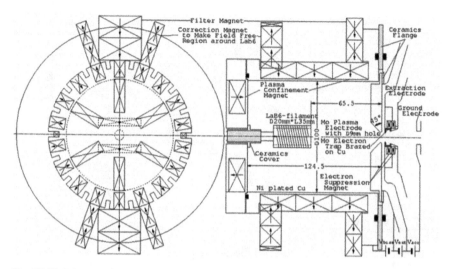

Fig. 5.201 Diagram of SPS at J-PARC with lanthanum hexaboride cathode. (Reproduced from Oguri et al. [191])

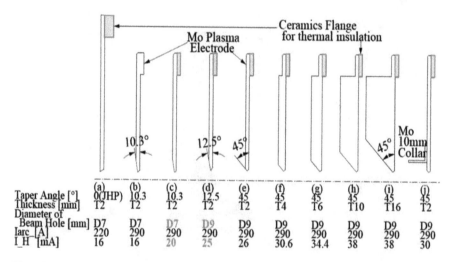

	(a)	(b)	(c)	(d)	(e)	(f)	(g)	(h)	(i)	(j)
Taper Angle [°]	0(JHP)	10.3	10.3	12.5	45	45	45	45	45	45
Thickness [mm]	T2	T2	T2	T2	T2	T4	T6	T10	T16	T2
Diameter of Beam Hole [mm]	D7	D7	D7	D9	D9	D9	D9	D9	D9	D9
Iarc [A]	220	290	290	290	290	290	290	290	290	290
I_H [mA]	16	16	20	25	26	30.6	34.4	38	38	30

Fig. 5.202 Various forms of plasma electrodes with negative ion currents. The maximum current was obtained with a plasma electrode in the shape of an "i." (Reproduced from Akira Ueno [192])

The results of testing the CERN magnetron at BNL are shown in Fig. 5.199. An H^- ion beam current of up to 130 mA was obtained at the required pulse repetition rate of 2 Hz.

Development of the CERN RF SPS was supported by computer simulation of plasma generation and negative ion beam formation [190]. The results of a meniscus simulation are shown in Fig. 5.200. The electric field has a deep penetration into the plasma because the positive ions have a low pressure.

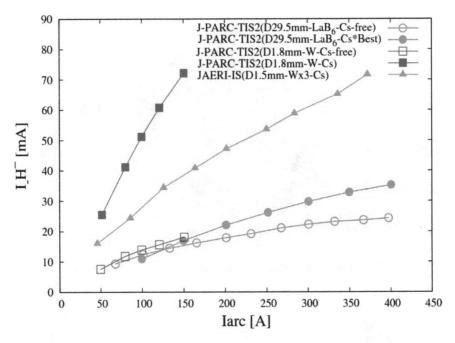

Fig. 5.203 Dependence of the intensity of H^- beams after cesium in the J-PARC SPS on the discharge current. (Reproduced from Akira Ueno [192])

5.26   Surface Plasma Sources at J-PARC, Japan

Development of an SPS at J-PARC (Japan Proton Accelerator Research Complex), Japan, was started to implement charge-exchange injection into a fast cyclic synchrotron—a booster at 3 GeV [191]. A schematic of the SPS at J-PARC with lanthanum hexaboride cathode is shown in Fig. 5.201. It consists of a gas-discharge chamber-anode of diameter 100 mm and length of 124 mm with a multipole magnetic wall, a lanthanum hexaboride cathode, a magnetic filter, a plasma electrode with a conical converter and an aperture diameter of 9 mm, an extraction electrode with permanent magnets and an electron trap, and a ground electrode. Studies have found that the negative ion beam intensity depends strongly on the shape of the cone of the plasma electrode.

Various forms of plasma electrodes and the negative ion current produced using them are shown in Fig. 5.202. The maximum current of 38 mA was obtained with a plasma electrode shaped as an "i." For production of maximum H^- current, the plasma electrode should be heated to a high temperature of 450 °C. The beam intensity was proportional to the length of the perimeter of the emission aperture rather than its area. From this, it was concluded that negative ions are formed on the surface of the conical converter without cesium. Chemical analysis of the elemental composition of the cone revealed a high content of lanthanum, which reduces the work function of the cone.

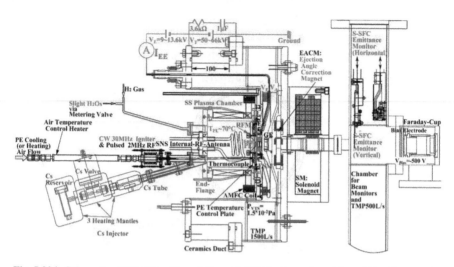

Fig. 5.204 Schematic of RF SPS with cesiation at J-PARC. (Reproduced from Akira Ueno [192])

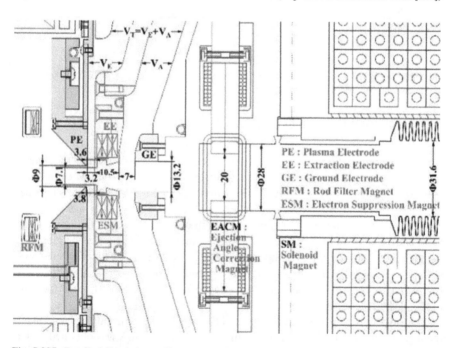

Fig. 5.205 Detailed dimensions of beam extraction region of J-PARC RF-driven SPS with cesiation H⁻ ion source test stand

The attempt for cesiation of this SPS was unsuccessful. After cesiation, the intensity of the beam in the SPS with lanthanum hexaboride cathode increased insignificantly. On the other hand, cesiation of the SPS with tungsten cathode was successful. The dependence of the intensity of H^- beams after cesium in these SPSs on the discharge current is shown in Fig. 5.203. In the SPS with tungsten cathode, the H^- ion current reached 70 mA at a discharge current of 150 A, whereas with lanthanum hexaboride cathode, a discharge current of 400 A was required to obtain 30 mA.

In order to increase the service life, J-PARC switched to plasma generation by an RF discharge with internal antenna, as for the SNS [192]. The cross-sectional view of the experimental setup of the SPS test stand, which consists of a stainless steel (SS) plasma chamber (PCH) unitized from an end-flange to the PE, extraction and ground electrodes (EE and GE), an ejection angle correction magnet (EACM), a vacuum chamber for differential pumping by a 1500 L/s turbo-molecular-pump (TMP) with a ceramic insulator duct for the terminal voltage (V_T) of 50 ~ 66 kV insulation, a solenoid magnet (SM) and a vacuum chamber for beam monitors, and a 500 L/s TMP, is shown in Fig. 5.204. The detailed dimensions of the beam extraction region are shown in Fig. 5.205 [193]. The SS PCH with 18 plasma confinement cusp magnets on the outer wall has two SS pipes to install rod-filter magnets (RFMs) cooled by water flowing inside. The SPS also has an end-flange with four plasma confinement cusp magnets on the outer wall, an internal RF antenna, a cesium (Cs) injector composed of a Cs-reservoir, a remotely controlled Cs-valve, and a Cs-tube, each of which is temperature controlled by using a thermocouple and a heating mantle attached to it; the PE is made of molybdenum, and a PE temperature control plate is made of oxygen-free copper (OFC). An axial magnetic field correction (AMFC) coil located around the downstream flange of the PCH is operated with a coil current of −4.38 A (403.4 AT) for all beam measurements presented in this paper. The AMFC has been shown to increase the RF efficiency by several %. The RF SPS was operated with a rather low H_2 flow rate of 17 sccm for a larger PE aperture of 9 mm by using a CW 30-MHz-RF plasma igniter with 50 W of power. The top of Fig. 5.205 shows the schematic circuit diagram of the high-voltage power supplies to produce the extraction and terminal voltages (V_E and V_T), plus the ampere meter to measure the extraction electrode current (IEE) directly. In order to reduce the V_T droop during the beam pulse, a 1 μF capacitor with a 3.6 kV overcurrent-suppression register was installed in parallel with the VT power supply. The H^- ion beam extracted by the V_E (9 ~ 13.6 kV) between the PE and the EE is accelerated by the VA between the EE and the GE, which is the difference of the V_T (50 ~ 66 kV) and V_E ($V_A = V_T - V_E$). The electrons co-extracted with the beam are bent by four electron suppression magnets (ESMs) onto a tungsten electron dump which is brazed on the OFC EE. The ejection angle of the H^- ion beam—caused by the rod-filter and electron-suppression fields—is corrected with an alignment error of about 50 μm by the EACM. The ejected H^- ion beam is focused by the SM into the beam monitor chamber. The horizontal and vertical emittances are measured using two sets of emittance monitors. Each monitor consists of a movable slit and movable slit-with-Faraday-cup (S-SFC). The measured emittance is visualized by randomly plotting dots in each mesh cell defined by the movement steps (typically 0.2 mm and 2 mrad). The number of dots per cell is proportional to the voltage signal and is normalized to make the total number 400,000. The I_{H^-} is measured with a Faraday cup attached on the downstream flange of the chamber, whose cup is grounded with a 50 Ω resistor. The I_{H^-} is calculated by converting the voltage across the resistor

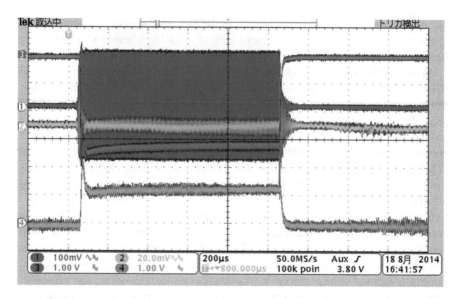

Fig. 5.206 Oscillograms illustrating the operation of the high-frequency SPS at J-PARC. (Reproduced from Akira Ueno [192]). 1, RF forward power; 2, RF reflected power; 3, H^- ion beam current at 20 mA/div, 66 mA; and 4, extraction gap current at 100 mA/div, 120 mA

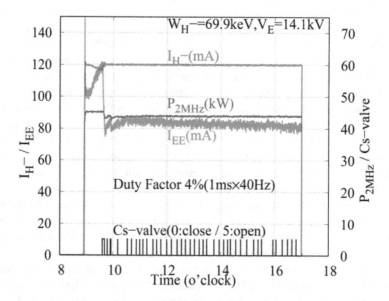

Fig. 5.207 Trend graph of IH-, IEE, P2MHz, and Cs-valve close/open during an 8-h operation of well-conditioned J-PARC-IS, in which feedback of P2MHz was used to control I_{H^-} to 120 mA. WH- and VE were fixed at 69 keV and 124.1 kV, respectively

with a coefficient of 20 mA/V. The coil current of the SM (I_{SM}) was set to 325 or 350 A (45,500 or 49,000 AT) for the I_H^- of 72 and 80 mA or 110 mA, respectively, in order to avoid too small beam size causing sputtering problems on the emittance monitor slits and tungsten Faraday cup. After a suitable conditioning procedure, it was confirmed that the lower plasma electrode temperature (T_{PE}) operation produced a beam with smaller transverse emittances, if the T_{PE} was higher than 50° C. All beam measurements presented in this paper were performed with a 2 MHz RF duty factor of 4.5% (1 ms u 45 Hz) in order to satisfy the radiation safety permission of the averaged I_H^- of 5 mA (110 mA u 4.5% = 4.95 mA). Although the feedback target value of T_{PE} was always set to 50 °C, the stationary state T_{PE} varied between 50 and 70 °C for the 4.5% duty factor operation. This was due to the insufficient PE air cooling capacity of 3600 L/h compared to the RF plasma power flow to the PE.

Oscilloscope traces illustrating the operation of the RF SPS at J-PARC are shown in Fig. 5.206. The forward RF power is trace 1, the reflected RF power is trace 2, the H^- ion beam current of 66 mA is trace 3, and the extractor current of 120 mA is trace 4. Operation of this RF SPS at a current of 60 mA for 30 days is ensured. The normalized beam emittance with 60 mA beam current is 1.5 π mm mrad. Further improvement of the J-PARC RF SPS is described in Refs. [194, 195].

Figure 5.207 shows the trend graph of a one-day (8 h) operation of the well-conditioned J-PARC IS after about an 88-h shutdown (vacuum pumping only). The beam energy, current, and duty factor are 69 keV, 120 mA, and 5% (1 ms, 40 Hz), respectively. The I_{H-}, I_{EE}, 2 MHz RF power (P_{2MHz}) and Cs-valve close/open (0/5) are shown with red, green, blue, and black lines, respectively. Feedback of the P_{2MHz} was used to control I_{H-} at 120 mA immediately after the operation was started. At first, the P_{2MHz} and I_{EE} gradually increased to 46.1 kW and 285 mA (out of range), respectively. Then, the Cs-valve was manually opened for 460 s at about 09:30. After the 2 MHz decreased lower than 40 kW, the Cs feedback was initiated. Cs feedback involves the control sequence to open the Cs-valve automatically for 24 s when the P_{2MHz} was increased higher than 40 kW. For this operation, the Cs- reservoir temperature was set to 180 °C. The injected Cs amount is estimated with the coefficient of 0.97 Pg/sec derived from previous measurement (23.3 Pg/opening). For the stationary condition after 11:30, there were ten openings over the course of 6.5 h, which correspond to a Cs injection rate of 35.8 Pg/h. For this Cs injection condition, the P_{2MHz} fluctuated between 38.8 and 40 kW according to the Cs density on the PE surface [196].

The AMFC effect for the I_{H-} enhancement with an AMFC coil was examined. The I_{H-} was optimally enhanced by the AMFC effect. At the I_{AMFC} of 994.4 AT in the case of the Isol of 400 A (59,040 AT), the I_{H-} with a flat top duty factor of 2% (800 μs × 25 Hz) of 77.5 mA was successfully extracted. The measured horizontal and vertical normalized rms emittances for the I_{H-} of 77.5 mA (I_{AMFC}: 994.4 AT, Isol: 400 A) were 0.45 and 0.44 π mm mrad, respectively. In that case, the I_{H-} of 69 mA and 70 mA in the horizontal and vertical plane were included within the normalized emittance of 1.5 π mm mrad, respectively. It satisfied the J-PARC second-stage requirements. It was considered that the I_{H-} enhancement was caused by the increase

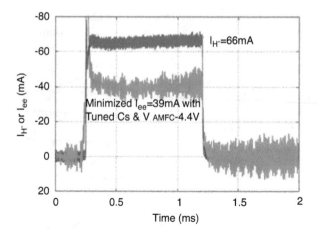

Fig. 5.208 H^- beam current, with co-extracted electron current less than the beam current, in the J-PARC RF SPS. (Reproduced from Akira Ueno [192])

of the low temperature plasma density due to the increase of the AMF. The I_{H^-} fluctuation was also affected by the AMF. In the near future, the optimization of the AMF to reduce the I fluctuation with keeping high I_{H^-} will be examined. An enhancement of H^- beam current by solenoidal magnetic field before was produced by Dudnikov [179, 180, 182, 184] and presented in Figs. 5.174, 5.182, 5.184, 5.187, and 5.193.

The beam brightness of the J-PARC cesiated RF-driven H^- SPS has been increased by using several unique measures, such as low (70 °C) plasma electrode temperature operation, a small H_2O feed for hydrogen plasma, and so on. The source allows injection of a 60 mA peak intensity beam into the high energy linear accelerators. In order to use the source stably for LINACs operated at a duty factor higher than the present value, 2.5%, 1 ms pulses at a repetition rate of 25 Hz—conditions to minimize the extraction electrode current (I_{EE}), whose main component is the co-extracted electron current—were investigated. The 66 mA H^- ion beam with low I_{ee} of 39 mA as shown in Fig. 5.208 (about one-fourth of that in ordinary operation) was stably operated by optimizing the rod-filter-field (RFF), cesium density, and axial magnetic field correction (AMFC). Notably, an AMFC of only −30 Gauss had the largest I_e reduction by about one-third. This decrease of I_e with increase of longitudinal magnetic field was recognized before by Dudnikov [184]; see Fig. 5.187. The corresponding 95% beam transverse normalized rms emittances were degraded by about 24% due to Cs density and RFF higher than those in ordinary operation.

The J-PARC cesiated RF-driven H^- SPS has been stably operated for about 4 years. The J-PARC RFQ LINAC successfully accelerated the required intensity of 60 mA, when a 70 mA beam was injected from the source. A high intensity beam with transverse emittance suitable for the RFQ is produced using several unique measures, such as slight water molecule (H_2O) addition into hydrogen plasma, low temperature (about 70 °C) operation of the 45°-tapered plasma electrode with

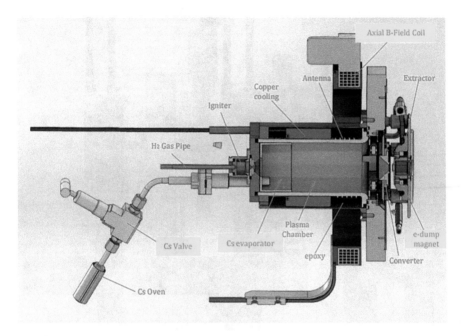

Fig. 5.209 A schematic of CSNS RF SPS with cesiation

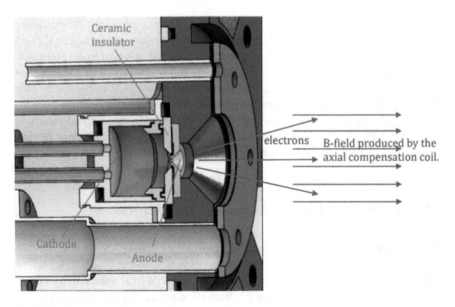

Fig. 5.210 Design of pulsed glow discharge igniter

16 mm thickness, and argon and/or nitrogen elimination in the hydrogen plasma, along with filter field optimization and so on. In order to identify the source beam

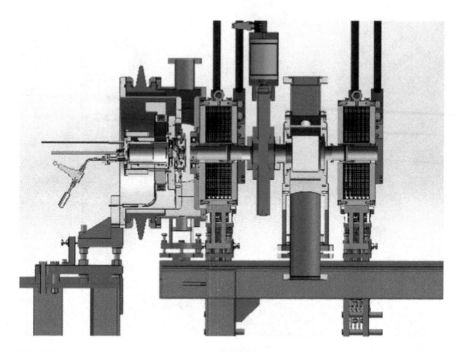

Fig. 5.211 A schematic of LEBT for CSNS RF SPS

intensity bottleneck and develop the optimal plasma electrode, extraction electrode, and ground electrode shapes, higher extraction and acceleration voltages (V_{ext} and V_{acc}) were examined. A 110 mA beam, of which about 103 mA is of transverse emittance appropriate for common RFQ designs, was stably operated with a duty factor of 5% (1 ms at 50 Hz) by using V_{ext} and V_{acc} of 12 kV and 50 kV, respectively. This substantial progress, with important information on the space charge limited bottlenecks in the extraction and acceleration gaps, will provide the optimal electrode shapes for J-PARC operation and realize the next-generation benchmark H^- ion source for high intensity and high energy LINACs.

5.27   RF Surface Plasma Source in Chinese SNS

Development of RF surface plasma H^- ion source for Chinese SNS is described in [197]. A schematic of CSNS RF SPS with cesiation is shown in Fig. 5.209.

It consists of silicon-nitride plasma chamber, high thermal conductivity epoxy insulation, glow discharge igniter in gas line, RF antenna, copper cooling, plasma electrode with magnetic filter and conical converter, extractor electrode with one pair permanent magnet for e-dumping, grounded electrode, Cs evaporator, axial B-field coil, and Cs injection system with Cs valve (which remove cusp magnets since 2020).

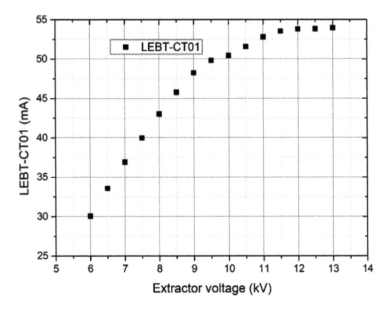

Fig. 5.212 Dependence of H^- beam current vs extraction voltage (RF power 30–31 kW, $H_2 = 21$ sccm)

Design of glow discharge igniter is shown in Fig. 5.210. It consists of ceramic body, hollow cathode with gas injection, ceramic insulator with small aperture ~1 mm diameter, and an anode. A pulsed 150 μs, 560 V voltage is applied to cathode. Electrons injected to main chamber to ignite main ~2 MHz RF discharge,

H^- beam pulse up to 1 ms, repetition 25 Hz. A schematic of LEBT is shown in Fig. 5.211.

It consists of double solenoids, optimized emittance growth, higher transmission efficiency, less than 0.75 m length, and new double-slit scanner.

A dependence of H^- beam current vs extraction voltage (RF power 30–31 kW, $H_2 = 21$ sccm) is shown in Fig. 5.212.

The RF-driven H^- ion source with external antenna and Si3N4 chamber runs successfully in CSNS with H^- beam current up to ~60 mA (35 mA after RFQ); it is able to operate more than 310 days (even longer). Vacuum condition and H_2 purifier are very important to increase the performance of the beam and to minimize the usage of cesium. Dark current is efficiently removed by biasing the extractor negatively and chopping the rising/fall fringe of beam. Cusp field is removed from the H^- ion source since 2021.

Last review of negative ion sources for accelerators was presented in [198]. High brightness, negative hydrogen ion sources are used extensively in scientific facilities operating worldwide. Negative hydrogen beams have become the preferred means of filling circular accelerators and storage rings as well as enabling efficient extraction from cyclotrons. Several well-known facilities now have considerable experience operating a variety of sources such as RF-, filament-, magnetron-, and

Penning-type H^- SPS. These facilities include the US Spallation Neutron Source (SNS), Japan Proton Accelerator Research Complex (J-PARC), Rutherford Appleton Laboratory (RAL-ISIS), Los Alamos Neutron Science Center (LANSCE), Fermi National Accelerator Laboratory (FNAL), Brookhaven National Laboratory (BNL), numerous installations of D-Pace (licensed by TRIUMF) ion sources used mainly on cyclotrons, and, most recently, the CERN-LINAC-4 injector. This book first summarizes the current performance of these ion sources in routine, daily operations with attention toward source service periods and availability metrics. Sustainability issues encountered at each facility are also reported and categorized to identify areas of common concern and key issues. Recent ion source improvements to address these issues are also discussed as well as plans for meeting future facility upgrade requirements.

5.28 Surface Plasma Sources for Low Energy Neutral Beam Production

One of the spectacular manifestations of the surface plasma mechanism of generation of negative ions should be the appearance of intense fluxes of neutral particles with energy hundreds of eV obtained when negative ions are destroyed in the discharge plasma. The fluxes of such particles have been investigated [199–201]. In the SPS, negative ions are formed when electrodes are bombarded with plasma particles due to the capture of electrons from the electrodes to the electron affinity levels of the sputtered and reflected particles. The probability of particles leaving the surface in the form of negative ions increases by hundreds of times with a decrease in the work function of the surface by the adsorption of cesium fed into the discharge. The resulting negative ions are accelerated in a thin near-electrode sheath with a potential difference between the electrode and the plasma. To reduce the destruction of negative ions in the SPS, the thickness of the plasma and gas layer between the emitting surface and the beam formation system is carefully minimized. If, however, the thickness of the plasma and gas layer exceeds the mean free path of negative ions before destruction, the accelerated negative ions lose the extra electron and become a stream of energetic neutral atoms that accumulate a significant fraction of

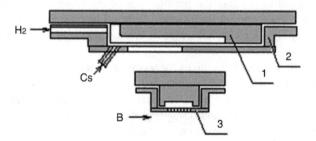

Fig. 5.213 Schematic of SPS for obtaining energetic neutral atoms. (Reproduced from Dudnikov and Fiksel [200]). 1, cathode; 2, anode; and 3, anode plate with emission grid

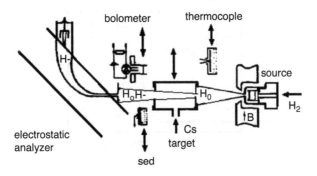

Fig. 5.214 Diagnostics for investigation of neutral beam production. (Reproduced from Dudnikov and Fiksel [200])

the power deposited in the discharge. The energy of accelerated negative ions is determined by the potential difference between emitting electrode and plasma. In most cases, the cathode of the gas-discharge device serves as the emitting electrode.

The emission of energetic atoms from the discharge in an SPS with semiplanotron electrode configuration has been investigated; a schematic of the setup used is shown in Fig. 5.213.

A high-current glow discharge is ignited between cathode 1 with a rectangular groove, with a notch at one end, and with an anode plate 3 covering the body of the plasma cell. Hydrogen and cesium vapor are fed into the groove with a notch for ignition of the discharge. An adjustable magnetic field $B \sim 500$ Gauss is oriented so that electrons oscillating between the cathode groove walls drift in crossed fields from the beginning of the groove to its closed end. Further, the plasma enters a narrow gap between the anode plate and the cathode, in which fast electrons are not retained. Electrons escape to the walls of the anode along the magnetic field lines of force and the plasma recombines. The length of the groove is 40 mm, the width is 8 mm, the depth is 3 mm, and the area of the working part of the cathode is 5.5 cm^2. In the first experiments, a stream of accelerated atoms emerged through an emission slit in the anode plate measuring 1×40 mm^2 and later through a grid of six slits, 0.7 mm wide and 2 mm deep, occupying an area of 6×20 mm^2. The total area of the emission surface was 84 mm^2.

The source operated in pulsed mode with half-sinusoidal current pulses of amplitude up to 100 A, duration 1 ms, and repetition rate up to 10 Hz. The discharge voltage was controlled by varying the cesium supply. Without cesium supply, the discharge voltage was 600–800 V; with increase in the cesium supply, the voltage decreased to 200 V. Note that over the wide operating range of magnetic field 500–1000 Gauss, there were no observed fluctuations in discharge parameters. The transition to a regime with "noise" occurred only with further increase in the magnetic field.

The beam diagnostics are shown in Fig. 5.214.

The charged component of the beam emerging from the SPS was bent in a magnetic field and was not measured. The neutral component was registered with a pulsed bolometer. A photodiode detects the radiation of a thin (5 microns) titanium

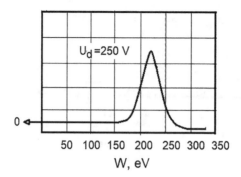

Fig. 5.215 Oscillograph of the energy spectrum of energetic neutral atoms from the SPS. (Reproduced from Dudnikov and Fiksel [200])

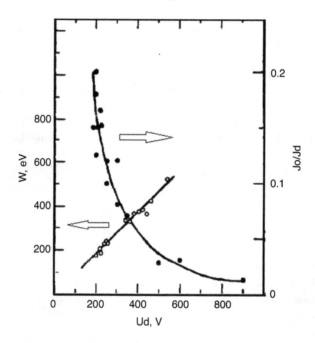

Fig. 5.216 Dependence of the average energy, W, of neutrals from the SPS, and the ratio J_o/J_d of emission current density of neutrals to discharge current density, on the discharge voltage U_d. (Reproduced from Dudnikov and Fiksel [200])

foil, heated by current to a temperature of 800–1000 K. The pulse heating of the foil by the energetic atom stream increases the heat flux and hence the photodiode current. (Constant heating of the foil is introduced to shift the spectrum of the pulsed radiation to the sensitivity region of the photodiode.) The bolometer was calibrated by passing a short current pulse through the foil. The sensitivity was 1 mA/J. As a secondary diagnosis, a secondary emission detector ("sed" in the figure) was used.

One of the main issues in the diagnosis of neutral beams is the measurement of its energy spectrum. The usual method—ionization of the neutral beam by an electron beam—has a low efficiency of 10^{-4}–10^{-5}. The efficiency of ionization in a gas target at energies of hundreds of eV is also small because of low ionization cross sections and strong scattering. However, for the diagnostics of neutral hydrogen atoms, they can be changed to negative H^- ions in a cesium target. It is known that the equilibrium yield of H^- in a cesium target is 20% at energies of 200 eV, so that even on targets with a comparatively low cesium density, it is easy to obtain an output of 1%, which is quite sufficient for diagnostic purposes. The setup for recording the atomic energy spectrum is shown in Fig. 5.214. The charge-exchange cesium target had a length of 15 cm. The target utilized a heat pipe approach, which allowed increasing the service life of the target by recirculation of cesium. The resulting negative ions are introduced at an angle of 45° to a parallel plate analysis system and after deflection through 90° are accepted by the collector. The measured current is amplified and fed to the oscilloscope. A triangular pulse of a regulated voltage of 30 μs duration is applied to the deflection plates and fed to the horizontal axis of the oscilloscope. The analyzer was calibrated by a monoenergetic electron beam. The measured energy broadening coincided with that calculated and amounted to ~10% of the energy at the base of the spectrum.

Figure 5.215 shows an oscillogram of the energy spectrum of negative ions at a discharge voltage of 250 V. Since the charge-exchange cross section in the energy range 200–600 eV is approximately constant, and the measured spectrum has a relatively small width, it can be asserted that it corresponds to the true energy spectrum of the energetic neutral atoms.

Figure 5.216 shows the dependence of the average energy ε of neutrals from the SPS as a function of discharge voltage U_p. It is seen that over the whole range $\varepsilon < eU_p$.

The width of the energy spectrum is ~50 eV at a discharge voltage $U_d = 200$ V and ~ 120–150 eV at $U_d = 600$–800 V. A broadening of the spectrum is observed when the discharge voltage is noisy.

Based on the measured energy spectrum of the neutrals and the energy content of the beam measured by the bolometer, it is possible to calculate the current of neutral atoms. The neutral current density was calculated by the formula $J_o = \pi W/2\varepsilon\tau S$, where W is the energy content of the beam recorded by the bolometer, ε is the average energy of the atoms, τ is the discharge time, and S is the area of the bolometer foil. The factor $\pi/2$ occurs when time dependence is taken into account.

In special experiments, it was found that about 90% of the neutral component is indeed in neutral hydrogen atoms with energy hundreds of volts. So when hydrogen was replaced by helium, which does not form negative ions, the signal dropped by a factor of 30. When the polarity of the voltage was changed in the hydrogen discharge, the signal fell by a factor of 400. Closing the emission slit of the LiF plate, transparent to ultraviolet, reduced the bolometer signal by a factor of 40 and the scattering helium target by a factor of 10.

It is assumed that the beam energy is completely absorbed in the bolometer, which lowers the beam current. The coefficient of reflection from titanium for hydrogen atoms with energy 200 eV is 20%.

The distribution of neutral current density normalized per unit of 10 cm from the source is shown in Fig. 5.217. When the discharge voltage was changed, the

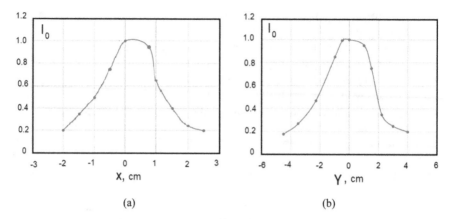

Fig. 5.217 Distribution of neutral current density, normalized to unity, at 10 cm from the source: (**a**) across the mesh plates and (**b**) along the mesh plates. (Reproduced from Dudnikov and Fiksel [200])

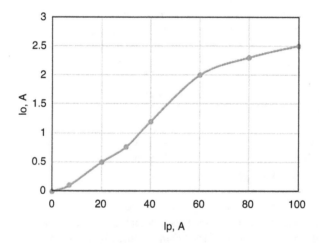

Fig. 5.218 Dependence of total neutral beam current I_o on discharge current I_p, for a fixed discharge voltage of 200 V. (Reproduced from Dudnikov and Fiksel [200])

distributions changed insignificantly. The distributions measured by the bolometer and the secondary emission detector coincided. The calculated secondary electron emission coefficient for atoms with energy ~200 eV is 0.2. The angular divergence at half-maximum was ~10°.

The total current of the neutral beam was calculated by integrating the current density over space. Figure 5.216 shows the dependence of the ratio of neutral beam emission current density to discharge current density on the discharge voltage (regulated by the cesium feed) at a fixed discharge current of 40 A. The neutral beam emission density was calculated as I_o/S_e and the discharge current density at the cathode as I_p/S_c, where I_o is the equivalent neutral current, I_d is the discharge current, S_e is the emission aperture area, and S_c is the cathode surface area. We see a sharp increase in the generation of energetic atoms when cesium is supplied to the discharge and a corresponding decrease in the discharge voltage. A similar dependence of negative ion generation is observed in the SPS. The fairly large spread of the data, especially at low discharge voltages, is explained by the fact that the discharge voltage, generally speaking, is not a one-to-one function of the state of the surface, on which the negative ion generation efficiency depends very strongly. It should also be noted that due to the inhomogeneity of the discharge, the true value of the discharge current density can differ from the calculated value.

Figure 5.218 shows the dependence of the total neutral beam current on discharge current for a discharge voltage of 200 V. The deviation from a linear dependence of the initial section was repeatedly checked. This can be explained by the fact that the negative ion generation is proportional to the discharge current, and their destruction in the initial section is also proportional to the discharge current. Therefore, $I_o \sim I_p^2$. The monotonic intensity ramp at high discharge current indicates the preservation of the high negative ion formation efficiency in the SPS at high discharge current. Beam of neutral atoms H^o with energy ~200 eV and intensity up to 2.5 equivalent ampere was produced.

Developments of surface plasma method of negative ion production are described in reviews [202–204].

5.29   Conclusion

During first 50 years of surface plasma sources development, an intensity of negative ion beams was increased more than 10^4 times. Long-time reliable operation of surface plasma sources was reached in many big accelerator and nuclear fusion facilities.

References

1. V. Dudnikov, *Negative Ion Sources* (NSU, Novosibirsk, 2018). В. Дудников, *Источники отрицательных ионов*. НГУ, Новосибирск, 2018
2. V. Dudnikov, *Development and Applications of Negative Ion Sources* (Springer, 2019)
3. V.G. Dudnikov, Charge exchange injection into accelerators and storage rings. Phys.-Uspekhi **62**, 405 (2019)
4. V. Dudnikov, Method of negative ion obtaining, Patent cccp, 411542, 10/III. 1972; http://www.findpatent.ru/patent/41/411542.html В.Г. Дудников, "Способ получения отрицательных ионов", Авторское свидетельство, М. Кл.Н 01 J 3/0,4, 411542, заявлено 10/III, 1972
5. V.G. Dudnikov, *Surface Plasma Method of Negative Ion Production*, Dissertation for doctor of Fis-Mat. nauk, INP SBAS, Novosibirsk 1976. В.Г. Дудников, *Поверхностно-плазменный метод получения пучков отрицательных ионов*, Диссертация на соискание учёной степени доктора физ.-мат. Наук, ИЯФ СОАН СССР, Новосибирск, 1976
6. B.P. Murin, Proceedings of VII international conference on high energy accelerators, CERN, p. 540 (1971)

7. M.I. Avramenko, G.I. Bazkikh, V.P. Golubev, G.I. Klenov, Proceedings of Allunion workshop on particle accelerators, M., Nauka, v. 1, p. 261 1973. М.И. Авраменко, Г.И. Бацких, В.П. Голубев, Г.И. Кленов и др. Труды Всесоюзного совещания по ускорителям заряженных частиц, М., Наука, , т. 1, стр. 261 (1973)

8. V.G. Dudnikov, Hydrogen negative ion source with Penning Geometry, in *Proceedings of IV Allunion conference for charged particles accelerators*, M. Nauka, v. 1, p. 323 (1975). English translation, V.G. Dudnikov, *Surface-plasma source of negative ions with Penning geometry*, Los Alamos, LA-TR--75-4 (1975). В. Г. Дудников, "Источник отрицательных ионов водорода с Пеннинговской геометрией, Труды IV Всесоюзного совещания по ускорителям заряженных частиц", М. Наука, т. 1, стр. 323 (1975)

9. V. Dudnikov, R.P. Johnson, Cesiation in highly efficient surface plasma sources. Phys. Rev. Accel. Beams **14**, 054801 (2011)

10. G.E. Derevyankin, V.G. Dudnikov, P.A. Zhuravlev, Electromagnetic shutter for a pulsed gas inlet into vacuum units. Pribory i Tekhnika Eksperimenta **91975**, 168–169

11. V.M. Neslin, Plasma Phys. **10**, 337 (1968)

12. C. Lejnne, Proceedings of Second Symposium on ion sources and ion beams, Berkeley, LBL Report No3399, p. I-1(1974)

13. G.I. Dimov, G.E. Derevyankin, V.G. Dudnikov, A 100 mA negative hydrogen-ion source for accelerators. IEEE Trans. Nucl. Sci. **24**(3), 1545 (1977)

14. V.G. Dudnikov, Review of high brightness ion sources for microlithography. Rev. Sci. Instrum. **67**(3), 915–920 (1996)

15. S.K. Guharay, W. Wang, V.G. Dudnikov, M. Reiser, J. Orloff, J. Melngailis, High brightness ion source for ion projection lithography. J. Vac. Sci. Technol. B. **14**, 3907 (1996)

16. V.G. Dudnikov, G.E. Derevyankin, D.V. Kovalevsky, V.Y. Savkin, et al., Surface plasma source to generate high-brightness H− beams for ion projection lithographya. Rev. Sci. Instrum. **67**(4), 1614–1617 (1996)

17. V. Dudnikov, Y. Belchenko, Preprint, INP 78-95, Novosibirsk, 1978, V. Dudnikov, Yu. Belchenko. J. Phys. **40**, 477 (1979)

18. Y.I. Belchenko, High current quasistacionary and pulsed surface plasma sources of Hydrogen negative ions, Doctor thesis, Novosibirsk, 1991. Ю.И. Бельченко, Сильноточные квазистационарные и импульсные поверхностно-плазменные источники отрицательных ионов водорода, Doctor thesis, Novosibirsk, (1991)

19. Y.I. Belchenko, G.I. Dimov, V.G. Dudnikov, A.S. Kupriyanov, Negative ion surface-plasma source development for fusion in Novosibirsk. Rev Phys Appl **23**(11), 1847 (1988)

20. Y.I. Belchenko, V.G. Dudnikov, Surface plasma source with increased efficiency of H− generation, Preprint INP 80–30, Novosibirsk, 1980, Proc. XV International conference on phenomena in ionized gases, Minsk, p. II, p. 1504 (1981). Ю. И. Бельченко, В. Г. Дудников, "Поверхностно-плазменный источник с повышенной эффективностью генерации ионов Н-"Препринт ИЯФ 80–30, Новосибирсу, 1980. Труды ХУ Международной конференция по явлениям в ионизированных газах, Минск, 1981, часть П, с.-1504 (1981)

21. Y.I. Belchenko, Fiz. Plazmy. **9**, 1219 (1983). Ю.И. Бельченко Препринт ИЯФ СО АН СССР, № 52-82, Новосибирск, 1982;Физика плазмы, 9, № 2 (1983).Yu. I. Belchenko, Preprint INP 52-82, Novosibirsk 1982

22. I. Belchenko Yu, A. Kupriyanov, S Rev. Phys. Appl. **23**, 1889 (1988)

23. A.S. Kupriyanov, Thesis for Cand. Phys.-Math. Sci. (Novosibirsk, Budker Institute of Nuclear Physics, SB RAS, 1994)

24. Y.I. Belchenko, A.S. Kupriyanov, High current surface plasma negative ion sources with geometrical focusing. Rev. Sci. Instrum. **61**, 484 (1990)

25. Y.I. Belchenko, G.E. Derevyankin, V.G. Dudnikov, Patent SSSP, Negative ion source, 854197 (1980), http://www.findpatent.ru/patent/85/854197.html, Бельченко Ю.И., Деревянкин Г.Е., Дудников В.Г., Источник Отрицательных Ионов, Авторское свидетельство, 854197, (1980)

26. Y.I. Belchenko, G.E. Derevyankin, V.G. Dudnikov, Proc. Simp. production and neutralization of negative hydrogen ions and beams, BNL 50727, 74 (1977)

27. G.E. Derevyankin, Surface plasma sources for accelerators, Ph. D. thesis, Novosibirsk, (1987)

28. G.E. Derevyankin, V.G. Dudnikov, Production of high brightness H⁻ beams in surface plasma sources. AIP Conf. Proc. **111**(1), 376–397 (1984)

29. A.A. Bashkeev, V.G. Dudnikov, Continuously operated negative ion surface plasma source, AIP CP 1097. NIBS Conf. Proc. **210**(1), 329–339 (1990)

30. V. Dudnikov, C.W. Schmidt, R. Hren, J. Wendt, Direct current surface plasma source with high emission current density. Rev. Sci. Instrum **73**(2), 989 (2002)

31. V. Dudnikov, C.W. Schmidt, R. Hren, J. Wendt, High current density negative ion source for beam line transport studies, PAC 2001, Chicago (2001)

32. V. Dudnikov, J.P. Farrell, Compact surface plasma sources for heavy negative ion production. Rev. Sci. Instrum. **75**, 5 (2004)

33. V. Dudnikov, J. Paul Farrell, Rev. Sci. Instrum. **75**(5), 1732 (2004)

34. K. Prelec, T. Sluyters, Proceedings of 2nd Symposium on ion sources and formation of ion beams, Berkeley, Ca., VIII-6, LBL-3399 (1974)

35. K. Prelec, T. Sluyters, A pulsed negative hydrogen source for currents up to one ampere. IEEE Trans Nucl Sci **22**(3), 1662 (1975)

36. K. Prelec, T. Sluyters, M. Grossman, High currant negative ion beams. IEEE Trans Nucl Sci **24**(3), 1521 (1977)

37. M. Kobayashi, Studies of the hollow discharge duoplasmatron operating in pure hydrogen mode and in hydrogen-cesium mode, BNL-21220, 1976

38. J.G. Alessi, T. Sluyters, Regular and asymmetric negative ion magnetron sources with grooved cathodes. Rev. Sci. Instrum **51**(12), 1630 (1980)

39. J. Alessi, A. Hershcovitch, K. Prelec, T. Sluyters, Development of multiampere negative ion sources. IEEE Trans. Nucl. Sci. **28**(3), 2652 (1981)

40. K. Prelec, Progress in the development of high-current, steady-state H-/D- sources at BNL, in *Proc. Second Symp. on the Production and Neutralization of Negative Hydrogen Ions and Beams*, ed. by K. Prelec, (BNL 51304, BNL, Upton, NY, 1080), p. 145

41. C. Schmidt, C. Curtis, Negative hydrogen-ion program at fermilab, in *Proceedings of the 1976 proton linear accelerator conference*, Chalk River, Ontario, Canada (1976)

42. C.W. Schmidt, C.D. Curtis, A 50-mA negative hydrogen-ion source. IEEE Trans. Nucl. Sci **26**(3), 4120 (1979)

43. C.W. Schmidt, C.D. Curtis, Proceedings of Symposium on the production and neutralization of negative hydrogen ions and beams, BNL, Sept. 26–30, 1977, BNL50727, p. 123, (1977)

44. A. Zelenski, G. Atoian, T. Lehn, D. Raparia, J. Ritter, High-intensity polarized and un-polarized sources and injector developments at BNL linac. AIP Conf. Proc. **2373**, 070003 (2021) https://doi.org/10.1063/5.0057677

45. H. Pereira, J. Lettry, J. Alessi, T. Kalvas, Estimation of sputtering damages on a magnetron H- ion source induced by Cs+ and H+ ions. AIP Conf. Proc. **1515**, 81 (2013). https://doi.org/10.1063/1.4792773

46. Y. Belchenko, V.I. Davydenko, G.E. Derevyankin, A.F. Dorogov, V.G. Dudnikov, Sov. Tech. Phys. Lett. **3**, 282 (1977)

47. C.D. Curtis, C.W. Owen, C.W. Schmidt, Factors affecting H-beam performance in the Fermilab Linac, in *Proceedings of the 1986 international linac conference*, Stanford, California, USA (1986)

48. V. Stipp, A. Dewitt, J. Madsen, A brighter H⁻ source for the intense pulsed neutron source accelerator systex. IEEE Trans. Nucl. Sci. **30**(4), 2743 (1983)

49. D.S. Barton, R.L. Witkover, Negative ion source tests for H- injection at the Brookraven AGS. IEEE Trans. Nucl. Sci. **28**(3) (1981)

50. J. Peters, The status of DESY H- sources. Rev. Sci. Instrum. **69**(2), 992 (1998)

51. J. Peters, The HERA magnetron: 24 years of experience, a world record run and new design, CP1097, 236, (2009)

52. J.G. Alessi, J.M. Brennan, A. Kponou, K. Prelec, H⁻ source and beam transport experiments for a new RFQ, PAC 1987, (1987)
53. D.S. Bollinger, A. Sosa, Overview of recent studies and design changes for the FNAL magnetron ion source. AIP Conf. Proc. **1869**, 030054 (2017)
54. P.W. Allison, E.A. Meyer, D.W. Mueller, R.R. Stevens, Jr, Status of The Lampf. H- Injector, in *Proceedings of the 1972 proton linear accelerator conference*, Los Alamos, New Mexico, USA (1972)
55. C. Robinson, *Aviation Week & Space Tech.*, p. 42, 1978; Rev. Mod. Phys., 59(3), Part II (1987)
56. Report to the APS of the Study Group on Science and Technology of Directed Energy Weapons, Executive Summary and Major Conclusions, Physics Today, (1987)
57. P.W. Allison, Experiments with Dudnikov type H- source, preprint, LA-UK -77-2113, Los Alamos, (1977)
58. P.W. Allison, Experiments with a Dudnikov Type H- Ion Source, in *Proceedings of symposium on the production and neutralization of negative hydrogen ions and beams*, Upton, New York, September 26–30, 1977, Brookhaven National Laboratory report BNL-50727, p. 119 (1977)
59. H. Vernon Smith, Jr., P. Allison, H- beam emittance measurements for the penning and the asymmetric, grooved magnetron surface-plasma source, in *Proceedings of the 1981 linear accelerator conference*, Santa Fe, New Mexico, USA (1981)
60. P. Alisson, J. Sherman, Operation experience with 100 keV, 100 mA injector. AIP Conf. Proc. **111** (1981)
61. H.V. Smith Jr., J.D. Sherman, C. Geisik, P. Allison, H temperature dependences in a Penning surface plasma source. Rev. Sci. Instrum. **63**, 2723 (1992)
62. J.D. Sherman, W.B. Ingalls, G. Rouleau, H.V. Smith Jr., Review of scaled penning H- surface plasma source with slitwithslit emitters for high duty factor linacs, CP642, in *High Intensity and High Brightness Hadron Beams: 20th ICFA Advanced Beam Dynamics Workshop on High Intensity and High Brightness Hadron Beams*, ed. by W. Chou, Y. Mori, D. Neuffer, J.-F. Ostiguy, (American Institute of Physics 0-7354-0097-0, 2002)
63. H. Vernon Smith Jr., P. Allison, J.D. Sherman, H⁻ and D⁻ scaling laws for Penning surface-plasma sources. Rev. Sci. Instrum. **65**(I), 123 (1994)
64. H. Vernon Smith Jr., P. Allison, J.D. Sherman, The 4x source. IEEE Trans. Nucl. Sci. **32**(5), 1797 (1985)
65. J.D. Sherman, P. Allison, H.V. Smith Jr., Beam potential measurement of an intense H- beam by use of the emissive probe technique. IEEE Trans. Nucl. Sci. **32**(5), 1973 (1985)
66. H. Vernon Smith, Jr., Emission Spectroscopy of the 4X Source Discharge with and without N₂ Gas, AT-10 Technical Note: 89–07 (1989)
67. V. Dudnikov, D. Bollinger, D. Faircloth, S. Lawrie, Potential for improving of the compact surface plasma sources. AIP Conf. Proc. **1515**, 369 (2013)
68. K.N. Leung, G.J. DeVries, K.W. Ehlers, L.T. Jackson, J.W. Stearns, M.D. Williams, M.P. Allison, Operation of a Dudnikov type Penning source with LaB6 cathodes, AIP Conf. Proc. **158**, 356 (1987); Y.D. Jones, R.P. Copeland, M.A. Parman, E.A. Baca, Xenon mixing in a cesium-free Dudnikov-type hydrogen ion source, in *Proceedings Volume 1061, Microwave and Particle Beam Sources and Directed Energy Concepts*; (1989). https://doi.org/10.1117/12.951842
69. P.E. Gear, R. Sidlow, Proceedings of 2nd International Conference on Low energy Ion Beams, Bath, Inst. Phys. Con. Ser No 54, p. 284 (1980)
70. R. Sidlow, P.J.S. Barratt, A.P. Letchford, M. Perkins, C.W. Planner, Operational Experience of Penning H- Ion Sources at ISIS, PAC (1996)
71. O. Tarvainen, et al., Caesium balance of the ISIS H- penning ion source in long pulse operation, NIBS 2022, Padova, Italy, 2022
72. D.C. Faircloth, A.P. Letchford, C. Gabor, M.O. Whitehead, T. Wood, Understanding extraction and beam transport in the ISIS H⁻ Penning surface plasma ion source. Rev. Sci. Instr. **79**, 02B717 (2008)

73. D.C. Faircloth, S.R. Lawrie, et al., High current results from the 2X scaled penning source, AIP Conf. Proc. **2052**, 050004-1–050004-7; 2018

74. D.C. Faircloth, S.R. Lawrie, J. Sherman, et al., 2X scaled penning source developments. AIP Conf. Proc. **2011**, 050028 (2018) https://doi.org/10.1063/1.505332

75. L. Sheng-Jin, H. Tao, O. Hua-Fu, et al., Status of CSNS H⁻ ion source. Chinese Phys. C. **39**(5), 057008 (2015)

76. S. Liu, H. Ouyang, T. Huang, et al., The improvements at CSNS ion source. AIP Conf. Proc. **2011**, 050022 (2018) https://doi.org/10.1063/1.5053320

77. S. Lawrie, A. Letchford, C. Gabor, et al., Developing the RAL front end test stand source to deliver a 60 mA, 50 Hz, 2 ms H– beam. AIP Conf. Proc. **1515**, 359 (2013)

78. W.K. Dagenhart, W.L. Stirling, H.H. Haselton, G.G. Kelley, J. Kim, C.C. Tsai, J.H. Whealton, Modified calutron negative ion source operation and future plans, in *Proceedings of the Second International Symposium on the Production and Neutralization of Negative Hydrogen Ions and Beams 1980*, Brookhaven National Laboratory, Upton, New York, p. 217 (1980)

79. Y. Belchenko, V. Savkin, Direct current H⁻ source for the medicine accelerator. Rev. Sci. Instrum. **75**(5), 1704 (2005)

80. Y. Belchenko, A. Sanin, A. Ivanov, 15 mA CW H- source for accelerators. AIP Conf. Proc. **1097**, 214–222 (2009)

81. Y. Belchenko, A. Sanin, V. Savkin, Study of fluctuations in the CW penning surface-plasma source of negative ions. AIP Conf. Proc. **1390**, 401–410 (2011). https://doi.org/10.1063/1.3637411

82. Y.I. Belchenko, A.I. Gorbovsky, A.A. Ivanov, A.L. Sanin, V.Y. Savkin, et al., Upgrade of CW negative hydrogen ion source. AIP Conf. Proc. **1515**, 448 (2013)

83. Y. Belchenko, A. Gorbovsky, A. Sanin, V. Savkin, The 25 mA continuous-wave surface-plasma source of H- ions. Rev. Sci. Instrum. **85**, 02B108 (2014)

84. Y. Belchenko, A. Sanin, O. Sotnikov, Comparative analysis of continuous-wave surface-plasma negative ion sources with various discharge geometry. Rev. Sci. Instrum. **85**, 02B116 (2014)

85. Y. Belchenko, A. Sanin, O. Sotnikov, Comparative analysis of continuous wave surface plasma sources of negative ions with various discharge geometries, Nucl. Phys. Eng. **6**(12), 81 (2015); Ю. И. Бельченко, А. Л. Санин, О. З. Сотников, "Сравнительный Анализ Стационарных Источников Отрицательных Ионов Водорода с Различной Геометрией Разряда", ЯДЕРНАЯ ФИЗИКА И ИНЖИНИРИНГ, том 6, № 1–2, с. 81–85. (2015)

86. A.A. Ivanov, A. Sanin, Y. Belchenko, I. Gusev, I. Emelev, V. Rashchenko, V. Savkin, I. Shchudlo, I. Sorokin, S. Taskaev, P. Zubarev, A. Gmyrya, Recent achievements in studies of negative beam formation and acceleration in the tandem accelerator at Budker Institute. AIP Conf. Proc. **2373**, 070002 (2021). https://doi.org/10.1063/5.0057441

87. K.N. Leung, K.W. Ehlers, Self-extraction negative ion source. Rev. Sci. Instrum. **53**, 803 (1982)

88. R.L. York, R.R. Stevens, Jr., A cusped field H- ion source for LAMPF, IEEE Trans. Nucl. Sci. **30**(4), (1983)

89. A. Takagi, Y. Mori, K. Ikegami, S. Fukumoto, KEK multicusp negative hydrogen ion source, PAC 1985. IEEE Trans. Nucl. Sci. **32**(5), 1782 (1985)

90. A.B. Wengrow, K.N. Leung, M.A. Leitner, et al., Development of a high duty factor, surface conversion H- ion source for the LANSe facility, PAC 1997, (1997)

91. R. Thomae, R. Gough, R. Keller, K. Leung, et al., Measurements on the LANSCE Upgrade H- Source, in *Proceedings of the 1999 particle accelerator conference*, New York, (1999)

92. M. Williams, R. Gough, K. Leung, et al., Design of ion source for lansce upgrade, in *Proceedings of the 1999 Particle Accelerator Conference*, New York, (1999)

93. V. Dudnikov, *Improvement of Converter Surface Plasma Sources*, ArXiv 5fe9f56d45851553a001595a.pdf (2020)

94. Y. Mori, A. Takagi, A. Ueno, S. Fukumoto, A versatile high intensity plasma sputter heavy negative ion source. Nucl. Instrum. Methods Phys. Res. **A270**, 194 (1988)

95. Y. Mori, G.D. Alton, A. Takagi, A. Ueno, S. Fukumoto, Further evaluation of the high intensity plasma sputter heavy negative ion source. Nucl. Instrum. Methods Phys. Res. **A273**, 5 (1988)

96. G.D. Alton, Y. Mori, A. Takagi, A. Ueno, S. Fukumoto, A high brightness plasma sputter heavy negative ion source. Nucl. Instrum. Methods Phys. Res. **B40/41**, 1008 (1989)

97. G.D. Alton, R. Lohwasser, B. Cui, et al., A high-intensity, rf plasma-sputter negative ion source (1994)

98. G.D. Alton, G.D. Mills, J. Dellwo, Design features of a high-intensity, cesium-sputter/plasma-sputter negative ion source. Rev. Sci. Instrum. **65**, 2006 (1994)

99. H. Tsuji, J. Ishikawa, Y. Gotoh, Y. Okada, RF plasma sputter-type DC-mode heavy negative ion source. AIP Conf. Proc. **287**, 530 (1992)

100. O. Koneko, Y. Takieri, K. Tsumori, et al., Proceedimgs of 16 International Conference on Fusion Energy, Montreal, **3**, 539 (1996)

101. Y. Okumura, M. Hanada, T. Inoue, et al., Proceedings of 16 Symposium On Fusion Technology, London, **2**, 1026 (1990)

102. K. Tsumori, A. Ando, T. Okuyama, et al., Improvement of the large current negative hydrogen ion source for neutral injection in the large helical device. Fusion Eng. Des. **26**, 473–483 (1995)

103. K. Tsumori, K. Ikeda, M. Kisaki, et al., Challenges toward improvement of deuterium-injection power in the Large Helical Device negative-ion-based NBIs. Nucl. Fusion **62**, 056016 (2020)

104. M. Hanada, A. Kojima, T. Inoue, et al., Development of the JT-60SA neutral beam injectors. AIP Conf. Proc. **1396**, 536 (2011)

105. ITER. [Online]. Available from: www.iter.org (Accessed April 15, 2021)

106. R.S. Hemsworth, D. Boilson, P. Blatchford, M.D. Palma, G. Chitarin, H.P.L. de Esch, et al., Overview of the design of the ITER heating neutral beam injectors. New J. Phys. **19**, 025005 (2017). https://doi.org/10.1088/1367-2630/19/2/025005

107. A. Chakraborty, C. Rotti, M. Bandyopadhyay, M.J. Singh, R. Gangadharan Nair, S. Shah, et al., Diagnostic Neutral Beam for ITER-Concept to Engineering. IEEE Trans. Plasma Sci. **38**, 248–253 (2010). https://doi.org/10.1109/tps.2009.2035809

108. Y. Belchenko, G. Dimov, V. Dudnikov, Physical principles of surface plasma source operation, in *Symposium on the Production and Neutralization of Negative Hydrogen Ions and Beams, Brookhaven, 1977* (Brookhaven National Laboratory (BNL), Upton, NY, 1977), pp. 79–96. Бельченко, Г. Димов, В. Дудников, «Физические основы поверхностно плазменного метода получения пучков отрицательных ионов», препринт ИЯФ 77–56, Новосибирск 1977. http://irbiscorp.spsl.nsc.ru/fulltext/prepr/1977/p1977_56.pdf

109. М.Е. Кишиневский, ЖТФ, **48**, 73 (1978). Препринт ИЯФ, 76-18, Новосибирск (1976). M.E. Kishinevskii, Zh. Tekh. Fiz. 48, 1281 (1978); translated in Sov. Phys. - Tech. Phys. 20, 799 (1975). М. Е. Кишиневкий, "К вопросу о вторичной отрицательно-ионной эмиссии", Взаимодействие атомных частиц с твердым телом, с. 22, ХАРЬКОВ 7-9 ИЮНЯ 1976 г. M. E. Kishinevskii, Interraction of atomic particles with solid, p. 22, Kharkov, (1976)

110. M. Hanada, A. Kojima, T. Inoue, K. Watanabe, M. Taniguchi, M. Kashiwagi, et al., Development of the JT-60SA neutral beam injectors. AIP Conf. Proc. **1390**, 536 (2011)

111. Y. Takeiri, O. Kaneko, K. Tsumori, Y. Oka, K. Ikeda, M. Osakabe, et al., High power and long-pulse injection with negative-ion-based neutral beam injectors in the large helical device. Nucl. Fusion **46**, S199–S210 (2006). https://doi.org/10.1088/0029-5515/46/6/s01

112. W. Kraus, M. Bandyopadhyay, H. Falter, et al., Progress in the development of rf driven H⁻ and D⁻ sources for neutral beam injection. Rev. Sci. Instrum. **75**, 1832 (2004)

113. U. Fantz et al., Plasma Phys. Control. Fusion **49**, B563 (2007)

114. SPIDER- The most powerful negative ion beam source. https://www.youtube.com/watch?v=NYC4zjmwCbo

115. N. den Harder, G. Orozco, R. Nocentini, et al., Computational design of magnetic beamlet deflection correction for NNBI. Fusion Eng. Des. **173**, 112837 (2021)

116. C. Wimmer, et al., BATMAN Upgrade: General results from beam optics studies, NIBS 2022, Padova, Italy, (2022)

117. D. Wünderlich, R. Riedl, F. Bonomo, et al., Long pulse operation at ELISE: Approaching the ITER parameters. AIP Conf. Proc. **2052**, 040001 (2018). https://doi.org/10.1063/1.5083735

118. Heinemann et al., Fusion Eng. Des. **146**, 455 (2019)

119. U. Fantz, RF-driven ion sources for fusion large and powerful ion sources for H⁻ and D⁻, NIBS 2022, Padova, (2022)

120. D. Wünderlich, I.M. Montellano, M. Lindqvist, A. Mimo, S. Mochalskyy, U. Fantz, Effects of the magnetic field topology on the co-extracted electron current in a negative ion source for fusion. J. Appl. Phys. **130**, 053303 (2021). https://doi.org/10.1063/5.0054949

121. R. Nocentini, U. Fantz, P. Franzen, et al., Toward a large RF ion source for the ITER neutral beam injector: Overview of the ELISE test facility and first results. IEEE Transa. Plasma Sci. **42**(3) (2014)

122. D. Yordanov, D. Wunderlich, C. Wimmer, U. Fantz, On the effect of biased surfaces in the vicinity of the large extraction area of the ELISE test facility. J. Phys. Conf. Ser. **2244**, 012050 (2022)

123. R.S. Hemsworth et al., Overview of the design of the ITER heating neutral beam injectors. New J. Phys. **19**, 025005 (2017)

124. V. Toigo, R. Piovan, S. Dal Bello, et al., The PRIMA Test Facility: SPIDER and MITICA test-beds for ITER neutral beam injectors. New J. Phys. **19**, 085004 (2017)

125. M. Boldrin, The 1MV MITICA power supply beyond the modern technologies limits: First experience during integration phas, NIBS 2022, Padova, Italy, (2022)

126. M. Agostini, M. Brombin, N. Marconato, E. Sartori, R. Pasqualott, G. Serianni, Co-extracted electrons and beam inhomogeneity in the large negative ion source SPIDER. Fusion Eng. Des. **168**, 112440 (2021)

127. A. Ivanov, G. Abdrashitov, V. Anashin, Y. Belchenko, A. Burdakov, V. Davydenko, P. Deichuli, G. Dimov, A. Dranichnikov, V. Kapitonov, V. Kolmogorov, A. Kondakov, A. Sanin, I. Shikhovtsev, N. Stupishin, A. Sorokin, S. Popov, M. Tiunov, V. Belov, A. Gorbovsky, V. Kobets, M. Binderbauer, S. Putvinski, A. Smirnov, L. Sevier, Rev. Sci. Instrum **85**, 02B102 (2014). https://doi.org/10.1063/1.4826326

128. D. Wünderlich et al., NNBI for ITER: status of long pulses in deuterium at the test facilities BATMAN Upgrade and ELISE. Nucl. Fusion **61**, 096023 (2021)

129. D. Wünderlich, R. Riedl, M. Fröschle, U. Fantz, et al., Operation of Large RF Driven Negative Ion Sources for Fusion at Pressures below 0.3 Pa. Plasma **4**, 172–182 (2021). https://doi.org/10.3390/plasma4010010

130. D. Wünderlich, R. Riedl, M. Fröschle, U. Fantz, B. Heinemann, Operation of large RF driven negative ion sources for fusion at pressures below 0.3 Pa. Plasma **4**, 172–182 (2021). https://doi.org/10.3390/plasma4010010

131. E. Sartori et al., First operations with cesium of the negative ion source SPIDER. Nucl. Fusion **62**, 086022 (2022)

132. K. Pandya, et al., Source performance and optimization in cesiated mode in ROBIN, NIBS 2022, Padova, Italy, (2022)

133. G. Abdrashitov, Y. Belchenko, A.A. Ivanov, Negative ion production in the RF surface-plasma source. AIP Conf. Proc. **1515**, 197 (2013)

134. Y. Belchenko, A. Gorbovsky, A. Ivanov, S. Konstantinov, A. Sanin, I. Shikhovtsev, M. Tiunov, Multiaperture negative ion source. AIP Conf. Proc. **1515**, 167 (2013). https://doi.org/10.1063/1.4792783

135. A.A. Ivanov, G.F. Abdrashitov, V.V. Anashin, et al., Development of a negative ion-based neutral beam injector in Novosibirsk. Rev. Sci. Instrum. **85**, 02B102 (2014)

136. O.Z. Sotnikov, Studies of Hydrogen negative ion sources for injector of high energy neutrals, PhD thesis, Novosibirsk, 2018. О.З. Сотников, «Исследование источника отрицательных ионов водорода для инжектора высокоэнергетичных нейтралов», Кандидатская диссертация, Новосибирск, (2018)

137. O. Sotnikov, A. Ivanov, Y. Belchenko, et al., Development of high-voltage negative ion based neutral beam injector for fusion devices, Nucl. Fusion **61**, 116017 (2021). V. Dudnikov, R. Johnson, Advanced large volume surface plasma H-/D- source for neutral beam injectors, Open System 2010, Novosibirsk, Russia, July, 2010

138. Y. Belchenko, A.A. Ivanov, A. Sanin, O. Sotnikov, Extracted and electrode currents in the inductively driven surface-plasma negative hydrogen ion source, NIBS 2016, Oxford (2018)

139. O.Z. Sotnikov, Y.I. Belchenko, A.A. Ivanov, A.L. Sanin, *Study of Negative Ion Based Injector Prototype at Budker Institute of Nuclear Physics*, ICIS 2021 (2021). 2021-ICIS-Sotnikov.pdf

140. Y. Xiea, C. Hu, J. Wei, et al., Research activities of RF-based negative ion source in the ASIPP, AIP Conf. Proc. **2373**, 030001 (2021); https://doi.org/10.1063/5.0057461. Published Online: 30 July 2021. Y. Xiea, et al., First results of negative ion extraction with Cs for CRAFT prototype negative beam source, Plasma Sci. Technol. **23** 012001 (2021)

141. K.N. Leung, G.J. DeVries, W.F. DiVergilio, R.W. Hamm, C.A. Hauck, W.B. Kunkel, D.S. McDonald, M.D. Williams, Rev. Sci. Instrum **62**, 100 (1991)

142. K.N. Leung, W.F. DiVergilio, C.A. Hauck, W.B. Kunkel, D.S. McDonald, Proceedings of 1991 IEEE Particle Accelerator Conference, San Francisco, CA (May 1991)

143. K.N. Leung, D.A. Bachman, D.S. McDonald, Production of H⁻ ions by an RF driven multi-cusp source. AIP Conf. Proc. **287**, 368 (1992)

144. J.R. Alonso, High-current negative-ion sources for pulsed spallation neutron sources: LBNL Workshop, October 1994 (invited). Rev. Sci. Instrum. **67**(3), 1308 (1996)

145. R. Thomae, P. Bach, R. Gough, J. Greer, R. Keller, K.N. Leung, Measurements on the H- ion source and low energy beam transport section for the SNS front-end system, in *XX International Linac Conference*, Monterey, California, p. 233 (2005)

146. S.K. Mukherjee, D. Cheng, M.A. Leitner, K.N. Leung, P.A. Luft, R.A. Gough, R Keller, M.D. Williams, Mechanical design of the prototype H- ion source for the spallation neutron source, in *Proceedings of the 1999 Particle Accelerator Conference*, New York, report WEA14 (1999)

147. M.A. Leitner, D.W. Cheng, R.A. Gough, R. Keller, K.N. Leung, S.K. Mukherjee, P.K. Scott, M.D. Williams, High-current, high-duty-factor experiments with the H- ion source for the spallation neutron source, in *Proceedings of the 1999 Particle Accelerator Conference*, New York, (1999)

148. M.A. Leitner, R.A. Gough, K.N. Leung, M.L. Rickard, P.K. Scott, A.B. Wengrow, M.D. Williams, D.C. Wutte, Development of the radio frequency driven H⁻ ion source for the National Spallation Neutron Source. Rev. Sci. Instrum. **69**, 962 (1998)

149. D.W. Cheng, R.A. Gough, M.D. Hoff, R. Keller, M.A. Leitner, K.N. Leung, J.W. Staples, M.D. Williams, Design of the prototype low energy beam transport line for the spallation neutron source, in *Proceedings of the 1999 Particle Accelerator Conference*, New York, (1999), p. 1958

150. R.F. Welton, M.P. Stockli, S.N. Murray, T.A. Justice, R. Keller, Operation of the SNS Ion Source At High Duty-Factor, in *Proceedings of EPAC 2004*, Lucerne, Switzerland (2004)

151. R.F. Welton, M.P. Stockli, S.N. Murray, R. Keller, Recent advances in the performance and understanding of the SNS ion source. AIP Conf. Proc. **1515**, 292 (2013)

152. M.P. Stockli, B. Han, S.N. Murray, et al., Ramping up the Spallation Neutron Source beam power with the H⁻ source using 0 mg Cs/day. Rev. Sci. Instrum. **81**, 02A729 (2012)

153. M.P. Stockli, B.X. Han, S.N. Multay, D. Newland, T.R. Pennisi, M. Santana, R.F. Welton, Ramping up the SNS Beam Power with the LBNL Baseline H⁻ Source, in *Negative Ions, Beams and Sources: 1st International Symposium*, ed. by E. Surrey, A. Simonin, vol. 223, (American Institute of Physics, 2009)

154. M.P. Stockli, B.X. Han, S.N. Murray Jr., T.R. Pennisi, M. Santana, C.M. Stinson, J. Tang, R.F. Welton, Recent performance of and extraction studies with the spallation neutron source H⁻ injector. AIP Conf. Proc. **1869**, 030010 (2017)

155. M.P. Stockli, Plasma-wall interactions in the cesiated SNS H⁻ ion source, in *26th Summer School and International Symposium on the Physics of Ionized Gases (SPIG2012)*, IOP

Publishing, J. Phys.: Conf. Ser. **399**, 012001 (2012). M. P. Stockli, Pulsed, high-current H- Ion Sources for Future Accelerators, p. 144, ICFA Beam Dynamics Newsletter No. 73 Issue Editor: G. Machicoane and P. N. Ostroumov Editor in Chief: Y. H. Chin, icfa-bd.kek. jp/Newsletter73.pdf

156. B.X. Han, M.P. Stockli, R.F. Welton, et al., Recent performance of the SNS H- ion source with a record long run. AIP Conf. Proc. **2373**, 040004 (2021). https://doi.org/10.1063/5.0057408

157. M.P. Stockli, B.X. Han, V. Dudnikov, et al., Operating the SNS RF H - ion source with a 10% duty factor, IPAC 2019, Melbourne, Australia (2019)

158. V. Dudnikov et al., Carbon Film in RF Surface Plasma Source with Cesiation, ICIS 2017, Geneva, Svitzerland (2017)

159. V. Dudnikov, A. Dudnikov, et al., Carbon Film in Radio Frequency Surface Plasma Source with Cesiation, NIBS 2018, Novosibirsk, Russia (2018)

160. V. Dudnikov, *Carbon Film in Radio Frequency Surface Plasma Source with Cesiation*, ArXiv, 1808.06003 (2018)

161. L. Osterlund, D.V. Chakarov, B. Kasemo, Potassium adsorption on graphite (0001). Surf. Sci. **420**(2), 174 (1999)

162. S.Y. Davydov, Alkali metal adsorption on graphite: Calculation of a work function variation in the Anderson–Newns–Muscat model. Applied Surface Science **257**, 1506 (2010)

163. E.V. Rut'kov, A.Y. Tontegode, Pis'ma Zh. Tekh. Fiz. **7**, 1122 (1981) [Sov. Tech. Phys. Lett. **7**, 480 (1981)]

164. A.Y. Tontegode, E.V. Rut'kov, Usp. Fiz. Nauk **163**, 57 (1993) [Phys.-Usp. **36**, 1053 (1993)]

165. N.R. Gall, E.V. Rut'kov, A.Y. Tontegode, Int. J. Mod. Phys. B **11**, 1865 (1997)

166. A.Y. Tontegode, Prog. Surf. Sci. **38**, 201 (1991)

167. J. Algdal, T. Balasubramanianb, M. Breitholtz, T. Kihlgren, L. Wallden, Thin graphite over-layers: Graphene and alkali metal intercalation. Surf. Sci. **601**, 1167–1175 (2007)

168. R. Gutser, C. Wimmer, U. Fantz, Work function measurements during plasma exposition at conditions relevant in negative ion sources for the ITER neutral beam injection. Rev. Sci. Instrum. **82**, 023506 (2011)

169. M. Breitholtz, J. Algdal, T. Kihlgren, S.-Å. Lindgren, L. Walldén, Alkali-metal-deposition-induced energy shifts of a secondary line in photoemission from graphite. Phys. Rev. B **70**, 125108 (2004)

170. R. Schletti, P. Wurz, T. Fröhlich, Rev. Sci. Instrum. **71**, 499 (2000)

171. V.Z. Kaibyshev, V.A. Koryukin, V.P. Obrezumov, Atom. Energiya **69**(3), 196–197 (1990)

172. G. Cartry, L. Schiesko, C. Hopf, A. Ahmad, M. Carrere, J.M. Layet, P. Kumar, R. Engeln, Production of negative ions on graphite surface in H_2/D_2 plasmas: Experiments and SRIM calculations. Phys. Plasmas **19**, 063503 (2012)

173. M.A. Gleeson, A.W. Kleyn, Effects of Cs-adsorption on the scattering of low energy hydro-genions from HOPG. Surf. Sci. **420**, 174 (1999)

174. R. Souda, E. Asari, H. Kawanowa, T. Suzuki, S. Otani, Capture and loss of valence electrons during low energy H^+ and H^- scattering from LaB6(100), Cs/Si (100) graphite and LiCl. Surf. Sci. **421**, 89 (1999)

175. J. Peters, *Review of Negative Hydrogen Ion Sources high Brightness/High Current* (Linac 98, Chicago, 1998)

176. R.F. Welton, V.G. Dudnikov, K.R. Gawne, B.X. Han, S.N. Murray, T.R. Pennisi, R.T. Roseberry, M. Santana, M.P. Stockli, M.W. Turvey, H^- radio frequency source development at the Spallation Neutron Source. Rev. Sci. Instrum **83**, 02A725 (2012)

177. R.F. Welton, V.G. Dudnikov, B.X. Han, S.N. Murray, T.R. Pennisi, R.T. Roseberry, M. Santana, M.P. Stockli, AIP Conf. Proc. **1515**, 341–348 (2013)

178. R.F. Welton, A. Aleksandrov, V.G. Dudnikov, B.X. Han, S.N. Murray, T.R. Pennisi, M. Piller, Y. Kang, M. Santana, M.P. Stockli, A look ahead: Status of the SNS external antenna ion source and the new RFQ test stand. AIP Conf. Proc. **1655**, 030002 (2015)

179. V. Dudnikov et al., AIP Conf. Proc. **925**, 153 (2007)

180. V. Dudnikov, R.P. Johnson, M. Stockli, R. Welton, H^- ion sources for high intensity proton drivers, report MO6RFP036, in *Proceedings of PAC09*, Vancouver, BC, Canada (2009)

181. D. Curreli, F. Chen, Equilibrium theory of cylindrical discharges with special application to helicons. Phys. Plasmas **18**, 113501 (2011)

182. V. Dudnikov, B. Han, R.P. Johnson, S.N. Murray, T.R. Pennisi, M. Santana, M.P. Stockli, R.F. Welton, Surface plasma source electrode activation by surface impurities. AIP Conf. Proc. **1390**, 411 (2011)

183. V. Dudnikov, R.P. Johnson, B. Han, et al., Features of radio frequency surface plasma sources with a solenoidal magnetic field. AIP Conf. Proc. **1869**, 030023 (2017)

184. V. Dudnikov, R.P. Johnson, G. Dudnikova, Negative ion radio frequency surface plasma source with solenoidal magnetic field. AIP Conf. Proc. **2052**, 050018 (2018)

185. V. Dudnikov, A. Dudnikov, Radio frequency discharge with control of plasma potential distribution. Rev. Sci. Instrum **83**, 02A720 (2012)

186. J. Lettry, D. Aguglia, Y. Coutron, A. Dallochio, H. Perreira, E. Chaudet, J. Hansen, E. Mahner, S. Mathot, S. Mattei, O. Midttun, P. Moyret, D. Nisbet, M. O'Neil, M. Paoluzzi, C. Pasquino, J. Sanchez Arias, C. Schmitzer, R. Scrivens, D. Steyaert, J. Gil Flores, H⁻ Ion sources for CERN's Linac4. AIP Conf. Proc. **1515**, 302 (2013)

187. R. Scrivens, M. Kronberger, D. Küchler, J. Lettry, et al., Overview of The Status and Developments On Primary Ion Sources at CERN, report THPS025, in *Proceedings of IPAC2011, San Sebastián, Spain* (2011)

188. J. Lettry, D. Aguglia, J. Alessi, P. Andersson, et al., CERN's Linac4 H⁻ sources: Status and operational results. AIP Conf. Proc. **1655**, 030005 (2015)

189. J. Lettry, S. Bertolo, U. Fantz, et al., Linac4 H⁻ source R&D: Cusp free ICP and magnetron discharge. AIP Conf. Proc. **2052**, 050008 (2018). https://doi.org/10.1063/1.5083762

190. S. Mochalskyy, J. Lettry, T. Minea, Beam formation in CERNs cesiated surfaces and volume H− ion sources. New J. Phys. **18**, 085011 (2016)

191. H. Oguri, A. Ueno, K. Ikegami, Y. Namekawa, K. Ohkoshi, Phys.Rev. ST Accel. Beams **12**, 010401 (2009)

192. A. Ueno, Cesiated surface H⁻ ion source: optimization studies. New J. Phys. **19**, 015004 (2017)

193. A. Ueno, K. Ohkoshi, K. Ikegami, et al., 110 mA operation of J-PARC cesiated RF driven H− ion source. AIP Conf. Proc. **2373**, 040002 (2021). https://doi.org/10.1063/5.0057552

194. A. Ueno et al., How to make extraction electrode current lower than beam and corresponding beam qualities in J-PARC cesiated RF-driven H⁻ ion source 66 mA operation. AIP Conf. Proc. **2011**, 050002 (2018)

195. A. Ueno, et al., Beam intensity bottleneck specification and 100 mA operation of J-PARC cesiated RF-driven H⁻ ion source, NIBS 2018, Novosibirsk, (2018)

196. A. Ueno, et al, 120 mA operation of J-PARC cesiated RF-driven H⁻ ion source, NIBS 2022, Padova, Italy, (2022)

197. W. Chen et al., Over 7200 hours commissioning of RF-driven negative hydrogen ion source developed at CSNS, NIBS 2022, Padova, Italy, (2022)

198. R. Welton, et al., Negative hydrogen ion sources for particle accelerators: Sustainability issues and recent improvements in long-term operations. J. Phys.: Conf. Ser. **2244**, 012045 (2022)

199. V.G. Dudnikov, G.I. Fiksel, Surface ftasma source of hydrogen atoms with an energy of hundreds eV. J. Phys. **40**, C7–479 (1979)

200. V. Dudnikov, G. Fiksel, Surface plasma source of intense flux of accelerated atoms, preprint INP 80–44, Novosibirsk (1980). В. Дудников, Г. Фиксель, Поверхностно-плазменный источник интенсивных потоков ускоренных атомов, Препринт ИЯФ 80–44, Новосибирск (1980)

201. V. Dudnikov, G. Fiksel, Fisika Plasmy **7**, 283–288 (1981)

202. V.G. Dudnikov, Surface-plasma method for the production of negative ion beams. Phys. Uspekhi **62**, 1233 (2019)

203. Y.I. Belchenko, V.I. Davydenko, P.P. Deichuli, et al., Studies of ion and neutral beam physics and technology at the Budker Institute of Nuclear Physics, SB RAS. Phys. Uspekhi **61**(6), 531–581 (2018)

204. V. Dudnikov, Development of a surface plasma method for negative ion beams production. J. Phys.: Conf. Ser. **2244**, 012034 (2022). https://doi.org/10.1088/1742-6596/2244/1/012034

Chapter 6
Transport of High Brightness Negative Ion Beams

Abstract High brightness negative ion beam transport, beam-plasma interaction, beam space charge compensation, and beam instability excitation are discussed, and designs of low-energy beam transport (LEBT) system are considered.

6.1 Instabilities in High Brightness Beam Transportation

H$^-$ ion beams with parameters enough to meet the needs of same important applications [1–3] were obtained in the early stage of SPS (surface plasma source) investigation [4, 5].

The transport of high brightness beams of negative ions has been considered in [6–8]. A setup for measuring the beam parameters after transport is shown in Fig. 5.4. It consists of an SPS with beam formation system and an absolute beam brightness detector [9]. The local normalized brightness of the beam B_n depends on the local emission current density J and the local effective transverse temperature of the ions T_i,

$$B_n = JMc^2 / \pi^2 T_i,$$

where M is the ion mass and c the speed of light.

Fluctuations in the parameters of the gas-discharge plasma cause fluctuations in the intensity of the extracted H$^-$ ion beams. Figure 6.1 illustrates the ion-optical characteristics of beams obtained from discharges with chaotic fluctuations [10].

Ions of energy $E = 20$ keV and current after magnetic analyzer $I_2^- = 0.1$ A for this case are characterized by values $2\Delta x = 3$–3.5 cm, $2\Delta\alpha_x = 7 \times 10^{-3}$ rad, $2\Delta y = 3$–3.5 cm, and $2\Delta\alpha_y = 2.5$–3 $\times 10^{-2}$ rad (see Fig. 6.1). The ordered beam divergence is preserved at a level of 10^{-2} radians per cm of transverse displacement. The recorded values of normalized emittance $E_{nx} = 0.5$ π mm.mrad and $E_{ny} = 2$ π mm.mrad correspond to a half-width of transverse energy spread $\Delta W_{ox} = 1.5$ keV and $\Delta W_{oy} = 80$ eV,

© The Author(s), under exclusive license to Springer Nature Switzerland AG 2023   407
V. Dudnikov, *Development and Applications of Negative Ion Sources*, Springer
Series on Atomic, Optical, and Plasma Physics 125,
https://doi.org/10.1007/978-3-031-28408-3_6

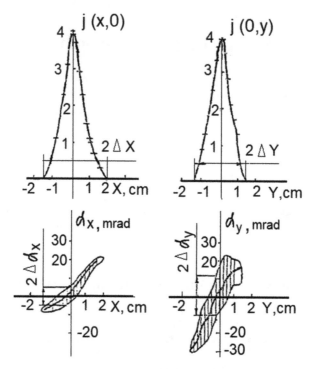

Fig. 6.1 Distribution of current density and beam emittance from discharges with chaotic fluctuations. (Reproduced from Dudnikov [5])

but at a discharge voltage $U_p = 100$ V, ions in the plasma cannot produce such high transverse energies in any way. It is natural to assume that such a high transverse energy spread is produced by the action of transverse fluctuating electric fields caused by broadband intensity fluctuations. Changes in the orientation of the emitting surface cannot yield such high values because of the limited aperture of the extraction system. These results relate to the formation of beams at an average residual gas pressure $p = 2$–5×10^{-5} Torr. In this case, through a magnet with an aperture along the magnetic field of 3 cm, 90% of the ions passed through collector 5 in Fig. 5.4. The impurity level of heavy ions can be 5–10%.

By increasing the average pressure in the chamber to $p = 10^{-4}$ Torr by reducing the pumping speed or by increasing the gas flow to the transport channel, the emittance of the beams generated was reduced by a factor of 1.5–2, and despite the decrease in the ratio of currents I_2/I_1 (see Fig. 5.4) to 0.8, the maximum H⁻ ion current density at the collector after the bending magnet increases. Apparently, in this case, the decrease in fluctuating fields in the beam leads to faster accumulation of space-charge compensating particles, and overcompensation up to positive potential occurs, with the accumulation of electrons.

Note that the generation of secondary particles in the transport beam and the interaction of the transport beam with the secondary ion beam-plasma should have a strong influence on the characteristics of the negative ion beams, as is the case for positive ion beams. Investigation of these questions with respect to beams of negative ions has only recently begun. The need for further investigation of this problem is beyond doubt.

When fluctuations are eliminated, the ordered beam divergence along the magnetic field after the extractor decreases from 0.2 to 0.05 radians. In this mode, for a beam of current $I_2 = 0.1$ A at energy $eU_o = 20$ keV, the values are $2\Delta x = 1\text{–}1.5$ cm, $2\Delta \alpha x = 2 \times 10^{-3}$ rad, $2\Delta y = 1$ cm, and $2\Delta \alpha_y = 2 \times 10^{-2}$ rad. These parameters correspond $E_{nx} = 0.03$ mm.mrad, $E_{ny} = 0.2$ mm.mrad, $\Delta W_x = 5$ eV (approx.), and $\Delta W_y = 1$ eV. Such transverse energy spread values are close to the expected values for H⁻ plasma ions with energy ~1 eV. The effect of the optics of the forming system is manifested by a significant excess of ΔW_{ox} over ΔW_{oy}. Because of this, the emittances E_{nx} and E_{ny} differ only by a factor of 6, while the dimensions of the emission slit in the corresponding directions differ by a factor of 20. When fluctuations are eliminated, the average normalized brightness of the formed beams, $B = I_2/\pi^2 E_{nx} E_{ny}$, increases from $B = 1 \times 10^{-2}$ A/mm²mrad² to $B = 2$ A/mm²mrad². The maximum brightness in the center of the beam is greater than this by about an order of magnitude.

In this case, beam transport is possible only with good space charge compensation. It is known that at a sufficient positive ion production rate, because of the faster escape of electrons, gas focusing of the negative ion beam is possible. When beam intensity fluctuations are eliminated, the conditions for gas focusing become particularly favorable.

When the beam current and the extraction voltage are coordinated, the beam divergence is small, and the beam current density is high, as shown in Fig. 6.2 for beam currents of 40 mA and 60 mA. But for a current of 90 mA, the beam spreads sharply, and the current density drops sharply because of a change in the curvature of emission meniscus. The dependence of the H⁻ ion current density in the beam on the emission current density is shown in Fig. 6.3.

For each extraction voltage, there is an emission current density for which the current density in the beam is maximum, and after which, the current density in the beam decreases. This is due to a change in the shape of the emission surface of the plasma (the "plasma meniscus") from concave to convex, away from the optimum shape for optimum beam formation optics.

Strong instability can be excited by secondary electrons entering the beam from a surface bombarded by the beam. The excitation of this instability is demonstrated by Fig. 6.4. Upper trace (a) is the signal from the collector (without secondary electron suppression) across a resistance of 100 Ohms; the lower trace (b) is the signal from a side collector at negative potential shielded by a grid. At the beginning, the negative potential of the beam stops secondary emission, but the signal oscillations are present. After compensation of space charge and overcompensation, a strong instability develops which leads to the ejection of compensating particles from the beam to the lateral collector. This stops the instability and again allows accumulation of space charge-compensating particles, which again initiates instability. This instability also leads to strong heating of the beam.

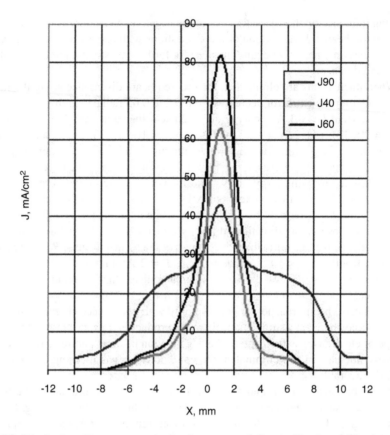

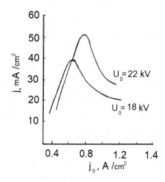

Fig. 6.2 Distribution of beam current density for currents of 40 mA, 60 mA, and 90 mA, at the same extraction voltage. (Reproduced from Dudnikov [5])

Fig. 6.3 Dependence of H⁻ ion current density on emission current density, for two different extraction voltages. (Reproduced from Dudnikov [5])

A noise-free beam can be stable if secondary electron emission is suppressed and the gas density is not very high. The energy dissipation of the oscillating compensating particles is very small, and all perturbations (e.g., micro-breakdowns in the extractor) can induce a long "ringing." Only after suppression of all perturbations

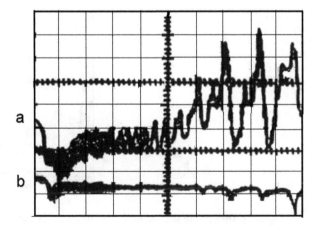

Fig. 6.4 Instability of an H⁻ ion beam caused by electron counterflow from the collector: (a) ion beam collector signal and (b) signal from a side collector shielded by a grid. (Reproduced from Dudnikov [5])

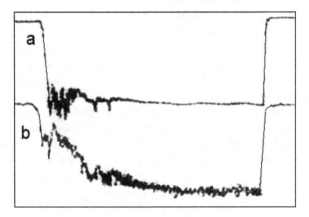

Fig. 6.5 Oscillogram of a quiet beam with a modulated current density. (a) Signal from the collector—the total beam current. (b) Signal from a collector behind a slit. (Reproduced from Dudnikov [5])

can we obtain a stable noise-free beam with high brightness and preservation of the ion temperature at a level below 1 eV. In this case, it is important to have an optimal relationship between the emission current density and the extraction voltage in order to prevent "overheating of ions" due to "overcooling" caused by beam extension—during beam extension along the magnetic field, the transverse ion temperature should decrease to the meV level (overcooling) but is held at the eV level by beam intensity fluctuations (overheating).

A noise-free cold beam can have a noisy microstructure. Measurements of the current density by a collector with a small collimating hole show high-frequency fluctuations, as shown in Fig. 6.5. The relative level of this noise is higher for a

Fig. 6.6 Oscillograms of
H⁻ ion beam current.
(From the Institute of
Nuclear Research,
Moscow). Upper signal (a)
with noise and bottom (b)
without noise

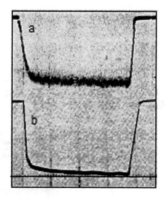

smaller slit and the noise grows along the beam. Noise in current density is lower if the slit used to measure the current density is oriented perpendicular to the magnetic field. This looks like breaking the beam into thin layers perpendicular to the magnetic field.

An example of the successful production of a noise-free beam was obtained at the Institute of Nuclear Research in Moscow [11]. A photograph of the beam current with and without noise is shown in Fig. 6.6.

Strong instability of an intense H⁻ ion beam has been observed and reported [12, 13]. The installation setup for observing this effect is shown in Fig. 6.7. It consists of a SPS with cesiation for H⁻ ion beam production 1; bending/focusing magnet 2; long vacuum tube for beam transport, with Rogowski coils to measure the beam current 3; secondary ion collector-analyzers 4; quadrupole lenses 5; and ion collectors 6. The beam potential was measured by capacitive probes consisting of a small thin-walled cylinder mounted on a quartz tube with a thin metal rod connected to a high capacitance cable. Since the coefficient of secondary emission of the metal is greater than one, the cylinder rapidly acquires the beam potential and monitors it without inertia. The H⁻ ion beam obtained from the SPS, with current up to 80 mA and current density up to 50 mA/cm², was investigated. The ion energy was up to 15 kV, the pulse duration was up to 0.6 ms, the gas pressure was 2×10^{-6} to 1×10^{-4} Torr, and the initial fluctuations in beam current are ~1%. The beam current was measured by Rogowski belts and the current density by ion collectors.

Figure 6.8 shows oscilloscope signals of the beam potential at different distances z of the capacitive probe from the beginning of the chamber. The large negative potential at the beginning of the pulse for a small z corresponds to the uncompensated space charge of the beam. During this period, the beam becomes very diffuse and part of it is lost to the walls of the chamber. Thereafter, the potential, which is proportional to the beam current at a given location, decreases substantially with increasing z.

Subsequent to the time for beam space charge compensation, $\tau_k = (n_a \sigma_i v^-)^{-1}$, where n_a is the density of gas molecules in the chamber, σ_i is the cross section for the formation of slow positive ions in collisions of H⁻ ions with molecules, and v^- is the velocity of beam ions, during which positive ions accumulate in the beam, for

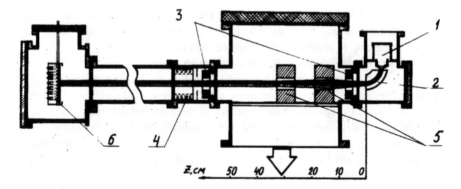

Fig. 6.7 Diagram of the installation for studying H⁻ ion beam transport. (Reproduced from Gabovich et al. [14])
(1) SPS for H⁻ ion beam production; (2) bending and focusing magnet; (3) Rogowski coil; (4) secondary ion collector-analyzers; (5) quadrupole lenses; and (6) ion collectors

Fig. 6.8 Development of instability and decompensation of H⁻ ion beam during transport. (Reproduced from Gabovich et al. [14]) Oscillograms of beam potential for left-hand column, slow oscilloscope sweep speed (100 μs time bar), and right-hand column, fast oscilloscope sweep speed (1 μs time bar). $z =$ (a) 10 cm, (b) 20 cm, (c) 30 cm, (d) 35 cm, and (e) 45 cm, $P = 2.5 \times 10^{-6}$ Torr

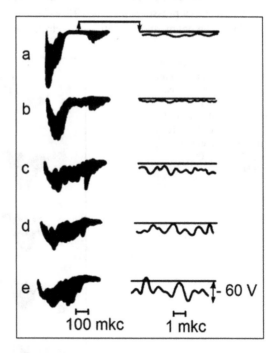

small z good beam space charge compensation is observed and an absence of large static and alternating fields as well as fluctuations in beam current density. However, already at $z = 20$–30 cm, compensation is disturbed, with fluctuations in the negative beam potential appearing, the amplitude of which increases to values that are tens of percent of the completely uncompensated space charge potential of the beam, $\Delta\varphi = I_{H^-}/v^-$, where I_{H^-} is the beam current at a given cross section and v^- is ion velocity. Figure 6.8 shows the dependence of the current density at the center of the beam on distance along the beam. Fluctuations in current density with frequency 1 MHz increase up to distances of 27 cm and then decrease.

The observed phenomenon is associated with beam ion-ion instability. The amplitude of fluctuations increases in the direction of beam propagation. The Langmuir frequency of the compensating positive ions calculated from the beam density is close to the characteristic frequency of the observed fluctuations in potential ~1 MHz. Signals proportional to the current density, taken from collectors positioned symmetric with respect to the axis and spaced by 20 mm, are in antiphase; this indicates that the oscillations occur in a direction perpendicular to the beam propagation direction.

For large z, strong potential fluctuations are accompanied by almost 100% fluctuation in the beam current density with significantly smaller total current fluctuations.

An important characteristic of the conditions for the onset of oscillations is the dependence of their amplitude on the gas pressure, shown in Fig. 6.9. Estimates show that at low pressures $p = 5 \times 10^{-6} - 4 \times 10^{-5}$ Torr, the electrons that form rapidly leave the beam, and the ratio of electron density to positive ion density is very

Fig. 6.9 Fluctuations in beam potential at different pressures. (Reproduced from Gabovich et al. [14]) $P =$ (**a**) 5×10^{-6} Torr, (**b**) 1.4×10^{-5}, (**c**) 2.8×10^{-5}, (**d**) 4×10^{-5}, (**e**) 6×10^{-5}, and (**f**) 6.8×10^{-5}. $z = 50$ cm

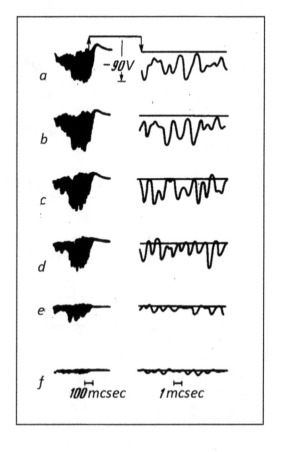

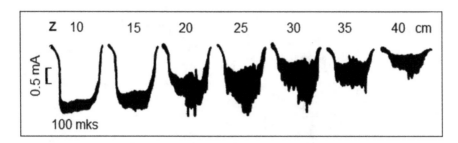

Fig. 6.10 Dependence of the current density in the center of the beam on distance. (Reproduced from Gabovich et al. [14])

small, $n_e/n_i \sim 5 \times 10^{-4} - 5 \times 10^{-3}$. This explains the possibility of the existence of observed fluctuations of large amplitude and their independence from the pressure in this interval; see Fig. 6.9a–d. With further increase in gas pressure, a known transition from negative to a small positive potential occurs, at which the electron density becomes comparable to the positive ion density; this is accompanied by a significant decrease in the amplitude of the potential oscillations, as in Fig. 6.9e, f.

The dependence of current density in the center of the beam on distance along the beam is shown in Fig. 6.10 [14]. The fluctuation level increases up to a distance of 27 cm and then decreases.

It is shown in this paper [14] that even in the absence of significant initial current fluctuations, strong fluctuations in negative beam potential arise as a result of the formation of a beam ion-ion instability in the compensated beam of negative ions as it propagates. The associated fields can lead to beam divergence and emittance growth, which can prevent the transport of an intense beam of negative ions at low gas pressure.

A general theory of extraction and transport of H⁻ ion beams has been developed by M. Reiser [15], who predicts the possibility of H⁻ ion beam extraction with current up to 200 mA and normalized brightness of ~1 A/(mm.mrad)². It is predicted that the equilibrium transverse dimension of the beam for gas focusing is determined by the emittance and does not depend on the intensity of the beam.

Experimental studies of gas focusing have been carried out and reported [16, 17]. A four-grid energy analyzer of secondary particles ejected from the beam was used. A schematic of the experiment is shown in Fig. 6.11. The construction of the four-grid energy analyzer of secondary particles is shown.

An SPS with Penning discharge was used, with emission slit 8×0.7 mm², which generates a beam of H⁻ ions of 80–90 mA current and energy 21 keV, with a noise level of ~7% as measured by a Faraday cup 3.5 cm from the emission aperture. The SPS operated at a frequency of 5 Hz with a pulse duration of 1 ms. The current density in the beam was 30 mA/cm² at 11 cm from the emission slit as measured by the emittance analyzer. A four-grid analyzer was located at a distance of 22 cm from the emission slit at a distance of 6.5 cm from the beam. A variety of potentials could be applied to the beam collector located at 55 cm. The hydrogen density at the operating source was 5×10^{12} cm⁻³. He or Xe gas was supplied to the transport area. The

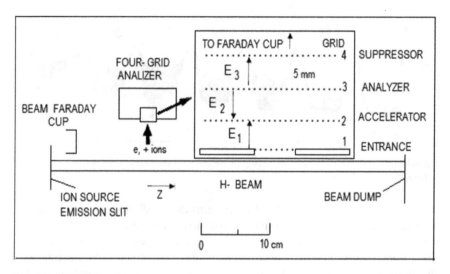

Fig. 6.11 Schematic of an experiment to measure the beam potential in gas focusing. The construction of the four-grid energy analyzer of secondary particles is shown. (Reproduced from Sherman et al. [16])

analyzer was used to measure the dependence of total ion current (at the analyzer) on the total cross section for the formation of secondary ions by the H$^-$ ion beam and also the energy spectra of the ions and electrons. Measurable ion currents at the analyzer appear when the threshold density of neutralizing gas exceeds 32×10^{12} cm^{-3} for He and 0.5×10^{12} cm^{-3} for Xe, which coincide with the calculation. At the same time, 57% and 11% (for He and Xe, respectively) of the negative ion beam was neutralized (stripped). From the energy spectra of ions leaving the beam at positive potential, we can determine the potential at the center of the beam and the potential drop between the center and the edge of the beam. The dependence of the potential of the center of the beam and the potential drop in the beam on xenon density are shown in Fig. 6.12. For Xe addition, the potential can reach up to 8 V and the potential drop in the beam up to 6 V. The potential for He addition is six times smaller. The electron current at the analyzer exceeds the calculated current by a factor of 100.

The dependence of the emittance of an H$^+$ ion beam with current 75 mA on density of added gas is shown in Fig. 6.13 [18]. Note the impressive decrease in emittance by a factor of 3 with increased krypton density in the transport path by just 1 m Pascal. Good space charge compensation and overcompensation can be very important for preservation of low emittance negative ion beams.

The impact of beam-plasma interaction on the transport at of negative ions in a cyclotron has been considered [19] and the choice of methods for transporting negative ion beams in [8].

The transport of H$^-$ ion beams obtained from the 4X SPS with Penning discharge at LANL (Los Alamos National Laboratory) has been considered and reported [20, 21]. This SPS was used with a distance between working cathode surfaces of 20 mm

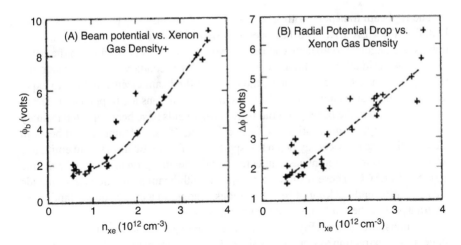

Fig. 6.12 Dependence of (**a**) potential of the center of the beam and (**b**) potential drop in the beam, on xenon density. (Reproduced from Sherman et al. [17])

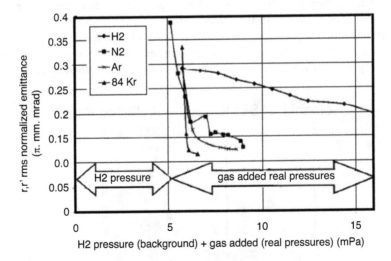

Fig. 6.13 Dependence of emittance on added gas density, for an H^+ ion beam with current 75 mA. (Reproduced from Beauvais et al. [18])

and an anode window 12×16 mm^2 with a circular emission aperture 5.4 mm in diameter. From a discharge with a $\pm 20\%$ noise level, a beam of current 100 mA was obtained with a normalized emittance of 0.23 π mm.mrad. From a discharge without noise (<1%) and with a larger gas supply, a beam of current of 67 mA with a normalized emittance of 0.12 π mm.mrad was obtained. This latter emittance is two times larger than the theoretical value for an ion temperature of 2 eV and an aperture of 5.4 mm.

The H⁻ beam transport, with space charge neutralization by positive ions, in the LEBT of the CERN RF SPS has been reported [22]. The space charge of intense unbunched negative ion beams can be compensated by ions created when the beam ionizes the residual gas, which creates a source of secondary particles inside the beam pipe. For negative ion beams, the effect of the beam electric field is to expel the electrons to the beam pipe walls, while the positive ions are trapped and start to be accumulated in the beam potential well. Experiments have been reported to study this space charge compensation (SCC) in a 35 mA, 45 keV H⁻ unbunched beam in the CERN Linac4 low energy beam transport (LEBT) line. Beam size and emittance were measured for different gases injected into the beam region to control the degree and speed of SCC. These results are compared with beam simulations that include the generation and tracking of secondary ions, leading to a unique understanding of the transport of the ion beam in some specific cases.

The transport system (Fig. 6.14) used for the experiments consists of a multi-stage 45 kV extraction system, one solenoid with a nominal focusing strength of 9 T².mm and total length of 304 mm, and horizontal and vertical trajectory correction magnets. Beam properties are measured using a Faraday cup for beam current, grid profile measurement for beam sizes, and a slit-grid emittance meter for the beam phase space. The beam signals from these instruments are sampled with a resolution of 6 μs, allowing the measurements of intensity and phase space to be made as a function of time. The emittance meter slit is grounded to suppress the secondary electrons generated when the beam hits the slit, forcing the secondary electrons to

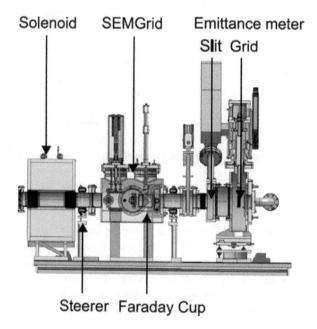

Fig. 6.14 Layout of the LEBT of the CERN RF SPS as installed at the 3 MeV test stand. (Reproduced from Valerio-Lizarraga et al. [22])

come back immediately to the slit and not disturb the beam. The motion of secondary positive ions in the beam potential was simulated. The results of emittance simulation and emittance measurement are represented in Fig. 6.17. In both cases, one can see a significant increase of emittance in the middle part of the emittance diagram, connected with positive ion oscillations. This effect can be eliminated by using heavier gases such as Xe with overcompensation of space charge and accumulation of electrons.

The flow diagram of the beam-plasma model in IBSimu for space charge compensation simulations using secondary particles instead of a reduced beam net current is shown in Fig. 6.15.

The interaction between the beam and the secondary ions created by the beam collision with the background gas along the trajectories of the beam is added to the model via their additional space charge contribution (this model is called the beam-plasma model). Starting from the solution of the full intensity net current model, we calculate the SCC effect in four steps as follows:

1. Track the beam through the electric and magnetic fields from the last potential solution, and the space charge density distribution ρ_b is calculated.
2. Use a Monte Carlo generator to create the secondary ions along the trajectories of the primary beam randomly but taking into account the mean free path of the beam with the residual gas. The mean free path calculation includes both residual H_2 and the injected gas.
3. Track the secondaries through the electric and magnetic field of the last solution for the time; evaluate the space charge density distribution contribution of the

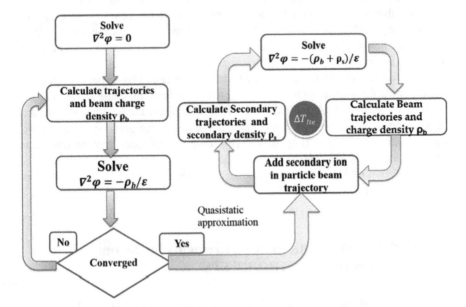

Fig. 6.15 Flow diagram of the beam-plasma model in IBSimu for space charge compensation simulations using secondary particles instead of a reduced beam net current

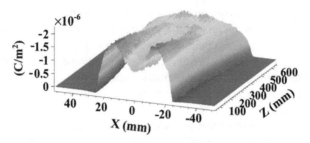

Fig. 6.16 Total space charge density (beam and secondary ions) during the SCC buildup. $t = 0.4\tau$

secondary particles (ρ_s) along their full trajectory. The final positions of the secondary ions are saved in order to reintroduce these particles together with the
new particles created in the next iteration in step 2.

4. Create a new electric potential by solving $\nabla^2\phi = -(\rho_s + \rho_b)/\epsilon_0$, and return to step
   1 until the total beam pulsed length has been simulated by the number of itera-
   tions times the secondary tracking time.

The grounded beam pipe and the negative beam potential trap the ions in the
system. The ions must fulfil the condition $E_i > q[\phi_B(r) - \phi_B(R)]$ to reach the vacuum
pipe at radius R. For $t \ll \tau$, only some ions created in the beam edges are able to
escape, whereas the ions created at the radius r that do not fulfil this condition will
oscillate around the beam center until the SCC reduces the potential sufficiently to
allow the ions to escape radially. During the SCC buildup, the beam potential distribution leads to the highest secondary ion concentration at the beam center, meaning
the space charge is not compensated uniformly across the beam volume in this
period. Figure 6.16 shows the sum of the beam and secondary ion space charge at a
time during the buildup of SCC; this hollow density is the cause of the emittance
growth in the first 60 μs.

The agreement between the simulated and experimental phase space results can
be seen by the unusual features that can be created in the beam phase space in some
cases. By increasing the H_2 pressure to 1×10^{-6} mbar above the baseline, we can see
the appearance of two distinct components in the transverse phase space when the
beam is strongly focused (Fig. 6.17). This type of emittance feature has been
observed previously [23, 24].

A simulation of H− extraction systems for CERN's Linac4 H− ion source is
presented in [25].

6.2 High-Speed Emittance Measurements for Beams
Extracted from J-PARC RF SPS

Oscillation of emittance and Twiss parameters in the negative ion (H-) beam from
the J-PARC (Japan Proton Accelerator Research Complex) 2 MHz RF SPS with
cesiation is measured by application of a double-slit emittance monitor located at

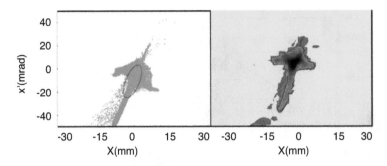

Fig. 6.17 Phase space simulation (left) and measurement (right) for 1.25×10^{-6} mbar H_2. (Reproduced from Valerio-Lizarraga et al. [22])

the RFQ entrance. The emittance monitor is equipped with a newly developed 60 MS/s data acquisition system, so that beam current oscillations at a few MHz can be observed with enough time resolution. From the measurement, it is shown that the beam phase space consists of (1) a direct current component in the beam core, (2) a 2 MHz oscillating component which takes place in the diverging halo, and (3) a doubled RF frequency (4 MHz) oscillation which faintly exists in the beam halo. The major component is the 2 MHz component, which resultantly decides the beam emittance oscillation frequency. A typical value of the beam emittance in the present experiment is 0.33 π mm-mrad, while the amplitude of the 2 MHz oscillation is around 0.04 π mm-mrad. The results indicate that the high-frequency oscillation component occupying about ten percent of the beam from the RF SPS travels a few meters passing through a magnetic lens focusing system [26]. RF beam intensity oscillation decreases space charge compensation and increases beam divergency.

A superposition of the beam phase spaces for phases 1 and 2 is shown in Fig. 6.18. The blue dots are the particle plots at phase 1, and the red dots are the beam orbits at phase 2, which is plotted behind the blue dots. As shown in Fig. 6.18, the phase space consists of three components: (1) beam core, (2) diverging halo, and (3) converging halo. Also, (4) a small asymmetric halo component takes place at the position around −3.4 mm and the angle 80 mrad in phase 2, while no difference is seen in the opposite position and angle. The difference of the beam phase space in phases 1 and 2 mainly lies in the outer part of (2) the diverging halo and (4) in the asymmetric halo component. The beam current wave forms (WF) at several points in these components are compared to understand the dominant beam components and the oscillation frequencies. Together with the components, above WFs are plotted also for the beam core and the converging halo.

Negative ion beam focusing is a key element for advanced applications of negative ion beams such as accelerators for particle physics, compact accelerators for medical fields, and plasma experiments for nuclear fusion because complicated magnetic fields exist both inside of the source plasma and the grid system. In order to understand the beam focusing, phase space structure measurements for a single beamlet have been performed with a research and development negative ion source

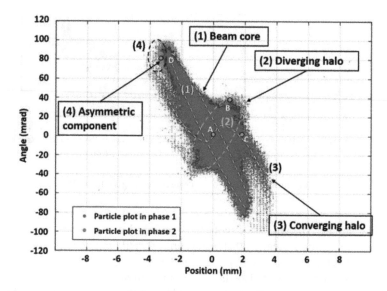

Fig. 6.18 The beam phase space at phases 1 and 2. The blue dots are the particle plots for phase 1 and the red dots are the particle plots for phase 2. The beam phase space in phase 1 is plotted in front of the one in phase 2. Indicators to show each of the components of the phase space are shown: (1) beam core, yellow dashed oval; (2) diverging halo, light blue dashed oval; (3) converging halo, yellow S-shaped curve; and (4) asymmetric halo, black dashed oval

at the National Institute for Fusion Science. A complicated phase space structure is observed in the direction parallel to the filter magnetic field in the vicinity of the plasma grid, while a single Gaussian beamlet structure is observed in the direction perpendicular to the filter field. Detailed analyses for the phase space structure of the single beamlet reveal that the complicated structure can be identified as a combination of three beam components with different beam axes. The shifts of each axis are also observed to depend on the ratio of the acceleration voltage for the extraction voltage, which may significantly degrade the beamlet focusing [27].

6.3 Study of Continuous Wave 33 keV H– Beam Transport Through the Low Energy Beam Transport Line

A transportation of intense H- beam trout low energy transport line of BINP was investigated [28]. The schematic of the injector test stand is shown in Fig. 6.19. It includes the negative ion SPS with cesiation, the bending magnet chamber for beam deflection, and an ~0.8-m-long transport line with an inner diameter of 10 cm. The Penning-type negative ion SPS producing a continuous wave H– beam with the current up to 15 mA and emission current density up to 150 mA/cm^2 was used. The beam 90° turn was performed by the peripheral magnetic field of the source (15° turn) and

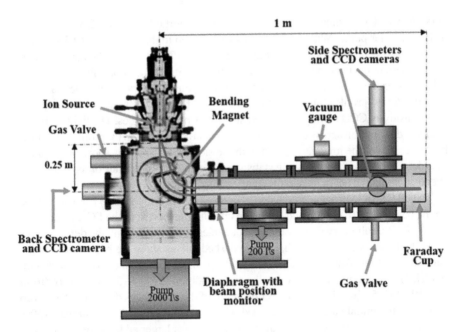

Fig. 6.19 Schematic of test stand for transportation of intense H- beam

by the dipole 75° bending magnet, made of permanent magnets with an iron yoke and equipped with water-cooled coils for the magnetic field adjustment. The beam focusing during the 75° turn, schematically shown in Fig. 6.19, was provided by the radial gradient of the dipole magnetic field and by the edge angle of the magnet poles used. To prevent the magnet pole overheating by the beam halo, the protective plates and temperature gauges were installed. The beam, transported to the tube end, was measured by the water-cooled Faraday cup (FC), equipped with the dipole magnet for preventing the escape of secondary electrons from the FC assembly. The adjustable bias voltage was applied to the FC plate to control its effect on the transported beam. High-resolution spectrometers, installed at the transport tube sides and at the view port, observing the beam along the axis from the back were used to measure the main and Doppler shifted Hα lines of the light emitted from the beam area. The focusing lenses and fiber optic cables were used for collecting beam light to the spectrometers and cameras. The gas puffing system consists of two leak valves, providing the continuous gas seed to the magnet chamber and the transport tube. Two vacuum gauges were installed to control the pressure in the bending magnet chamber and in the transport tube. The pressure readings of the vacuum gauges were corrected after compensating the sensitivities with a calibrated ion gauge. The diaphragm with ⌀30 mm opening was installed between the bending magnet chamber and the transport tube to provide the differential pumping of the magnet chamber and transport tube volumes. The beam position monitor consisting of four thermistors and four secondary emission sensors was installed at the diaphragm periphery to align the beam position with the transport tube axis. The source and bending magnet chamber is supplied with a

2000 l/s (for H_2) turbomolecular pump and the transport tube—with a 200 l/s (for H_2) turbomolecular pump. With the typical hydrogen feed to the source ~0.1 l Torr/s, the background vacuum in the transport tube was about ~$2 \cdot 10^{-5}$ Torr. The correlation of gas pressure in the bending chamber vs gas pressure in the transport tube for the typical hydrogen feed to the source and xenon addition to the transport tube was controlled. Beam profiles and widths were determined by taking an image with high-sensitivity CCD cameras, installed at the transport tube sides. The light emission intensity distribution across the beam axis in the vertical direction (Y-profile), obtained by processing of the image, recorded with the side CCD camera is shown controlled. In order to increase the processing statistics, the elemental Y-profiles of the light were averaged along the beam axis in the image central part.

The effects of hydrogen, argon, and xenon gas addition to the LEBT transport tube on the beam sizes (FWHM) and the beam current transported to the FC are shown in Fig. 6.20. The solid curves show the dependencies of the FC current, the dotted curves show the beam horizontal width, and the dashed curves show the beam width in the vertical direction. An appreciable ~33% drop of the beam size from 9 to 6 mm (FWHM) was recorded with a small xenon addition $\Delta P_{Xe} \sim 3 \cdot 10^{-6}$ Torr (from $2.2 \cdot 10^{-5}$ to $2.5 \cdot 10^{-5}$ Torr). A further decrease of the beam sizes down to 4.5 mm (up to 50% of the initial value) was recorded with larger Xe addition. The similar ~30% decrease of the beam size was obtained with 10 times larger addition of argon (with pressure growth $\Delta P_{Ar} \sim 3 \cdot 10^{-5}$ Torr) and with ~40 times larger addition of hydrogen (with pressure growth $\Delta P_{H2} \sim 1.3 \cdot 10^{-4}$ Torr). The addition of Xe to the LEBT causes an increased H− beam stripping during transport as compared with that caused by Ar or H_2 addition. Namely, the ~30% drop in the beam size recorded with Xe pressure

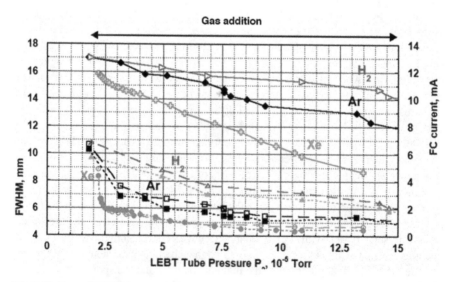

Fig. 6.20 Beam FWHM (full width at half maximum) (left scale, dotted and dashed curves) and FC current at the LEBT exit (right scale, solid curves), measured with various gas additions to the transport tube. Green (triangles), H_2 addition; black (diamonds or squares), argon addition; and pink (crosses or circles), xenon addition. Filled markers (dotted lines), horizontal FWHM, and empty markers (dashed lines), vertical FWHM. The initial H_2 pressure 1.7–2.2 $\cdot 10^{-5}$ Torr is produced by the H_2 flow from the ion source

increment $\Delta P_{Xe} \sim 3 \cdot 10^{-6}$ Torr produces $\sim 5\%$ of H– beam stripping (Fig. 6.20), while the $\sim 50\%$ decrease in the beam size with Xe addition $\Delta P_{Xe} \sim 3 \cdot 10^{-5}$ Torr is accompanied by $\sim 20\%$ stripping of the H– beam. It is important to note that the measured FC current shown in Fig. 6.20 consists mainly of negative ions with a small fraction of positive ion current, produced due to H– ion stripping in the transport tube, so the transported H– beam current I_{H-} is slightly larger than the FC current, as shown in Fig. 6.20 (positive ion current $I^+ \sim 0.07 \, I_{H-}$ at $\Delta P_{Xe} = 10^{-4}$ Torr). The obtained values of H– ion stripping with gas addition to the LEBT corresponded to the calculated values. An integral cross section $\sigma_{-0} + \sigma_{-+}$ of 33 keV H – ion stripping in gases 7 used for calculation is shown in the last column of Table 6.1. The addition of gases to the bending magnet chamber was studied as well. It was recorded that gas addition to the bending magnet chamber produces a larger decrease of the beam height ΔY as compared with that for the beam width ΔX. The similar 30% reduction of the beam height ΔY from 9 to 6 mm was recorded with xenon addition $\Delta P_1 \sim 10^{-5}$ Torr to the bending magnet chamber (and accompanying gain $\Delta P_2 \sim 3 \cdot 10^{-6}$ Torr of the transport tube pressure). But the beam horizontal width ΔX (FWHM) has decreased only 15% in this case. The gradual reduction of the beam height and width was recorded with further xenon addition up to $P_1 \sim 18 \cdot 10^{-5}$ Torr (and accompanying growth of $P_2 \sim 6 \cdot 10^{-5}$ Torr). The oscillations of the FC current in the 40 kHz–12 MHz frequency range were measured with the Tektronix oscilloscope. The rms of integral power of oscillations was calculated from the signal Fourier spectrum. The standard rms of FC current oscillations was about 10–20% of the average beam current, but no correlation to the beam oscillation frequency or the amplitude with the additional gas puffing to the transport tube was detected. The negative ion beam dump potential and injection of secondary electrons, emitted from the dump to the beam well, could change the beam-plasma dynamics and affect the space charge compensation. To control this effect, a dipole magnetic filter at the FC back was used, and the experiments with the FC biasing to potential ± 100 V with respect to transport tube walls were performed. A little influence on the beam size and no detectable change in the fluctuation level were recorded during this FC potential change. It shows a good suppression of the secondary electron escape from the FC by the magnetic filter. The three-dimensional structures of the magnetic and electrical fields, as well as the trajectories of the negative ion beam path through the bending magnet and the transport chamber, were simulated using the COMSOL software. No space charge effects were taken into account. The calculated trajectories for the 35 keV beam are schematically shown in

Table 6.1 Data for H_2, Ar, and Xe addition to the LEBT for 12–13 mA, 33 keV beam transport. Accompanied H– beam stripping (measured and calculated) is also shown

	Gas addition	33% beam shrink		H– beam stripping	
	$\Delta P2$, 10^{-5} Torr	(FWHM) (mm)	Measured (%)	Calculated (%) $\sigma_{-0} + \sigma_{-+}$, 10^{-15} cm^2	
ΔXe	0.3	$9 \rightarrow 6$	5	3.5	4.2
ΔAr	2	$10.5 \rightarrow 7$	9	10	1.8
ΔH_2	7	$10.5 \rightarrow 7$	12	15	0.75

Fig. 6.19. The beam initial diameter 3.5 mm and angular divergence ±100 mrad were taken in this case. The calculated radial profile of the calculated radial profile of the transporting beam has the FWHM of 30 mm after the bending magnet and of 13 mm at the LEBT exit. The value at the exit is only ~15% higher than the measured FWHM in the case with no gas addition. The coincidence of the measured beam FWHM with the "ballistic" FWHM calculated by COMSOL for the beam with no space charge permits to state that the space charge of the transported beam is compensated at the residual hydrogen pressure of $2 \cdot 10^{-5}$ Torr in the LEBT. The further gas addition stipulates the overcompensation of the continuous wave H− beam space charge. The overcompensation of the beam is supported by accumulation of positive ions and by reversing of the beam potential. The positively charged "ridge" is formed in the beam area. The electric field of the overcompensated beam with a positive charge increases to the beam periphery, and it focuses the peripheral negative ions to the beam axis. The ratio of positive ion production and removal from the beam governs the beam ridge potential height, and it is remarkably higher in the case of heavy gas with a large cross section of ionization like xenon. The beam focusing with heavy gas addition was experimentally observed and studied before for the intense pulsed beams. The visible effect of H− ion focusing needs relatively large Xe addition of $\Delta P_{Xe} \sim 6 \cdot 10^{-5}$ Torr in the case of secondary electron emission from the beam dump. This value of $\Delta P_{Xe} \sim 6 \cdot 10^{-5}$ Torr is inappropriate for H− ion transport purpose due to strong H− beam stripping. Much lower addition of krypton $\Delta P_{Kr} \sim 2 \cdot 10^{-6}$ Torr was necessary to focus the pulsed beam in the case of secondary electron trapping near the grounded emittance meter. As it was observed in our experiments, the 33% compression of the CW H− beam could be obtained with the low addition of xenon $\Delta P_{Xe} \sim 3 \cdot 10^{-6}$ Torr to the transport tube if the secondary electrons entering from the FC to the beam area are suppressed. No effect of the FC plate ±100 V biasing on the H− beam transport and beam oscillations show the importance of secondary electron blocking. The xenon addition is more effective for beam focusing and transport. Namely, the H− beam with an intensity up to 13 mA was transported and focused to the LEBT exit under a decreased level of the beam losses (<5%).

6.4   Low Energy Beam Transport System Developments

For high brightness beam production, it is important to preserve the brightness in the low energy transport system (LEBT) used to transport and match the ion beams to the next stage of acceleration, usually a radio frequency quadrupoles (RFQs) [8]. One recent review of LEBTs for positive and negative ion beams is presented in [29]. While electrostatic focusing can be problematic for high-current beam transport, reliable electrostatic LEBT operation has been demonstrated with H− beams up to 60 mA [30]. Now, however, it is commonly accepted that an optimal LEBT for high-current accelerator applications consists of focusing solenoids with space charge compensation (SCC). Two-solenoid magnetic LEBTs are successfully used for high-current (>100 mA) proton beam [31]. Preservation of low emittances (~0.15 π mm mrad) requires the addition of a heavy gas (Kr, ~1 mPa or Ar ~ 4 mPa) which causes ~5% of proton loss in ~1 m long MLEBT. Adding of N_2 or H_2 up to necessary density increases beam loss up to 15% [18]. An impressive illustration of rms emittance reduction up to three times by a small admixture of heavy gases is shown in Fig. 6.13, adopted from [30]. A significant advantage of SCC by heavy gases is obvious. Similar gas density (Xe, ~3 10^{12} cm^{-3}) is required to preserve low emittances of H$^-$ beams, but such gas densities cause unacceptably high H$^-$ beam losses in the long MLEBT [10, 11, 32, 33] because H$^-$ stripping cross sections are much higher than cross sections for proton neutralization at LEBT energies. For this reason, the first experiments of high brightness H$^-$ beam formation/transportation in surface plasma sources (SPS) with cesiation [10, 31, 34] used space charge compensation only in the first short magnetic focusing/separation section directly after extraction where the beam current density is very high (>1 A/cm^2). This was followed by fast beam acceleration up to higher energy. An example of such a system with H$^-$ beam extraction from a Penning discharge SPS is shown in Fig. 5.1. This system is used in the linac pre-injector of the Moscow Institute of Nuclear Research (MINR) [11]. A fast electromagnetic valve [34] was used for pulsed gas injection into the SPS and into the bending magnet aperture for space charge compensation with low gas pumping. Similar versions of short LEBTs were used in other high-voltage pre-injectors with compact SPS in FNAL (Fermi National Accelerator Laboratory) [35], ANL (Argonne National Laboratory), BNL (Brookhaven National Laboratory), RAL (Rutherford-Appleton *Laboratory*), and DESY (Deutsches Elektronen-Synchrotron). An LEBT with beam brightness preservation by SCC in a short bending magnet can be followed by beam focusing to an RFQ either by a short solenoid or by an electrostatic Einzel lens. Using a bending/focusing magnet with SCC after extraction was proposed also to protect downstream high-voltage structures from escaping cesium and fast neutrals. However, after optimization of compact SPS operation, the escaping cesium was decreased significantly, and it became possible to use a further acceleration and focusing by electrostatic systems without bending magnets (small angle sources) [6, 7, 34, 36], which can deliver the beams with higher intensity and brightness.

A strong electrostatic focusing LEBT has been successfully adopted for reliable transportation of high-current H$^-$ beams in the SNS front end [27, 28, 37]. Some modifications of such electrostatic LEBTs can be expected to improve the reliable transport of more intense positive and negative ion beams without greatly degrading their brightness and intensity.

Now the most commonly accepted LEBTs for high-current accelerator applications use magnetic solenoids and space charge compensation (SCC). Reviews of magnetic LEBTs with SCC descriptions and simulations are presented in [27, 29, 30]. For a discussion of magnetic LEBT features, we will use the MLEBT design prepared for the SNS front end [38] which is topologically similar to other MLEBTs, including the first H$^-$ RFQ pre-injector at BNL [39].

Figure 6.21 shows a schematic view of the designed two-solenoid MLEBT. It consists of two solenoids S1 and S2 that focus the H$^-$ beam into the RFQ entrance at the right, separated by 50 cm used for pumping and beam diagnostics. The chopper located near the entrance to the RFQ will operate only at low voltage for chopping and steering, because no high voltage is required for electrostatic focusing.

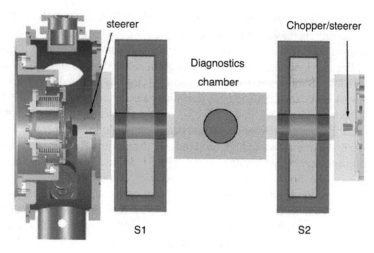

Fig. 6.21   A schematic view of the two-solenoid S1 and S2 MLEBT

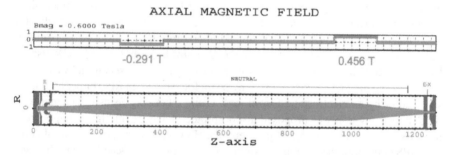

Fig. 6.22   PBGUNS simulation of beam transport using the prototype two-solenoid MLEBT

This eliminates high-voltage sparks and their harmful transients. The beam can be characterized before being injected into the RFQ to simplify the beam diagnostics. The simplified simulations were carried out using PBGUNS (particle beam gun simulation) [40]. The dimensions for the two solenoids were decided to be the same with an effective length of 135 mm. Figure 6.22 shows a PBGUNS simulation of the beam transport in the prototype two-solenoid MLEBT design. A 60 mA beam is extracted from the RF SPS, accelerated up to 65 keV, and transported through the two-solenoid MLEBT with ("manual" defined level of SCC) compensated space charge after extraction because negative ion beams rapidly accumulate the secondary ions produced in collisions with the residual gas. However, the rapidly changing electric fields in the chopper prohibit a substantial accumulation of neutralizing ions, and therefore the beam inside the chopper is simulated with full space change and with solenoid fields of -0.291 T and 0.456 T. The calculated emittances at the RFQ entrance exceed $0.28\ \pi$ mm mrad, with the required Twiss parameters $\alpha = 1.6$ and

$\beta = 0.06$ mm/mrad produced after ELEBT, but are still within the 0.35 π mm mrad acceptance limit of the SNS RFQ and linac. Here we will not discuss well-known integral behavior of beams in MLEBT with SCC such as ion accumulation time, average compensation levels, and beam defocusing, described in [27, 29, 30, 41], but will concentrate mainly on processes determining the beam brightness degradation and on this degradation suppression. The possible sources of beam halo formation and emittance growth in high intensity injectors are aberrations due to the ion source extraction optics, optical aberrations of the focusing elements of the LEBT, beam density fluctuations due to ion source instability or power modulation as in RF ion sources, and nonlinearity of the electric fields created by the space charge of the beam and compensating particles.

For high intensity beams at low energy, space charge forces are particularly strong. The electric field created by the space charge tends to defocus the beam and is strongly nonlinear as it is induced by the nonuniform charged particle distributions of the beam and compensating particles. The defocusing effect of the space charge can be compensated by transporting the low energy beam in the space charge compensation (SCC) regime, but the partial SCC can be the main driving force of the beam brightness degradation (beam heating) through nonlinear fluctuating electric fields created by complex motion of compensating particles (electrons and ions) in the beam's potential well. The strong beam instability connected with secondary particle oscillations in the beam potential was observed and described in [41–43]. A first theory of this instability was developed by Chirikov [44], and a kinetic description with quadrupole oscillations was presented in [45]. Reviews with further development in this field, named now e-p instability (or electron cloud effect), were presented in [46–48]. Now this instability is a main limiting factor of beam performance in the Large Hadron Collider (LHC) in CERN, including some beam transfer lines and in many other accelerators and storage rings. Simulations of SCC for protons and H$^-$ beams in MLEBT with secondary particle motion in the beams were conducted in [49, 50].

For the first electric field scale estimation, it is possible to use a simple model of a cylindrical ion beam. The electrostatic potential $\varphi(r)$ created by the beam with current I_b, uniform current density, and compensation level fn, inside radius r_b surrounded by tube with radius r_w, is $\varphi(r) = Vs\,[1 + 2 \ln (r_w/r_b)]$, where $Vs = I_b(1 - f_n)/(4\pi\varepsilon_o v_b) > 0$ (< 0) for positive (negative) ion beams. Note that $\varphi\,(r = r_b) = 2Vs\,\ln(r_w/r_b)$ and $\varphi(r = r_w) = 0$. Here, $v_b = \beta c$ is the particle velocity and r_w is the radius of the conducting cylindrical wall.

A potential difference between a beam axis ($r = 0$) and beam boundary ($r = r_b$) is $V_b = I_b(1 - f_n)/(4\pi\varepsilon_o c\,\beta)$, where $I_b/(4\pi\varepsilon_o c) = 30$ V for $I_b = 1$ A. For proton (or H$^-$) with energy W and mass M, we have $\beta = (2\,W/Mc^2)^{1/2} = 4.7\,10^{-3}$, at $W = 10$ keV; we have $V_b = 6.3\,10^3$ V. The electric field on the boundary of this beam is E [V/cm] $= 30\,I_b[A]/(\beta\,r_b[cm])$. For $I_b = 0.1$ A, $E(r) = 30\,I_b$ [A] $r/\beta\,(r_b)^2 = 6.3\,10^2$ V/cm for $r_b = 1$ cm and $E = 6.3\,10^3$ V/cm for beam focused into the RFQ aperture up to $r_b \sim 0.1$ cm. For beam focusing to a small spot, it is difficult to keep symmetry, and in the focused beam, the electric field can have unpredictable directions. On an emission surface of the high brightness ion source, the ions have a transverse ion temperature ~ 1 eV. After beam extraction and expansion up to 10 times from ~1 mm in the compact SPS up to ~1 cm, the local energy spread of ions should be decreased ~100 times up to 10^{-2} eV. In this situation, the fluctuating nonlinear electric fields can induce significant local energy spread with a strong increase in local ion temperature and related brightness degradation. The intrabeam electric fields can be compensated up to much lower level by secondary particles with opposite charges: electrons and negative ions for positive ion beams and positive ions for negative ion beams. For proton beams, the compensation degree f_n can be as high as 98%. For high beam brightness preservation, a high density of heavy gases (~10^{12} cm^{-3} for Kr) is needed. For protons with energy 50–100 keV, the beam attenuation in the MLEBT is relatively low (~5%) because cross sections for electron capture are relatively low and the gas ionization cross sections are relatively high [6, 7, 51, 52]. However, for low energy positive ions for low emittance preservation, very high gas density is needed, which causes more significant beam attenuation because the ionization cross sections become very low [51] but electron capture cross sections are high. In this situation, for more efficient compensating particle generation, it is better to use secondary emission of electrons and negative ions or active plasma generation by a plasma gun or by electron beams.

Simulation of SCC for positive ions in MLEBTs with analysis of compensating particle motion was conducted, for example, in Refs [29, 51, 52]. that demonstrated qualitative agreement with experiments [30], which exhibit more complex behavior of SCC with beam focusing to a small focusing waist. The usual location of the emittance scanner and other diagnostics between the solenoids is not very useful because most significant emittance growth arises in the beam focus at the entry to the RFQ. Beam intensity fluctuations increase the emittance of the extracted beam and can be further amplified during transport in the MLEBT with very strong beam brightness degradation. Transport efficiency and beam quality of positive ions in the RFQ can be improved by shortening the MLEBT. Some designs of shortened MLEBT were proposed in [52], including one solenoid MLEBT, movable for Twiss parameter optimization. This problem is not of high priority for high energy positive ions but is more important for low energy positive [53] and negative ion beams.

6.5   Features of Negative Ion Beam Transportation in MLEBTs

Negative ion beam transport with SCC in the MLEBT is more complex and more difficult for optimization. Positive ions compensating the negative ion beam have much higher mass and low initial energy (~ 0.1 eV). The positive ion radial distribution which neutralizes the negative ion beams is very sharp on the axis because each ion trajectory, created with zero initial velocity, crosses the beam center. This behavior is very different compared to the electron distribution for positively charged beams because electrons have a large initial energy ~5–10 eV and relative angular momentum, preventing electrons from reaching the beam axis. The ions produced by ionization on different distances from the beam axis can have high potential energy (~ 10^2 eV) and in the parabolic potential of quasiuniform beam density can collapse to the axis at the same time forming a very peaked "Christmas tree" density distribution. A high concentration of secondary ions on the beam axis leads to very fast compensation of the beam center. A positive field appears when the secondary ion density exceeds the beam density producing "overcompensation" and enables some electron trapping. If the axial symmetry of beam or secondary ion generation is broken, the peak of the ion distribution can be off axis, which can cause beam deflections and dipole oscillations. In more detailed simulations of SCC in MLEBT with secondary particle dynamics presented in [29, 51], the SCC features discussed above were reproduced.

However, in experimental investigations of H⁻ beams, SCC with beam potential and secondary particle monitoring [9, 12, 14, 16, 17, 53–55] more complex behavior was observed with longitudinal fluxes of secondary particles, strong influence of electrode biasing, direct registration of transverse oscillations, and brightness degradation. In the compensated negative ion beam, the light positive ions can have very nonuniform distributions with very nonlinear electric fields, leading to beam brightness degradation and some exotic phase space distributions. The beam brightness degradation can be suppressed by using SCC with fast generation of heavy positive ions. In this situation, the positive ions cannot collapse to the center of the beam, and electrons can compensate the transverse electric fields. Unfortunately, for this, the necessary high gas density was too high, leading to significant H⁻ beam attenuation in existing long MLEBTs [56–59]. Thus, it is difficult to use long MLEBTs for production of H⁻ beams with highest intensity ~0.1 A without brightness degradation. Transport efficiency and beam quality of H⁻ ions in the RFQ can be improved by shortening the MLEBT. Some designs of shortened MLEBT were proposed in [54], including one solenoid MLEBT, movable for Twiss parameter optimization. The emittance scanner located between solenoids is not useful because the main brightness degradation arises after the second focusing lens. An absolute beam brightness detector (ABBD) [9] can be useful in the focus position for beam brightness optimization. Beam intensity fluctuations increase the emittance of the extracted beam and can be further amplified during transport in the MLEBT with very strong beam brightness degradation. This was observed in the RF SPS for CERN Linac 4 [55] where the beam intensity modulation (the beam current issued from the CERN (DESY type) plasma generator showed a very clear 2 MHz ripple (±10%) that is correlated to the reversal of the axial electrical field of the solenoid antenna) can be a reason of additional brightness degradation.

The program of SCC optimization for positive and negative ion beams was established in the Fermi National Laboratory [60]. Using the advanced test stand and the PIC code WARP, it is possible to produce a better understanding of complex SCC processes and improve MLEBT performance. Active generation of compensating plasma can be useful to produce the highest beam performance [53]. Another possibility for a predictable LEBT is an electrostatic LEBT (ELEBT).

6.6   Features of Electrostatic LEBT (ELEBT)

In an ELEBT, the beam is transported without any SCC because the compensating particles are quickly removed by the electric field induced by the focusing elements. The ELEBT is compatible with fast beam chopping as there is no transient time for the SCC. Furthermore, the design of ELEBTs is simplified by the fact that no repelling electrode for compensating particle trapping is needed. So the beam lines can be very compact and transparent for gas, which tends to minimize the beam losses by stripping (but can be useful to decrease the flux of back accelerated positive ions to the ion source). As an example, Fig. 5.102 shows a schematic of the LBNL designed version of the SNS ion source with the 12-cm-long ELEBT equipped with two Einzel lenses [61]. One possible disadvantage of the ELEBTs is vulnerability to beam losses that can lead to high-voltage breakdowns and beam trips. The Einzel lenses have some optical aberrations that create beam halo and emittance growth. To limit this effect, the beam radius should not exceed two-thirds of the lens aperture radius. Finally, the design of the ELEBTs is intensity limited. Design of this ELEBT is shown in Fig. 5.103. The design features are very important for reliable ELEBT operation with intense beam. With this design, the commissioning of the SNS accelerator chain was accomplished at low duty factor. A history of SNS commissioning and RF ion source and ELEBT modification is presented in [27, 28, 40, 42, 62].

Now the SNS uses an electrostatic, two-lens (G3 and G5) ~12-cm-long ELEBT, which focuses the 65 keV, ~ 6% duty factor, high-current H^- ion beam (up to 60 mA) into the radio frequency quadrupole (RFQ) accelerator with the required Twiss parameters $\alpha = 1.7$ and $\beta = 0.06$ mm/mrad. The second lens (G5) is split into four electrically isolated segments, which allow small superimposed voltages to steer, and rapidly switched superimposed ±2.5 kV voltages to chop the beam [29]. The ELEBT's compactness prohibits any beam characterization before it is accelerated to 2.5 MeV by the RFQ. The first beam current monitor is located near the exit of the RFQ.

Some problems arise during the start of the ELEBT operation in SNS with high beam intensity and df. High-voltage pulses from lens 2 sparks occasionally break the chopper electronics. Uncontrolled beam losses and beam-induced and corona-induced secondary ions heat the lenses and the extractor. Sometimes lens 1 (G3) reaches temperatures where thermionically enhanced field emissions make the lens inoperable. To reduce the operational risks associated with the ELEBT, and to improve functionality of the ion source and LEBT, the two-solenoid MLEBT described above (shown in Fig. 6.19) was developed, designed, and fabricated with a hope to match or exceed the requirements for the present ELEBT, including the fast beam chopping at ~1 MHz. Fortunately, the performances of RF ion source and

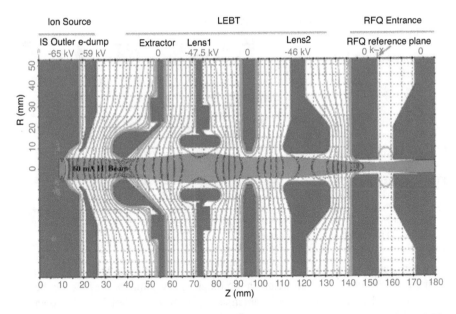

Fig. 6.23 A drawing of modified ELEBT with PBGUNS simulation of beam transport to the RFQ entrance

ELEBT were improved, and necessity to replace the ELEBT by a MLEBT was avoided. The improved ELEBT can reliably keep 65 kV of extraction voltages with H⁻ beam ~ 60 mA and lenses 1 and 2 can be conditioned up to 50 kV. A schematic drawing of the modified ELEBR with PBGUN simulated 60 mA H⁻ beam is shown in Fig. 6.23.

Similar ELEBTs are used in the SNS front end to supply SNS accelerators with H⁻ beams for change exchange injection into the compressing storage ring, in test stand for ion source testing, and in the accelerator test facility for ion source testing with a spare RFQ. With this ELEBT and beam diagnostics, a saddle antenna RF surface plasma source (SA RF SPS) was tested with production of 67 mA of pulsed H⁻ beam with efficiency of 1.6 mA/kW [63, 64].

This ELEBT will be used for beam transport with the newly modified SA RF SPS [65, 66] in which the efficiency of negative ion generation is improved up to 5 mA/kW. In this SA RF SPS, a transverse magnetic field in the extractor and inside the collar is created by permanent magnets inserted into a water-cooled extractor as shown in Fig. 6.24.

H⁻ ions are formed on the cesiated conical collar surface and extracted by an electric field, created by the potential difference between the collar and the extractor (e dump). The H⁻ beam is accelerated by the electric field between the extractor and the accelerating electrode. Electrons escaping from the collar are intercepted by a strong transverse magnetic field and collected at the extractor. The transverse component of magnetic field B_x is zero between centers of the permanent magnets and increases up to $B_{x0} = 1$ kG at a distance $L = 8$ mm. This dependence can be

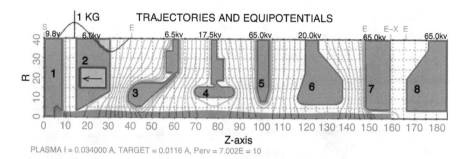

PLASMA I = 0.034000 A, TARGET = 0.0116 A, Perv = 7.002E = 10

Fig. 6.24 Diagram for axisymmetric simulation of the ion beam extraction and ELEBT for the saddle antenna RF SPS with SNS ELEBT without magnetic field (electrodes from left to right: (1) plasma electrode with a conical collar; (2) extraction electrode with permanent magnet (e dump); (3) accelerator electrode; (4) first electrostatic lens; (5) grounded electrode; (6) second electrostatic lens; (7) RFQ plate; and (8) RFQ entrance)

approximated by formula, $B_x = B_{x0} \sin(\pi Z/2L)$ for $-2L < Z < 2L$, from the center of permanent magnets in the extractor. In the diagram for simulation of the ion beam extraction and ELEBT for saddle antenna RF SPS with SNS LEBT shown in Fig. 6.24, the centers of magnets have a coordinate $Z_m = 23$ mm. For these parameters,

$$B_x = B_{x0} \sin\left[\pi\left(Z - Z_m\right)/2L\right] = 1000\sin\left[\pi\left(Z - 23\right)/16\right]\left[\text{Gauss}\right]$$
$$\text{for } 7\,\text{mm} < Z < 39\,\text{mm}.$$

The beam current over 1-m-long 6-mm-wide slit is equivalent to ~20 mA beam from a 6 mm diameter circular aperture. The electrons drift in crossed ExB fields reproduced. To estimate the H⁻ beam deflection by the transverse magnetic field, the extraction and transport of a ribbon beam with equivalent emission current density were simulated. The output from a simulation including the transverse magnetic field is shown in Fig. 6.25.

It seems that the needed tilt and offset of the ion source are small if the voltages on the extractor and lenses are properly chosen as in this example. Sometimes the ELEBT center ground (G4) on the SNS test stand was overheated, causing it to buckle excessively, snapping screws and insulators. Nonuniform heating creates nonuniform local expansion stresses and deformation preventing normal ELEBT operation. To prevent such situations, the mounts were reengineered to double their heat sinking, and a thermo couple was added to understand the issues and reduce the likelihood of future failures. For further improving the heat sinking, it is possible to use insulators from AlN with much higher thermal conductivity. In city, heating of electrode in good vacuum is helping for preventing high-voltage arcing and for corona current (field emission). To prevent the deformation during the heating, it is possible to fabricate the electrodes' supports from material with very small thermal extension such as invar. Well surface processed Mo is one of the best materials for high-voltage electrodes. Surface melting by low energy electrons can be used to

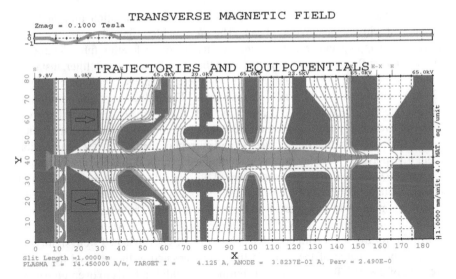

Fig. 6.25 Simulation of the beam deflection with extraction voltage 8 kV and acceleration voltage 65 kV in the SNS LEBT

Fig. 6.26 Design of recent version of SNS ELEBT

increase the arcing voltage [67]. The design of a recent version of the SNS ELEBT is shown in Fig. 6.26.

Such an ELEBT can be used for proton beam transport by changing the electrode voltages. The beam intensity is scaled as $(V)^{3/2}$. An increase of electrode voltages by 1.5 times can increase the transported beam intensity by 1.8 times.

An accelerator-based neutron source using a vacuum-insulated tandem accelerator (VITA) has had more than a decade of operation at the Budker Institute of Nuclear Physics.

Studies of H- beam production, acceleration, and transport at different vacuum conditions in the transport channel were performed [68]. About 95% beam transmission was achieved. The presence of secondary electrons with a current of

~0.4 mA was recorded, which were produced in the acceleration tube and co-accelerated with the negative ion beam. The presumable origin of secondary electrons is H- stripping and secondary emission from the acceleration tube electrodes. The secondary electrons were deflected from the beam by a magnetic filter, installed at the transport tube sides. Reliable operation of the ion injector with ion energy of 133 keV and current of 14.5 mA was achieved. The influence of gas addition to compensate the ion beam space charge was studied extensively earlier. In the present case, hydrogen, argon, and xenon injections into the 0.8-m-long transport tube were tested. The resultant transverse size and current of the DC 33 keV negative ion beam transported to Faraday cup were studied using optical diagnostics. A drop of the beam size from 9 to 6 mm was recorded with small ~3 • 10^{-6} Torr addition of xenon. A similar decrease of the beam size was obtained with 10 times larger addition of argon and ~40 times larger addition of hydrogen. The measured beam full-width at half maximum (FWHM) was compared with the value calculated by COMSOL for the beam with no space charge effect. They matched well, signifying that the space charge of the beam is fully compensated at the residual hydrogen pressure of 2 • 10^{-5} Torr in the LEBT. Further gas addition leads to overcompensation of DC H− beam space charge and to beam focusing. Xenon addition is more effective for beam focusing and transport. The H− beam with intensity up to 13 mA was transported and focused to the LEBT exit under a decreased level of the beam losses (5%).

A possible scheme is presented for the injection of high power atomic beams within the project of the tokamak with reactor technologies (TRT), which is being developed [69] (Fig. 6.27). In each of the two TRT injection ports, it is proposed to install the two-beam injection complex consisting of two high energy (500 keV) injectors arranged vertically one above the other. A prototype of such an injector is being developed at the Budker Institute of Nuclear Physics of the Siberian Branch of the Russian Academy of Sciences (BINP). For each of the BINP-developed high-voltage injectors, it is proposed to separately form the beam of negative hydrogen ions, accelerate it using the separate single-aperture accelerating tube, and neutralize it in the efficient plasma neutralizer. The power of each two-beam complex will initially be 7 and ~5.7 MW for hydrogen and deuterium atoms, respectively. In the future, it is planned to increase the total injection power up to 20 MW by increasing the energy and current of the beams used in the injection complexes. For the TRT facility, a possible scheme is discussed for the injectors of fast deuterium atomic beams with energies of up to 200 keV based on positive ions.

6.7   Recuperation of Positive and Negative Ion Beams

The neutral ion beam injection will be used to further heat the plasma temperature to reach the ignition of the fusion reaction to producing energy in ITER project, and some energy efficiency is required. Recently a very simple beam energy recovery based on space charge effect has been proposed as an alternative to the electrostatic

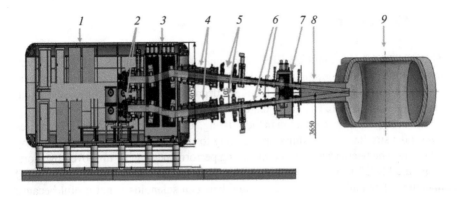

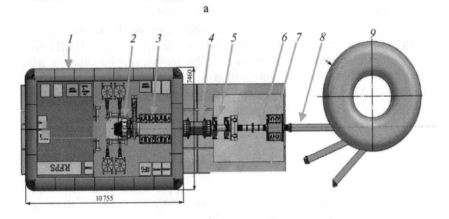

Fig. 6.27 (**a**) Schematic of neutral beam injection into TRT (side view): (1) high-voltage platform, (2) source of ions, (3) LEBT tank, (4) accelerating tubes, (5) quadrupole magnets, (6) neutralizers, (7) separator, (8) TRT port, and (9) TRT. (**b**) Top view

ion dump since it had some advantages in removing the residual ion after the negative beam neutralization [70]. Preliminary simulation results showed that the proposed device was able to remove all the residual ions by collecting them at very low energy on proper electrodes. Further simulations with more accurate space charge calculations were needed to confirm the high residual ion collection efficiency obtained in the preliminary simulations. A new simulation result with a code that performs more accurate space charge calculations is presented. The more accurate space charge calculations presented here suggest some modification on the previous collector model to increase the ion collection efficiency. Further ion recovery simulations with the modified collectors have been also presented and discussed. The proposal of the experimental test foreseen for that device on a scaled ion beam source will be also updated.

6.8   Conclusion

The existing two-solenoid MLEBT is adequate for high intensity H^+, D^+ ion beam transportation and matching with an RFQ. By optimal injection of heavy gases, the emittance of beams after MLEBT can be decreased up to three times, and beam brightness can be increased up to ten times with low intensity loss.

Beam brightness preservation in the existing, long two-solenoid MLEBT requires a high-density heavy gas admixture, leading to significant H^- beam attenuation by stripping. The beam transport and matching performance can be improved by shortening the MLEBT or by using a one-solenoid MLEBT, which is movable for RFQ matching. The emittance scanner located between solenoids is not useful because the main brightness degradation arises after the second focusing lens. An absolute beam brightness detector (ABBD) [9] can be useful in the focus position for beam brightness optimization. Beam intensity fluctuations increase the emittance of the extracted beam and can be further amplified during transport in the MLEBT with very strong beam brightness degradation. For this reason, it is important to have noiseless beams without intensity modulation.

The improved ELEBT design may be the best solution for reliable transport and matching of very intense, high brightness negative and positive ion beams.

References

1. V. Dudnikov, *Negative Ion Sources* (NSU, Novosibirsk, 2018). В. Дудников, *Источники отрицательных ионов*. НГУ, Новосибирск, 2018
2. V. Dudnikov, *Development and Applications of Negative Ion Sources* (Springer, 2019)
3. V.G. Dudnikov, Charge exchange injection into accelerators and storage rings. Physics-Uspekhi **62**(4), 405 (2019)
4. V. Dudnikov, Method of negative ion obtaining, Patent cccp, 411542, 10/III. 1972; http://www.findpatent.ru/patent/41/411542.html. В.Г. Дудников, "Способ получения отрицательных ионов", Авторское свидетельство, М. Кл.Н 01 J 3/0,4, 411542, заявлено 10/III, 1972
5. V.G. Dudnikov, Surface Plasma Method of Negative Ion Production, Dissertation for doctor of Fis-Mat. nauk, INP SBAS, Novosibirsk, 1976. В.Г. Дудников, "Поверхностно-плазменный метод получения пучков отрицательных ионов", Диссертация на соискание учёной степени доктора физ.-мат. Наук, ИЯФ СОАН СССР, Новосибирск, 1976
6. V.G. Dudnikov, G.E. Derevyankin, The art of the high brightness ion beam production. AIP Conf. Proc. **287**, 239 (1992)
7. G.E. Derevyankin, V.G. Dudnikov, Production of high brightness H^- beam in surface plasma source. AIP Conf. Proc. **111**, 376 (1984)
8. V. Dudnikov, B. Han, M. Stockli, R. Welton, G. Dudnikova, Low energy beam transport system developments. AIP Conf. Proc. **1655**(1), 050003 (2015)
9. V. Dudnikov, Absolute beam brightness detector. Rev. Sci. Instrum. **83**, 02A714 (2012)
10. G.Y. Derevyankin, V.G. Dudnikov, V.S. Klenov, Ion-optical characteristics of H- ion beam produced by surface -plasma sources. Sov. Phys. Tech. Phys. **48**, 404 (1978)
11. A.S. Belov, V.S. Klenov, V.P. Yakushev, Rev. Sci. Instrum. **63**, 2422 (1992)
12. D.G. Dzhabbarov, A. Naida, Spatial development of the instability of a dense beam of negative ions in a rarefied gas. Sov. Phys. JETP **51**(6), 1132 (1980)

13. D.G. Dzhabbarov, A.P. Naida, Motion of positive ions in the plasma produced by a dense beam of negative ions in a gas. Sov. J. Plasma Phys. **6**(3), 316 (1980)
14. M.D. Gabovich, D.G. Jababbarov, A.P. Naida, Effect of decompensation high dense beams of negative ions. Pisma JETF **296**, 536 (1979). М. Д. Габович, Д. Г. Джаббаров, А. П. Найда, «Эффект декомпенсации плотного пучка отрицательных ионов», Письма в ЖЭТФ, 29б, 536 1979
15. M. Reiser, Formanion and transport of high-brightness H⁻ beams, in *Proceedings of the 1988 Linear Accelerator Conference*, Williamsburg, Virginia, USA (1988)
16. J. Sherman, E. Pitcher, P. Allison, H⁻ beam neutralization measurements with a gridded-energy analyzer, A noninterceptive beam diagnostic, in *Proceedings of the 1988 Linear Accelerator Conference*, Williamsburg, Virginia, USA (1988)
17. J.D. Sherman, P.V. Allison, H.V. Smith, Beam potential measurement of an intense H⁻ beam by use of the emissive probe technique. IEEE Trans. Nucl. Sci. **32**(5), 1973 (1985)
18. P.Y. Beauvais et al., Rev. Sci. Instrum. **71**, 1413 (2000)
19. S.V. Grigorenko, S.Y. Udovichenko, Influence of beam plasma interaction for transportation of ion in cyclotron injectors. ZTF **73**, 119 (2003). С.В. Григоренко, С. Ю. Удовиченко, «Влияние пучково-плазменного взаимодействия на транспортировку ионов в инжекторе циклотрона», Ж.Т.Ф. 73, 119 2003
20. H. Vernon Smith Jr., P. Allison, J.D. Sherman, The 4X source. IEEE Trans. Nucl. Sci. **NS-32**(5), 1985 (1797)
21. H. V. Smith, Jr., P. Allison, and J. J. Sherman, The 4X Source, LA-UR-84-1843 (1984)
22. C.A. Valerio-Lizarraga, I. Leon-Monzon, R. Scrivens, Negative ion beam space charge compensation by residual gas. Phys. Rev. Spec. Top. Accel. Beams **18**, 080101 (2015)
23. G. Derevyankin, V. Dudnikov, V.S Klenov, Formation of H⁻ beams in surface plasma sources for accelerators, Preprint INP, 79–17, Novosibirsk (1979). Г. Е. Деревянкин, В. Г. Дудников, В. С. Кленов, Формирование пучков ионов Н- в поверхностно плазменных источниках для ускорителей, Препринт ИЯФ 79-17 Новосибирск (1979)
24. G. Derevyankin, V. Dudnikov, M.L. Throshkov, Features of formation beams of H- ions in surface plasma sources for accelerators, Preprint INP 82–110 (1982). Г. Е. Деревянкин, В. Г. Дудников, М. Л. Трошков, «Особенности формирования пучков ионов Н- в поверхностно-плазменных источниках для ускорителей», Прерипринт ИЯФ СОРАН, 82–110 1982
25. D.A. Fink, T. Kalvas, J. Lettry, Ø. Midttun, D. Noll, H⁻ extraction systems for CERN's Linac4 H⁻ ion source, Nucl. Inst. Methods Phys. Res. A **904**, 179–187 (2018)
26. T. Shibata, K. Shinto, M. Wada, H. Oguri, K. Ikegami, K. Ohkoshi, K. Nanmo, High-speed emittance measurements for beams extracted from J-PARC RF ion source. AIP Conf. Proc. **2373**, 050002-1–050002-9 (2021). https://doi.org/10.1063/5.0057418
27. Y. Haba, K. Nagaoka, K. Tsumori, et al., Characterisation of negative ion beam focusing based on phase space structure. New J. Phys. **22**, 023017 (2020). https://doi.org/10.1088/1367-2630/ab6d41
28. A. Sanin, Y. Belchenko, S. Popov, et al., Study of continuous wave 33 keV H− beam transport through the low energy beam transport section. Rev. Sci. Instrum. **90**, 113323 (2019). https://doi.org/10.1063/1.5128591
29. M.P. Stockli, Ion injectors for high-intensity accelerators, in *Reviews of Accelerator Science and Technology*, vol. 6, (World Scientific Publishing Company, 2013), pp. 197–219
30. M.P. Stockli, B.X. Han, S.N. Murray, et al., Ramping up the SNS beam power with the LBNL baseline H− source, in *AIP Conference Proceedings CP1097*, (2009), pp. 223–235
31. N. Chauvin, O. Delferrière, R. Duperrier et al., Source and injector design for intense light ion beams including space charge neutralization, in *TH302, Proceedings of Linear Accelerator Conference LINAC2010*, Tsukuba, Japan, 2010
32. V. Dudnikov, Proceedings 4th All-Union Conference. on Charged Particle Accelerators, Moscow **1**, 323 (1974)

33. G.I. Dimov, G.Y. Dereviankin, V.G. Dudnikov, 100-mA negative hydrogen-ion source for accelerators. IEEE Trans. Nucl. Sci. **NS-24**(3), 1545–1547 (1977)
34. G. Derevyankin, V. Dudnikov, P. Zhuravlev, Prib. Tekh. Eksp. **5**, 168–169 (1975)
35. C. Schmidt, C. Curtis, "A 50-mA negative hydrogen-ion source," PAC 1979. IEEE Trans. Nucl. Sci. **NS-26**(3) (1979)
36. V.G. Dudnikov, Rev. Sci. Instrum. **67**(3), 915–920 (1996)
37. M.P. Stockli et al., Rev. Sci. Instrum. **85**(2), 02B137 (2014)
38. B. Han, D.J. Newland, W.T. Hunter, M.P. Stockli, Physics design of a prototype 2-solenoid LEBT for the SNS injector, in *WEP038, Proceedings of 2011 Particle Accelerator Conference*, New York, 2011
39. J.G. Alessi et al., The new AGS H⁻ RFQ preinjector, EPAC88, 1988; J. Alessi, et al., H⁻ source and low energy transport for the BNL RFQ preinjector, AIP Conf. Proc. **210**, 711 (1990)
40. PBGUNS., http://www.far-tech.com
41. G. Budker, G. Dimov, V. Dudnikov, Sov. At. Energy **22**, 384 (1967)
42. G. Budker, G. Dimov, V. Dudnikov, Experiments on producing intensive proton beams by means of the method of charge-exchange injection, in *Proceedings of the International Symposium on Electron and Positron Storage Rings, Saclay, France, 1966*, (Saclay, Paris, 1966) Article No. VIII-6-1
43. G. Budker, G. Dimov, V. Dudnikov, V. Shamovsky, Experiments on electron compensation of proton beam in ring accelerator, in *Proceedings of the sixth Internatinal Conference on High energy Accelerators*, Harvard University (1967)
44. B.V. Chirikov, Sov. At. Energy **19**, 1149 (1965)
45. D. Koshkarev, P. Zenkevich, Part. Accel. **3**, 1–9 (1972)
46. V. Dudnikov, Some features of transverse instability of partly compensated proton beams, in *Proceedings of the Particle Accelerator Conference, Chicago, 2001*, (IEEE, Piscataway, 2001)
47. M. Reiser, *Theory and Design of Charged Particle Beams*, vol 567, second edn. (Wiley-VCH, Weiheim, 2008)
48. F. Zimmermann, Phys. Rev. Spec. Top. Accel Beams **7**, 124801 (2004)
49. A. Bensmail et al., Phys. Rev. Spec. Top. Accel. Beams **10**, 070101 (2007)
50. L. Neri, L. Celona, S. Gammino, et al., Rev. Sci. Instrum. **85**, 02A723 (2014)
51. A. Dudnikov, V. Dudnikov, Rev. Sci. Instrum. **73**, 723 (2002).; A. Dudnikov, Simplified beam line with space charge compensation of low energy ion beam, in *TUPPB033, Proceedings of RUPAC2012*, Saint-Petersburg, Russia, 2012
52. J. Nipper, G. Flanagan, R.P. Johnson, M. Popovic, High temperature superconducting magnets for efficient low energy beam transport systems, in *THPPD045, Proceedings of IPAC2012*, New Orleans, Louisiana, USA, 2012; M. Popovic, A low energy beam transport system for the superconducting proton linac, FNAL, January 9, 2006
53. R. Scrivens, G. Bellodi, O. Crettiez, V. Dimov, et al., Rev. Sci. Instrum. **85**, 02A729 (2014)
54. J. Sherman, E. Pitcher, P. Allison, in *Proceeding of the International Linear Accelerator Conference (LINAC 88)*, Williamsburg (CEBAF Report 89-001, Newport News, VA, 1989), pp. 155–157
55. R. Ferdinand, J.D. Sherman, R.R. Stevens, Jr., T. Zaugg, in *Proceedings of the 1997 Particle Accelerator Conference*, Vancouver, Canada, 1997
56. D. Raparia, J. Alessi, B. Briscoe et al., Reducing losses and emittance in high intensity linac at BNL, in *MOPD12, Proceedings of HB2010*, Morschach, Switzerland, 2010
57. C. Gabor, D.C. Faircloth, D.A. Lee, et al., Rev. Sci. Instrum. **81**, 02B718 (2010).; J. Back, J. Pozimski, P. Savage et al., Commissioning of the low energy beam transport, in *MOPEC078, Proceedings of IPAC'10*, Kyoto, Japan, 2010
58. C.Y. Tan, D.S. Bollinger, K.L. Duel et al., The FNAL injector upgrade status, THPPP065, IPAC12, New Orleans, LU, 2012
59. H. Oguri, A. Ueno, K. Ikegami, Y. Namekawa, K. Ohkoshi, Phys. Rev. Spec. Top. Accel. Beams **12**, 010401 (2009)

60. M. Chung, L. Prost, V. Shiltsev, Space-charge compensation for high-intensity linear, and circular accelerators at Fermilab, in *TUOBB1, Proceedings PAC2013*, Pasadena, CA, 2013
61. D.W. Cheng, IPAC12, New Orleans, LU, 2012; H. Oguri, A. Ueno, K. Ikegami, Y. Namekawa, K. Ohkoshi, Phys. Rev. Spec. Top. Accel Beams **12**, 010401 (2009); V. Dudnikov, Rev. Sci. Instrum. **83**, 02A714 (2012); M. Chung, L. Prost, V. Shiltsev, Space-charge compensation for high-intensity linear, and circular accelerators; M.D. Hoff, K.D. Kennedy et al., Design of the prototype low energy beam transport line for the spallation neutron source, in *PAC 1999*, New York, 1999
62. B.X. Han, T. Kalvas, O. Tarvainen, et al., Rev. Sci. Instrum. **83**, 02B727 (2012)
63. V. Dudnikov, R.P. Johnson, S. Murray et al., RF H$^-$ ion source with saddle antenna, in *THPEC073, Proceedings of IPAC'10*, Kyoto, Japan, 2010
64. V. Dudnikov, B. Han, R.P. Johnson, et al., Surface plasma source electrode activation by surface impurities. AIP Conf. Proc. **1390**, 411 (2011)
65. V. Dudnikov, R.P. Johnson, S. Murrey, et al., Rev. Sci. Instrum. **85**, 02B111 (2014)
66. V. Dudnikov, R. Johnson, S. Murrey et al., Improving efficiency of ions production in ion source with saddle antenna, in *TUPSM22, Proceedings of NA PAC2013*, Pasadena, CA, USA, 2013
67. E. Ozur, D.I. Proskurovsky, K.V. Karlik, A wide-aperture, low-energy, and high-current electron beam source with a plasma anode based on a reflective discharge. Instrum. Exp. Tech. **48**(6), 753–760 (2005)
68. A.A. Ivanov, A. Sanin, Y. Belchenko, Recent achievements in studies of negative beam formation and acceleration in the tandem accelerator at Budker institute. AIP Conf. Proc. **2373**, 070002 (2021). https://doi.org/10.1063/5.0057441
69. Y.I. Belchenko et al., Possible scheme of atomic beam injector for plasma heating and current drive at the TRT tokamak. Plasma Phys. Rep. **47**(11), 1151–1157 (2021) © Pleiades Publishing, Ltd., 2021. Russian Text © The Author(s), 2021, published in Fizika Plazmy, 2021, Vol. 47, No. 11, pp. 1031–1037
70. V. Variale et al., Rev. Sci. Instrum. **91**, 013516 (2020)

Chapter 7
Development of Conversion Targets for High-Energy Neutral Beam Injectors

Abstract Conversion targets for high-energy neutral beam injectors are discussed. Gas targets, plasma targets, and photon targets for efficient conversion of high-energy negative ion beams to neutral beams are considered.

7.1 Gas Targets

In first development of charge exchange injection into accelerator and storage rings, a conversion of high-energy negative ions into neutral beam [1–3] was used. For this, a measurement of stripping cross sections σ_{-10} and σ_{01} in different gases for H- ions with energy up to 1.5 MeV was provided [4]. Neutralization efficiency up to 52% was reached (see Fig. 2.3).

The injection of beams of fast hydrogen isotope atoms provides an important tool for plasma maintenance and heating in plasma traps with magnetic confinement. Beam energies required for this purpose amount to hundreds of keV and may reach a few MeV in the future [5, 6]. High-energy neutral beams can be obtained only by neutralization of accelerated ions in special conversion targets. A variety of gas or metal vapor targets were examined to convert a beam of positive ions of hydrogen isotopes into an atomic beam. However, the efficiency of neutralization decreases to below 0.2 at H^+ ion energies in excess of 100 keV (for D^+, over 200 keV). Therefore, the use of hydrogen negative ions is preferred, starting from an energy of 75 keV per nucleon [7]. In this case, the yield of atoms in metal vapor targets can be as high as 0.65 in the chosen particle energy range.

Particles exiting the accelerator include electrons and D^o and D^+ as well as $D-$. Fast D^o and D^+ are created by stripping of the $D-$ in the extractor and accelerator. The D^o that are not so divergent that they hit one of the grids will exit the accelerator, as will the D^+ with sufficient energy to overcome the retarding field downstream of their birth point. The D^o and D^+ that exit the accelerator will continue along the beamline and enter the neutralizer, with some D^o being converted to D^+ by collisions

V. Dudnikov, *Development and Applications of Negative Ion Sources*, Springer Series on Atomic, Optical, and Plasma Physics 125, https://doi.org/10.1007/978-3-031-28408-3_7

with D_2 in the gap between the accelerator and the neutralizer. The most dangerous particles exiting the accelerator are electrons as they will carry a substantial power. The backscattering of electrons from metal surfaces is significant, and after >2 backscattering events, some electrons could reach the thermal load-sensitive panels of the cryopumps. In the HNBs (heating neutral beams), the electrons exiting the accelerator are deflected downward by the long-range magnetic field from the beam source so that most of them are lost on a cooled electron dump below the entrance to the neutralizer and on the bottom of the neutralizer. Other electron dumps are deployed at each side of the neutralizer entrance and on the vessel walls to catch most of the backscattered electrons. The neutralizer and the electron dumps are water cooled, with the cooling channels being created in the various panels by deep drilling. Where the channel length is >1 m, the channels are drilled from both ends. The expected maximum drill deviation of 1 mm is fully compatible with the thickness of the panels. If enhanced cooling is required, and the channel is created by drilling from each end, a section of twisted tape is inserted from each end of the channel. That is the case for the leading edge elements of the neutralizer. Neutralization is achieved via collisions with D_2 inside the "neutralizer," the relevant reactions being

$$D - + D_2 \Rightarrow D^\circ + \ldots$$

$$D - + D_2 \Rightarrow D^+ + \ldots$$

$$D^\circ + D_2 \Rightarrow D^+ + \ldots$$

where … represents all other possible products of the reaction. To reduce the neutralizer length while keeping the gas flow into the neutralizer needed for the optimum neutralization of the beam, the neutralizer is divided into four adjacent vertical channels, and the gas is added midway along the neutralizer channel from five vertically disposed nozzles. Each channel is of rectangular cross section, 3 m long, 1.7 m high, and 105 mm wide at the beam entrance and 90 mm wide at the exit of the channel; see Fig. 7.1. The channel walls are formed by five panels each made up of three 1-m-long panels of oxygen-free copper that is cooled by water flowing in deep drilled channels in the walls (a total flow of 55 kg s^{-1}), and the maximum temperature of the Cu is kept below 150 °C. The out-of-plane deformation caused by the thermal loads is calculated to be <2 mm. At the entrance to the neutralizer, the channel walls are protected from the beam ions and neutrals and the fast electrons by "leading edge elements" (LEs). The LEs will intercept power densities of <5 MW m^{-2} and a maximum power of 0.3 MW. The actively cooled LEs consist of CuCrZr bars with a rounded shape on the upstream side with an 18 mm diameter, 1.8-m-long vertical cooling channel. Twisted tapes with a 70 mm pitch inside the cooling channels promote water turbulence. Finite element simulations show that with subcooled nucleate boiling conditions, the critical heat flux condition is 17 MW m^{-2} assuming the Yagov correlation [8].

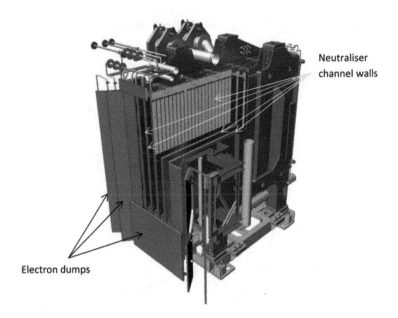

Fig. 7.1 Section view of the neutralizer and the electron dumps for ITER (International Thermonuclear Experimental Reactor) neutral beam injector. (Reproduced from Hemsworth [8])

7.2 Plasma Targets

The electrical efficiency of the entire beamline system is a key parameter for the economic application of neutral beams to fusion power plants [9]. A steady-state 1 GW thermal (2.7 GW fusion) power device might require some 200 MW of injected current drive power, so minimizing the power re-circulating in the system is of great significance. The system codes that are used to determine the operating envelope of the power plant usually adopt a figure of merit, F, for current drive systems defined as

$$F = \gamma\varepsilon \geq 0.25$$

where γ is the current drive efficiency and ε is the electrical efficiency of the driving system. The value of γ depends on the plasma scenario but is generally in the range 0.4–0.6 $Am^{-2} W^{-1}$ for neutral beam systems; thus, electrical efficiency in the range 0.625–0.4 is required. The electrical efficiency is calculated through a neutral beam system code that includes all power consumption aspects of the beam system. The reduction in the target density will influence the beam losses and the pumping requirements. Values for the efficiency are given in Table 7.1 together with those for ITER, a powered plasma neutralizer, and a photoneutralizer at 90% neutralization fraction for comparison.

G. Dimov at BINP (Budker Institute of Nuclear Physics) proposed to use a plasma target for negative ion beam conversion into neutrals [10]. In this case, the

Table 7.1 Electrical efficiency for gas, plasma, and photoneutralizers

Parameter	Gas	Driven plasma	Plasma	Photo
Electrical efficiency	0.35	0.49	0.52	0.58
Neutralization fraction	0.58	0.8	0.8	0.9
Beam losses	0.42	0.36	0.36	0.31
Auxiliary power	6 MW	6 MW	5 MW	4 MW

Beam parameters: Energy 1.0 MeV, current 40 A, and divergence 5 mrad

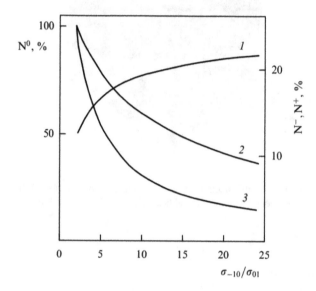

Fig. 7.2 $\gamma = \sigma_{-10}/\sigma_{01}$ dependence of beam charge component fractions at optimal target thickness: (1) atoms; (2) negative ions; and (3) protons. (Reproduced from Dimov [10])

atom yield resulting from the conversion of a negative ion beam into an atomic beam depends on the ratio between the detachment cross section of a weakly bound electron from a negative ion and the ionization cross section of the resultant atom: $\gamma = \sigma_{-10}/\sigma_{01}$. Then, the cross section of double ionization of a negative ion in a single collision is considered to be much smaller than cross sections of above processes $\sigma_{-11} \ll \sigma_{01}, \sigma_{-10}$. Therefore, the atom output from a target with optimal linear density is $N^0_{opt} = 1/\gamma^{(1/\gamma - 1)}$. The theoretical dependence of charge component fractions of the beam on γ is plotted in Fig. 7.2. For the majority of gases, the ratio of cross sections is 2.5–3.5 and atom yield ranges 0.52–0.55 [11, 12].

The involvement of alkali metal vapor targets allows reaching $N^0_{opt} = 0.6$ [13]. For a fully ionized plasma, the cross section of electron detachment from a negative ion markedly increases due to the absence of shielding of the long-range Coulomb potential for impact parameters of collisions smaller than the Debye radius. In the Born approximation, the dependence of the electron loss cross section on the relative velocity of colliding particles has the form $\sigma \sim \ln(mu^2/S)/Su^2$, where m is the electron mass and S is the electron affinity of the atom or the atomic ionization potential. In other words, the cross section ratio for targets from completely ionized hydrogen plasma weakly (logarithmically) depends on the collision energy as a result of fulfillment of the condition $mu^2/Su^2 \gg 1$; in fact, the cross section ratio is roughly 20. For this cross section ratio, atomic yield at an energy of several hundred keV or higher (when the applicability condition for the Born approximation is certainly satisfied) must be ~0.85. In the experiments of G I Dimov and G. V. Roslyakov with lithium and magnesium plasma jets [14, 15], the yield of atoms amounted to almost 0.8 at a negative ion beam energy of 0.5–1 MeV. The yield proved to be somewhat higher for a lithium plasma target, and its variation over the entire energy range was ~1%. Atomic yield for the magnesium plasma target was significantly smaller, and it tended to decrease with increasing beam energy.

The difference between atomic yields for the energies of 970 and 370 keV was ~2%. The lower atomic yield in magnesium plasma was attributable to the fact that the impact parameters of beam atom-ion collisions were commensurate with the size of the electron shell of Mg ions; it resulted in an appreciable increase in the atom ionization cross section upon collision [14]. Direct measurements of cross section ratios σ_{-10}/σ_{01} and $\sigma_{-11}/\sigma_{-10}$ in the lithium and magnesium plasmas [15] fully confirmed this conjecture. It appears that a completely ionized hydrogen plasma provides the best working medium for plasma conversion targets (disregarding the possibility of using a plasma with multiply charged ions). For the purpose of calculations, cross section σ^e_{-10} was assumed to be equal to σ^p_{-10}. A certain nonmonotone behavior of the calculated target thickness and atomic yield arises from the scatter of experimental values of the respective cross sections. The departure of target thickness from optimal values and the presence of unionized impurities in the plasma are the main causes behind the reduction in atomic yield. In the case of deviation of the target thickness from an optimal value x, a change in the yield of atoms is expressed as

$$\Delta N^0_{opt} = N^0_{opt} / 2 \left[\left(\Delta \xi / \xi \right)^2 \gamma \ln^2 \gamma / \left(\gamma - 1 \right)^2 \right].$$

For hydrogen plasma, the target thickness inhomogeneity may be significant: a 20% inhomogeneity leads to a 1% decrease in the atomic yield. Decreased atomic yield in the case of incomplete ionization of the target material is associated with lowering an effective cross section ratio taking into account beam particle collisions with atoms and molecules. Figure 7.3 presents the calculated dependence of the maximum atomic yield in the target on the content of unionized impurity.

Evidently, the impurity being molecular hydrogen, the plasma yield decreases but insignificantly, even at a relatively low degree of ionization. When an electron is detached from a negative ion, the resulting atom may be either in the ground or in one of the excited states. It does not lead to further appreciable ionization of beam excited atoms, because the deactivation cross section of metastable atoms in collisions under these conditions exceeds the ionization cross section by an order of magnitude. Experiments with a jet of hydrogen plasma [16] gave evidence of 84.5 ± 0.5% atomic yield at the energy of the hydrogen negative ion beam of around

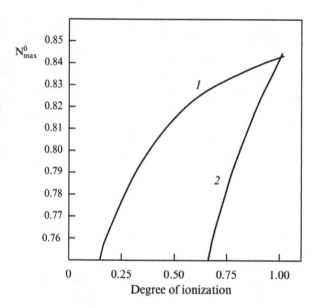

Fig. 7.3 Calculated dependence of maximum atom yield from the target on its degree of ioniza-tion: (1) molecular hydrogen admixture, and (2) molecular nitrogen admixture. (Reproduced from Dimov [10])

500 keV. A schematic representation of the experimental design is given in Fig. 7.4. Plasma source 1 produced a hydrogen plasma jet moving along the lines of the magnetic field created by curved solenoidal plasma driver 2. The straight section of the driver served to pump out the accompanying gas and purify plasma from molec-ular ions produced via dissociative recombination and ionization. After the plasma passed the turn, it entered the next 80-cm-long straight section of the driver and functioned as the target for the negative ion beam undergoing conversion. The coil placed at the end of the driver created a field opposite in sign to the solenoid field, resulting in the formation of a cusp configuration, and the plasma jet was sent along the radius into the beam collector with a powerful pumping system. The solenoid magnetic field was roughly ~0.1 T and did not appreciably deflect the primary nega-tive ion beam with an energy of several hundred keV.

The plasma was generated in a discharge for which an LaB_6 disk 50 mm in diam-eter served as the cathode heated up to 1400–1500 °C by an indirect heating spiral. The role of anode was played by a stainless steel grid installed approximately 35 cm from the cathode. The gas was puffed into the discharge chamber through a pulse valve at a rate of 10^{18} molecules per pulse. Discharge current varied in the range from 0.2 to 1 kA, and cathode voltage from 75 to 150 V, depending on cathode temperature and amount of hydrogen introduced into the discharge. Measurements with a mass spectrometer installed at the exit from the plasma source showed that plasma contained 85% H^+ ions and roughly 15% H^+_2 ions in the routine operating regime (with a plasma density of 10^{13} cm^{-3}). Plasma density in the discharge and plasma driver was measured by using mobile Langmuir probes and estimated from

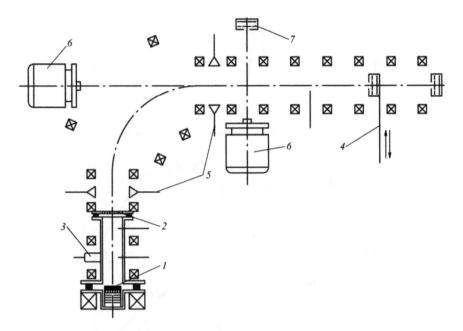

Fig. 7.4 Schematic of an experiment [13]: (1) plasma source; (2) plasma driver; (3) gas feed valve; (4) diagnostic beam attenuation detector; (5) microwave probe antenna; (6) diagnostic beam injector; and (7) beam detector. (Reproduced from Dimov [10])

the weakening of the atomic beam created by diagnostic injector 4 (see Fig. 7.4) and from the cutoff of microwave radiation with a wavelength of 7–11 mm. The section average density of the ionized gas across the jet was determined based on the weakening of the proton beam from the diagnostic injector. Moreover, there were no appreciable losses during jet transport from the source with the aid of the curved solenoid (at densities above 2×10^{13} cm^{-3}). Plasma density after the turn decreased by approximately 20% due to recombination of molecular ions in the jet. Stabilization of the flute instability during plasma jet propagation in the curved solenoid took place due to the plasma jet freezing into the butt end, where the cathode was located, as shown in special experiments in which the jet was cut off the cathode upon switching on the pulsed coil installed between the plasma source and the curved part of the solenoid to create a field opposite the solenoid field. Figure 7.4 shows the characteristic shape of the signals from Langmuir probes placed outside the plasma at the outer and inner radii of the turn and of the current in the coil that cuts off the jet. When the coil field coincided with the direction of the solenoid field, the signals from both probes decreased similarly, perhaps due to a decline in the plasma flow passing through the local magnetic mirror (Fig. 7.5b). If the cutoff coil induced a counter-propagating field, an ejection of plasma occurred along the radial direction at the jet turning site (Fig. 7.5c, d).

Further measurements demonstrated that the jet moved in the curved solenoid field without appreciable losses or radial dispersion. The degree of plasma

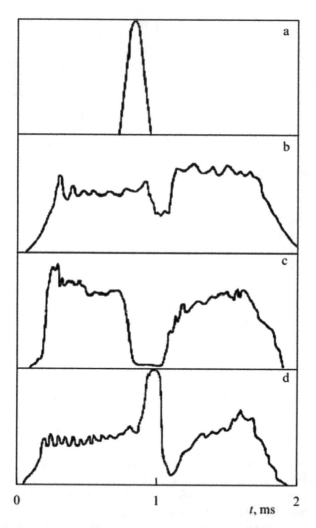

Fig. 7.5 (**a**) Cutoff coil current, characteristic shape of time dependences of probe signals, (**b**) cutoff coil field direction coincides with that of solenoid field, and (**c, d**) cutoff coil field pointed away from the solenoid field. (Reproduced from Dimov [14])

ionization after the turn was close to 100%. Measurements of the conversion coefficient in the target were performed for a negative ion beam with a pulse duration of 0.3 ms and an energy of 500 keV coming from a Van de Graaff accelerator (Fig. 7.6). The negative ion beam from the accelerator passed through analyzing magnet 1 and collimator 2.

After passage through plasma target 3 in solenoid 4, the beam diverged in the magnetic field of the bent section into charge components H-, H^0, and H$^+$. Their currents were measured using Faraday cups 6 after stripping by 3500 A$^\circ$ thick lavsan (Dacron in the USA) films. The coefficient of conversion of all charge components

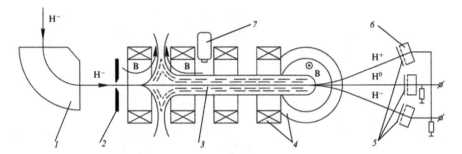

Fig. 7.6 Schematic diagram of target conversion coefficient measurements: (1) analyzing magnet; (2) collimator; (3) plasma target; (4) solenoid; (5) stripping films; (6) Faraday cups; and (7) gas valve. (Reproduced from Dimov et al. [15])

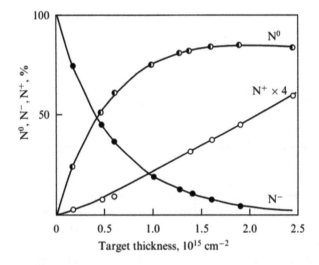

Fig. 7.7 Beam charge components plotted vs plasma target thickness. (Reproduced from Dimov [15])

of the beam to protons was roughly 0.999, which made unnecessary a calibration of signals from the atomic beam detector.

The results of the measurement of the yield of charge components from the weakening of the negative ion beam depending on the target linear thickness are presented in Fig. 7.7. The fraction of atoms reaches a maximum value of $(84.5 \pm 0.5)\%$ at a 2×10^{15} cm^{-2} plasma target thickness, in agreement with the theory. (Reproduced from Dimov et al. [15]).

The development of an alternative variant of the plasma target with plasma confinement in the magnetic field of a multipole configuration is currently underway at BINP [17, 18]. This variant, unlike the jet flow target, may help to reduce energy expenditures for plasma generation.

The prototype of the plasma neutralizer was constructed and studied at the separate test bench [19]. The prototype scheme is shown in Fig. 7.8a. The magnetic field

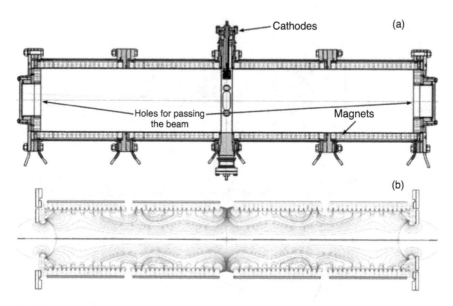

Fig. 7.8 A schematic of the plasma target prototype: (**a**) scheme and (**b**) the magnetic field of the target (the iron yoke, position of the magnets, and magnetic field lines are shown). (Reproduced from Emelev [17])

of the target, the iron yoke, position of the magnets, and magnetic field lines are shown in Fig. 7.8b.

It has a cylindrical chamber 120 cm long and 20 cm in diameter. The ends of the cylinder have apertures 10 cm in diameter for beam passing through. The multipole axially symmetric magnetic field of the trap was generated by permanent magnets installed from the outside on the thin-walled vacuum chamber. The field was strengthened by an iron yoke. The field change period was 1.5 cm, and its strength on the wall of the vacuum chamber was 7 kG. The trap consisted of two parts, in each of which, a 100-G paraxial longitudinal magnetic field was formed oppositely directed in its right and left halves. Annular magnets generated a magnetic field in the holes at the ends of the trap opposite that in the central region of the respective part of the device. It was expected that opposite magnetic fields would effectively curb the escape of plasma particles into the end holes due to generalized angular momentum conservation in the axially symmetric magnetic field.

Plasma is generated by the arc discharge with LaB_6 cathodes that are installed at the periphery of the central plane. Working gas (hydrogen) is injected at the chamber center. The complex magnetic field configuration of the neutralizer is produced by an array of circular permanent magnets. Discharge power is 220 kW. This configuration with multicusp at the walls and inversed end mirrors provides the plasma confinement. Schematic of a plasma target test bench (top view) is shown in Fig. 7.9a. A photograph of H- beam neutralization test bench with plasma neutralizer prototype is shown in Fig. 7.9b. The plasma parameters in the neutralizer are measured using the probes and diagnostic neutral beam. At the power of ~150 kW,

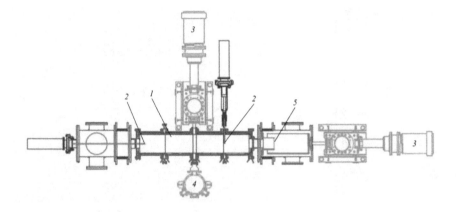

a

b

Fig. 7.9 (**a**) Schematic of a plasma target test bench (top view): (1) plasma target; (2) probes; (3) diagnostic neutral beam injector; (4) passed beam analyzer; and (5) plasma flow density meter. (**b**) Photograph of H- beam neutralization test bench with plasma neutralizer prototype. (Reproduced from Dimov [17])

the plasma density of $n_i \sim 1.8 \times 10^{13}$ cm^{-3} is reached. At discharge power 220 kW, the mean plasma density $n_i = 1.5 \times 10^{13}$ cm^{-3} and degree of ionization $\sim 50\%$ in the center of the neutralizer are obtained. The spatial profile of plasma parameters in the neutralizer is studied as well. Figure 7.10 shows the plasma density profiles measured by the probes in the central cross section of the neutralizer and at a distance of one-quarter of the neutralizer length from the center. The plasma density in the beam area ($r = 5$ cm) at these cross sections has an inhomogeneity of $\pm 23\%$ and $\pm 10\%$ correspondingly.

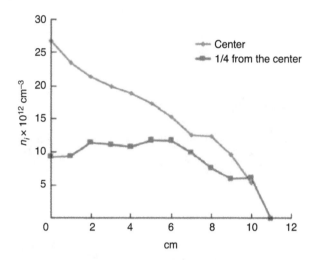

Fig. 7.10 The radial profile of plasma density in the plasma neutralizer. Discharge power 220 kW. (Reproduced from Dimov [17])

The multipole plasma target was tested on a bench schematically depicted in Fig. 7.9a. A photograph of H- beam neutralization test bench with plasma neutralizer prototype is shown in Fig. 7.9b [20]. Vacuum tanks were attached on either side of the trap for gas pumping and installation of diagnostic equipment. Plasma density and its profile were determined with the aid of movable Langmuir probes; its density and degree of ionization were measured by probing plasma with a 5–10 keV atomic beam. The beam to be injected had to pass through the target plasma and reach a magnetic analyzer, where it was divided into three charge components, namely, hydrogen atoms, protons, and hydrogen negative ions.

Charged beam components were registered by Faraday cups and the atomic component by a secondary emission detector. The diagnostic beam and analyzer can be arranged both at the ends of the trap and along the radius in its central part.

Plasma density in the trap and the degree of its ionization could be determined independently based on variations in charge component currents. One of the vacuum tanks contained a specially designed collector to measure the plasma flow. Plasma density showed a linear dependence on the discharge power. At 200 kW, the plasma density in the central region of the trap amounted to $n_i = 2.2 \ 10^{13}$ cm^{-3}. The electron temperature was equal to 3–5 eV. Plasma density profile in the radial directions is shown in Fig. 7.10, suggesting that the plasma was confined in the central region of the trap. Its density fell dramatically on approaching the walls of the vacuum chamber (Fig. 7.9a) but remained virtually constant along the trap axis. Plasma densities measured by charge exchange of an atomic beam fairly well agree with probing data. Plasma density in the central section of the trap is $n_i > 2 \ 10^{13}$ cm^{-3}. The degree of plasma ionization in the central section measured by the charge-exchange method as applied to an atomic beam is 50%. Measurements of the full plasma flow outgoing from the hole intended for the beam to be neutralized demonstrated the efficiency of bounding the flow into the hole by the oppositely directed magnetic field. The plasma flow into the hole was almost 30 times smaller than the unbounded one. Further experiments are designed to enhance plasma ionization in the target and increase the efficiency of its generation.

The paper [8] describes a method for using the plasma created by the passage of the beam itself through a neutral gas background to enhance the neutralization efficiency with no additional power input. A zero-dimensional model of the beam-gas-plasma system in the neutralizer is derived for the case where magnetic confinement is applied to the neutralizer walls. It is shown that this method can confine the electrons generated by stripping from the beam particles and by ionization of the background gas. Confining these electrons enhances the plasma production in the neutralizer to densities close to the optimum required for maximum plasma neutralization efficiency. Although the economy of power is marginal compared to the driven plasma concept, there is significant value in reduced complexity and increased reliability.

7.3 Photon Targets

One promising neutralization technique for negative ion beams involves the use of a photon target based on the electron photodetachment reaction [21]. Cross section for destruction of H⁻ ions by photons as a function of electron energy or photon wavelength is shown in Fig. 3.2. The quantum energy is higher than the electron affinity energy but lower than the ionization potential of the hydrogen atom, which allows, in principle, coefficients of negative ion beam conversion close to 100% to be obtained. The degree of neutralization of a negative ion beam in the photon target is given by the expression [21].

$$\eta = J^0 / J^- = 1 - \exp\left(-\sigma c P / \hbar \omega d \mathrm{Vi}\right) \qquad (7.1)$$

where J^0 and J^- are the neutral atom and negative ion currents, respectively, c is the speed of light, σ is the photodetachment cross section, P is the radiation power density inside the trap, Vi is the particle velocity, $\hbar \omega$ is the photon energy, and d is the length of the photon-occupied region. In view of the small photodetachment cross section, photons have to cross the negative ion beam many times to ensure a high-degree neutralization (above 90%). It is achieved by the employment of a mirror system confining radiation within itself. A variety of scenarios have been proposed for the purpose [14, 22] based on different variants of the Fabry-Perot cavity. Modern technologies make it possible, in principle, to create such photon targets, but a large number of challenging technical problems remain to be solved. Specifically, high-reflectivity (above 0.999) mirrors resistant to incident radiation of several hundred kilowatts per cm² and requiring intense cooling are needed. Moreover, thermal stabilization of the entire mirror system must be ensured, and rigorous demands on the quality of laser radiation met to fulfill the phase synchronism condition for a large number of passes. The current state of research on resonant photon targets is overviewed in Refs. [14, 23]. An alternative approach covers nonresonant photon accumulation [24, 25]. Such a concept of the photon trap reduces to a system of reflective surfaces capable of multiple beam reflection. The basic principle behind the device is well described in terms of a mathematical billiard system [26] with a wide enough range of billiard trajectories stable in the phase space. The energy density in this system increases in proportion to the beam lifetime. The integral lifetime in a photon trap is largely determined, as it uses a resonant photon accumulator, by photon losses due to reflections and photon escape time from the system, because the mirrors cannot form a closed surface. The essential difference between this and resonant accumulators is the absence of stringent conditions imposed on the phase relationships between the large number of rays in its interior and radiation input through a highly reflective surface, rather than through one or several small holes. Long-term retention of photons is ensured by conservation of certain adiabatic invariants. The efficiency of photon accumulation is virtually independent of the quality of injected radiation, which allows considering a relatively cheap commercial high-efficiency optical fiber laser a potential radiation source. The nonresonant photon accumulation concept proposed in Ref. [17] suggests conservation inside the trap of some adiabatic invariants bounding the photon-occupied region instead of exact phase relationships between the rays at different stages of evolution. Such accumulation is analogous to the confinement of charged particles in open magnetic traps, as proposed by G I Budker [5]. Let us consider the two-dimensional configuration of the mirrors. Figure 7.10 shows that a photon acquires an additional horizontal momentum in the direction of the height maximum between the mirrors after each new reflection. At small deviations from vertical motion, the photon tends toward the central "equilibrium" position. Let us define the photon position immediately after its nth reflection by the abscissa of the reflection point x_n, its height $F(x_n)$, and the angle β_n between the vertical line and photon velocity (see Fig. 7.11). The horizontal movement is then described by the following set of equations:

$$x_{n+1} - x_n = \left(F\left(x_{n+1}\right) + F\left(x_n\right)\right) \tan \beta_n; \tag{7.2}$$

$$\beta_{n+1} - \beta_n = 2\ dF\left(x_{n+1}\right) / dx. \tag{7.3}$$

Linearization of the (7.2) and (7.3) system to study stability gives

$$x_{n+1} - x_n = 2F\left(0\right)\beta_n; \tag{7.4}$$

$$\beta_{n+1} - \beta_n = 2\left(d^2 F\left(0\right) / dx^2\right) x_{n+1}. \tag{7.5}$$

Combining expressions (7.4) and (7.5) leads to the following linear recurrent relationship:

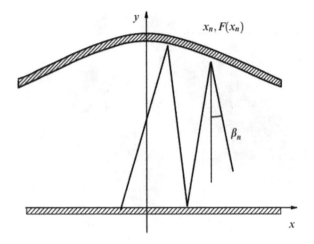

Fig. 7.11 Schematic diagram of a quasiplanar photon trap. (Reproduced from Emelev [19])

$$x_{n+2} - 2x_{n+1} + x_n = 4\left(F(0)d^2F(0)/dx^2\right)x_{n+1} = -4F(0)x_{n+1}/R; \qquad (7.6)$$

where R is the radius of curvature of the upper mirror. The stability condition of Eq. (7.6) is easy to find as

$$F(0) < R. \qquad (7.7)$$

Evidently, the fulfillment of inequality (7.7) conserves the adiabatic invariant

$$F(x)\cos \beta = \text{const}; \qquad (7.8)$$

limiting the region occupied by photons.

The possible geometry of a photon accumulator based on this principle is presented in Fig. 7.12. Each mirror consists of a cylindrical part conjugated with spherically shaped mirrors docked at the ends.

Clearly, the radii of both cylindrical and spherical mirrors are equal. The end mirror may have a different shape provided that it ensures a gradual decrease in the distance between the upper and lower mirrors during movement from the center of the trap. Photon motion along the axis in such a system is essentially anharmonic. It was shown in [30] that the lifetime of radiation injected into such an accumulator is largely determined by the reflectivity of the mirrors. Radiation storing efficiency thus achieved proved sufficient for experiments on neutralization of hydrogen and deuterium negative ion beams with a particle energy of 10 keV [27, 28]. The experimental setup is schematically depicted in Fig. 7.13. Laser radiation 1 passes through beam splitter 2 and is focused by lens 3 onto the inlet of photon accumulator 4 inside vacuum chamber 5. The input angular spread determined mostly by the focal length is about 3.

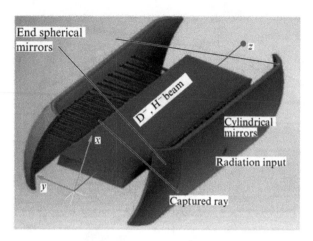

Fig. 7.12 Schematic of an elongated adiabatic photon trap. (Reproduced from Emelev [19])

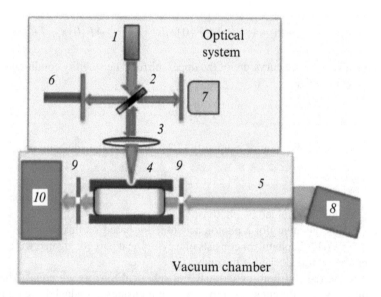

Fig. 7.13 Schematic diagram of measuring the negative ion beam neutralization coefficient: (1) laser radiation; (2) beam splitter; (3) lens; (4) photon accumulator; (5) vacuum chamber; (6) light guide; (7) CCD; (8) magnet; (9) diaphragms; and (10) analyzer. (Reproduced from Popov [30])

Neutralization coefficient plotted vs laser radiation power P_L is shown in Fig. 7.14. Curves 1 and 2 correspond to hydrogen and deuterium ions, and curve 3 is horizontally compressed curve 1.

An antireflective beam splitter directs a minor part of the injected radiation to light guide 6 for monitoring input radiation power. The use of a charge-coupled device (CCD) 7 ensures precise control of the laser ray and its direction exactly into the inlet hole. H- and D- beams were generated using the DINA-4A injector [29].

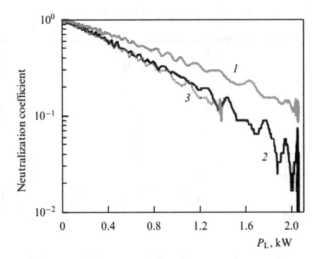

Fig. 7.14 Neutralization coefficient plotted vs laser radiation power P_L. Curves 1 and 2 correspond to hydrogen and deuterium ions, and curve 3 is $(U_D m_H / m_D U_H)^{1/2}$ times horizontally compressed curve 1. (Reproduced from Popov [30])

The ion beam from the injector passed through the gas target after which negative ions were separated from other charge fractions by magnet 8. A narrow negative ion beam cut out by diaphragms 9 was sent to the photon target and reached magnetic analyzer 10 at the exit from it. The two-channel design made possible simultaneous registration of positive and negative ions in the beam behind the photon target. The beam energy varied from 6 to 12 keV. Pulse duration was 150 ms.

Typical oscillograms obtained in these experiments are presented in Fig. 7.15a. In Fig. 7.15b, curve 1 at the time interval of 45–85 ms and curve 3 extended in the vertical direction with the aid of an ad hoc scaling factor are practically superimposed. The best coincidence of the curves was obtained with the scaling factor of 50 ± 4 corresponding to a $98\% \pm 0.2\%$ degree of neutralization. The fraction of positive particles at the exit from the neutralizer was below the noise level equivalent to 0.1% of the signal amplitude.

Results of the above experiments provide convincing evidence of the possibility of efficient radiant energy accumulation from a low-quality photon beam unsuitable for traditional approaches with the use of Fabry-Perot cells. Reference [30] presents a conceptual project of a nonresonant neutralizer for higher than 90% neutralization at a total injected radiation power of 310 kW for the powerful neutral injection systems currently under design and construction [31]. Industrial high-performance optical fiber lasers may serve as radiation sources [32]. The biggest technological challenge to be addressed is the manufacture of highly reflective (0.9995) mirrors having a large $(0.5 \times 8 \text{ m}^2)$ area. Such mirrors can be composed of smaller segments with minimal gaps between them. An advantage of such an approach is that it does not require a powerful narrowband master laser emitting ultrahigh-quality radiation nor does it need an expensive complex system for stabilizing and adjusting resonator elements.

Review of development conversion target is presented in [33].

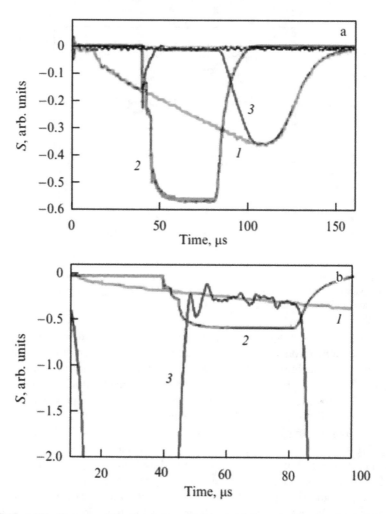

Fig. 7.15 (a) Oscillograms of the signals from negative ion beam neutralization. Curve 1 is variation of D- current without neutralization, curve 2 is data from the laser pumping monitor, and curve 3 is D- current at laser switch-on. (Reproduced from Popov [30])

References

1. V. Dudnikov, *Negative Ion Sources* (NSU, Novosibirsk, 2018). В. Дудников, *Источники отрицательных ионов*. НГУ, Новосибирск, 2018
2. V. Dudnikov, *Development and Applications of Negative Ion Sources* (Springer, 2019)
3. V.G. Dudnikov, Charge exchange injection into accelerators and storage rings. Physics-Uspekhi **62**(4), 405 (2019)
4. G.I. Dimov, V.G. Dudnikov, Cross sections for stripping of 1 Mev negative hydrogen ions Inоb certain gases. Soviet Physics Technical Physics-USSR **11**(7), 919 (1967). Г.И.Димов, В.Г. Дудников, «Сечения обдирки отрицательных ионов водорода с энергией 1 МэВ в некоторых газах», ЖТФ 11 (7), 919 (1967)

5. G.I. Budker, *Plasma Physics and the Problem of Controlled Thermonuclear Reactions*, vol 3 (Ed. M A Leontovich) (Pergamon Press, New York, 1959), p. 1; Translated from Russian: in Fizika Plazmy i Problema Upravlyaemykh Termoyadernykh Reaktsii Vol. 3 (Ed. M A Leontovich) (Moscow: Izd. AN SSSR, 1958) p. 3

6. G.I. Dimov, V.V. Zakaidakov, M.E. Kishinevskii, Fiz. Plazmy **2**, 597 (1967)

7. H.S. Staten, in Proceedings of the 3rd International Symposium on Production and Neutralization of Negative Ions and Beams, Brookhaven, p. 587 (1983)

8. R.S. Hemsworth et al., Overview of the design of the ITER heating neutral beam injectors. New J. Phys. **19**, 025005 (2017)

9. E. Surrey, A. Holmes, The beam driven plasma neutralizer. AIP Conf. Proc. **1515**, 532 (2013). https://doi.org/10.1063/1.4792825

10. G.I. Dimov, Charge exchange injection method of protons into accelerators and storage rings, Preprint 304, INP SBAS, Novosibirsk, 1969. Г.И. Димов, "Перезарядный метод инжекции протонов в ускорители и накопители", Препринт 304 ИЯФ СО АН СССР, Новосибирск, 1969

11. S.K. Allison, Rev. Mod. Phys. **30**, 1137 (1958)

12. H. Tawara, A. Russek, Rev. Mod. Phys. **45**, 178 (1973)

13. B.A. D'yachkov, Zh. Tekh. Fiz. **38**, 1259 (1968)

14. G.I. Dimov, G.V. Roslyakov, Nucl. Fusion **15**, 551 (1975)

15. G.I. Dimov, A.A. Ivanov, G.V. Roslyakov, Zh. Tekh. Fiz. **50**, 2300 (1980)

16. G.I. Dimov, A.A. Ivanov, G.V. Roslyakov, Sov. J. Plasma Phys. **6**, 513 (1980).; Fiz. Plazmy 6 933 (1980)

17. I.S. Emelev, "Генератор плазмы с инверсным магнитным полем для тандемного источника отрицательных ионов и других применений", Dissertaciya kandidat Fis. Mat. Nauk, BINP (2020)

18. I.S. Emelev, A.A. Ivanov, A plasma target for neutralization of the negative ion beam. AIP Conf. Proc. **2052**, 070005 (2018). https://doi.org/10.1063/1.5083785

19. O. Sotnikov, A. Ivanov, Y. Belchenko, et al., Development of high-voltage negative ion based neutral beam injector for fusion devices. Nucl. Fusion **61**, 116017 (2021)

20. O. Sotnikov, A. Ivanov, G. Abdrashitov, Y. Belchenko, et al., Development of high-voltage negative ion based neutral beam injector for fusion devices. Nucl. Fusion **61**, 116017. http://iopscience.iop.org/article/10.1088/1741-4326/ac175a (2021)

21. W. Chaibi et al., AIP Conf. Proc. **1097**, 385 (2009)

22. M. Kovari, B. Crowley, Fusion Eng. Design **85**, 745 (2010)

23. A. Simonin et al., AIP Conf. Proc. **1390**, 494 (2011)

24. S.S. Popov et al., *Tezisy XLII Mezhdunarodnoi Konf. po Fizike i UTS (Theses of the XLII Intern. Conf. on Plasma Physics and Controlled Thermonuclear Reactions)* (ZAO NTTs PLAZMAIOFAN, IOF RAN, Moscow, 2015), p. 395

25. S.S. Popov, et al.., arXiv:504.07511

26. V.V. Kozlov, D.V. Treshchev, *Billiardy. Geneticheskoe Vvedenie v Dinamiku System s Udarami (Billiards. Genetic Introduction to Dynamics of Systems with Strokes)* (Izd. MGU, Moscow, 1991), p. 168

27. S.S. Popov et al., AIP Conf. Proc. **1869**, 050005 (2017)

28. M.G. Atlukhanov et al., AIP Conf. Proc. **1771**, 030024 (2016)

29. Y.I. Belchenko et al., Rev. Sci. Instrum. **61**, 378 (1990)

30. S.S. Popov et al., AIP Conf. Proc. (2018)., in print

31. R. Hemsworth et al., Nucl. Fusion **49**, 045006 (2009)

32. IPG.IRE-POLUS., http://www.ipgphotonics.com/ru/products/lasers/nepreryvnye-lazery-vysokoy-moshchnosti/1-mikron/yls-sm110-kvt

33. Y.I. Belchenko, V.I. Davydenko, P.P. Deichuli, et al., Studies of ion and neutral beam physics and technology at the Budker Institute of Nuclear Physics, SB RAS. Physics - Uspekhi **61**(6), 531–581 (2018)

Chapter 8
General Remarks on Surface Plasma Sources

Abstract The development of the surface plasma method of negative ion beam production, classification of surface plasma sources, neutral beams in space (SDI), H^- injector for ESS storage ring, and recent developments in surface plasma sources are reviewed.

8.1 Introduction

Soviet academician N. Semenov (Nobel prize winner for discovery of chain reactions) proposed in 1945 to suppress the explosions of nuclear bombs by irradiating them with neutrons [1]. A large accelerator development program was established in the USSR. V. Teplyakov has invented radio frequency quadrupole (RFQ, Teplyakov Accelerator) within the framework of this program [2]. A similar proposal was put forward by R. Wilson in 1953 [3].

After experimental development of charge exchange injection with the conversion of negative ion into neutrals [4] (my PhD thesis and doctoral dissertation by G.I. Dimov) [5, 6], the director of the Institute of Nuclear Physics (Novosibirsk, Russia), G. Budker, suggested using high energy neutral beams in space to influence space objects (inspecting satellites for the presence of nuclear materials, suppressing nuclear explosions, etc.) [7].

In the Korolev Space Central Design Burro (already without Korolev), after the closure of the Soviet lunar program, unused funds remained, and in 1969, a contract was signed with the INP to develop a neutral injector for an ion beam current of H^- 10 mA for 2 Mrub [8] (at that time in USSR, Novosibirsk, a nine-story building for 286 apartments was cost 1 Mrub, 1 \$ = 0.86 rubles). The Strategic Defense Initiative using neutral particle beams was established in the USA in 1976 [9]. Then three groups were organized in INP, which were engaged in different methods of negative ion beams obtaining. For charge exchange injection into the Dimov's lab, charge exchange and plasma sources of negative ions (of the Ehlers type) were developed. In the first year of Novosibirsk State University, I read Gaponov's textbook

© The Author(s), under exclusive license to Springer Nature Switzerland AG 2023
V. Dudnikov, *Development and Applications of Negative Ion Sources*, Springer
Series on Atomic, Optical, and Plasma Physics 125,
https://doi.org/10.1007/978-3-031-28408-3_8

Electronics and then reread it. In it, I found a mention of the secondary emission of negative ions, little known at that time. I began to study secondary ion-ion emission in more detail. There was a lot of data, but they were very contradictory, and everywhere the probabilities of secondary emission were very small, and the theoretical foundations of this phenomenon were absent. It was known that the deposition of alkali metals increases the coefficient of secondary ion-electron emission.

The works of Ayukhanov of 1961 [10] and of Kron of 1962 [11] were discovered, in which it was shown that when deposited alkali metals, the coefficient of secondary emission of negative ions also increases, but the resulting currents of H^- ions were at a sub-microampere level.

I suggested that Dimov take up the secondary emission method for obtaining negative ions. He enthusiastically accepted this proposal, and a group was organized to obtain H^- ions by bombarding a surface with cesium ions under his leadership. I took an active part in this work. A stand was set up and rather soon H^- beams with currents up to -2.5 mA were obtained [12], but the brightness of these beams was very low and the service life of these devices was very short.

G. Roslyakov's group was engaged in charge exchange sources [13].

My student Y. Belchenko and I worked on plasma sources. The groups worked intensively, but they were far from reaching the required parameters. Budker gathered weekly meetings where various proposals were discussed, but no solutions were found. By the end of the contract, everyone despaired of the possibility of obtaining the necessary currents and went on vacation for the summer. Yura Belchenko left with a building team in Bilibino in Kolyma (to build the Bilibino nuclear power plant) and took our laboratory assistant with him. The mechanics also went on vacation.

8.2 Discovery of Cesiation Effect

I stayed alone for the summer and continued to work with a plasma source with a planotron geometry. It yielded up to 4.5 mA H^- in 1 ms pulses, but with an electron current to the extractor 30 times higher, a discharge current of 100 A, and a discharge voltage of 600 V (a power of 60 kW put into a device with a volume of ~ 1 cm^3). Taking into account the acquired "negative" experience gained after a number of trial experiments, the design of a new source was brought to the form depicted schematically in Fig. 8.1. The body of plasma cell 3 was fixed on insulators 2 of plexiglass in the gap between the pole pieces 1. The plates of extraction electrode 14 were welded to special protrusions from the pole pieces 1, creating a magnetic field in the high-voltage gap (but not a Penning trap configuration). A pair of pole pieces with the source was installed between the grounded poles of the electromagnet. A plasma cell with planotron configuration is formed by the cathode, consisting of the central plate of cathode 10 and cathode side shields 11 and a cathode-enclosing anode formed by parts of the plasma cell body 3 and anode insert 5. A cathode made of 0.2-mm-thick molybdenum foil was attached to tantalum

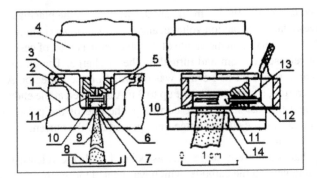

Fig. 8.1 Schematic of planotron plasma configuration for producing a negative ion beam. 1, electromagnet pole tips; 2, high-voltage insulators; 3, body of plasma chamber; 4, pulsed gas inlet valve; 5, anode insert; 6, screen of emission slit; 7, emission slit; 8, collector; 9, negative ion beam; 10, central cathode plate; 11, cathode side shields; 12, cathode holders; 13, cathode insulators; and 14, extractor plates

Fig. 8.2 Photograph of the first planotron with numeration of parts

current leads 12 passing through the wall of anode insert 5 and insulated from it by ceramic tubes 13. The volume of the plasma cell was minimized as much as possible. Gaps between the cathode and the anode, in which a discharge should not burn, were reduced to 1 mm. Hydrogen was supplied to the plasma cell through a short channel by a pulsed electromagnetic valve 4. Emission slit 7 with dimensions 0.5 × 10 mm^2 oriented across the magnetic field was cut in the thin-walled body of the plasma cell. From the discharge region, particles could pass to the emission slit through the gap between the anode projections 6 shielding the emission slit from the dense, high-current plasma. A photograph of the first planotron is shown in Fig. 8.2.

The old power supply systems were used for pulsed hydrogen gas injection, plasma ignition and support of the discharge, and ion extraction. A beam collector

8 was installed to monitor the beam current 9. The body of the plasma cell was held at the negative-polarity extraction voltage, and the collector was held at low voltage.

On July 1, 1971, I fixed on the anode of the planotron a tablet of cesium chromate with 1 mg of released cesium and turned on the discharge. The emission slit was shielded and an ion current of 1.5 mA was recorded on the collector. After several minutes of operation, a current surge of up to 3 mA appeared at the end of the pulse. After optimizing the gas supply, the current at the end of the pulse increased to 4 mA, but after 20 mi, the current surge disappeared and the current to the collector became 1 mA again.

Deciding that the current surge was associated with the release of cesium, I placed several tablets on the cathode, which was heated more strongly, and covered them with a nickel mesh.

In this configuration, the collector current quickly increased to 12.5 mA, and after optimizing the gas supply and discharge, a rectangular pulse of 15 mA was obtained.

The discharge voltage dropped from 600 V to 100 V. After that, within a week, various configurations of discharges were tested, it was verified that these are mainly H^- ions, and the currents of electrons and heavy ions to the collector are small [14]. After that, I left everything in working order and went on vacation to my village Gunda in Buryatia.

After my return, the source worked as it should, the H^- current was quickly increased to 100 mA, by optimization of design and mode of operation, and then to 300 mA, with emission current density up to 4 A/cm^2, and then to 0.9 A from a source the size of a cigarette lighter [15].

Under the contract with the Central Design Bureau, we successfully reported and concluded two new contracts of 5 Mrub each for the manufacture of a surface plasma sources (SPSs) for 100 mA for the linear accelerator of the meson factory of the Institute of Nuclear Research (Moscow) and for the Institute High Energy Physics (Protvino).

Then the results with cesiation were recognized as deep secret and were not allowed to be published. But many delegations from the Soviet Union (Central Design Bureau, NIIEFA, Kurchatov Institute, Institute of Nuclear Research, Institute of High Energy Physics) and from the USA began to visit INP. Budker allowed the SPS (surface plasma source) to be shown to high-ranking US visitors as objects for bidding. International publication [11] was permeated only in 1974.

One practical result of the development of high-brightness SPSs with cesiation is the wide use of charge exchange injection in circular accelerators for routine operation. Now SPSs are "sources of life" and "working horses" for large accelerator complexes: SNS spallation neutron source at the Oak Ridge National Laboratory (ORNL), Fermi National Accelerator Laboratory (FNAL), Brookhaven National Laboratory (BNL), Los Alamos Neutron Scientific Center (LANSCE), ISIS at the Rutherford Appleton Laboratory (RAL), Large Hadron Collider (LHC) in CERN (European Council for Nuclear Research), Japan Proton Accelerator Research Complex (KEK-J-PARC), Deutsche Electron-Synchrotron (DESY), and Chinese Spallation Neutron Source (CSNS). Charge exchange injection is used on a

CELSIUS storage ring (Uppsala, Sweden) and in the COZY storage ring (Research Center Juelich, Germany). Charge exchange injection was used in the synchrotron of the Institute of Theoretical and Experimental Physics (ITEP, Moscow, Russia) for the accumulation of carbon ions. A transition to charge exchange injection in the CERN booster (Geneva, Switzerland) is finished and being prepared charge exchange injection at the IHEP booster (Protvino, Russia). It is planned to use charge exchange injection in storage ring of European Spallation Source (ESS).

The efficiency and operational reliability of these SPSs have determined the productivity of these laboratories and their big machines. Many discoveries in high-energy physics were made using SPSs.

The development of high-brightness SPSs was first stimulated by the successful accumulation of intense proton beam using charge exchange injection [6, 16–20] and subsequently to some extent by interest in particle beam weapons in space (e.g., as an element of the US Strategic Defense Initiative (SDI) or "Star Wars" program in the late 1980s [21]). The testing of the acceleration and neutralization of H^- ions in space is described and (the Beam Experiment Abroad Rocket (BEAR)) is shown in Fig. 8.3 [22, 23]. Developed by LANL (Los Alamos National Laboratory) with McDonnel Douglas, Westinghouse, and GRUMMAN Corp, it was cost 794 M\$.

Military applications and secret work led to a big delay in the first publications, but informal contacts were relatively fast. Until 1971, the main attention was concentrated on charge exchange negative ion sources because there was no hope to extract more than 5 mA of H^- ions directly from the plasma.

The cesiation effect, a significant enhancement of negative ion emission from the gas discharge with simultaneous decrease of co-extracted electron current to less than the negative ion current, was observed for the first time on July 1, 1971, at the Institute of Nuclear Physics (Novosibirsk, Russia) by introducing into the discharge in the planotron a compound containing 1 mg of cesium [24–26]. This observation, recounted in a review [27, 28], was further developed and soon recognized as a prime new surface plasma method of negative ion production. The patent application [13] stated as follows: "a method of negative ion production in gas discharges, comprising adding into the discharge an admixture of substance with a low ionization potential such as cesium, for example, for enhancement of negative ion

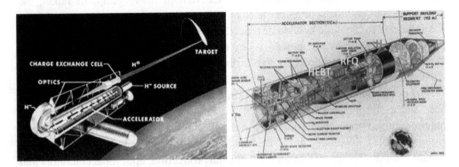

Fig. 8.3 Schematic of BEAR experiment. The testing of the acceleration and neutralization of H^- ion beam in space

formation." Subsequent experiments demonstrated that cesium adsorption decreases the surface work function from 4–5 eV to ~1.2–1.3 eV, which enhances secondary emission of negative ions caused by the interaction of the plasma with the electrode surface and thereby enhances surface plasma generation (SPG) of negative ions. Ion sources based on this process have been named surface plasma sources (SPSs). The theoretical explanation of this enhancement of negative ion emission by cesiation was presented by Kishinevskii [29–31]. Further development of the SPS was conducted by the Belchenko, Dimov, Dudnikov coalition at the Budker Institute of Nuclear Physics. The planotron was invented by the author [13]. The development of the high-brightness SPS with Penning discharge has been described by the author [32]. The semiplanotron SPS with effective geometric focusing was developed and reported by the author [33].

Development and adaptation of SPSs were then further carried out at many laboratories around the word from 1972.

A small admixture of cesium or other impurity with low ionization potential (ILIP) in the gas discharge significantly improves H^- production [10, 11, 13]. When done correctly, a cesiated SPS works very well. However, improper cesiation can complicate ion source operation. For example, injection of too much cesium can cause the discharge to become unstable, and sparking occurs in the extractor with loss of stable ion source operation. With low cesium concentration, the efficiency of negative ion production is too low. With "proper" cesiation, the efficiency of negative ion production is high, and extended ion source operation is stable with low cesium consumption.

Brookhaven National Laboratory (BNL) symposiums and European conferences on the production and use of light negative ions were established [34] later replaced by the International Symposium of Negative Ions, Beams and Sources (NIBS) [35].

The physical principles of SPS operation have been published in [36] and have been reproduced in many reviews and texts. A good overview of SPSs for accelerators has been presented in Peters' reports [37–39]. A review of the early work on SPSs has been given by N. Wells [40]. The work of the Allison group in Los Alamos is reported in [41]. A major development program for high-current SPSs (tens of amperes of H^- or D^- ions) for thermonuclear research was carried out at the Lawrence Berkeley National Laboratory (LBNL) for many years. In Japan, the work continues and the sources are used to inject high energy neutrals into the JT-60 Tokamak and to the large helicon device (LHD) stellarator. The author proposed the preparation of polarized negative ions by resonant charge exchange on slow negative deuterium ions obtained by the surface plasma method and was carried out collaboratively with Belov [42].

The development of negative *heavy* ion SPSs for technological applications has been successful, but improvement of dc sources is still needed to meet the broad requirements of many industrial users. A dc SPS for long-term operation with accelerators has been developed at the Budker Institute of Nuclear Physics (BINP) team at Novosibirsk. Cesiated SPSs with RF plasma production, both for accelerators and for neutral beam injectors, have been significantly improved by the SNS ion source group at Oak Ridge and by the IPP (Max-Planck-Institut für Plasmaphysik,

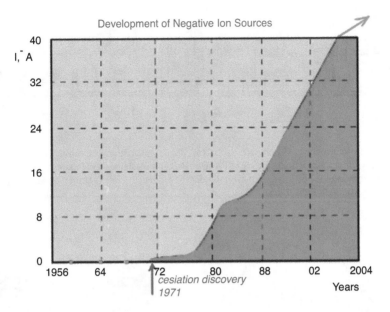

Fig. 8.4 History of development of negative ion sources. Growth of beam intensity in time. (Reproduced from Vadim Dudnikov [44], with permission from AIP Publishing)

Garching, Germany) team [43]. The historical development of negative ion sources—the growth of beam current with time—is shown in Fig. 8.4.

Many versions of the SPS were developed and optimized for a variety of applications. The addition of cesium increases the formation of negative ions in all discharges. But the most efficient production of beams of negative ions with high brightness is obtained in special SPSs optimized for various applications [44–46]. Some basic configurations of compact high-brightness SPSs (CSPSs) are shown in Fig. 8.3.

Figure 8.5 shows (a) plain magnetron (planotron), (b) magnetron with geometric focusing, (c) Penning discharge SPS (Dudnikov-type source) adapted for injection into ISIS, (d) semiplanotron, and (e) hollow cathode SPS.

The compact SPSs shown in Fig. 8.5 use glow discharges with cold cathodes in crossed ExB fields. They have a high plasma density of up to 10^{14} cm^{-3}, a high of negative ion emission current density of up to 8 A/cm^2, and a small gap of 1–5 mm between the emitter cathode and the emission aperture in the anode. They are simple, have high energy efficiency of up to 100 mA/kW, and have high gas efficiency of up to 30% with pulsed gas feed. CSPSs are excellent for pulsed mode operation, although CSPSs have been successfully used for continuous (dc) operation with an emission current density of up to 300 mA/cm^2 [47] and up to 1 A/cm^2 [48].

A different situation holds for large volume SPSs (LV SPSs) with discharge volume up to hundreds of liters, as shown in Fig. 8.6. The first LV SPS, Fig. 8.6a, was developed at LBNL, Berkeley. The gap between the emitter (5) and extractor aperture is very large (8–12 cm) and the plasma and gas density must be kept low to

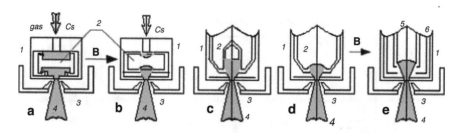

Fig. 8.5 Basic versions of compact SPSs. (**a**) Planar magnetron-planotron; (**b**) magnetron-planotron with geometric focusing; (**c**) SPS with Penning discharge, Dudnikov-type source; (**d**) semiplanotron; and (**e**) SPS with hollow cathode. (Reproduced from Vadim Dudnikov [44], with permission from AIP Publishing)

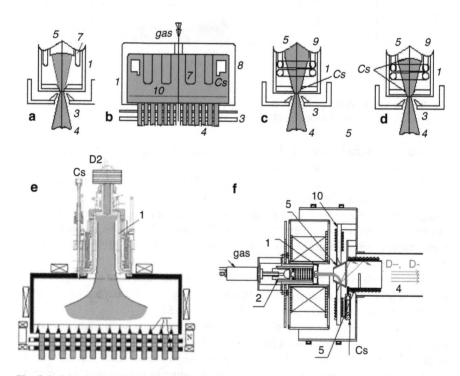

Fig. 8.6 Schematics of the main versions of large volume SPS (LV SPS). (Reproduced from Vadim Dudnikov [44], with permission from AIP Publishing). (**a**) SPS with multipole magnetic wall and a converter, (**b**) SPS with formation of negative ions on the plasma electrode, (**c**) RF SPS with anodic generation of negative ions, (**d**) RF SPS with emitter, (**e**) RF SPS with external antenna, and (**f**) SPS for polarized negative ions. The main components of these sources are 1, anode gas discharge chamber; 2, cold cathode emitter; 3, extractor with a magnetic system; 4, ion beam; 5, independent emitter; 6, hollow cathode; 7, heated cathode; 8, multipole magnetic wall; 9, RF antenna; and 10, magnetic filter

prevent severe loss of negative ions. In LV SPSs, hot filaments (7 in Fig. 8.6b), RF coils (9 in Fig. 8.6c and d), and high power microwaves, together with multicusp magnets (8 in Fig. 8.6e), are used for plasma generation at low gas density. LV SPSs have low power density and can be operated in dc mode. The ion beam current density is typically less than ~100 mA/cm^2 and the beam brightness is modest. LV SPSs with negative ion production on the plasma grid surface (anode production), as in Fig. 7.3b, have been adapted for high current (up to 70 A) negative ion beam production for plasma heating by neutral beam injection. A typical LV SPS as used for accelerator injection using RF plasma generated by an internal antenna is shown in Fig. 8.6c. Some versions of LV SPS, as in Fig. 7.3d with emitter (5), have been adapted for negative heavy ion beam formation. LV SPSs for neutral beam injection using an RF discharge generated by an external antenna or by a saddle antenna with longitudinal magnetic field are shown in Fig. 8.6e [49]. Version (f) of Fig. 8.6 is used in polarized H^-/D^- sources [28].

Table 8.1 lists some of the characteristics of modern high-current sources of H^- ions for accelerator injection. The most intense H^- ion beams are obtained from the BNL magnetron SPS (up to 100–120 mA) with a service life of up to 3 A.hours. The longest service life is with the RF SPS of the spallation neutron source (SNS) at Oak Ridge, with service life of up to 7 A.hours (19 weeks with H^- beam current of 60 mA at a duty cycle of 6%) [50].

A semiplanotron SPS with improved cathode cooling and geometric beam focusing has been proposed [51]. Cross-sectional schematic of a magnetron SPS is shown in Fig. 8.7 and a three-dimensional image in Fig. 8.8.

Table 8.1 Modern high-current H^- ion sources of for accelerators

Source	Type	Arc or RF and repetition rate	Beam duty factor %	Pulsed current mA	Extraction aperture mm	Lifetime hours	Lifetime A.hours
BNL	Magnetron	12A, 130 V, 7.5 Hz	0.5	120	2.8⌀	5760	3.0
FNAL	Magnetron	15A, 180 V, 15 Hz	0.35	80	3.2 ⌀	6500	3.2
RAL	Penning	55A, 70 V, 50 Hz	3.75	55	0.6 × 10	840	0.51
CSNS	Penning	50A, 100 V, 25 Hz	1.5	50	0.6 × 10	720	0.46
INR RAS	Penning	50A, 120 V, 50 Hz	1.0	40	1 × 10		
LANSCE	Converter	35A, 180 V, 120 Hz	10	18	9.8 ⌀	670	0.87
SNS	RF internal	60 kW, 2 MHz, 60 Hz	6.0	60	7 ⌀	4400	7
J-PARC	RF internal	30 kW, 2 MHz, 25 Hz	2.0	90	9 ⌀	1850	2.0
CERN	RF external	40 kW, 2 MHz, 0.8 Hz	0.07	45		1200	0.026

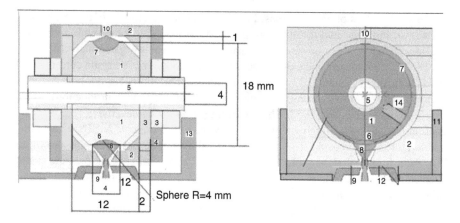

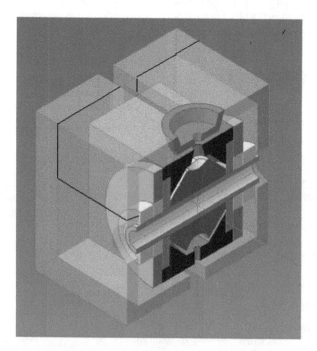

Fig. 8.7 Cross-sectional schematics of magnetron SPS with forced cathode cooling and geometric focusing. Left: cross section along the magnetic field. Right: cross section across the magnetic field. (Reproduced from V. Dudnikov and G. Dudnikova [51], with permission from AIP Publishing). Some major components are 1, cathode disk; 2, anode; 3, insulator; 4, magnetic poles; 5, cooling tube; 6, spherical dimple emitter of negative ions; 7, semicylindrical groove discharge channel; 8, flux of focused negative ions; 9, beam of negative ions extracted through the emission hole; 10, gas supply; 11, cesium feed; 12, extractor; 13, magnet; and 14, hollow cathode

Fig. 8.8 Three-dimensional image of a magnetron SPS with cooled cathode. (Reproduced from V. Dudnikov and G. Dudnikova [51], with permission from AIP Publishing)

An advanced design of magnetron SPS with the spherical focusing of emitted negative ions and forced cathode and anode cooling is shown in Fig. 8.7. This new magnetron SPS is capable for DC operation with high average negative ion current generation. This H^- SPS could be used as injector to ESS storage ring. H^- charge exchange (stripping) injection [52] into the European spallation neutron source (ESS) storage ring requires ~80 mA H^- ion source that delivers 3 ms pulses at 14 Hz repetition rate (duty factor ~ 4%) [53, 54] that can be extended to 28 Hz (df 8%). This can be achieved with a magnetron surface plasma H^- source (SPS) with active cathode and anode cooling. The Brookhaven National Laboratory (BNL) magnetron SPS can produce an H^- beam current of 100 mA with about 2 kW discharge power and can operate up to 0.7% duty factor (average power 14 W, energy efficiency up to 67 mA/kW) without active cooling [55]. An RF SPS in SNS has energy efficiency ~ 1 mA/kW [56].

We describe how active cathode and anode cooling can be applied to the BNL source to increase the average discharge power up to 140 W (df 8%) to satisfy the needs of the ESS [52, 53]. We also describe the use of a short electrostatic LEBT as is used at the Oak Ridge National Laboratory Spallation Neutron Source to improve the beam delivery to the RFQ.

Cross sections of new magnetron are shown in Fig. 8.7 [57]. A disk shape cathode (1) has 18 mm diameter D and 12 mm thickness H. A surrounded anode (2) is separated from the cathode by insulators (3). A vacuum gap between cathode and anode is d ~ 1 mm. Cathode is cooled by liquid or gas flux flowing through the cooling tube (5) with OD ~ 4 mm. The magnetron is compressed by ferromagnetic poles (4).

A working gas is injected to the discharge chamber through a channel (10). Cesium is added to discharge through second channel (11). Magnetic field, created by magnet (13) and formed by magnetic poles (4), has direction along axis of cooling tube (5).

The discharge in the crossed ExB fields is localized in the cylindrical grove (7) as in the semiplanotrons SPS. The cylindrical grove focus emitted negative ions to the anode surface and fast particles keep anode surface clean by sputtering the flakes and deposit. A plasma drift in the discharge can be closed around the cathode perimeter or can be bracket by shallow cylindrical grove. For beam formation, negative ions emitted from the spherical dimple (6) are used, geometrically focused to the emission aperture made in anode (2). These ions are extracted by electric field applied between anode (2) and extractor (12). Emission aperture of ~2 mm diameter has a conical shape. The spherical dimple with a curvature radius $R \sim 4$ mm has a working surface $S \sim 12$ mm^2. For the emission current H^- of 0.1 A, it is necessary to have the emission current density on the cathode surface $J_e \sim 1$ A/cm^2, which is acceptable for pulsed operation. The emission current density of H^- ~0.1 A/ cm^2 necessary for 10 mA extraction is acceptable for DC operation. Anode (2) is cooled by gas or liquid flow flowing through the cooling tube attached to the anode front. Material of cathode and anode for H^- beam production is molybdenum. The surface of spherical dimple should by mirror smooth for efficient negative ion emission and sharp focusing into the emission aperture. For heavy negative ion production, it is

Fig. 8.9 Photographs of cathode and anode of magnetron SPS with active cooling

possible to use some compound with necessary elements and necessary emission and discharge properties such as LaB$_6$ [58]. Two-stage extraction/acceleration is preferable for operation with high average beam current for collection of co-extracted electrons to the electrode with low potential. Gas valve [59] can be used for pulsed operation.

Photograph of cathode and anode of magnetron SPS with active cooling is shown in Fig. 8.9.

For H^- charge exchange (stripping), injection [60, 61] to ESS storage ring is necessary H^- ion source with current ~90 mA, pulses 3 ms and repetition of 14 Hz (duty factor ~ 4%) [62] extendable to repetition up to 28 Hz (df 8%). Possible solution can be magnetron surface plasma H^- source with active cathode and anode cooling. BNL magnetron SPS can produce H^- beam current of 100 mA at ~2 kW (energy efficiency of H^- generation up to 67 mA/kW, in SNS Rf SPS the energy efficiency ~1 mA/kW) discharge power and can operate up to duty factor 0.7% (average power ~ 14 W) without active cooling [55]. With active cathode and anode cooling, it is possible to increase average discharge power up to 140 W (df 8%).

For beam delivery to RFQ, it is possible to use a short electrostatic LEBT as in SNS.

8.3   A Proposed ESS Injector

A schematic of proposed ESS injector is shown in Fig. 8.10 [63]. It consists of surface plasma negative ion source with magnetron configuration comprising of cathode 1 and anode 2 with emission aperture, extractor electrode 3, and magnetic pole

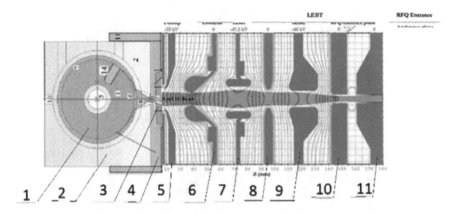

Fig. 8.10 A schematic of proposed ESS injector. 1, cathode; 2, anode; 3, extractor; 4, magnetic pole; 5, electron damp; 6, grounded electrode; 7, lens 1; 8, grounded electrode; 9, lens; 2, corrector; 10, RFQ wall; and 12, RFQ van

4. Extracted ion beam is accelerated to grounded electrode 6. Co-extracted electrons are collected by electron damp 4. The accelerated beam is focused by electrostatic Einzel lens 1 (7) and lens 2 (9) into RFQ wall aperture 10 and focused by RFQ vanes 12.

A design of electrostatic LEBT is shown in Fig. 8.11. It operates well with H^- beam current 60 mA at 65 kV with df up to 10%.

A joint of beams from two ion sources is presented in Ref. [51]. But it is more practical to have two separate RFQ for protons and for H^- and joint both beams after RFQ. Figure 8.12 shows preferable schematic of joint p and H^- beams.

Figure 8.13 shows the erosion of material on a BNL magnetron SPS that successfully operated for 2 years: the cathode has a hole of 1.8 mm^2 close to the center of its spherical focusing dimple, and the anode cover plate shows marks in the vicinity of the extraction hole spread in an area of 6.2 mm^2, which is not influent for magnetron operation. This damage is produced by back accelerated positive ions of Cs^+ and H_2^+. Estimation of sputtering of cathode and anode magnetron SPS was presented in [64].

But estimation of cesium density was incorrect, because during discharge, cesium is strongly ionized and cannot escape the discharge chamber as shown in Fig. 7 from [25, 65]. Figure 6 shows a typical oscillogram of the cesium ion current from the collector of the mass spectrometer, illustrating changes in the cesium atoms flux from the source in time at a high (~1000 K).

planotron cathode temperature, in conjunction with oscillograms of discharge current I_p and discharge voltage U_d. One can see that cesium atoms leave the source mainly after the end of the discharge pulse. Cesium release during the pulse is small, since cesium is highly ionized and the extraction voltage blocks the escape of cesium ions.

An SPS with plasma accelerator has been proposed [66]. A schematic of this SPS is shown in Fig. 8.14a. It consists of a plasma accelerator with an anode layer that

Fig. 8.11 Construction of the LEBT for transporting the *H⁻* beam in RF SPS for SNS to the RFQ

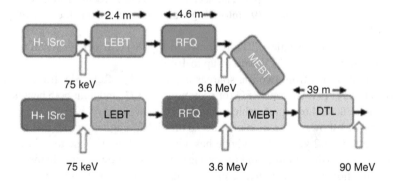

Fig. 8.12 The layout merges the two species in the MEBT from [52]

accelerates the flow of plasma 6 to converter 7, focusing a stream of emitted negative ions into an emission aperture with a transverse magnetic field to suppress the flow of accompanying electrons.

An SPS with plasma accelerator and multiaperture beam formation system is shown in Fig. 8.14b. In this SPS, a stream of ions 6 accelerated by the plasma accelerator with anode layer is directed to plasma electrode 3 with a multiaperture beam forming system coated with a thin cesium film. The emitted negative ions are pulled by extractor voltage 8. The generated beam 10 is accelerated by the potential between the pulling electrode 8 and grounded electrode 4. The magnetic field produced by magnet 9 blocks the flow of accompanying electrons.

An SPS with plasma accelerator with an extended acceleration zone is shown in Fig. 8.15. It consists of a plasma accelerator with an extended acceleration zone and

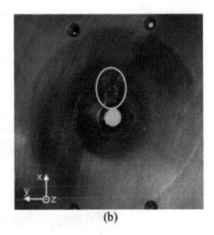

Fig. 8.13 Wear traces on the (**a**) cathode and (**b**) the anode cover plate of BNL's magnetron. The location of the traces on the cathode and anode cover plate is indicated by a circle and an ellipse (red and green), respectively

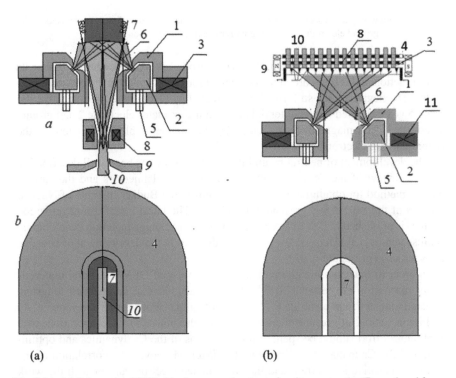

Fig. 8.14 (**a**) Schematic of SPS with plasma accelerator and emitter-converter. (Reproduced from Vadim Dudnikov [66], with permission from AIP Publishing). 1, magnetic pole-cathode; 2, anode; 3, permanent magnet; 4, magnetic pole; 5, anode cooling; 6, accelerated ion flow; 7, negative ion emitter; 8, electron suppressor; 9, extractor; and 10, negative ion beam. (**b**) Schematic of SPS with plasma accelerator and multiaperture beam formation system. 1, magnetic pole-cathode; 2, anode; 3, plasma electrode; 4, ground electrode; 5, anode cooling; 6, accelerated ion flow; 7, negative ion emitter; 8, extractor; 9, electron suppressor; 10, negative ion beam; and 11, permanent magnet

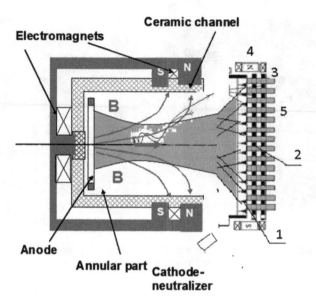

Fig. 8.15 SPS with plasma accelerator with an extended zone of acceleration. 1, plasma electrode; 2, extractor; 3, ground electrode; and 4, magnet for electron suppression

a multiaperture ion-optical system. Plasma electrode 1 is cesium coated to reduce its work function and is bombarded by a stream of ions accelerated in the plasma accelerator. The negative ions so formed are extracted by electrode 2. The negative ion beam 5 is accelerated by the potential between the extractor electrode 2 and ground electrode 3. The magnetic field created by the magnet 4 blocks the flow of the accompanying electrons.

The development of neutral beam injectors for ITER has been well described in [67]. A review of the developments of ion and neutral injectors and the surface plasma method for obtaining negative ion beams at the Budker Institute of Nuclear Physics of the SB RAS is presented in [68, 69]. The latest review on negative ion sources is published in [70, 71]. A review of charge exchange injection development is presented in [13]. An overview of recent advances in the development of negative ion sources is presented in [72].

In order to find possible alternatives to Cs in view of DEMO, several materials under discussion are tested [73, 74]. Hence, LaB_6, MoLa, lanthanated tungsten, bariated tungsten, and electride C12A7 are not valid options for H^- ion sources.

In conclusion, at present, there is no valid alternative to Cs evaporation; thus continuous effort should be spent in investigations of the Cs dynamics and optimization of the Cs management in view of DEMO. Moreover, the correlation of the extracted negative ion current and the co-extracted electron current with the work function of a cesiated surface in controlled conditions as in a laboratory experiment should be the next step for new investigations, in order to understand the underlying mechanisms of negative ion extraction at ion sources.

8.4 Conclusion

Since the discovery of the cesiation effect and the development of the surface plasma mechanism of negative ion formation with cesiation, many modifications of the SPS have been developed, and many improvements have been introduced. Now SPSs have become a highly reliable technology. Thousands of highly qualified scientists, engineers, highly skilled workers, and administrators around the world are engaged in the development and use of SPSs with cesiation. The intensity of negative ion beams has increased by a factor of 10^4, from a record 3 mA to more than 40 A. The cost of advanced injectors has increased from k\$ to M\$. Large projects are now ongoing for the development of SPSs for the Large Hadron Collider and for ITER. The development and manufacture of SPSs with cesiation have become a business with turnover of order US\$ billions [75, 76].

Last review of Physics and application of hydrogen negative Ion Sources is presented in book of Marthe Bacal (ed.), "Physics and Application of Hydrogen Negative Ion Sources," Springer Cham, Switzerland, 2023. https://doi.org/10.1007/978-3-031-21476-9.

References

1. V. Matyushkin, *Everyday life of Arzamas-16*, Molodaya Gvardia, 2007 (in Russian); Малышкин В. *Каждодневная жизнь Арзамаса 16*, Молодая Гвардия (2007)
2. M. Kapchinskii, V.A. Teplyakov, Prib. Tekh. Eksp., No. 4 (1970) p. 19 and No. 4 (1970) p. 17
3. S.S. Schweber, R. Wilson Defending against nuclear weapons: A 1950s proposal. Phys. Today **60**(4), 36 (2007). https://doi.org/10.1063/1.2731971
4. G.I. Dimov, V.G. Dudnikov, Cross sections for stripping of-1-MeV negative hydrogen ions in certain gases. Sov. Phys. Tech. Phys.-USSR **11**(7), 919 (1967).; Г. Димов, В. Дудников, «Измерение сечений обдирки отрицательных ионов водорода с энергией 1 МэВ в некоторых газах», ЖТФ, 36, 1239 (1966)
5. G.I. Budker, G.I. Dimov, V.G. Dudnikov, Experiments on producing intensive proton beams by means of the method of charge-exchange injection, in *Proc. of the Intern. Symp. on Electron and Positron Storage Ring*, France, Sakley, 1966, Rep. VIII, 6.1 (1966)
6. G.I. Budker, G.I. Dimov, V.G. Dudnikov, Experiments on producing intensive proton beams by means of the method of charge-exchange injection. Sov. At. Energy **22**(5), 441 (1967)
7. V. Dudnikov, Development of a surface plasma method for negative ion beams production. J. Phys.: Conf. Ser. **2244**, 01203 (2022)
8. R.Z. Sagdeev, *The Making of a Soviet Scientist* (John Wiley & Sons, 1994)
9. C. Robinson, Aviation Week&Space Tech., p. 42, Oct., 1978; Rev. Mod. Phys., 59(3), Part II (1987)
10. У.А. Арифов, А.Х. Аюханов, Изв. АН Уз. ССР, серия физ-мат. наук, **6**, 34 (1961). U.A. Arifov, A.H. Ayukhanov, Isvestiya AN Us. CCR, seriya Fis-mat. nauk, **6**, 34 (1961)
11. V.E. Kron, J. Appl. Phys. **34**, 3523 (1962)
12. E.D. Bender, M. Kishinevskii, I.I. Morozov, Proceedings of Second International Symposium on Production and Neutralization of Negative Hydrogen Ions and Beams, Brookhaven, NY. Preprint BNL 51304, 60 (1980)
13. G.I. Dimov, G.V. Roslyakov, ZTF, **42**, 1186 (1972)

14. V. Dudnikov, Method of negative ion obtaining, USSR Patent 411542, 10/III. 1972; http://www.findpatent.ru/patent/41/411542.html, V.G. Dudnikov, Technique for producing negative ions, https://inis.iaea.org/search/search.aspx?orig_q=RN:9355182; В. Дудников, Метод получения отрицательных ионов, Патент СССР , 411542 заявлено 10/ Ш/1972

15. Y.I. Belchenko, G.I. Dimov, V.G. Dudnikov, Powerful injector of neutrals with a surface-plasma source of negative ions. Nucl. Fusion **14**(1), 113–114 (1974)

16. G. Budker, G. Dimov, V. Dudnikov, Proc. Int. Symp. on Electron and Positron Storage Ring, France, Sakley, 1966, rep. VIII, 6.1 (1966)

17. V. Dudnikov, Production of intense proton beam in storage ring by charge exchange injection method, Dissertation for candidate of Fis-Mat. Nauk, INP SBAS, Novosibirsk, 1966. В. Дудников, "Получение интенсивного протонного пучка в накопителе методом перезарядной инжекции", Диссертация, представленная на соискание учёной степени кандидата физ.-мат. Наук, ИЯФ СОАН СССР, Новосибирск, (1966)

18. G.I. Dimov, Charge exchange Injection in accelerators and storage rings, Dissertation for Doctor of Fis-Mat. Nauk, INP SBAS, Novosibirsk 1968. Г. И. Димов, Перезарядная инжекция в ускорители и накопители, докторская диссертация, ИЯФ, (1968)

19. G.I. Dimov, V.G. Dudnikov, Determination of current and its distribution in storage ring. Instrum. Exp. Tech. **3**(553) (1969)

20. A. Rumolo, Z. Ghalam, T. Katsouleas, et al., Electron cloud effects on beam evolution in a circular accelerator. Phys. Rev. Special Topics. Accel. Beams **6**, 081002 (2003)

21. A. Gsponer, Physics of high-intensity high-energy particle beam propagation in open air and outer-space plasmas, arXiv.org > physics > arXiv: physics/0409157

22. D. Schrage, L. Young, B. Campbell, et al., Flight-qualified RFQ for the Bear project, in *Proceedings of the 1988 Linear Accelerator Conference*, Williamsburg, Virginia, USA (1988)

23. P.G. O'Shea, T. Butler, L.D. Hansborough, M.T. Lynch, K.F. McKenna, D.L. Schrage, M.R. Shubaly, J.E. Stovall, T.J. Zaugg, The Bear Accelerator. PAC 89, (1989)

24. V. Dudnikov, *Negative Ion Sources* (NSU, Novosibirsk, 2018). В. Дудников, *Источники отрицательных ионов*. НГУ, Новосибирск, 2018

25. V. Dudnikov, *Development and Applications of Negative Ion Sources* (Springer, 2019)

26. V.G. Dudnikov, Charge exchange injection into accelerators and storage rings. Phys.-Uspekhi **62**(4), 405 (2019). В. Дудников, Перезарядная инжекция в ускорители и накопители, Успехи физических наук, 184, 433 (2019)

27. V. Dudnikov, Method of negative ion obtaining, Patent cccp, 411542, 10/III. 1972; http://www.findpatent.ru/patent/41/411542.html. В.Г. Дудников, Способ получения отрицательных ионов, Авторское свидетельство, М. Кл.Н 01 J 3/0,4, 411542, заявлено 10/III,1972,

28. V.G. Dudnikov, Surface Plasma Method of Negative Ion Production, Dissertation for doctor of Fis-Mat. nauk, INP SBAS, Novosibirsk 1976. В.Г. Дудников, Поверхностно-плазменный метод получения пучков отрицательных ионов, Диссертация на соискание учёной степени доктора физ.-мат. Наук, ИЯФ СОАН СССР, Новосибирск, 1976

29. M.E. Kishinevskii, Z. Tekh. Fiz. **45**, 1281 (1975).; translated in Sov. Phys. – Tech. Phys. 20, 799 (1975). М. Е. Кишиневский, ЖТФ, т. 45, в., 6, 1281, 1975, препринт ИЯФ 116-73, Новосибирск (1973)

30. M.E. Kishinevskii, Z. Tekh. Fiz. **48**, 1281 (1978).; translated in Sov. Phys. – Tech. Phys. 20, 799 (1978). М. Е. Кишиневский, ЖТФ, 48, 73 (1978). Препринт ИЯФ, 76-18, Новосибирск (1978)

31. M.E. Kishinevskii, Interraction of atomic particles with solid, p. 22, Kharkov, 1976. М. Е. Кишиневкий, К вопросу о вторичной отрицательно-ионной эмиссии, Взаимодействие атомных частиц с твердым телом, с. 22, ХАРЬКОВ 7–9 ИЮНЯ 1976 г

32. V.G. Dudnikov, Hydrogen negative ion source with Penning Geometry, M. Nauka, vol. 1, p. 323 (1975). English translation, V.G. Dudnikov, Surface-plasma source of negative ions with Penning geometry, Los Alamos, LA-TR--75-4 (1975). В. Г. Дудников, Источник отрицательных ионов водорода с Пеннинговской геометрией, Труды IV Всесоюзного совещания по ускорителям заряженных частиц, М. Наука, т. 1, стр. 323 (1975)

33. V. Dudnikov, Y. Belchenko, Preprint, INP 78-95, Novosibirsk, 1978, V. Dudnikov, Y. Belchenko, J. Phys. **40**, 477 (1979)

34. Proc. of the 15th International Conference on Ion Sources, 9–13 September 2013, Chiba, Japan; Rev. Sci. Instrum. 85, No2I, part II, (2014). Proc. of the 14th Internat. Conf. on Ion Sources, 11–16 September 2011, Giadian Naxos, Italy; Rev. Sci. Instrum. 83, No. 2, Part 2, (2012). Proc. of the 13th Internat. Conf. on Ion Sources, 20–25 September 2009, Gatlinburg, TN, USA; Rev. Sci. Insrum. 81, No2, part II, (2010)

35. AIP 1655: Fourth International Symposium on Negative Ions, Beams and Sources (NIBS 2014) Edited by Werner Kraus, Paul McNeely, published (2015). Proceedings, 3rd International Symposium on Negative Ions, Beams and Sources (NIBS 2012): Jyväskylä, Finland, September 3–7, 2012, Olli Tarvainen (ed.), Taneli Kalvas (ed.) 2012AIP Conf.Proc. 1515 (2012). AIP CP 1097; Negative Ions, Beams and Sources, 1st International Symposium, Edited by E. Surrey and A. Simonin. AIP CP 925, 11th International Symposium on the Production and Neutralization of Negative Ions and Beams, Santa Fe, NM, USA, 13–15 September 2006. Edited by M. Stockli. AIP CP 763, 10th International Symposium on the Production and Neutralization of Negative Ions and Beams, Kiev, Ukraine, 14–17 September 2004. Edited by. J. Sherman and Yu. Belchenko. AIP CP 639, Ninth International Symposium on the Production and Neutralization of Negative Ions and Beams, Gif-sur-Yvette, France, 30–31 May 2002. Edited by M. Stockli. Proc. Second Symp. Production and Neutralization of Negative Hydrogen Ions and Beams, Brookhaven, 1980 (BNL, Upton, NY, 1980), BNL- 51304, edited. By Th. Sluyters. Proc. Symp. Production and Neutralization of Negative Hydrogen Ions and Beams, Brookhaven, 1977 (BNL, Upton, NY, 1977), BNL- 50727, edited. By Th. Sluyters and C. Prelec. C. Schmidt, "Production and neutralization of negative ions and beams", (Report on the 5th International Symposium, Upton, NY, USA, 30 October – 3 November 1989). 162 AIP CP158, Production and neutralization of negative ions and beams (Report on the 4th International Symposium, Upton, NY, USA, 1986) edited by. J. Alessi

36. Y. Belchenko, G. Dimov, V. Dudnikov, Physical principles of surface plasma source operation, Symposium on the Production and Neutralization of Negative Hydrogen Ions and Beams, Brookhaven, 1977 (Brookhaven National Laboratory (BNL), Upton, NY, 1977), pp. 79–96. Ю. Бельченко, Г. Димов, В. Дудников, «Физические основы поверхностно плазменного метода получения пучков отрицательных ионов», препринт ИЯФ 77–56, Новосибирк 1977. http://irbiscorp.spsl.nsc.ru/fulltext/prepr/1977/p1977_56.pdf; Yu Belchenko, G. Dimov, V. Dudnikov, Physical principles of surface plasma source method of negative ion production, Preprint IYaF 77–56, Novosibirsk 1977

37. J. Peters, Review of negative hydrogen ion sources high brightness/high current, in *2002 Proc. 21st Linear Accelerator Conference (LINAC 2002)* (Gyeongju, Korea) p. 42 (www.jacow.org)

38. J. Peters, The status of DESY H- sources. Rev. Sci. Instrum. **69**(2), 992 (1998)

39. J. Peters, *Review of Negative Hydrogen Ion Sources High Brightness/High Current* (Linac 98, Chicago, 1998)

40. N. Wells, The Development of High-Intensity Negative Ion Sources and Beams in the USSR, Rand Corp, Report R-No.2816-ARPA (1981)

41. J.D. Sherman, W.B. Ingalls, G. Rouleau, H. Vernon Smith, Jr. Review of Scaled Penning H-Surface Plasma Source with SlitwithSlit Emitters for High Duty Factor Linacs, CP642, in *High Intensity and High Brightness Hadron Beams: 20th ICFA Advanced Beam Dynamics Workshop on High Intensity and High Brightness Hadron Beams*, ed. by W. Chou, Y. Mori, D. Neuffer, and J.-F. Ostiguy 2002 American Institute of Physics 0-7354-0097-0 (2002)

42. A.S. Belov, V.G. Dudnikov, S.K. Esin, et al., Rev. Sci. Instrum. **67**, 1293 (1996)

43. V. Dudnikov, Development of a surface plasma method for negative ion beams production. J. Phys.: Conf. Ser. **2244**, 012034 (2022)

44. V. Dudnikov, Forty years of surface plasma source development. Rev. Sci. Instrum. **83**, 02A708 (2012)

45. V. Dudnikov, Progress in the Negative Ion Sources Development, Report FRBOR01, in *Proceedings of RUPAC2012*, Saint-Petersburg, Russia (2012)

46. V. Dudnikov, Forty-five years with cesiated surface plasma sources. AIP CP **1869**, 030044 (2017)
47. Y. Belchenko, A. Gorbovsky, A. Sanin, V. Savkin, The 25 mA continuous-wave surface-plasma source of H ions. Rev. Sci. Instrum. **85**, 02B108 (2014)
48. V. Dudnikov, C.W. Schmidt, R. Hren, J. Wendt, High current density negative ion source for beam line transport studies, PAC 2001, Chicago 2001
49. V. Dudnikov, R. Johnson, Advanced large volume surface plasma H-/D- source (lv sps) for neutral beam injectors, OS 2010 Novosibirsk, (2010)
50. M.P. Stockli, B.X. Han, S.N. Murray, T.R. Pennisi, M. Santana, C.M. Stinson, R.F. Welton, Record performance of and extraction studies with the Spallation Neutron Source Hinjector. AIP Conf. Proc. **1869**, 030010 (2017)
51. V. Dudnikov, G. Dudnikova, Compact surface plasma H- source with geometrical focusing. Rev. Sci. Instrum. **87**, 02B101 (2016)
52. G. Budker, G. Dimov, V. Dudnikov, Experimental investigation of the intense proton beam accumulation in storage ring by charge-exchange injection method. Sov. At. Energy **22**(384), 441 (1967)
53. A. Alekou, E. Baussano, A.K. Bhattacharyyaj , et al., The European Spallation Source neutrino Super Beam Conceptual Design Report, arXiv:2206.01208v1 [hep-ex] 2 Jun 2022
54. A. Alekoue, E. Baussano , N. Blaskovic Kraljevici, et al., The European Spallation Source neutrino Super Beam, arXiv:2203.08803v1 [physics.acc-ph] 15 Mar 2022
55. A. Zelenski, G. Atoian, T. Lehn, D. Raparia, J. Ritter, High-intensity polarized and unpolarized sources and injector developments at BNL Linac. AIP Conf. Proc. **2373**, 070003 (2021). https://doi.org/10.1063/5.0057677
56. R.F. Welton, V.G. Dudnikov, B.X. Han, et al., Improvements to the internal and external antenna H− ion sources at the Spallation Neutron Source. Rev. Sci. Instrum. **85**(2), 02B135 (2014)
57. V. Dudnikov, G. Dudnikova, Compact surface plasma H⁻ ion source with geometrical focusing. Rev. Sci. Instrum. **87**, 02B101 (2016). https://doi.org/10.1063/1.4931700
58. V. Dudnikov, J. Paul Farrell, Compact surface plasma sources for heavy negative ion production. Rev. Sci. Instrum. **75**(5) (2004)
59. G.E. Derevyankin, V.G. Dudnikov, P.A. Zhuravlev, Electromagnetic shutter for a pulsed gas inlet into vacuum units. Pribory i Tekhnika Eksperimenta **5**, 168–169 (1975)
60. G. Budker, G. Dimov, V. Dudnikov, Experimental investigation of the intense proton beam accumulation in storage ring by charge-exchange injection method. Sov. At. Energy **22**, 384 (1967)
61. V.G. Dudnikov, Charge exchange injection into accelerators and storage rings. Phys. Uspechi **62**(4) (2019)
62. A. Alekou, E. Baussano, N. Blaskovic Kraljevici, et al., The European Spallation Source neutrino Super Beam conceptual design report, arXiv:2203.08803v1 [physics.acc-ph] 15 Mar 2022; A. Alekou, et al., The European Spallation Source neutrino super-beam conceptual design report, Eur. Phys. J. Spec. Top. (2022); https://doi.org/10.1140/epjs/s11734-022-00664-w
63. V. Dudnikov, An H- Surface Plasma Source for the ESS storage ring, NIBS 2022, Padua, Italy, 2022
64. H. Pereira, J. Lettry, J. Alessi, T. Kalvas, Estimation of sputtering damages on a magnetron H- ion source induced by Cs+ and H+ ions. AIP Conf. Proc. **1515**, 81 (2013). https://doi.org/10.1063/1.4792773
65. Y. Belchenko, V.I. Davydenko, G.E. Derevyankin, A.F. Dorogov, V.G. Dudnikov, Sov. Tech. Phys. Lett. **3**, 282 (1977)
66. V. Dudnikov, Surface plasma source with anode layer plasma accelerator. Rev. Sci. Instrum. **83**, 02A713 (2012)
67. R.S. Hemsworth et al., Overview of the design of the ITER heating neutral beam injectors. New. J. Phys. **19**, 025005 (2017)
68. Y.I. Belchenko, V.I. Davydenko, P.P. Deichuli, et al., Stadies of ion and neutral beam physics and technology at the Budker Institute of Nuclear Physics SB RAS. Uspekhi Fizicheskikh Nauk **188**(6), 595 (2018)

69. Y.I. Belchenko, A.A. Ivanov, A.L. Sanin, O.Z. Sotnikov, Development of surface-plasma negative ions sources at the Budker Institute of nuclear physics. AIP Conf. Proc. **2052**, 030006 (2018)

70. V. Dudnikov, Modern high intensity H⁻ accelerator sources, http://arxiv.org/abs/1806.03391

71. V. Dudnikov, Surface plasma method of negative ion beams production. Phys. Uspekhi **6**(5) (2019). В. Г. Дудников, Поверхностно плазменный метод получения пучков отрицательных ионов, УФН, 189 (2019)

72. U. Fantz, J. Lettry, Focus on sources of negatively charged ions. New J. Phys. **20**, 060201 (2018)

73. S. Cristofaro, Work function of caesiated surfaces in H_2/D_2 low temperature plasmas correlated with negative ion formation, IPP 2019–05 März 2019, Dissertation zur Erlangung des akademischen Grades

74. M. Kobayashi, M. Sasao, M. Kisaki, T. Eguchi, M. Wada, Study of the beam extraction system of a negative ion source with a C12A7 electride plasma electrode. AIP Conf. Proc. **2052**, 020003 (2018).; https://doi.org/10.1063/1.5083721. Published Online: 28 December (2018)

75. ITER Neutral Beam Test Facility – 2017; https://www.youtube.com/watch?v=DKOHJYxK15o

76. https://www.iter.org/construction/NBTF

Index

V. Dudnikov, *Development and Applications of Negative Ion Sources*, Springer
Series on Atomic, Optical, and Plasma Physics 125,
https://doi.org/10.1007/978-3-031-28408-3

CPSIA information can be obtained
at www.ICGtesting.com
Printed in the USA
LVHW050752300623
751144LV00001B/5